VOLUME ONE HUNDRED AND FIFTY SIX

# PROGRESS IN MOLECULAR BIOLOGY AND TRANSLATIONAL SCIENCE

## Gangliosides in Health and Disease

Edited by

RONALD L. SCHNAAR
*Johns Hopkins University School of Medicine, Baltimore, MD, United States*

PABLO H. H. LOPEZ
*Instituto de Investigación Médica Mercedes y Martin Ferreyra, INIMEC-CONICET-Universidad Nacional de Córdoba, Córdoba, Argentina*

ELSEVIER

ACADEMIC PRESS
An imprint of Elsevier

Academic Press is an imprint of Elsevier
125 London Wall, London EC2Y 5AS, United Kingdom
525 B Street, Suite 1800, San Diego, CA 92101-4495, United States
50 Hampshire Street, 5th Floor, Cambridge, MA 02139, United States
The Boulevard, Langford Lane, Kidlington, Oxford OX5 1GB, United Kingdom

First edition 2018



ISBN: 978-0-12-812341-6
ISSN: 1877-1173

For information on all Academic Press publications
visit our website at https://www.elsevier.com/books-and-journals



*Publisher:* Zoe Kruze
*Acquisition Editor:* Ashlie M. Jackman
*Editorial Project Manager:* Alina Cleju
*Production Project Manager:* James Selvam
*Cover Designer:* Matthew Limbert

Typeset by Thomson Digital

# CONTENTS

# CONTRIBUTORS

**José Abad-Rodríguez**
Hospital Nacional de Parapléjicos (SESCAM), Toledo, Spain

**Bárbara Beatriz Báez**
Instituto de Investigación Médica Mercedes y Martin Ferreyra, INIMEC-CONICET-Universidad Nacional de Córdoba, Córdoba, Argentina

**Joseph T. Barbieri**
Medical College of Wisconsin, Milwaukee, WI, United States

**Elena Chiricozzi**
University of Milano, Milano, Italy

**Duncan Hieu M. Dam**
Northwestern University, Chicago, IL, United States

**Jose L. Daniotti**
National University of Cordoba, Ciudad Universitaria, Córdoba, Argentina

**Philippe Delannoy**
University of Lille, Villeneuve d'Ascq, France

**Justine H. Dewald**
University of Lille, Villeneuve d'Ascq, France

**Koichi Furukawa**
Department of Biomedical Sciences, Chubu University College of Life and Health Sciences, Kasugai, Japan
Department of Biochemistry II, Nagoya University Graduate School of Medicine, Nagoya, Japan

**Keiko Furukawa**
Department of Biomedical Sciences, Chubu University College of Life and Health Sciences, Kasugai, Japan

**Hans-Joachim Gabius**
Ludwig-Maximilians-University Munich, Munich, Germany

**Shinji Go**
Institute of Molecular Biomembrane and Glycobiology, Tohoku Medical and Pharmaceutical University, Sendai, Miyagi, Japan

**John A. Goodfellow**
College of Medical, Veterinary and Life Sciences, University of Glasgow, Glasgow, Scotland

**Sara Grassi**
University of Milano, Milano, Italy

**Sophie Groux-Degroote**
University of Lille, Villeneuve d'Ascq, France

**Noboru Hashimoto**
Department of Biochemistry II, Nagoya University Graduate School of Medicine, Nagoya, Japan

**Kei-ichiro Inamori**
Institute of Molecular Biomembrane and Glycobiology, Tohoku Medical and Pharmaceutical University, Sendai, Miyagi, Japan

**Jin-ichi Inokuchi**
Institute of Molecular Biomembrane and Glycobiology, Tohoku Medical and Pharmaceutical University, Sendai, Miyagi, Japan

**Yutaka Itokazu**
Department of Neuroscience and Regenerative Medicine, Medical College of Georgia, Augusta University, Augusta, GA, United States
Charlie Norwood VA Medical Center, Augusta, GA, United States

**Kazuya Kabayama**
Osaka University, Toyonaka, Osaka, Japan

**Hirotaka Kanoh**
Institute of Molecular Biomembrane and Glycobiology, Tohoku Medical and Pharmaceutical University, Sendai, Miyagi, Japan

**Koichi Kato**
Nagoya City University, Nagoya, Japan
Okazaki Institute for Integrative Bioscience and Institute for Molecular Science, National Institutes of Natural Sciences, Okazaki, Japan

**Yuji Kondo**
Department of Biochemistry II, Nagoya University Graduate School of Medicine, Nagoya, Japan

**Jürgen Kopitz**
Ruprecht-Karls-University Heidelberg, Heidelberg, Germany

**Robert W. Ledeen**
Department of Pharmacology, Physiology & Neuroscience, Rutgers New Jersey Medical School, Newark, NJ, United States

**T. August Li**
Johns Hopkins University School of Medicine, Baltimore, MD, United States

**Pablo H.H. Lopez**
Instituto de Investigación Médica Mercedes y Martin Ferreyra, INIMEC-CONICET-Universidad Nacional de Córdoba, Córdoba, Argentina

**Katsumi Matsuzaki**
Kyoto University, Kyoto, Japan

**Laura Mauri**
University of Milano, Milano, Italy

**Taeko Miyagi**
Miyagi Cancer Center Research Institute, Natori, Japan

**Masakazu Nagafuku**
Institute of Molecular Biomembrane and Glycobiology, Tohoku Medical and Pharmaceutical University, Sendai, Miyagi, Japan

**Yuki Ohkawa**
Department of Biomedical Sciences, Chubu University College of Life and Health Sciences, Kasugai, Japan

**Yuhsuke Ohmi**
Department of Biomedical Sciences, Chubu University College of Life and Health Sciences, Kasugai, Japan

**Isao Ohno**
Center for Medical Education, Tohoku Medical and Pharmaceutical University, Sendai, Miyagi, Japan

**Amy S. Paller**
Northwestern University, Chicago, IL, United States

**Alessandro Prinetti**
University of Milano, Milano, Italy

**Simona Prioni**
University of Milano, Milano, Italy

**Macarena Rodríguez-Walker**
National University of Cordoba, Ciudad Universitaria, Córdoba, Argentina

**Konrad Sandhoff**
Life and Medical Sciences (LIMES) Institute, University of Bonn, Bonn, Germany

**Roger Sandhoff**
Lipid Biochemistry Group, German Cancer Research Center, Heidelberg, Germany

**Heike Schulze**
Life and Medical Sciences (LIMES) Institute, University of Bonn, Bonn, Germany

**Ronald L. Schnaar**
Johns Hopkins University School of Medicine, Baltimore, MD, United States

**Kazuhiro Shiozaki**
Kagoshima University, Kagoshima, Japan

**Fumi Shishido**
Institute of Molecular Biomembrane and Glycobiology, Tohoku Medical and Pharmaceutical University, Sendai, Miyagi, Japan

**Ji Shuting**
Department of Biochemistry II, Nagoya University Graduate School of Medicine, Nagoya, Japan

**Sandro Sonnino**
University of Milano, Milano, Italy

**Akemi Suzuki**
Institute of Molecular Biomembrane and Glycobiology, Tohoku Medical and Pharmaceutical University, Sendai, Miyagi, Japan

**Orie Tajima**
Department of Biomedical Sciences, Chubu University College of Life and Health Sciences, Kasugai, Japan

**Kohta Takahashi**
Tohoku Medical and Pharmaceutical University, Sendai, Japan

**Satoshi Uemura**
Tohoku Medical and Pharmaceutical University, Sendai, Miyagi, Japan

**Jing Wang**
Department of Neuroscience and Regenerative Medicine, Medical College of Georgia, Augusta University, Augusta, GA, United States
Charlie Norwood VA Medical Center, Augusta, GA, United States

**Hugh J. Willison**
College of Medical, Veterinary and Life Sciences, University of Glasgow, Glasgow, Scotland

**Gusheng Wu**
Department of Pharmacology, Physiology & Neuroscience, Rutgers New Jersey Medical School, Newark, NJ, United States

**Kazunori Yamaguchi**
Miyagi Cancer Center Research Institute, Natori, Japan

**Koji Yamamoto**
Miyagi Cancer Center Research Institute, Natori, Japan
Saitama Medical University, Moroyama, Saitama, Japan

**Katsuhiko Yanagisawa**
Center for Advanced Medicine for Dementia, National Center for Geriatrics and Gerontology, Obu, Japan

**Robert K. Yu**
Department of Neuroscience and Regenerative Medicine, Medical College of Georgia, Augusta University, Augusta, GA, United States
Charlie Norwood VA Medical Center, Augusta, GA, United States

**Madison Zuverink**
Medical College of Wisconsin, Milwaukee, WI, United States

# PREFACE AND GANGLIOSIDE NOMENCLATURE

**Ronald L. Schnaar*, Pablo H.H. Lopez**[†]
[*] Johns Hopkins University School of Medicine, Baltimore, MD, United States
[†] Instituto de Investigación Médica Mercedes y Martin Ferreyra, INIMEC-CONICET-Universidad Nacional de Córdoba, Córdoba, Argentina

## Preface

Gangliosides, defined as glycosphingolipids containing one or more sialic acids, are broadly distributed in vertebrates[1] and common to all mammalian tissues[2] where they are most abundant at cell surfaces but are also expressed on intracellular membranes.[3] Mostly embedded in the outer leaflet of the plasma membrane, ganglioside glycans extend outward where they are part of the molecular "face" of the cell to the external environment.[4] Based on the characteristics of their ceramide lipid moieties, gangliosides associate laterally with each other and selected other membrane lipids and proteins into membrane microdomains.[5] The functions of gangliosides are based on transmolecular recognition by factors outside their cell of origin (lectins, pathogens, toxins) or by lateral (*cis*) molecular interactions with molecules on their own cells, including receptor tyrosine kinases and ion channels.[6]

In 1993, Prof. Lars Svennerholm, a pioneer of neurochemistry and the field of gangliosides, convened the 83rd Nobel Symposium on the "Biological Function of Gangliosides."[7] In the intervening 25 years, advances in molecular genetics and structural biology have driven a wave of additional discovery in this area with implications ranging from microbial pathogenesis to brain development and neurodegeneration. In this volume, we invited investigators who are leaders in ganglioside discovery to provide updates and insights on ganglioside metabolism, biophysics, recognition, and ganglioside biological roles in infection, immunity, cancer, diabetes, development and regeneration, wound healing, and neurodegenerative diseases. The editors express their profound thanks to all of the contributors for their generous efforts and excellent contributions.

The editors dedicate this volume to Dr. Sen-itiroh Hakomori, Pacific Northwest Research Institute, whose contributions to understanding the structures and functions of gangliosides and other glycosphingolipids have been legion. With >400 published papers on glycosphingolipids in his >65-year research career his many discoveries and enduring insights have touched all of the work described herein.

## Ganglioside Nomenclature

A discussion of ganglioside nomenclature is relevant, especially for investigators coming to gangliosides from diverse fields. By far, the most common nomenclature used in the field is that of Lars Svennerholm.[8] In June 1980, a few months after one of us (RLS) joined the faculty at Johns Hopkins, Dr. Svennerholm visited and gave a compelling series of lectures devoted to ganglioside structure, metabolism, distribution and function. In that series, he shared the history of his eponymous ganglioside nomenclature and his profound discontent with it. By then the Svennerholm ganglioside nomenclature had gained such wide acceptance that it was part of the scientific lexicon and could not be readily reversed. As he related it, he was studying extracted human brain gangliosides using adsorptive chromatography to separate monosialo and disialo species.[9] When he analyzed the monosialo ganglioside (GM) fraction by thin layer chromatography (TLC), three species appeared, which he numbered starting with the one nearest the TLC origin. This numbering system was experimental and was not informed by the structures, which were not yet determined. As more complex species migrated slower by TLC, this resulted in the most complex of the GM series having the name GM1 with the progressively less complex species named GM2 and GM3, respectively. Attempting to make Svennerholm ganglioside nomenclature systematic (after the fact) is a fool's errand, but as its' use has expanded over the years, an attempt to do so is presented in Fig. 1.

Whereas a ganglioside is any glycosphingolipid with a sialic acid (independent of its carbohydrate structure) the term "ganglio series" designates a specific neutral glycosphingolipid core of up to four sugars with the structure Galβ1-3GalNAcβ1-4Galβ1-4GlcβCer (the gangliotetraose core) abundant in the brain and several other tissues. In IUPAC[10] nomenclature (Table 1), each neutral glycan in the core is assigned a roman numeral starting with the glucose attached to the ceramide (Fig. 1). Molecules with this core (in full or truncated) are denoted by the capital letter "G" in Svennerholm nomenclature. This is followed by a capital letter designating the total number of sialic acids on the structure as mono (M), di (D), tri (T), tetra (Q), or pentasialo (P). This is followed by an Arabic number designating the length of the

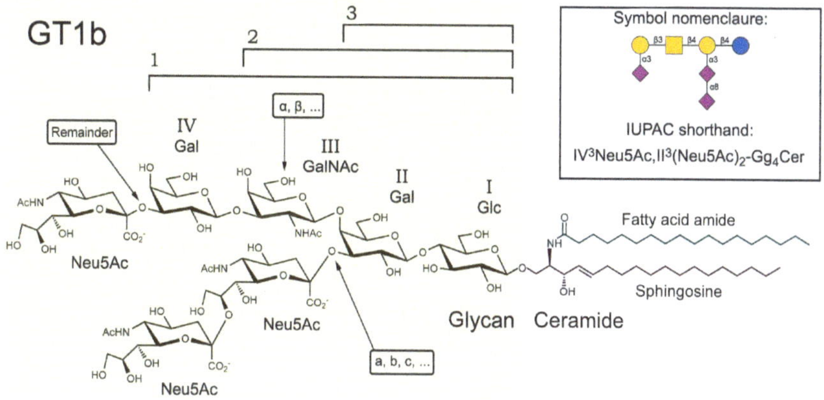


**Fig. 1** Ganglioside nomenclature. Ganglioside GT1b is shown as an example. The gangliotetraose core structure (Galβ1-3GalNAcβ1-4Galβ1-4GlcβCer) is shown with each glycan designated by a Roman numeral according to IUPAC[10] shorthand nomenclature (e.g., Table 1). In the widely used nomenclature of Svennerholm,[8] an Arabic numeral designates the length of the neutral glycan chain as shown in the brackets above the structure, with larger numbers designating shorter glycan chains. Common attachment positions of sialic acids (shown as Neu5Ac) are indicated: the number of sialic acids at Gal (II) is indicated by a lower case Latin letter, those at GalNAc (III) as lower case Greek letters, and any remaining sialic acids are attached to the terminal Gal (IV). Gangliosides are embedded into membranes by their variable ceramide structures with most of the glycan extending outward.[4]

gangliotetraose core as either all four neutral sugars (1), three sugars (2) or two sugars (3) as shown in Fig. 1. The number of sialic acid residues on the 3-hydroxyl of the innermost galactose (Gal II) is designated by a lower case letter for one (a), two (b), or three (c) sialic acids. For the cases of mono-sialogangliosides GM1, GM2, and GM3, the lower case "a" is optional and most often omitted. Rare "α-series" gangliosides have a single sialic acid on the 6-hydroxyl of the GalNAc residue. Any sialic acid not designated by a lower case Latin or Greek letter are attached to the 3-hydroxyl of the nonreducing terminal galactose (Gal IV). Thus, GT1b (shown in Fig. 1) has three sialic acids (T), the full tetrasaccharide neutral core (1), two of its three sialic acids on the innermost galactose (b), and the remaining one sialic acid on the terminal galactose. In mammals, multiple sialic acids are attached to one another via α2,8-linkage. Unfortunately, some Svennerholm-inspired designations are exceptions to the rules (e.g., see GM1b, Table 1).

Whereas Svennerholm nomenclature is inherently nonsystematic, and IUPAC nomenclature[10] tends to be too cumbersome for daily use, agreed

**Table 1** Nomenclature of some common gangliosides.

| Svennerholm | Glycan Symbol | IUPAC Shorthand[a] | IUPAC Structure |
|---|---|---|---|
| GM3 | 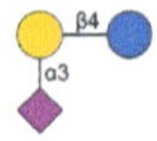 | II$^3$Neu5Ac-LacCer | Neu5Acα2-3Galβ1-4GlcβCer |
| GD3 | 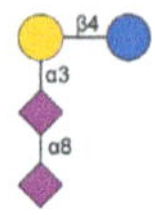 | II$^3$(Neu5Ac)$_2$-LacCer | Neu5Acα2-8Neu5Acα2-3Galβ1-4GlcβCer |
| GM2 | 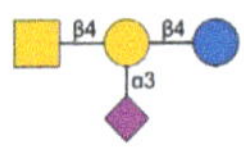 | II$^3$Neu5Ac-Gg$_3$Cer | GalNAcβ1-4(Neu5Acα2-3)Galβ1-4GlcβCer |
| GD2 | 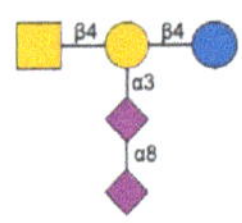 | II$^3$(Neu5Ac)$_2$-Gg$_3$Cer | GalNAcβ1-4(Neu5Acα2-8Neu5Acα2-3)Galβ1-4GlcβCer |
| GM1 (GM1a) | 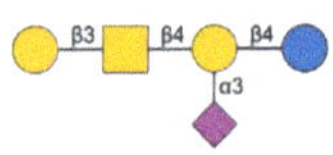 | II$^3$Neu5Ac-Gg$_4$Cer | Galβ1-3GalNAcβ1-4(Neu5Acα2-3)Galβ1-4GlcβCer |
| GD1b | 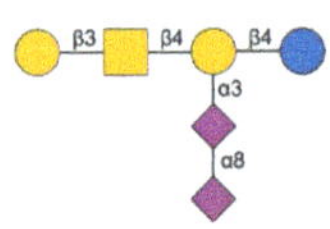 | II$^3$(Neu5Ac)$_2$-Gg$_4$Cer | Galβ1-3GalNAcβ1-4(Neu5Acα2-8Neu5Acα2-3)Galβ1-4GlcβCer |
| GD1a | 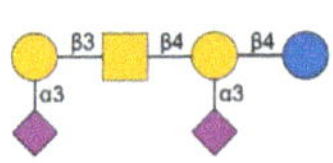 | IV$^3$Neu5Ac, II$^3$Neu5Ac-Gg$_4$Cer | Neu5Acα2-3Galβ1-3GalNAcβ1-4(Neu5Acα2-3)Galβ1-4GlcβCer |
| GT1b | 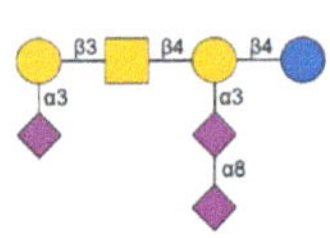 | IV$^3$Neu5Ac, II$^3$(Neu5Ac)$_2$-Gg$_4$Cer | Neu5Acα2-3Galβ1-3GalNAcβ1-4(Neu5Acα2-8Neu5Acα2-3)Galβ1-4GlcβCer |
| GQ1b | 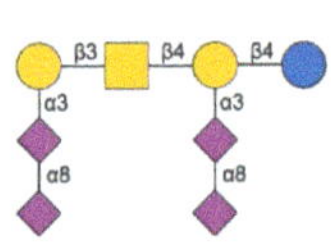 | IV$^3$(Neu5Ac)$_2$, II$^3$(Neu5Ac)$_2$-Gg$_4$Cer | Neu5Acα2-8Neu5Acα2-3Galβ1-3GalNAcβ1-4(Neu5Acα2-8Neu5Acα2-3)Galβ1-4GlcβCer |
| GM1b (*cis*GM1) | 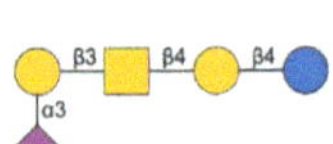 | IV$^3$Neu5Ac-Gg$_4$Cer | Neu5Acα2-3Galβ1-3GalNAcβ1-4Galβ1-4GlcβCer |
| GD1α | 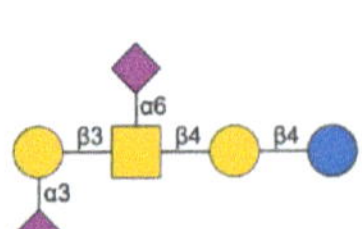 | IV$^3$Neu5Ac, III$^6$Neu5Ac-Gg$_4$Cer | Neu5Acα2-3Galβ1-3(Neu5Acα2-6)GalNAcβ1-4Galβ1-4GlcβCer |

[a] Sialic acids (Neu5Ac) are all considered to be in α linkage. Formal IUPAC nomenclature recommendations[10] include the anomeric linkage between hyphens in the shorthand version, such as II$^3$-α-NeuAc-LacCer for GM3.

upon symbol nomenclature[11] is helping to better convey glycan structures to a wider scientific audience. Some of the more common gangliosides discussed in this volume are shown in Table 1, along with Svennerholm (or Svennerholm-inspired), IUPAC, and symbol nomenclatures. A more complete table of structures can be found in the first chapter of this volume. More complete compendia of ganglioside glycan structures including variations in sialic acids and lipid moieties (sphingosine structure and length, fatty acid amide structure, and length) contain more than a thousand ganglioside entries.[12,13]

Several chapters in this volume include representations of gangliosides and their metabolic pathways relevant to consideration of their biological functions and medical significance. We have chosen to retain this volume-wide redundancy to provide ready access to this information in the context of each chapter and so that each chapter can be viewed as a stand-alone review of a particular aspect of ganglioside biology and/or pathophysiology.

## REFERENCES

1. Avrova, N. F. (1971). Brain ganglioside patterns of vertebrates. *J Neurochem, 18*, 667–674.
2. Puro, K., Maury, P., & Huttunen, J. K. (1969). Qualitative and quantitative patterns of gangliosides in extraneural tissues. *Biochim Biophys Acta, 187*, 230–235.
3. Keenan, T. W., Morre, D. J., & Huang, C. M. (1972). Distribution of gangliosides among subcellular fractions from rat liver and bovine mammary gland. *FEBS Lett, 24*, 204–208.
4. DeMarco, M. L., & Woods, R. J. (2009). Atomic-resolution conformational analysis of the GM3 ganglioside in a lipid bilayer and its implications for ganglioside-protein recognition at membrane surfaces. *Glycobiology, 19*, 344–355.
5. Aureli, M., Grassi, S., Prioni, S., Sonnino, S., & Prinetti, A. (2015). Lipid membrane domains in the brain. *Biochim Biophys Acta, 1851*, 1006–1016.
6. Todeschini, A. R., & Hakomori, S. I. (2008). Functional role of glycosphingolipids and gangliosides in control of cell adhesion, motility, and growth, through glycosynaptic microdomains. *Biochim Biophys Acta, 1780*, 421–433.
7. Svennerholm, L., Asbury, A. K., Reisfeld, R. A., et al. (1994). *Biological Function of Gangliosides*. Amsterdam: Elsevier.
8. Svennerholm, L. (1994). Designation and schematic structure of gangliosides and allied glycosphingolipids. *Prog Brain Res, 101*, xi–xiv.
9. Svennerholm, L. (1963). Chromatographic separation of human brain gangliosides. *J Neurochem, 10*, 613–623.
10. Chester, M. A. (1997). IUPAC-IUB Joint Commission on Biochemical Nomenclature (JCBN): nomenclature of glycolipids. *Eur J Biochem, 257*, 293–298.
11. Varki, A., Cummings, R. D., Aebi, M., et al. (2015). Symbol nomenclature for graphical representations of glycans. *Glycobiology, 25*, 1323–1324.
12. Sud, M., Fahy, E., Cotter, D., Dennis, E. A., & Subramaniam, S. (2012). LIPID MAPS-nature lipidomics gateway: an online resource for students and educators interested in lipids. *J Chem Educ, 89*, 291–292.
13. Fahy, E., Subramaniam, S., Murphy, R. C., et al. (2009). Update of the LIPID MAPS comprehensive classification system for lipids. *J Lipid Res, 50*, S9–S14.

CHAPTER ONE

# Ganglioside Metabolism in Health and Disease

**Roger Sandhoff*, Heike Schulze†, Konrad Sandhoff†,1**
* Lipid Biochemistry Group, German Cancer Research Center, Heidelberg, Germany
† Life and Medical Sciences (LIMES) Institute, University of Bonn, Bonn, Germany
[1] Corresponding author. E-mail address: sandhoff@uni-bonn.de

## Contents



## Abstract

Gangliosides (GGs) are cell type-specific sialic acid-containing glycosphingolipids (GSLs), which are enriched in mammalian brain. Defects in GSL metabolism cause fatal human diseases. GSLs are composed of a hydrophilic oligosaccharide linked in 1-*O*-position to a hydrophobic ceramide anchor, which itself is composed of a long-chain amino alcohol, the sphingoid base, and an amide-bound acyl chain. Biosynthesis of mammalian GGs and other GSLs starts with the formation of their hydrophobic ceramide anchor in the endoplasmic reticulum, followed by sequential glycosylation reactions along the secretory pathway, mainly at the luminal surface of Golgi and

*Progress in Molecular Biology and Translational Science*, Volume 156
ISSN 1877-1173
http://dx.doi.org/10.1016/bs.pmbts.2018.01.002

trans-Golgi network (TGN) membranes. Few membrane-anchored and often promiscuous glycosyltransferases allow the formation of cell type-specific glycolipid patterns in a combinatorial process. Inherited defects of these transferases therefore affect not only single structures but defined glycolipid series. GGs and other GSLs are thereafter transported by an exocytotic membrane flow to the plasma membrane where they are expressed in cell type-specific patterns, which can be modified by metabolic reactions at or near the cellular surface. Endocytosed (glyco)sphingolipids are degraded, together with other membrane lipids in a stepwise fashion by endolysosomal enzymes with the help of small lipid-binding proteins, the sphingolipid activator proteins (SAPs), at the surface of intraluminal lysosomal vesicles. Inherited defects in a sphingolipid-degrading enzyme or SAP cause the accumulation of the corresponding lipid substrates. Endolysosomal GSL degradation is strongly modified by the lipid components of the organelle microenvironments.

# 1. INTRODUCTION

## 1.1 Definition and Historical Aspects

Gangliosides (GGs) had been discovered by Ernst Klenk in brain extracts while studying amaurotic familial idiocy and Niemann–Pick disease.[1,2] In 1942, he created the term "gangliosides" to signify their predominant location in ganglion cells.[3,4] GGs are a large group of sialic acid-containing glycosphingolipids (GSLs) present in all classes of vertebrates.[5] They are cell type specifically expressed (Fig. 1A). GSLs are composed of a hydrophilic oligosaccharide linked in 1-*O*-position to a hydrophobic ceramide anchor, which itself is composed of a long-chain amino alcohol, the sphingoid base, and an amide-bound acyl chain. GM1 (Fig. 1B) (or GM1a), belonging to the a-series of neuronal GGs, was the first complex GG, for which the correct chemical structure was elucidated.[6] It contains *N*-acetyl neuraminic acid, which is among the most common of the about 50 different sialic acids that have been described to date, all of which are derivatives of the 5-amino-2-keto-3-deoxy nononic acid, the "neuraminic acid."[7] Sialic acids are essential constituents of glycolipids, oligosaccharides, and glycoproteins at the cell surface of mammalian and other cells, performing their biosynthesis.[7–9] GGs are amphipathic lipids and integral constituents of cell membranes. As a matter of the topology of their biosynthesis, GGs locate in the anticytosolic leaflets and concentrate on the cell surface. There, GGs contribute to cell surface $Ca^{2+}$-binding. They can participate in membrane microdomains,

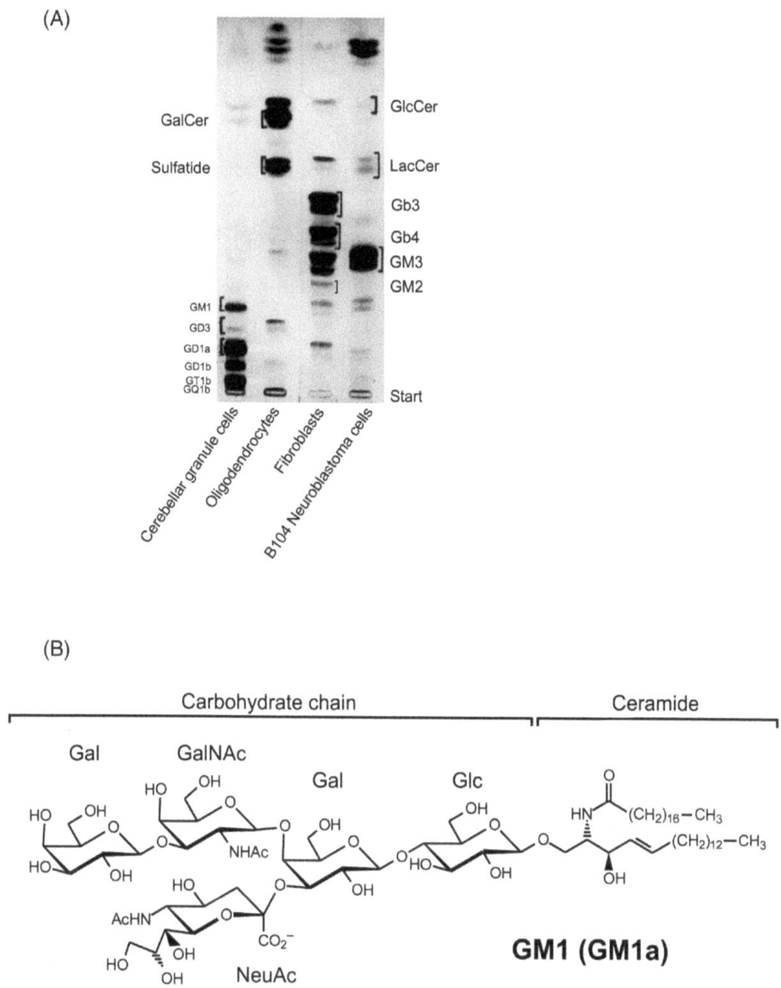


**Fig. 1** (A) Cell type specific incorporation of [$^{14}$C]-galactose into glycosphingolipids. Primary cultured cerebellar neurons (granule cells), oligodendrocytes, fibroblasts, and B104 neuroblastoma cells were metabolically labeled with [$^{14}$C]-galactose for 48 h and the radiolabeled GSLs were separated by thin layer chromatography, and visualized by fluorography (*modified after van Echten G, Sandhoff K. Modulation of ganglioside biosynthesis in primary cultured neurons. J Neurochem. 1989;52(1):207–214* [42]). (B) Chemical structure of GM1a, one of the major brain GGs and preferred ligand of cholera toxin. Whereas the oligosaccharide moiety is defined by the name, the sphingoid base and fatty acid of the hydrophobic ceramide moiety may vary. Here, it is presented with C18-sphingosine and stearic acid, which compose the main neuronal ceramide anchor.

serve as ligands of lectins, and modulate activity of membrane proteins such as the insulin receptor, leptin receptor, or the EGF receptor (EGFR).[10–18] These interactions depend, on one hand, on the structure of the glycan head group and, on the other hand, on the structural properties of the ceramide anchor, which may differ significantly from cell to cell type. Mature neurons, for example, are rich in a- and b-series GGs with a core tetraosyl moiety and a varying amount of sialic acids (GM1a, GD1a, GD1b, GT1b), which are anchored to a ceramide moiety containing almost only stearic acid[6,19] (Fig. 2). From these GGs, only the head groups of axonal GD1a and

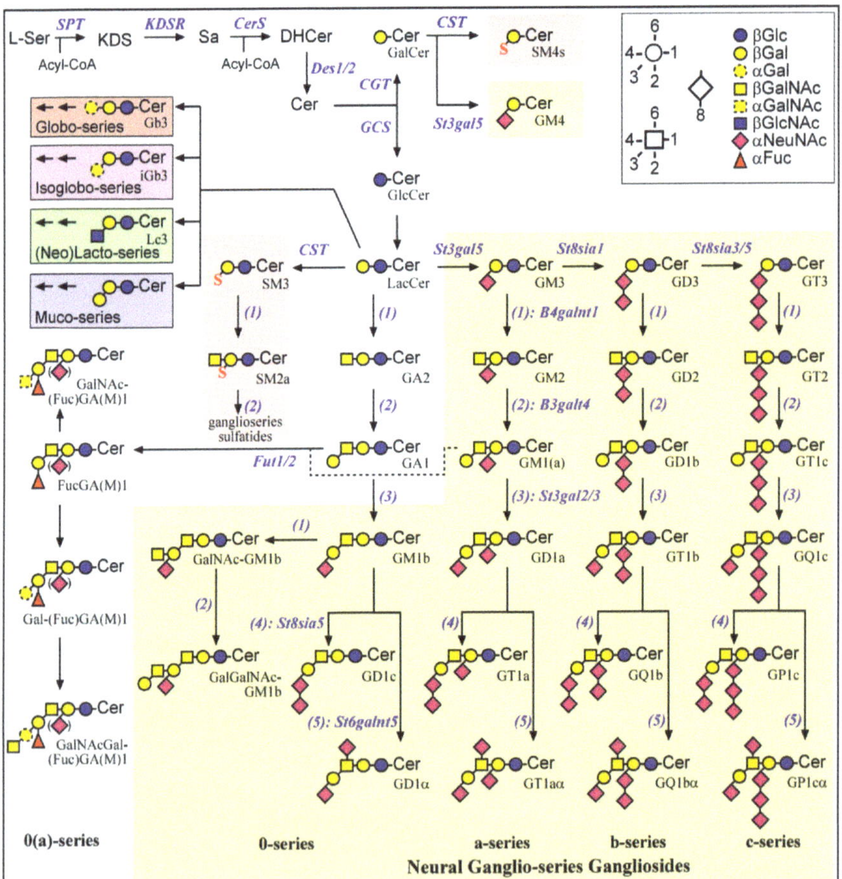


**Fig. 2** Pathways of the combinatorial ganglioside biosynthesis. *KDS*, 3-Ketodihydrosphingosine; *Cer*, ceramide; *DHCer*, dihydroceramide; *Sa*, sphinganine/dihydrosphingosine; *GalCer*, galactosylceramide; *GlcCer*, glucosylceramide; *L-Ser*, L-serine. Ganglioside names are abbreviated according to Svennerholm.[43]

GT1b efficiently bind myelin-associated glycoprotein (MAG) on the surrounding myelin sheaths.[20–22] Murine muscle cells in contrast are enriched in the small GG, GM3, but containing also mainly stearic acid in its ceramide anchor,[23] which is, like in brain, due to the expression of ceramide synthase 1.[24] GM3 interacts with and inactivates the insulin receptor in these cells.[10,25] GM3 and GGs of other organs, such as liver and intestine, endothelium or immune cells, prefer palmitic acid and very long-chain fatty acids over stearic acid in their ceramide anchor.[26–31] Interestingly, different acyl chains, synthetically incorporated into the ceramide moiety of GM1, have been shown to change the endosomal and intracellular trafficking of GM1 in cell culture.[32,33] In conclusion, the biological function of a GG depends on its structure and on its membrane microenvironment, both of which are cell-type dependent. Besides these normal biological functions, GGs are also ligands for lectin-type bacterial exotoxins such as cholera toxin, shiga/verotoxin, tetanus toxin, or botulinum toxins.[34–41] Biological implications of these properties are discussed in details in the other chapters of this issue.

## 2. ANABOLISM OF GANGLIOSIDES

### 2.1 Synthesis of the Lipophilic Glycosphingolipid-Anchor, Ceramide

#### *2.1.1 Endoplasmic Reticulum, Step 1, Serine Palmitoyl-CoA Transferase*

De novo biosynthesis of sphingolipids starts in and at the cytosolic membrane of the endoplasmic reticulum (ER)[44] by the condensation of an activated fatty acid (mainly palmitoyl-CoA) with the amino acid L-serine yielding 3-keto sphinganine/3-keto dihydrosphingosine (KdS) (Fig. 2). This rate limiting step is catalyzed by the serine palmitoyl-CoA transferase (SPT), which requires pyridoxal 5′-phosphate (PLP) as cofactor for the decarboxylative, Claisen-like condensation.[45–50] SPT is regulated in a feedback loop by inhibitory ORM proteins in yeast[51,52] and ORMDL proteins in mammals.[53,54] Mammalian SPTs are heterotrimeric proteins consisting of two large subunits, SPTLC1 combined with either SPTLC2 or SPTLC3, and one of two small subunits, SPTssa or SPTssb. Although the presence of SPTLC3 shifts substrate specificity toward myristoyl-CoA,[55] SPTssb shifts it toward stearoyl-CoA,[56] thereby modulating the length of the sphingoid base.

### 2.1.2 Endoplasmic Reticulum, Step 2, 3-Keto Dihydrosphingosine Reductase

The second step of sphingolipid synthesis is the reduction of KdS catalyzed by the KdS reductase on the cytosolic side of the ER, an enzyme encoded by the 3-ketodihydrosphingosine reductase (*KDSR*) (FVT-1) gene.[57] The enzyme oxidizes NADPH to produce sphinganine/dihydrosphingosine (Sa),[48] which is directly the substrate for the ER-located ceramide synthases (CerS).[58,59]

### 2.1.3 Endoplasmic Reticulum, Step 3, Ceramide Synthases

CerS convert sphinganine from de novo synthesis to dihydroceramides as well as sphingosine and other sphingoid bases from the salvage pathway directly yielding ceramides.[60] Although there is only one gene encoding a ceramide synthase in *Drosophilamelanogaster* (schlank),[61] there are three genes in *Caenorhabditis elegans* (*hyl-1*, *hyl-2*, and *lagr-1*)[62] and six genes in mammals (former *LASS1-6*, now *CERS1-6*)[63] leading to different CerS isoforms. Ceramide synthases are cell type specifically expressed, leading to a characteristic ceramide anchor profile of sphingolipids and GGs in each organ.[24,64] Due to high expression levels of *CERS1* in neuronal cells, the major ceramide anchor of mammalian GGs extracted from brain contains stearic acid.[65–67] A recent study showed that enzyme activities of CerS2-6 are regulated by phosphorylation in the C-terminal region and suggested that CerS phosphorylation may be a regulatory element in the control of the distribution of the acyl chain lengths in sphingolipids.[68] Interestingly, these CerS contain, in contrast to CerS1, also a HOX-domain,[69] which has been shown in drosophila to regulate triglyceride metabolism,[61,70] probably via nuclear DNA-binding.[71] Schlank, the only CerS of drosophila, can translocate to the nucleus and bind with its homeodomain promotor regions of lipase genes to regulate their transcription.[72]

### 2.1.4 Endoplasmic Reticulum, Step 4, Dihydroceramide Desaturase

Sphingolipid neo-biosynthesis results in the formation of dihydroceramides, which are subsequently converted to ceramides by the dihydroceramide desaturase 1 encoded by the *DEGS* (*DES1*) gene. DES1 requires the cofactor NAD(P)H and oxygen to introduce a trans-double bond in 4-position and generate *N*-acyl sphingosine.[73,74] Although sphingosine is the major sphingoid base of neuronal GGs, intestinal GSLs including GGs are rich in phytosphingosine.[75–77] Here, dihydroceramides are converted to phytoceramides by a dihydroceramide C-4 hydroxylase encoded by the *DEGS2* (*DES2*) gene.[78–80]

### 2.1.5 Endoplasmic Reticulum, 5a Generation of Ganglioside GM4

GG GM4 is the only GG derived from galactosylceramide (GalCer), which is synthesized by the β-galactosylceramide synthase (UGT8a) located in the ER with its active site facing the luminal side.[81] After vesicular transport to the Golgi, GM3-synthase (GM3S, St3gal5) transfers a sialic acid from CMP-sialic acid in α2,3 linkage to the galactosyl moiety of GalCer and forms the most simple GG GM4,[82] but eventually only when St3gal5 is not in complex with lactosylceramide (LacCer) synthase.[83] GM4 is a minor component of the brain and, otherwise, detected in low concentrations in erythrocytes, kidney, and intestine.[84–86] Little is known about its biological function.

### 2.1.6 Golgi, Step 5b, Glucosylceramide Synthase

Complex GGs and GSLs are synthesized by the stepwise addition of monosaccharides to ceramide. Corresponding glycosyltransferases use nucleotide-activated sugars and act in a combinatorial fashion.[46] Except for the synthesis of cerebrosides in the ER,[81] glycosylation of sphingolipids is located to the Golgi and trans-Golgi network (TGN).[46] Ceramides reach the Golgi by either vesicular transport or by protein-bound transfer involving ceramide transfer protein (CERT). CERT preferentially transports ceramides having C14 to C20 long acyl chains.[87] The initial step of complex GSL formation is the transfer of glucose (Glc) to ceramide by the UDP-glucose ceramide glucosyltransferase or glucosylceramide synthase (GlcCer-S) encoded by the *UGCG* gene.[88] The product β-glucosylceramide (GlcCer) is released into the cytosolic leaflet of the early Golgi membranes due to the location of the catalytic site and has to be transferred to the luminal side for further glycosylation.[89,90] Translocation of GlcCer may involve retrograde vesicular transport to the ER, ATP-independent ER phospholipid flippases, and subsequent vesicular transport back to the Golgi for further glycosylation.[91,92]

### 2.1.7 Golgi, Step 6, Lactosylceramide Synthase

In the lumen of the Golgi, one of the two LacCer synthases encoded by B4Galt5 and B4Galt6 can transfer a galactose from UDP-galactose in β-glycosidic linkage to C4′ of glucosylceramide. However, in mice, LacCer is predominantly synthesized by B4Galt5.[93,94] LacCer is at the main branching point in complex GSL synthesis. Different glycosyltransferases glycosylate LacCer and initiate different GSL series, that is, (iso)lacto-, globo-, isoglobo-, asialoganglio-, ganglio-, muco-, and sulfoganglio-series (Fig. 2, Table 1).

**Table 1** Schematic Formulas of Gangliosides in the Mammalian Nervous System

| | Ganglioside/GSL | Structure | Occurence |
|---|---|---|---|
| | Cer | Ceramide, *N*-acylsphingosine | Ubiquitous |
| | GalCer | Galβ1→1Cer | Myelin sheaths, kidney |
| | GM4 | Neu5Acα2→3Galβ1→1Cer | Kidney[a], myelin |
| | GlcCer | Glcβ1→1Cer | Ubiquitous |
| | LacCer | Galβ1→4Glcβ1→1Cer | Neutrophils (intracellular granules)[b] |
| | GA2 | GalNAcβ1→4Galβ1→4Glcβ1→1Cer | |
| | GA1 | Galβ1→3GalNAcβ1→4Galβ1→4Glcβ1→1Cer | Intestine (mouse) |
| 0-series | GM1b | Neu5Acα2→3Galβ1→3GalNAcβ1→4Galβ1→ 4Glcβ1→1Cer | Only trace amounts in adult mammalian brain.[a] |
| | GD1c | Neu5Acα2→8Neu5Acα2→3Galβ1→ 3GalNAcβ1→4Galβ1→4Glcβ1→1Cer | Only trace amounts in adult mammalian brain.[a] |
| | GD1α | Neu5Acα2→3Galβ1→3(Neu5Acα2→6)GalNAcβ1→ 4Galβ1→4Glcβ1→1Cer | Only trace amounts in adult mammalian brain.[a] |
| a-series | GM3 | Neu5Acα2→3Galβ1→4Glcβ1→1Cer | Relatively broad distributed, prominent GG in muscle endothelial cells |
| | GM2 | GalNAcβ1→4(Neu5Acα2→3)Galβ1→4Glcβ1→ 1Cer | Mouse liver, human tumors such as melanoma, neuroblastoma, gliobalstoma, and renal cell carcinoma |
| | GM1a | Galβ1→3GalNAcβ1→4(Neu5Acα2→3)Galβ1→ 4Glcβ1→1Cer | Major ganglioside in adult mammalian brain. Mouse intestine, testis |
| | GD1a | Neu5Acα2→3Galβ1→3GalNAcβ1→ 4(Neu5Acα2→3)Galβ1→4Glcβ1→1Cer | Major ganglioside in adult mammalian brain. |
| | GT1a | Neu5Acα2→8Neu5Acα2→3Galβ1→3GalNAcβ1→ 4(Neu5Acα2→3)Galβ1→4Glcβ1→1Cer | Major ganglioside in adult mammalian brain |
| | GT1α | Neu5Acα2→3Galβ1→3(Neu5Acα2→6)GalNAcβ1→ 4(Neu5Acα2→3)Galβ1→4Glcβ1→1Cer | Erythrocytes, spleen, bone marrow, testis, liver[c], adult brain.[c] |
| b-series | GD3 | Neu5Acα2→8Neu5Acα2→3Galβ1→4Glcβ1→1Cer | Embryonic neuronal tissue, lymphocytes |
| | GD2 | GalNAcβ1→4(Neu5Acα2→8Neu5Acα2→3)Galβ1→ 4Glcβ1→1Cer | Paediatric and adult solid tumors including neuroblastoma, glioma, retinoblastoma, ...[d] |
| | GD1b | Galβ1→3GalNAcβ1→4(Neu5Acα2→8Neu5Acα2→ 3)Galβ1→4Glcβ1→1Cer | Major ganglioside in adult mammalian brain |
| | GT1b | Neu5Acα2→3Galβ1→3GalNAcβ1→4(Neu5Acα2→ 8Neu5Acα2→3)Galβ1→4Glcβ1→1Cer | Major ganglioside in adult mammalian brain |
| | GQ1b | Neu5Acα2→8Neu5Acα2→3Galβ1→3GalNAcβ1→ 4(Neu5Acα2→8Neu5Acα2→3)Galβ1→4Glcβ1→1Cer | Major ganglioside in adult mammalian brain |
| | GQ1bα | Neu5Acα2→3Galβ1→3(Neu5Acα2→6)GalNAcβ1→ 4(Neu5Acα2→8Neu5Acα2→3)Galβ1→4Glcβ1→1Cer | |
| | GT3 | Neu5Acα2→8Neu5Acα2→8Neu5Acα2→3Galβ1→ 4Glcβ1→1Cer | Only trace amounts in adult mammalian brain. |
| | GT2 | GalNAcβ1→4(Neu5Acα2→8Neu5Acα2→ 8Neu5Acα2→3)Galβ1→4Glcβ1→1Cer | Formed during mammalian brain development.[a] |

| | Ganglioside/GSL | Structure | Occurence |
|---|---|---|---|
| | GT1c | Galβ1→3GalNAcβ1→4(Neu5Acα2→8Neu5Acα2→8Neu5Acα2→3)Galβ1→4Glcβ1→1Cer | |
| c-series | GQ1 | Neu5Acα2→3Galβ1→3GalNAcβ1→4(Neu5Acα2→8Neu5Acα2→8Neu5Acα2→3) Galβ1→4Glcβ1→1Cer | |
| | GP1c | Neu5Acα2→8Neu5Acα2→3Galβ1→3GalNAcβ1→4(Neu5Acα2→8Neu5Acα2→8Neu5Acα2→3)Galβ1→4Glcβ1→1Cer | |
| | GP1cα | Neu5Acα2→3Galβ1→3(Neu5Acα2→6)GalNAcβ1→4(Neu5Acα2→8Neu5Acα2→8Neu5Acα2→3)Galβ1→4Glcβ1→1Cer | |
| | Fuc-GM1 | Fucα1→2Galβ1→3GalNAcβ1→4(Neu5Acα2→3)Galβ1→4Glcβ1→1Cer | Mouse intestine and testis |
| Fucosylated | Fuc-Gal-GM1 | Fucα1→2Galα1→3Galβ1→3GalNAcβ1→4(Neu5Acα2→3)Galβ1→4Glcβ1→1Cer | |
| | Fuc-GD1b | Fucα1→2Galβ1→3GalNAcβ1→4(Neu5Acα2→8Neu5Acα2→3)Galβ1→4Glcβ1→1Cer | |
| | Fuc-Gal-GD1b | Fucα1→2Galα1→4Galβ1→3GalNAcβ1→4(Neu5Acα2→8Neu5Acα2→3)Galβ1→4Glcβ1→1Cer | |
| Galactosyl | Gal-GD1b | Galα1→4Galβ1→3GalNAcβ1→4(Neu5Acα2→8Neu5Acα2→3)Galβ1→4Glcβ1→1Cer | |
| | Gal-GalNAc-GM1b | Galβ1→3GalNAcβ1→4(Neu5Acα2→3)Galβ1→3GalNAcβ1→4Galβ1→4Glcβ1→1Cer | |
| | GalNAc-GM1 | GalNAcβ1→4Galβ1→3GalNAcβ1→4(Neu5Acα2→3)Galβ1→4Glcβ1→1Cer | Squamous cell cancer |
| | GalNAc-GM1b | GalNAcβ1→4(Neu5Acα2→3)Galβ1→3GalNAcβ1→4Galβ1→4Glcβ1→1Cer | |
| | GalNAc-GD1b | GalNAcβ1→4Galβ1→3GalNAcβ1→4(Neu5Acα2→8Neu5Acα2→3)Galβ1→4Glcβ1→1Cer | |
| | 9-*O*-Ac-GD3 | Neu5,9Ac$_2$α2→8Neu5Acα2→3Galβ1→4Glcβ1→1Cer | T and B lymphocytes, enhanced in ALL, glioblastoma, breast cancer, basaliomas, and other tumors. |
| | 9-*O*-Ac-GD1b | Galβ1→3GalNAcβ1→4(Neu5,9Ac$_2$α2→8Neu5Acα2→3)Galβ1→4Glcβ1→1Cer | Human peripheral nerve tissue, bovine brain[e] |
| Acetylated | 9-*O*-Ac-GT1b | Neu5Acα2→3Galβ1→3GalNAcβ1→4(Neu5,9Ac$_2$α2→8Neu5Acα2→3)Galβ1→4Glcβ1→1Cer | Brain[f] |
| | 9-*O*-Ac-GQ1b | Neu5Acα2→8Neu5Acα2→3Galβ1→3GalNAcβ1→4(Neu5,9Ac$_2$α2→8Neu5Acα2→3)Galβ1→4Glcβ1→1Cer | |
| | 9-*O*-Ac-GT3 | Neu5,9Ac$_2$α2→8Neu5Acα2→8Neu5Acα2→3Galβ1→4Glcβ1→1Cer | |
| | 9-*O*-Ac-GT2 | GalNAcβ1→4(Neu5,9Ac$_2$α2→8Neu5Acα2→8Neu5Acα2→3)Galβ1→4Glcβ1→1Cer | |

| Ganglioside/GSL | Structure | Occurence |
|---|---|---|
| GD1b-lactone | Galβ1→3GalNAcβ1→4(Neu5Acα2→8, 1→ 9 Neu5Acα2→3)Galβ1→4Glcβ1→1Cer | Adult human brain[g] |
| 3′isoLM1 | Neu5Acα2→3Galβ1→3GlcNAcβ1→3Galβ1→4Glcβ1→ 1Cer | Fetal brain[h] |
| 6′LM1 | Neu5Acα2→6Galβ1→4GlcNAcβ1→3Galβ1→4Glcβ1→ 1Cer | |
| 3′LM1 | Neu5Acα2→3Galβ1→4GlcNAcβ1→3Galβ1→4Glcβ1→ 1Cer | |
| 3′LD1 | Neu5Acα2→8Neu5Acα2→3Glcβ1→4GalNAcβ1→ 3Galβ1→4Glcβ1→1Cer | |
| Gb3Cer | Galβ1→4Galβ1→4Glcβ1→1Cer | Human erythrocytes, kidney, spleen, serum, and other extraneural tissues |
| Gb4Cer | GalNAcβ1→3-Galβ1→4Galβ1→4Glcβ1→1Cer | Human erythrocytes, kidney, spleen, serum, and other extraneural tissues |

Sialic acid containing glycosphingolipids are underlined in yellow.

*ALL*, Childhood acute lymphoblastic leukemia; *Cer*, ceramide; *Gal*, galactose; *GalNAc*, *N*-acetylgalactosamine; *Glc*, glucose, *GG*, ganglioside; *Neu5Ac*, *N*-acetylneuraminic acid. All human gangliosides and nonhuman brain gangliosides contain exclusively or predominantely Neu5Ac, whereas nonhuman extraneural mammalian gangliosides often contain Neu5Gc.

a: [95], b: [96], c: [86], d: [97], e: [98], f: [99], g: [100], h: [101].

## 2.1.8 Golgi, Combinatorial Ganglioside Biosynthesis

In neurons, the GM3/GM4-sialyltransferase, ST3GalT5, transfers sialic acid from CMP-sialic acid to the galactosyl moiety of LacCer in α2,3 linkage and forms the simple GG GM3.[102] Further sialylation of GM3 generates GD3, in which the second sialic acid is α2,8-linked to the first sialic acid. This step is catalyzed by the GD3-synthase (St8sia1) (Fig. 2). In contrast to downstream glycosyltransferases, GD3-, GM3-, and LacCer-synthases are highly specific for their glycolipid substrates. Additional sialylation of GD3 at the terminal sialic acid, resulting in GT3 formation, is catalyzed by two sialyltransferases, St8sia3 and St8sia5, both of which are promiscuous for various glycolipid substrates.[103,104] As suggested by Saul Roseman, biosynthesis of complex glycolipids and GGs is mediated by multiglycosyltransferase systems.[105–107] They are often arranged in a modular organization in homo- and hetero-multienzyme complexes in the membranes of the Golgi and TGN.[108] Cysteines of membrane spanning glycosyltransferases are targets of *S*-acylation and are also involved in the formation of homodimers.[109,110] Higher GGs are formed by stepwise glycosylation reactions at the luminal face of

Golgi and TGN membranes. Enzymatic steps for the formation of a- and b-series GGs, for example, GM1 and GD1a, had already been identified in the 1960s and 1970s.[111,112] It was expected that different glycosyltransferases were involved in the biosynthesis of a- and b-series GGs, respectively. However, our group failed to detect specific GalNAc-transferases for the biosynthesis of GM2 (belonging to the a-series) and GD2 (belonging to the b-series), respectively.[113] On the contrary, kinetic competition experiments between the lipid acceptors GM3 and GD3 proved that both competed for one and the same active site of a quite unspecific GalNAc-transferase.[113] All glycosyltransferases involved in the biosynthesis of complex GGs turned out to be rather promiscuous for their lipid substrates,[114–116] resulting in the assembly line given in Fig. 2.

LacCer/GA3, GM3, GD3, and GT3—and even the sulfated form of LacCer (SM3)—are all substrates for a single *N*-acetylgalactosaminyl transferase, *B4galnt1*, which elongates the lactosyl moiety with *N*-acetyl galactosamine (GalNAc) in β1,4-linkage yielding GA2/Gg3Cer, GM2, GD2, and GT2 as well as the sulfated SM2a, the first members of the 0-, a-, b-, and c-series of GGs[113,114] and the sulfoganglio-series.[31] Likewise, these products are all substrates of the galactosyl transferase B3galt4.[117] Its products—GA1, GM1a, GD1b, and GT1c—are, therefore, all substrates of the sialyltransferases St3gal2 and St3gal1.[116] Thus the major brain GGs—GM1a, GD1a, GD1b, and GT1b—are generated by combinatorial biochemistry. Initially, this concept was established in enzymatic competition experiments[113,118] and later was confirmed in mutant mouse models deficient in either of the following enzmyes: GalNAc-transferase (B4galnt1), GM3-synthase (St3gal5), GD3-synthase (St8sia1), and the sialyltransferases 1-4 (St3gal1-4).[25,119–123]

### 2.1.9 Sialic Acid Modifications

The physiological precursor of all sialic acids is *N*-acetyl D-mannosamine (ManNAc). Mammalian biosynthesis of the most common sialic acid, NeuNAc, begins with the formation of ManNAc from UDP-*N*-acetylglucosamine (UDP-GlcNAc), which is catalyzed by the UDP-GlcNAc 2-epimerase encoded by the *GNE* gene. Due to the hydrolytic release of UDP, the reaction is irreversible. The bifunctional enzyme then phosphorylates ManNAc with its kinase activity and generates ManNAc 6-phosphate (ManNAc-6P). The bifunctional enzyme catalyzes the rate-limiting step of NeuNAc synthesis and therefore functions as a key regulator of cell surface sialylation. Subsequently, sialic acid synthase/NeuNAc 9-P synthase, encoded by the *NANS* gene, catalyzes the condensation of MacNAc-6P with phosphoenolpyruvate to release NeuNAc-9P. NeuNAc-9P is then dephosphorylated to NeuNAc by the NeuNAc-9P phosphatase (*NANP* gene) and the activated form of sialic acid, CMP-sialic acid, is generated by consuming CTP catalyzed by the CMP-sialic acid synthetase (*CMAS* gene).[9]

#### 2.1.9.1 *N*-Glycolyl-Neuraminic Acid

A major modification of NeuNAc in mammals is the hydroxylation of the *N*-bound acetyl to form *N*-glycolyl-neuraminic acid (NeuGc), catalyzed by the CMP-NeuNAc hydroxylase (*CMAH* gene).[124] Men, however, have lost activity of this gene due to mutation.[125] Nevertheless, NeuGc is found in several tumor types, such as melanoma and breast cancer,[126] but rarely in normal tissue. It is presumed that humans acquire NeuGc from dietary sources. Metabolically processed and incorporated into glycoconjugates, it is then expressed on epithelial and endothelial cell surfaces.[124]

Although other mammals, including mice, horse, sheep, and goats, are able to synthesize NeuGc, it is always absent in brain.[124] One reason may be that NeuGc significantly alters the conformational shape of polysialic acids on neural cell adhesion molecules, thereby inhibiting substantially sialidase activity.[127] On the contrary, axonal GGs stabilize the binding of the axons to the innermost myelin sheath in a trans-lectin interaction with myelin associated glycoprotein (MAG/Siglec-4a),[22] which depends on the presence of NeuNAc, but is inhibited by terminal NeuGc on GGs.[21] Indeed, overexpression of a functional CMAH gene in mouse brain and incorporation of NeuGc in GGs disrupt the binding between axon surface GGs and the myelin sheets surrounding the axons, thus resulting in disturbed locomotor activity, impaired memory, and abnormal axon myelination.[128]

#### 2.1.9.2 *O*-Acetylation

Another frequent modification of sialic acids is *O*-acetylation,[129] which is observed at the C4, C7, C8, and C9 hydroxyl groups and is also found on GGs.[130,131] Influenza C viruses differ from other influenza viruses because their preferred sialic acid is *N*-acetyl-9-*O*-acetylneuraminic acid (Neu5,9Ac2),[132,133] which facilitates binding when present on GGs.[134] A preferred substrate of *O*-acetylation is GG GD3, where *O*-acetylation occurs on the glycerol side chain of terminal α-2,8-linked sialic acid.[135] 9-*O*-acetyl GD3(CD60b) and 7-*O*-acetyl-GD3 (CD60c) are differentially regulated during differentiation and maturation of human T and B lymphocytes.[136]

Modification by *O*-acetylation reduces the rate of hydrolysis by sialidases.[137,138] 9-*O*-acetyl-GD3 is considered an oncofetal marker.[139] Considered a biomarker of melanoma,[140] *O*-acetylated GD3 was found to be enhanced on childhood lymphoblastic leukemia,[141] glioblastoma,[142] breast cancer, basaliomas, and tumors of neuroectodermal origin.[131,143] *O*-acetylation is regulated by sialate *O*-acetyl-transferases (SOAT) and *O*-acetyl-esterases after sialic acid had attached to the glycoconjugate.[139] The mammalian AcCoA:sialate-7(9)-*O*-acetyl-transferase (E2.3.1.45) is thought to initially transfer the acetyl group to the 7-position from where it migrates nonenzymatically to the 9-position, possibly via the 8-position. The 9-*O*-acetyl-product then can be acetylated again in 7-position by the same enzyme, generating a 7,9-*O*-bisacetyl-sialate residue.[144,145] Recently, human CASD1 (CAS1 domain 1) was shown to transfer acetyl residues from acetyl-CoA via intermediate acetylation of a serine residue as part of the enzymes catalytic triad (Ser–His–Asp) to generate in combination with sialyltransferase St8sia1 GG 9-*O*-acetyl-GD3[146] and a potential transfer of the acetyl group to CMP-NeuNAc, used by St8sia1 to sialate GM3, was discussed. In a previous study, overexpression of CASD1 together with St8sia1 yielded 7-*O*-acetyl-GD3 in a different cell system.[147] In both studies CASD1 located to the Golgi. Eventually, additional cell-specific factors regulate the outcome of acetylation, which could explain the differential results. Another enzyme, the AcCoA:sialate-4-*O*-acetyltransferase (EC 2.3.1.44), was described in Golgi-enriched fractions of guinea pig liver[148] and its preferred substrate is GGs.

#### 2.1.9.3 Lactonization

Another intriguing modification of GGs at the plasma membrane is lactonization of disialio residues like in GD1b ($GD1b_L$) (Fig. 3), in which the carboxylate of the terminal sialic acid forms a lacton with the (7- or) 9-hydroxyl group of the inner sialic acid. As a minor component, $GD1b_L$ is found in mammalian brain and resists sialidase treatment.[100,149] The mechanisms of inducing the lactonization process are not clarified. Lactonization profoundly influenced the conformational, aggregational, and biological properties of GD1b and thus influences GG-mediated cellular processes.[150]

### *2.1.10 Cell Surface Modeling of Gangliosides*

Complex GGs and GSLs reach the plasma membrane by vesicular transport and are exposed to the extracellular leaflet where they can undergo lateral

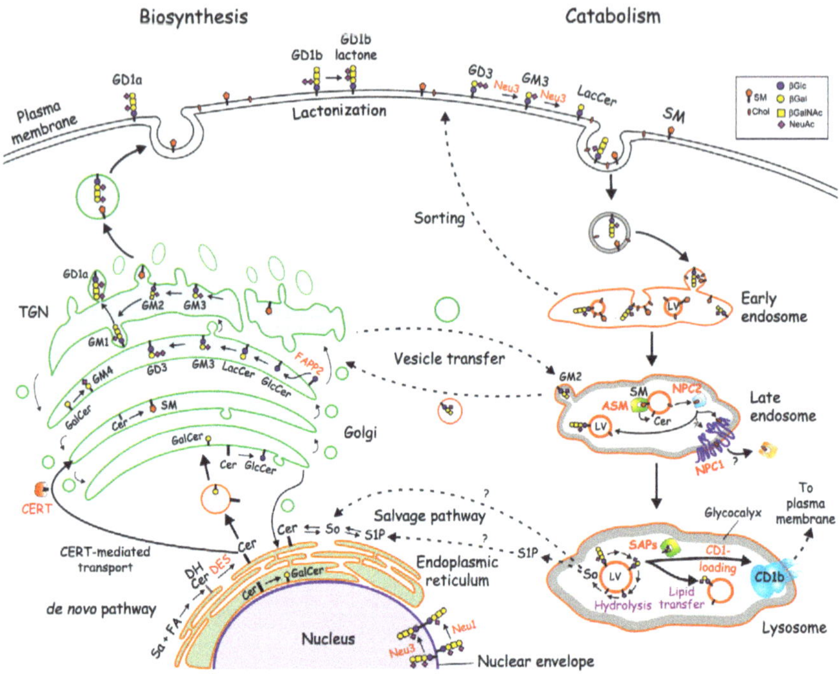


**Fig. 3** Subcellular compartmentalization and intracellular trafficking of ganglioside biosynthesis and secretion (left side) and of endocytosis and catabolism (right side). *ASM*, Acid sphingomyelinase; *Cer*, ceramide; *CERT*, ceramide transfer protein; *DES*, dihydroceramide desaturase; *DHCer*, dihydroceramide; *FA*, fatty acid; *GlcCer*, glucosylceramide; *LacCer*, lactosylceramide; *LV*, intraendolysosomal luminal vesicles; *Neu*, neuraminidase; *NPC*, Niemann–Pick disease type C protein; *Sa*, sphinganine; *SAP*, sphingolipid activator protein; *SM*, sphingomyelin; *So*, sphingosine; *S1P*, sphingosine-1-phosphate; *TGN*, trans Golgi network. *Modified after Breiden B, Sandhoff K. Ganglioside metabolism and its inherited diseases. Methods Mol Biol. 2018 [in press]*

associations and trans-interactions[22] (Fig. 3). In brain, GGs were found enriched in nerve endings.[152,153] Remodeling of GGs on synaptic surfaces may contribute to synaptic function and hitherto to their plasticity, which is considered to be of great importance for memory, behavior, and learning functions.[154] On postsynaptic membranes, GGs appear to regulate trafficking of receptors like the AMPA-type glutamate receptor (AMPAR). It was postulated that receptors and their trafficking complexes bind to different GGs (e.g., the AMPAR to GM1) and its trafficking complex to GT1b, thereby segregating into different membrane domains. ATP-dependent release of GG associations would then facilitate association of receptors with their trafficking complexes and corresponding endocytosis.[155] Such interactions may be regulated by plasma membrane-located sialidase and sialyltransferase activities, both of which are present in synaptosomal membranes.[156–160] Experiments with radiolabeled GGs and other sphingolipids demonstrated hydrolysis on isolated neuronal plasma membranes following a two-dimensional Michaelis–Menten kinetic and stimulation of activity with increasing membrane fluidity.[161,162] Neu3, which is one of four sialidases (Neu1-4), is mainly a plasma membrane-associated sialidase on the cell surface and has been shown to cleave sialic acids from complex GGs. The human, murine, and bovine Neu3 have a pH optimum around 4.6, and the human Neu3, in addition, has a second optimum between pH 6.0 and pH 6.5.[163] Local pH changes within the synaptic cleft, for example, upon hydrolysis of ATP, may increase Neu3 activity and quickly regulate changes in GGs compositions by desialylation of polysialo-GGs. Neu3 activity further is regulated by phosphatidic acid, which binds, activates (at neutral pH), and translocates Neu3 to the plasma membrane.[164] Whether GGs are degraded on neuronal plasma membranes down to ceramide in vivo, as proposed according to experiments in cultured cerebellar granula cells,[165] awaits further investigations.

### 2.1.11 Recycling of Gangliosides

If GG modifications contribute to regulated cell surface events like control of receptor surface expression, the recycling of desialated GGs would save energy and time to enable enhanced cycle time of the events. Radiotracer techniques and ESR spectroscopy experiments demonstrated that exogenously added GM2, incorporated into the plasma membrane of GM2 gangliosidosis cells with a full block of GM2 degradation, could still form radiolabeled GGs GM1 and GD1a, by trafficking directly from endosomes to the TGN compartment,[166,167] which presumably involves the retromer complex.[168] Similarly, generation of GD1a in part by direct sialylation of GM1, but not of GD1b and GT1b, was observed in rat liver.[169,170] These findings are corroborated by transcytosis or endosomal to TGN and ER trafficking of GM1 especially if containing unsaturated or shorter ceramide anchors,[32,33] an important aspect of cholera toxin metabolism in epithelial cell and hence its nocuous effect. Analyzing fluorescent GSL analogs, a recycling of labeled glucosylceramide from and back to the plasma membrane was observed[171] and a nondegradable glucosylceramide analogue can be converted to complex GGs involving trafficking from proximal to distal Golgi cisternae.[172] Vimentin-intermediates appear to support the recycling of GSLs from endosomes to Golgi and

TGN as well as of sphingoid bases from lysosomes to ER for ceramide synthesis.[173–175] Whether such a GGs recycling could play a role in synaptic plasticity is not known. According to the aforementioned studies, synaptic GGs (eventually desialylated at the synapse) would, however, need to travel back to the most proximal segment of a subset of dendrites for glycosylation/sialylation and recycling via the Golgi/TGN.[176]

#### *2.1.12 Nuclear and Mitochondrial Gangliosides*

Besides sphingomyelin (SM) and GSLs, small amounts of GGs have also been found in mitochondria, nucleus, and ER membranes.[177] Their origin is still poorly investigated. Nuclear envelope membranes contain GGs GM1 and GD1a, together with the sialidases Neu3 in the inner membrane and Neu1 in the outer nuclear envelope membrane. GM1 of the inner membrane stimulates the activity of a $Na^+/Ca^{2+}$ exchanger. The absence of GM1 in B4galNt1-knockout mice triggers an elimination of the cells by apoptosis. GD3 is a minor component of most tissues.[177] It is upregulated in many tumors,[178] and induces a loss of mitochondrial transmembrane potential, DNA fragmentation, and apoptosis.[179] Other sphingolipids also occur in the nucleus and the mitochondria. SM, for instance, can be degraded to ceramide to trigger apoptosis, and furthermore to sphingosine and free fatty acids.[177]

## 3. ANABOLISM AND DISEASE

### 3.1 Consequences of Dysfunctional Genes in Human Disease and Animal Models

#### *3.1.1 Endoplasmic Reticulum, Step 1, Serine Palmitoyl-CoA Transferase*

Upon aging, GGs of the central nervous system (CNS) are increasingly enriched with the longer C20-sphingosine (C20-GGs),[180–182] particularly in the molecular layer of the dentate gyrus.[183] Pathological accumulation of C20-GGs in a mouse model of mutated SPTssb leads to neurodegeneration.[184] The high specificity for L-serine as a substrate of SPT is lost by missense mutations in the SPTLC1 gene, which cause hereditary sensory and autonomic neuropathy type 1 (HSAN1). Increasing amounts of glycine and L-alanine are then converted by the $SPT^{HSAN1}$ to the 1-deoxy- and 1-deoxymethyl-sphingolipids,[185] which cannot be converted, neither to higher sphingolipids including GGs nor to sphingosine 1-phosphate. Thus they have to be degraded by a sphingosine 1-phosphate lyase 1 (SGPL1) independent pathway involving cytochrome P450 oxygenases leading to increased half-life of these toxic compounds.[186]

### 3.1.2 Endoplasmic Reticulum, Step 3, Ceramide Synthases

Deficiency of *CerS1* activity in mice leads to decreased levels of stearic acid-containing GGs and ceramides, but to an increase of free sphingoid bases followed by neurodegeneration.[187,188] This phenotype is correlated with decreased CerS1 activity due to mutations in CERS1 causing a progressive disorder with myoclonic epilepsy and dementia.[189] The major GG of mouse hepatocytes is GM2, and it contains mostly very long-chain fatty acids in its ceramide anchor.[27] Disruption of the Cers2 gene in mice causes depletion of ceramide anchors with very long acyl chains (mainly C22 and C24) and a compensatory increase of anchors with long chains (C16-fatty acids).[190,191] CerS2-depleted mice exhibit glucose intolerance and abrogation of insulin receptor and Akt phosphorylation in liver.[192] The decisive role of GG GM2 in this process, however, is unclear. GGs of most tissues are enriched in saturated fatty acids with a chain length between 16 and 24 carbon atoms, which is due to the preference of the CerS1, CerS2, and CerS4-6 for corresponding saturated acyl-CoA substrates. GGs and neutral GSLs found in differentiating male germ cells, however, are highly enriched in polyunsaturated and ultra-long acyl chains (3–6 double bonds and 28–32 carbon atoms in length).[193,194] These GSLs are important for the stability of intercellular bridges connecting clonal spermatids and loss of them leads to an arrest of spermatogenesis and male infertility.[195] The polyunsaturated anchor shifts these GSLs including GGs from the detergent insoluble to the detergent soluble fraction[193] and likewise modulates their participation into microdomains of cell membranes similar to corresponding ceramides and SMs.[196,197]

### 3.1.3 Golgi, Step 5, Glucosylceramide Synthase

Loss of glucosylceramide synthase activity due to a systemic disruption of the *Ugcg* gene arrests embryogenesis at E6.5 in mice.[198] Mice with cell-specific disruption of the Ugcg gene in neural cells using a Cre/loxP-system under control of the nestin promotor were apparently normal at birth, but died around 3 weeks after birth. All neuronal GGs are gone in these mice, without replacement by any glycolipid. Instead, SM accumulates, whereas ceramide levels stay more or less steady. The mice developed severe ataxia from 5 days after birth and frequently dropped sideways without external contact. Microscopically, GGs deficiency leads to reduced axon branching, best visible in Purkinje cells and instability of myelin sheath axon assembly.[199] However, it remains unclear, if a loss of GM1-induced integrin signaling via cross-linking with galectin-1 causes a reduced axonal branching similar to in vitro neuritogenesis.[200] Tackling Ugcg, specifically in Purkinje cells, induced disrupted myelin sheaths with detached paranodal loops, thus indicating that Purkinje cells GGs are required for axonal homeostasis and correct myelin sheath formation.[201] GG deletion in forebrain neurons only by disruption of the Ugcg gene under the control of the CamK promotor revealed GGs of the hypothalamic neurons to activate the leptin receptor (ObR), which appears to be important for corresponding neuronal activity in response to alterations in peripheral energy signals. Corresponding GG loss is associated with obesity, hypothermia, and lower sympathetic activity in mice.[18] Further studies suggested that complex GG GD1a interacts dynamically with the insulin receptor on adult hypothalamic neurons by modulating mediobasal hypothalamus insulin signaling and white adipose tissue function in fasted mice.[202] Investigating corresponding GG in a model of familial Alzheimer's disease suggested GGs to limit levels of functional dendritic insulin receptors on the cell surface of neurons by increasing caveolin-1-mediated insulin receptor internalization. Consequently, GGs decrease insulin receptor signaling in neurons exposed to oligomeric amyloid-β-derived diffusible ligand stress and may contribute to the progression of neurodegeneration in Alzheimer's disease.[203] Cell-specific disruption of the *Ugcg* gene in nonneuronal tissues leads to GG GM2 depletion in murine hepatocytes without detectable loss of major liver functions.[27] Likewise, depletion of GGs like GD1a did not affect Sertoli cell function for murine male fertility in a corresponding cell-specific model of *Ugcg*-depletion.[195] Cell-specific disruption of the Ugcg gene in mice depleted GGs together with abundant neutral GSL in intestinal mucosa cells indicating GSL involvement in endocytic function of enterocytes,[26] and in male germ cells where these compounds are important for the stability of intercellular bridges between sister spermatids.[195]

### 3.1.4 Golgi, GM3-Synthase (GM3S)/ST3GAL5

#### 3.1.4.1 Human Disorders Due to ST3GAL5 Mutations

A nonsense mutation (c.694C > T) in GM3-synthase (GM3S) with loss of GM3S activity causes recessive infantile-onset symptomatic epilepsy syndrome, developmental stagnation, and blindness. Seizure activity starts within the first year of life. In corresponding human plasma samples, GGs are gone and the precursor LacCers accumulates.[204] Vision loss in these children could be associated with optic nerve atrophy and possibly cortical visual impairment, but retinal function remained normal.[205] In a follow-up study, 20 out of 38 patients with GM3S deficiency developed progressively pigmentation abnormalities including bilateral freckle-like hyperpigmented macules on the extremities as well as depigmented macules and patches after the first years of life.[206] In addition, children with this nonsense mutation (ST3GAL5 c.694C>T) develop a neonatal onset hearing impairment likely due to an essential role of GG GM3 for the structural integrity and function of cochlear hair cells.[207] Similar symptoms were observed in children with a c.862C4T (p.Arg288*) mutation in the GM3S gene. GM3 was absent in patient fibroblasts, which showed enhanced expression of neutral globo-series GSLs. These changes appear to cause a decrease in the mitochondrial membrane potential and enhanced apoptosis in fibroblasts from these patients.[208] Recently, the "Salt and Pepper" syndrome, an autosomal recessive condition characterized by severe intellectual disability, epilepsy, scoliosis, choreoathetosis, dysmorphic facial features, and altered dermal pigmentation, was associated with a homozygous c.994G>A transition in the ST3GAL5 gene leading to an exchange of a glutamate by a lysine (p. E332K) within a highly conserved sialyltransferase domain. Patient fibroblasts were devoid of GG GM3 and displayed changes in protein glycosylation. The latter was corroborated in zebrafish with antisense morpholinos that targeted zebrafish st3gal5 expression. In addition, increased cell apoptosis in multiple brain regions of zebrafish embryos was observed.[209] In accordance with lysosomal degradation of GG fed to cells in culture as major pathway, the amount of GGs in breast milk and general nutritional intervention does not seem to alter the outcome of the human disease.[210]

#### 3.1.4.2 GM3S Mouse Models

Unlike man lacking GM3S activity, mice with a genetic deletion of the GM3S are viable and appear without major abnormalities, but have enhanced insulin receptor phosphorylation in skeletal muscle. Mutant mice are protected from high-fat diet-induced insulin resistance, supporting GG GM3 as a negative regulator of insulin signaling. Analysis of mutant mouse brain revealed expression of 0-series complex GGs (GM1b, GD1α) instead of a- and b-series GGs keeping up the lipid-bound sialic acid levels to normal.[25] Furthermore, GM1b and GD1α express just as GD1a and GT1b of normal brain the same terminal trisaccharide (NeuNAc2α3Gal1β3GalNAc1β-), which facilitates lectin-binding of MAG, which is in case of GD1α due to the sialic acid in 6-position of the

GalNAc-residue even enhanced.[211,212] Whether a similar replacement of neuronal a- and b-series by 0-series GGs takes place in human patients with GM3S deficiency remains so far unknown. More detailed investigations revealed a gender-specific impairment of neuropsychological behaviors in juvenile mice resembling attention-deficit hyperactivity disorder (ADHD).[213] In accordance with patients, mice lacking GG GM3S exhibit complete hearing loss due to selective degeneration of the organ of Corti. Apparently, GG GM3 is essential for the structural integrity and function of cochlear hair cells. In contrast to total brain, the cochlea of mice is rich in GM3 with complex GGs being less abundant. GM3 localizes to different regions than GT1b and is almost exclusively expressed during early postnatal maturation of the cochlea.[207,214] In this critical phase, loss of GM3S probably leads to a complete loss of GGs in outer and inner hair cells, which in part could explain deafness due to GM3S deficiency in mice and eventually in man.

GG GM3 is not only an inhibitor of insulin receptor signaling, but GM3S expression is also increased in human diabetic foot skin, ob/ob, and diet-induced obese diabetic mouse skin. Consequently, disruption of the St3gal5 gene in diet-induced obese mice prevents the diabetic wound-healing defect due to increased activation of insulin receptor and insulin growth factor-1 receptor in keratinocytes.[215] Furthermore, depletion of GM3S reverses neuropathic pain and small fiber neuropathy in diet-induced diabetic mice possibly by preventing increased calcium influx and neuronal hyperexcitability in dorsal root ganglia sensory neurons.[216]

GM3 is the most widely distributed GG among tissues and further nonneurological phenotypes should be expected in corresponding deficient mouse models.

Indeed, analysis of GM3S-deficient mice revealed severely compromised TCR-dependent activation of CD4$^+$ T cells, but not of CD8$^+$ T cell, which was not observed in GM2-synthase deficient mice. Correspondingly, CD4$^+$ T cells expressed relatively more a-series GG, but the portion of 0-series GG was higher in CD8$^+$ T cells.[217,218] GGs GM3 and more complex GGs are also expressed by human lymphocytes,[219,220] and human T and B cells partially convert it into to *O*-acyl-GD3,[136] making a transfer of corresponding function principally possible. CD4$^+$ T cells are a key component in primary inflammatory lesions. A subset of Th17 cells and regulatory T cells are suggested to mediate inflammation in rheumatoid arthritis.[221] Human articular cartilage contains mainly GM3 and GD3,[222] and mRNA levels of GM3S in the synovium are higher in patients with rheumatoid arthritis than in those with osteoarthritis. In a collagen-induced arthritis, mouse model GM3S deficiency, however, exacerbated inflammatory arthritis and induced T-cell activation prompting overproduction of cytokines involved in rheumatoid arthritis. These results indicate that GM3 may act as a regulatory molecule to suppress the immunologic response in $CD4^+$ T cells to stimulus of T-cell receptor antigen.[221] The apparent discrepancy of results to the aforementioned effects in $CD4^+$ cells may underlie different functions of GM3 and GGs in subsets of lymphocytes and local environments. In an age-associated and in an instability-induced model of osteoarthritis, the depletion GGs due to GM3S deficiency led to enhanced cartilage degradation in mice, which was in vitro associated by enhanced induction of metalloproteinase secretion from chondrocytes.[223] However, in a model of articular cartilage injuries, depletion of GGs due to GM3S deficiency enhanced articular cartilage repair. In this model, GM3/GGs appeared to play a role in the late repair phase and enhanced certain hypertrophy-related genes.[224] By comparing mouse embryonic fibroblasts (MEFs) from GM3S-deficient mice with control MEFs, a modulation of the glycoprotein *N*-glycan profile was observed,[225] implicating that changes in posttranslational protein modifications due to altered intracellular GSL profiles may contribute to one or the other phenotype observed in GM3S mice.[225] Also, other pathological processes such as nephronophthisis-related polycystic kidney disease appear to be enhanced by the presence of GGs and are diminished in GM3S-deficient mice.[226]

### *3.1.5 Golgi, GD3-Synthase/ST8SIA1*

Patients with mutations in the GD3-synthase have, so far, not been reported.

### *3.1.6 GD3 Deficiency in Mice*

GD3 and b-series GG expression change dramatically during neuronal development. However, mice and embryonic stem cells undergo neuronal differentiation in the absence of b-series GGs. Loss of b-series GGs in GD3S-deleted mice was compensated by a corresponding increase of a-series GGs, mainly GM1a and GD1a.[120,227,228] Likewise, b-series GG GQ1bα disappeared and corresponding a-series GT1aα increased in cholinergic neurons. GQ1bα is specifically elevated in a double-Tg (PAA/PSEN1) Alzheimer mouse model, and deletion of GD3S in this model reverses the neuropathology, indicating a destructive impact of b-series GGs in Alzheimer disease.[229,230] Initially, impaired regeneration of the lesioned hypoglossal nerve was observed in GD3S-deficient mice.[228] Follow-up studies revealed also morphological abnormalities in the sciatic nerve and neuronal disturbances during peripheral nerve regeneration,[231] likely due to a loss of 9-*O*-acetyl GD3 expression. The self-renewal ability of neural stem cells is diminished in

GD3S-KO mice, which was accompanied by reduced expression of EGFR, likely due to lost interaction of EGFR with GD3, thus inducing increased degradation.[232] In another study, progressive loss of hippocampal neural stem cells was observed, leading to impaired neurogenesis at the granular cell layer of the olfactory bulb and the dentate gyrus in adult; and loss of b-series GGs was associated with depression-like behavior.[233] However, under pathological conditions, when parkinsonism was induced by 1-methyl-4phenyl-1,2,3,6-tetrahydropyridine (MPTP), GD3S deficiency had neuroprotective properties leading to smaller lesions and less decrease in striatal dopamine reduction.[234] Phenotypically, increased sensory responses to thermal and mechanical stimuli was evaluated but mice showed decreased responses during the late phase of the formalin test, indicating b- and c-series GGs are critical components of the sensory nervous system responsible for the transmission of acute pain sensation and pain modulation. Moreover, they appear to play an important role in the development of hyperalgesia induced by inflammation.[235] On the molecular levels, loss of b- and c-series GGs could modulate activity of membrane receptor involved in these processes similar to EGF-receptor phosphorylation in these (GD3S-deficient) mice.[236] Investigation of GD3S-deficient mice further demonstrated positively regulated leptin/Ob receptor-mediated signals in hypothalamus, which was attributed to increased levels of a-series GGs[237] supporting previous findings.[18] Moreover, b-series GGs appear to stimulate the secretion rate of leptin in adipose tissue, which retains leptin in the absence of GD3S; the involvement of membrane microdomains was previously discussed.[238]

### *3.1.7 Golgi, GM2 Synthase/B4GALNT1*

#### 3.1.7.1 Human Disorders Due to B4GALNT1 Mutations

In contrast to ST3GAL5 mutations, humans with autosomal recessive B4GALNT1 (GM2 synthase) deficiency have no hearing loss, visual impairment, or epileptic encephalopathy, but suffer from progressive hereditary spastic paraplegia. Analysis of patient fibroblasts revealed lack of GM2 production and increased levels of the precursor GG GM3[239] confirming GM3 to be sufficient for postnatal maturation of the cochlea.

### *3.1.8 GM2S Mouse Models*

The GM2S-deficient mouse model was one of the first models established that affected GG biosynthesis. All complex GGs of the brain are gone and simple GGs GM3 and GD3 remain.[122,240] In addition, increased levels of 9-

*O*-acetyl GD3 were observed in the brain.[241] First investigations revealed surprisingly subtle defects in their nervous system including slight reduction in the neural conduction velocity.[122] Loss of this enzyme, however, led to an arrest in spermatogenesis.[242] Since GM3S-knockout mice are fertile, it was concluded that neutral GSLs of the ganglio-series, which also depend on GM2S, were important.[193,194] Follow-up studies demonstrated myelination defects and axonal degeneration[243] leading to motor deficits in 8-month-old mice.[244] Neuromuscular junctions stay fully functional in these mice, demonstrating redundancy of complex GGs with GM3 and GD3,[245] even in aged mice.[246] Furthermore, the rate of motor nerve terminal regeneration was not affected.[247] Analysis of cerebellar granule neurons from neonatal GM2S-knockout, however, revealed impaired capacity for calcium regulation and degeneration in the presence of depolarizing levels of potassium or glutamate.[248,249] GM2S-deficient mice suffer enhanced susceptibility to kainate-induced seizures, neuronal apoptosis, and death.[250] Similarly, hypoglossal nerve regeneration is impaired upon GM2S deficiency, indicating complex GGs to control the expression levels of neurotrophic factors and receptors.[251] The shift from complex to simple GGs in GM2-S-deficient mice enhances amyloid β-protein deposition[252] and accelerated the clinical disease and pathology in a mouse model of mucopolysaccharidosis type IIIB.[253] Furthermore, GM2S-deficient mice develop Parkinson disease symptoms, likely due to reduced glial cell-derived neurotrophic factor (GDNF) signaling, as a result of absent GM1–GDNF receptor interaction.[254]

Immunological analysis of GM2S-depleted mice revealed reduced size and cell number in thymus and spleen associated with reduced proliferation upon IL-2 treatment[255] and deficient $CD8^+$ T-cell activation.[217] To what extent the activity of membrane receptors, like the IL-2 receptor, are modulated by the presence of GGs is not known, but similar mechanisms as the modulation of EGFR phosphorylation in mice expressing different GGs[236] could be the basis of such findings.

### 3.1.9 GM3-Only Mice

wCombining loss of GD3-synthase (St8sia1) and GalNAc-transferase (B4galnt1) in mice restricts lipid-bound sialic acid to GM3 (and the minor GM4) in these "GM3-only" mice, which exhibit lethal audiogenic seizures. In contrast to the corresponding single-knockout mice expressing either a-series GG or simple GG GM3 and GD3, the lipid-bound sialic acid content was significantly reduced in brain of GM3-only mice.[120] Similar GM3-only mice, but established independently, displayed sudden death starting from 12 weeks on. However, it was not attributed to audiogenic seizures and differences were associated with the differences in background.[119] Further investigation of the latter mice revealed reduced motor and sensory functions as well as emotional response, progressing with aging[256]; and this was later attributed substantially to dysfunctional muscarinic acetylcholine receptors.[257] Wisp2/CCN5 (Wnt-1-induced signaling protein-2/cysteine-rich 61-connective tissue growth factor-nephroblastoma-overexpressed 5) was upregulated in the brain of these mice, likely as a protective mechanism against neurodegeneration[258]; the latter likely inducing the observed corresponding upregulation of complement-related genes, proliferation of astrocytes, and infiltration of microglia.[12,259,260] Likewise, GG deficiency causes inflammation and neurodegeneration via the activation of complement system in the spinal cord.[261] GM3-only mice displayed largely intact synaptic transmission at the neuromuscular junction, showing, however, some extra degree of rundown of acetylcholine release at high intensity use, indicating a modulating role of presynaptic GGs.[262] Initially, a refractory skin injury was observed in the second mouse GM3-only model, primarily on the face of the mutant mice at 25 weeks after birth and attributed to reduced sensory function.[119]

### *3.1.10 LacCer/SM3-Only Mice*

Combining loss of GM3-synthase (St3gal5) and GalNAc-transferase (B4galnt1) in mice depleted all main GGs of wild-type tissues. The neutral GSL lactosylceramide (LacCer), the precursor of GM3 synthesis, accumulated in the brain. Surprisingly, neuronal cells in the brain of these GG-null mice start to express sulfated LacCer, eventually in an attempt to compensate loss of lipid-bound negative charge. Loss of GGs resulted in CNS degeneration and altered axon–glial interactions[123] and the mice died almost as early as those depleted completely for GSL in neural cells,[199] although a few 10% survived until 3 months.

### *3.1.11 Sialyltransferase-Deficient Mice*

Lately, mutant mice with a deficiency of either of the four sialyltransferases St3gal1-4 were established. Only St3gal2 appears to be essential for homeostatic levels of brain GGs on its own, with about half the normal amount of GD1a and GT1b remaining in mutant brains. St3gal3 obviously was responsible for the synthesis of most of the residual GGs with terminal sialic acids as the St3gal2/3-double-null mice were almost completely depleted in GD1a and GT1b.[121]

In St3gal2/3-double-null mice however, the total protein-bound sialic acid content was also reduced to less than 50%, mainly due to loss of St3gal3, making it difficult to attribute the observed phenotype to either GGs or proteins.[263] In contrast, St3gal2 deficiency alone did not affect protein sialylation, but corresponding mice developed late-onset obesity, which was attributed to altered GG expression in adipose tissue resulting in diminished IR sensitivity.[264]

# 4. CATABOLISM

## 4.1 Endocytosis of Gangliosides, Catabolic, and Salvage Pathways

### *4.1.1 Topology of Ganglioside Uptake by Endocytosis*

The constitutive catabolic pathway of SLs, GSLs, and GGs occurs in the lysosomes, the digestive compartment of the cell. Metabolic studies and immunoelectron microscopy of endocytotic processes indicate that plasma membrane-bound radio- and biotin-labeled GGs reach intralysosomal luminal vesicles (LVs) for degradation. They are the major sites of SL, GSL, GG, and membrane lipid catabolism[265,266] (Fig. 3).

As proposed,[267] LVs are generated by an inward budding of the endosomal perimeter membrane and budding off into the luminal space is catalyzed by ESCRT proteins.[268] Components of the endosomal perimeter membrane, lipids and proteins, either remain in the perimeter membrane or are sorted into the LVs, where they can be attacked by the digestive juice in the lysosome. The degradation of their membrane components, lipids and proteins, destabilizes and finally will destroy the bilayer structure of LV membranes, collapses their topology, and generate transiently micellar and other amphiphilic aggregates.

### *4.1.2 Lysosomes are Intracellular Stomachs for the Digestion of Membrane Components and Macromolecules*

Lysosomes degrade a variety of macromolecules and complex lipids and release their components into the cytosol as nutrients for use in energy metabolism and biosynthetic salvage pathways. Macromolecules and complex lipids reach LVs by endocytotic pathways[266] (Fig. 3), phagocytosis and autophagy.[269] At the level of late endosomes, LVs mature by a lipid sorting process,[266,270] thus removing membrane stabilizing lipids of the plasma membrane, SM, and cholesterol. Acid sphingomyelinase (ASM) cleaves several phospholipids, mainly SM, thereby releasing ceramide.[271,272] The decrease of inhibiting SM levels and the concomitant increase of stimulating ceramide levels facilitate cholesterol secretion from the LVs by the transfer and sterol-binding glycoprotein NPC2 to the transmembrane spanning NPC1, a glycoprotein of the perimeter membrane.[273] Cholesterol secretion by NPC2 is further stimulated by bis(monoacylglycero)phosphate (BMP), which is a rather slowly catabolized intermediate of phosphatidylglycerol degradation in the LVs.[273,274]

Catabolism of GGs, SLs, and SM proceeds at the surface of LVs (Fig. 3),[266] whereas the lysosomal perimeter membrane components are protected by a luminally oriented thick glycocalix, containing digestion-resistant polylactosamine structures.[275,276]

Soluble lysosomal hydrolases can recognize soluble macromolecules such as glycoproteins and oligosaccharides directly, but due to the lipid phase problem, they hardly attack lipophilic membrane components of LVs and other membranes, such as GSLs and GGs. Their catabolism needs the help of membrane perturbing lipid-binding proteins, the sphingolipid activator proteins (SAPs) (saposins A, B, C, and D, and GM2AP), small glycoproteins with lipid transfer, and fusogenic functions at low pH values.[277] Their defects cause rare and fatal, often degenerative brain diseases[265,278,279] with a clinical picture often quite similar to that of the respective enzyme deficiency.[265,279,280] An inherited deficiency of prosaposin, the precursor of four SAPs, the saposins A, B, C, and D, triggers a perinatal fatal disease with accumulation of LVs, ceramide, and several GSLs, and a loss of the water permeability barrier of the skin.[281–283]

### 4.1.3 Endocytosis of Gangliosides, Their Catabolic and Salvage Pathways

Routes of GG metabolism have been analyzed in cultured cells, fibroblasts, neurons, and others. ESR spectroscopy and radiotracer techniques showed that exogenously added tritium- and nitroxide-labeled GGs bind as micelles to the surface of cultured cells, a binding that can be removed by trypsinisation. A small trypsin-resistant component of about 10–15 nmol labeled GG per mg cell protein, however, displayed characteristic order parameters and unisotropic ESR spectra. They proved that ceramide moieties of the labeled GGs were fully integrated into the plasma membrane.[166] With the help of radiolabeled GGs, a stepwise catabolic degradation along the endocytotic pathway was observed besides several salvage pathways.[167,168]

In GM2 gangliosidosis cells with a block in GM2 degradation, radiolabeled GM2 was directly glycosylated to form radiolabeled GM1 and GD1a, apparently after trafficking from endosomes to the biosynthetic TGN compartment,[167] most likely mediated by the retromer complex[168] (Fig. 3).

Direct glycosylation GGs recycling from the endosomes to the biosynthetic compartments, Golgi and TGN, was also demonstrated in rats,[169,170] and for fluorescent and nondegradable analogs of glucosylceramide also in cell culture.[171,172]

Cellular metabolic studies on GGs, radiolabeled in their sphingoid base, showed a release of labeled sphingoid base from the lysosomes and its incorporation into newly synthesized SM and GGs.[167,173–175]

Desialysation is a crucial and initial step in the constitutional degradation of glycoproteins and oligosialyl GGs in mammalian tissues,[284] which could have already taken place at the plasma membrane or later in the early and late endosomes. Membrane-bound sialidases, Neu1, Neu4, and also Neu3 are the major sialidases that catabolize oligosialyl GGs to generate GG GM1.[285,286] The lysosomal sialidase Neu1 is in a complex with two other hydrolases, cathepsin A (PPCA, a stabilizing protein for sialidase), and the GM1 cleaving β-galactosidase. Inherited defects of Neu1 cause a storage of sialylated metabolites in sialidosis, whereas genetic defects in PPCA trigger storage of GM1 besides an additional accumulation of oligosaccharides and glycolipids.[287,288]

As shown by electron dense tracers and immunoelectron microscopy, biotin-tagged GM1 can insert into the plasma membrane of cultured cells and travel to the LVs for constitutive degradation.[289] Catabolism of GSLs, SLs, and GGs internalized by endocytosis occurs at the LVs. It is blocked by genetic defects of lysosomal proteins[266] (Fig. 3) causing the accumulation of intralysosomal lipid and membrane storage. An inherited loss of all four saposins due to a genetic loss of their precursor protein, prosaposin, triggers a dramatic accumulation of LVs, GSLs, and SLs in late endosomes and lysosomes. Highly increased levels of LVs with an average diameter of 90 nm have been observed by electron microscopy in patients' skin samples and cultured fibroblasts.[290–293] Feeding the missing prosaposin (Sap precursors) to patients' fibroblasts at nanomolar concentrations restored the capability of the mutant cells to degrade the accumulated GSLs, ceramide, GlcCer, LacCer, and GG GM3, and this reversed the storage of LVs. LVs are obviously the major location of GG and sphingolipid degradation as well as of other membrane components.[291,294]

Membrane lipids entering normally functioning lysosomes are usually readily degraded. The catabolism of phosphatidylglycerol (PG), however, generates an intermediate, the anionic lysolipid BMP which is only slowly catabolized and therefore increases up to 40–70 mol% of the LV-phospholipids in the lysosomes.[270,297,298] BMP is negatively charged, even at pH values as low as 4, like other phospholipids (phosphatidic acid, PG, and phosphatidylinositol).[272,299] It is expected to generate a negative surface potential on LVs[272] which should electrostatically bind protonated lysosomal enzymes and SAPs in the acidic lysosol with pH values around 4–5.[272,300]

Surface-bound lysosomal enzymes like ASM,[301] acid ceramidase,[302] hexosaminidase A, hexosaminidase B, and other glycoproteins are partially protected against premature proteolytic digestion in the lysosome. Compensation of the negative surface charge of LVs by feeding cationic amphiphilic drugs (such as desipramine) to cultured fibroblasts favors their release and proteolytic digestion, triggering an induced phospholipid storage disease, a phospholipidosis.[300,301,303]

In conclusion, PL, SL, and membrane digestion take place at and in LVs, within the late endosomes and lysosomes.[266] Several catabolic blocks can trigger an accumulation of LVs as observed, for example, in GM1-gangliosidosis[304] and prosaposin-deficient human tissues.[290,291] In other sphingolipidoses, the intralysosomal storage material appears as membranous cytoplasmatic bodies (MCB) (GM2-gangliosidoses)[305] or as membrane-bound vesicles containing twisted structures (Gaucher disease),[306] presumably generated from LVs that change their structure during life due to accumulation of quite different lipids and proteins.

Whereas LVs are almost free of a protecting coat, the lysosomal perimeter membrane is covered by a thick glycocalix on its luminal surface. It is generated by integral lysosomal glycoproteins, which are heavily N-glycosylated with digestion-resistant polylactosamine units.[276] It forms an efficient barrier to prevent hydrolases and SAPs to reach the lysosomal perimeter membrane and to digest their lipid and protein components. It is also an efficient barrier for lipids like cholesterol, phospholipids, GSLs, SLs and most of their hydrophobic breakdown products to reach the perimeter membrane and to escape from the lumen of the lysosome.

Defects in the biosynthesis of the lysosomal perimeter membrane glycoproteins, however, increase their turnover[276] and may attenuate the barrier function of the membrane. Even small degradation products like sphingoid bases should be released from the lysosome either by a transmembrane transporter or by trapping sphingosine reaching the membrane by phosphorylation at the cytosolic surface.[307] NPC1 has been in the focus, but seems not to be relevant for sphingosine export from the lysosomes.[298]

### 4.1.4 Membrane Lipids Regulate Maturation of LVs in Late Endosomes

Organelles of eukaryotic cells are surrounded by a limiting membrane, each of them has an organelle-specific protein and lipid composition.[308] A high content of complex GGs, SM, and cholesterol is typical for neuronal plasma membranes. The high cholesterol content (about 40 mol% of membrane lipids) is required to keep the physiological functions of many membrane proteins, including those involved in the generation and maintenance of electrochemical gradients. On the other hand, cholesterol is inhibitory to some steps in the maturation of LVs,[272,273] and the SL degradation.[309,310] Therefore it is apparently removed along the endocytotic pathway. LVs mature mainly at the stage of late endosomes by an efflux of cholesterol, mainly mediated by two sterol-binding transfer proteins, the soluble glycoprotein NPC2 and the perimeter membrane spanning NPC1.[311]

Inherited defects of either protein cause a modest lysosomal cholesterol accumulation in patients with Niemann–Pick disease type C (NPC). SM also accumulates in NPC disease, thus strongly inhibiting the cholesterol transfer and secretion by NPC2,[272,273] the catabolism of GM2[310] and glucosylceramide.[309] SM hydrolysis by ASM[272] and the cholesterol transfer by NPC2 are strongly stimulated by BMP[272,273] and other anionic phospholipids such as PA, PG, and PI.[272]

Therefore SM accumulation in NP disease, type A and B, explains the secondary accumulation of cholesterol[312] and in part also the accumulation of GGs GM2 and GM3 and glycolipids like glucosylceramide in NPC disease. Inhibition of catabolic steps by increasing lysosomal cholesterol levels in NPC contributes to the secondary accumulation of small GGs and GSLs like GM2 and GM3 in NPC, respectively.[273,310,313]

Lowering SM and cholesterol levels during maturation of LVs in the late endosomes and the generation of anionic BMP seems to be essential to reach physiological rates of GSL and SL degradation in endosomes and lysosomes.[314]

### 4.1.5 Protracted Clinical Forms of Gangliosidoses and the Threshold Theory

Gene mutations causing an almost complete loss of catabolic activities result in excessive lysosomal lipid accumulation in infantile forms of diseases, and a slow spread of storage materials from lysosomes to nonlysosomal membranes and compartments, thus affecting widespread cellular functions.[314,315]

Lysosomal storage and dysfunction of infantile diseases cause:

- Inhibition of autophagy and expansion of lysosomal compartments at the expense of other organelles.

- Accumulation of cytotoxic cationic lipids like galactosylsphingosine in Krabbe disease,[316] glucosylsphingosine in Gaucher disease[273,309,310,317–319] and sphingoid bases in other lipidoses. They inhibit lysosomal digestion and trigger proteolysis of ASM and other lysosomal hydrolases.[272] Storage compounds may inhibit hydrolases and SAPs. For instance, accumulation of SM inhibits cholesterol secretion from late endosomes by NPC2 in NP diseases and increased lysosomal cholesterol levels trigger secondary storage of small GGs and GSLs in NPC disease
- The lysosomal storage process releases proinflammatory cytokines, activates macrophages and causes an influx of macrophages into the brain of Gaucher mice,[320] as well as an inflammation of the CNS in mouse models of gangliosidoses.[321]

In contrast, gene mutations causing protracted clinical forms of gangliosidoses often generate variant proteins, which retain some residual enzymatic activity and other functions. They can give rise to milder clinical forms, often described as late infantile, juvenile, adult, or chronic diseases[265] with a broad range of clinical symptomatology.

At the biochemical level, the protracted clinical forms often display strongly reduced levels of catabolic activities and a reduced extent of GSL and SL accumulation (Fig. 4B). These observations can be modeled on the basis of a simplified kinetic theory.[265,296] They reveal a close correlation between predicted and measured residual cellular turnover of accumulating GG GM2, and the clinical course of infantile, juvenile and adult forms of Tay–Sachs disease and protracted forms with hexosaminidase A deficiency.[295,296] About 10% of normal GM2-cleaving activity appears to be sufficient for normal life, as assayed by cellular turnover and the in vitro hydrolysis of radiolabeled GG GM2 by hexosaminidase A in the presence of the GM2 activator protein. Adult, juvenile and infantile forms display increasingly lower levels of GM2-degrading activities (Fig. 4B). Similar observations have been described for MLD, Gaucher disease, and ASM-deficient Niemann–Pick disease.[295,322,323]

## 5. CATABOLISM OF GANGLIOSIDES

Stepwise degradation of GGs starts with the removal of terminal sialic acid units from the oligosaccharide chain by neuraminidase(s) (Fig. 5). On the level of the mono-sialoGGs, GM1 degradation continues with the removal of terminal galactose by GM1-β-galactosidase, assisted by the GM2 activator

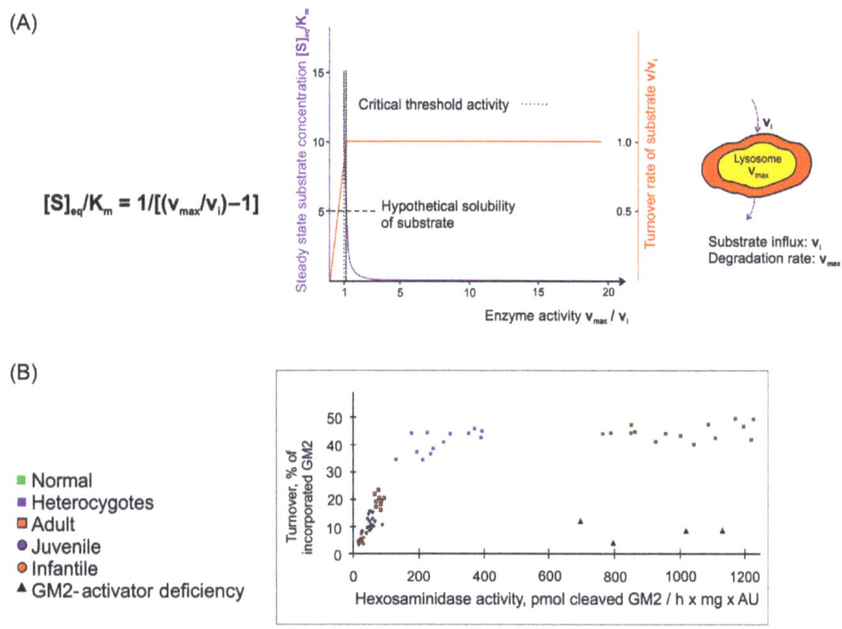


**Fig. 4** Residual catabolic activity correlates with clinical forms of GM2 gangliosidoses[265] (*modified after Leinekugel P, Michel S, Conzelmann E, Sandhoff K. Quantitative correlation between the residual activity of beta-hexosaminidase A and arylsulfatase A and the severity of the resulting lysosomal storage disease. Hum Genet. 1992;88(5):513–523*). (A) Steady-state substrate concentration and turnover as a function of enzyme activity.[295,296] The model underlying this theoretical calculation assumes influx of the substrate into the lysosomal compartment at a constant rate (vi) and its subsequent utilization by the catabolic enzyme. *Blue line* = [S]eq steady-state substrate concentration; . . . . . . . . . = theoretical threshold of enzyme activity; - - - - - - - = critical threshold value, taking limited solubility of substrate into account; *red line* = turnover rate of substrate (flux rate). (B) To verify the basic assumptions on which this model rests, experimental studies were performed in cell culture.[295] The radiolabeled substrate ganglioside GM2 was added to cultures of skin fibroblasts with different activities of β-hexosaminidase A and its uptake and turnover measured. The correlation between residual enzyme activity and the turnover rate of the substrate was essentially as predicted: degradation rate of ganglioside GM2 increased steeply with residual activity, to reach the control level at a residual activity of approximately 10%–15% of normal. All cells with an activity above this critical threshold had a normal turnover. Comparison of the results of these feeding studies with the clinical status of the donor of each cell line basically confirmed our notions but also revealed the limitations of the cell culture approach (*modified after Leinekugel P, Michel S, Conzelmann E, Sandhoff K. Quantitative correlation between the residual activity of beta-hexosaminidase A and arylsulfatase A and the severity of the resulting lysosomal storage disease. Hum Genet. 1992;88(5):513–523*).

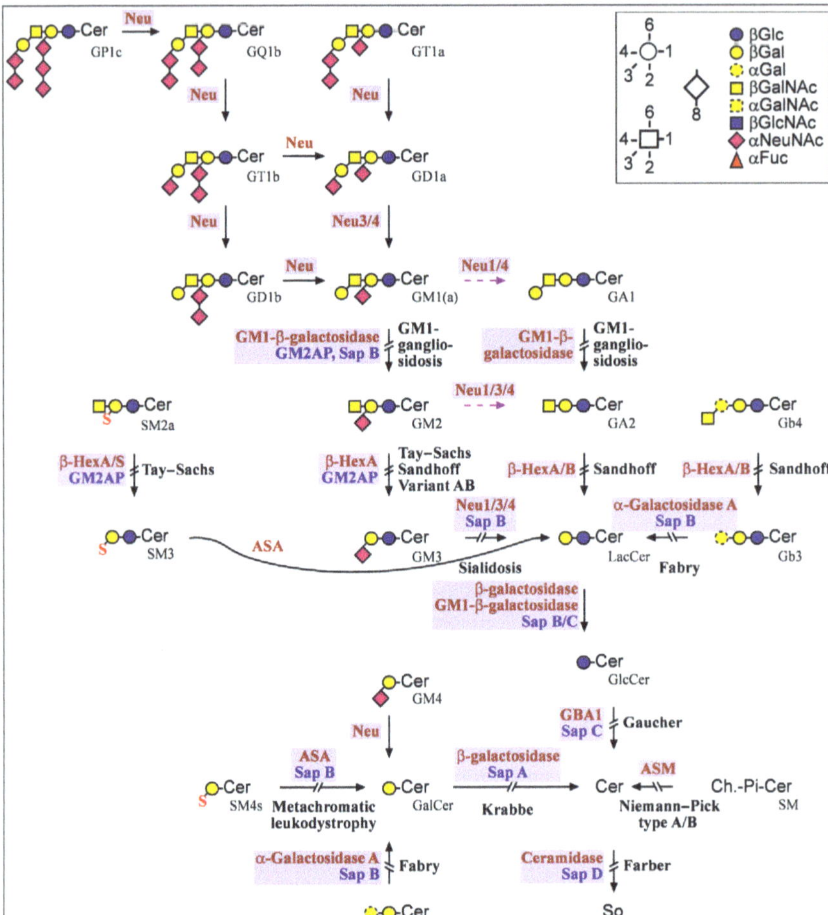


**Fig. 5** Lysosomal ganglioside catabolism. The names of hydrolases (red), activator proteins (blue), and metabolic diseases (black) are indicated in the scheme. *Broken purple arrows* indicate the bypass pathway of the ganglioside GM1 and GM2 degradation in mice. Sialidoses caused by different inherited sialidase (Neu) deficiencies are not marked in the figure due to their overlapping oligosialoganglioside storage pattern caused by promiscuous and overlapping substrate specificities of mammalian sialidases. *ASA*, Arylsulfatase A; *ASM*, acid sphingomyelinase; *GBA1*, lysosomal β-glucocerebrosidase; *Hex A*, β-hexosaminidase A; *Neu*, neuraminidase; *Sap*, saposin.

protein or Sap B to yield GM2.[324] Thereafter the terminal N-acetylgalactosamine residue is split off by β-hexosaminidase A with the help of the GM2 activator protein, to form GM3, which is then degraded to LacCer by α-sialidase and Sap B.[325] In mouse, the first two steps can be bypassed by the removal of sialic acid from GM1 and GM2, producing the corresponding asialo-derivates GA1 and GA2 respectively, from which the galactose residue and the N-acetylgalactosamine residue are removed by a β-galactosidase and β-hexosaminidase A and B, respectively, with the assistance of the GM2 activator protein to form LacCer. The Sap B or Sap C-assisted action of β-galactosidase degrades LacCer to GlcCer,[326] whose glucosyl residue is removed by Sap C-assisted β-glucosidase.[309] Finally ceramide is cleaved by acid ceramidase aided by Sap D to sphingosine and free fatty acid.[327]

Our focus is on the direct pathway of the degradation of GG GM1. Fig. 5 gives also a schematically overview of other degradation steps and inherited diseases, which are described elsewhere in more details.[278,328]

## 5.1 Catabolism and Disease

GG degradation at the surface of endo-lysosomal vesicles can be blocked at almost every step by an inherited deficiency either of the hydrolase or the respective activator protein. The gangliosidoses comprise GM1-, and the GM2-gangliosidoses. The disorders are characterized mainly by neuronal GG storage and by the accumulation of complex GSL in various tissues. They are rare and often fatal, in most cases of the neurodegenerative diseases.[278,279,329]

### *5.1.1 GM1 Gangliosidosis and Morquio Type B Disease*

GM1 gangliosidosis and Morquio type B disease are both caused by mutations in the β-galactosidase gene (GBL1) and resulting deficiency of lysosomal acid β-galactosidase (E.C.3.2.1.23).[330–332] Their mode of inheritance is autosomal recessive. GBL1 was mapped on chromosome 3 (3p21.33). It gives rise to two alternatively spliced mRNAs, one encoding for lysosomal β-galactosidase, the other for the elastin-binding protein (EBP) which is localized in the endosomal compartment.[333,334] β-Galactosidase is part of a lysosomal multienzyme complex, containing sialidase, cathepsin A, the so-called protective protein[288] and N-acetylaminogalacto-6-sulfate sulfatase.[335]

GM1 gangliosidosis is a progressive neurodegenerative disease. Three clinical forms have been described.[332,336] Type 1 is the severe infantile phenotype with clinical symptoms in early infancy or immediately after birth. Patients show progressive neurologic deterioration, visceromegaly, cherry red spot, and skeletal abnormalities. Death occurs within a few years. Type 2 is the later onset infantile/juvenile form with slowly progressive neurological signs, less severe skeletal changes, and an overall heterogeneous phenotypic expression. Type 3 (adult/chronic form) is the mildest phenotype. It is characterized by cognitive deterioration and progressive cerebellar ataxia. The estimated incidence of

GM1 gangliosidosis is 1:100,000–200,000 live births.[337] β-Galactosidase-deficient-knockout mouse develop progressive spastic diplegia within a few months after birth, and die of emaciation at 7–10 months of age.[338]

Morquio type B disease (Mucopolysaccharidosis type IVB) is caused by GBL1 gene mutations different from those of GM1 gangliosidosis, which lead to a changed substrate specificity of the enzyme, thereby resulting in major accumulation of galactose containing keratan sulfate and oligosaccharides.[339,340] The disease is characterized by generalized skeletal dysplasia with corneal clouding. The CNS is not involved and intelligence appears to be normal. Increased urinary keratan sulfate excretion is the hallmark of the disease.[332]

### 5.1.2 GM2 Gangliosidoses

The GM2 gangliosidoses are a group of three disorders that result from defects in degradation of GG GM2 and related glycolipids[341,342] by β-hexosaminidases (EC 3.2.1.52) and the GM2-activator protein. Three lysosomal β-hexosaminidases, which differ in the combination of their two subunits (α and β) and their substrate specificity have been described. β-Hexosaminidase A (αβ, consisting of α and β subunit) cleaves terminal β-glycosidically linked N-acetylglucosamine- and N-acetylgalactosamine residues from negatively charged and uncharged glycoconjugates. β-Hexosaminidase B (ββ) cleaves uncharged substrates such as glycolipids GA2 or Globoside ($Gb_4Cer$) and oligosaccharides with terminal N-acetylhexosamine residues. β-Hexosaminidase S (αα) contributes to the degradation of glycosaminoglycans, sulfated glycolipid SM2a and glycolipid GA2 and GM2. Inherited deficiency of α- or β-chain of the β-hexosaminidase isoenzymes or the GM2-activator protein leads to one of the three different variants of the disorder. They are named according to the isoenzyme that remains intact.

In contrast to the human clinical phenotype, mouse models for Tay–Sachs and Sandhoff disease differ severely in their phenotypes. In accordance with human patients with the infantile onset variant, the Sandhoff mouse, lacking hexosaminidases A and B, shows a severe neurological phenotype. In contrast to human patients, the Tay–Sachs mouse model lacking hexosaminidases A and S, shows no significant neurological phenotype. Mouse fibroblast sialidase has a higher activity than human, to accept GM2 as substrate, converting it slowly to GA2,[343] which is further degraded by the still intact β-hexosaminidase B in the Tay–Sachs mice.[341]

### 5.1.3 Tay–Sachs Disease (B-Variant) and Variant B1

The B-variant of the GM2 gangliosidoses is caused by an α-chain deficiency, and the subsequent deficiency of hexosaminidases A and S, with normal hexosaminidase B.[341] The disorder is subclassified into infantile, juvenile, chronic, and adult form, corresponding to increasing residual enzyme activity.[295] In some patients, an α-chain is synthesized, which has no catabolic activity against GG GM2, but is active against neutral substrates. This variant is also called B1-Variant.[344] In vitro studies demonstrated the ability of human hexosaminidases A and S, to degrade sulfatide SM2a in the presence of human GM2AP,[345] therefore SM2a accumulation was also observed in a human Tay–Sachs liver as well as in kidneys of the Tay–Sachs mouse model.[31]

### 5.1.4 Sandhoff Disease (0 Variant)

Sandhoff disease is caused by mutation in the beta subunit of hexosaminidase leading to a functional loss of both hexosaminidases A and B. Negatively charged glycolipids and uncharged glycolipids such as glycolipid GA2 are stored in the brain and globoside in visceral organs.[341,342] The clinical phenotype is similar to Tay–Sachs disease, with additional storage of globoside and oligosaccharides in visceral organs and mucopolysacchariduria.

### 5.1.5 GM2 Activator Deficiency (AB Variant)

The AB-variant is characterized by deficiency of the GM2 activator protein. β-Hexosaminidase A, B, and S activities are normal. The phenotype corresponds to that of Tay–Sachs disease.[346]

### 5.1.6 Further Degradation Steps

#### 5.1.6.1 Gaucher Disease

In Gaucher disease deficiency of glucosylceramide-β-glucosidase (EC 3.2.1.45, also called glucocerebrosidase) leads to accumulation of glucosylceramide.[317,347,348]

As in most sphingolipidoses three different types of the disorder are distinguished.

The attenuated form, Gaucher disease type I, has a nonneuropathic course and is the most frequent form of Gaucher disease, which is the most common sphingolipidosis. It has a frequency of 1: 50,000–200,000 births, but is higher among the Ashkenazi Jewish population (1:1000). It can be treated with enzyme replacement therapy.[349,350]

Gaucher disease type II, the acute form, is a very rare panethnic disease characterized by an additional storage of the toxic glucosylsphingosine and the involvement of the nervous system with early onset and a life expectancy of less than two years. The sub-acute or juvenile form, Gaucher disease type III, is an intermediate variant of the other two types. In all variants, patients may show hepatosplenomegaly, anemia, thrombocytopenia, and bone damage. The severity of these symptoms differs widely, but is inversely correlated to the residual enzyme activity determined in skin fibroblasts of Gaucher patients.[323] Complete glucosylceramide-β-glucosidase deficiency leads to a perinatal fatal form, the "collodion baby" phenotype with a severe impairment of skin function.[351]

### *5.1.7 Farber Disease*

Inherited deficiency of lysosomal acid ceramidase (EC 3.5.1.23, also called N-acylsphingosine amidohydrolase) leads to Farber disease. The enzyme is a heterodimer composed of two subunits,[352] which are derived from a common precursor that is processed within late endosomes and lysosomes.[353,354] It catalyzes the degradation of ceramide to sphingosine and fatty acid. In vivo, the enzyme requires the presence of Sap D.[327] Patients with Farber disease develop painful and progressive joint deformations, subcutaneous nodules (lipogranulomas), and progressive hoarseness.

### *5.1.8 Sphingolipid and Membrane Storage Caused by Defective Saposins*

#### 5.1.8.1 Prosaposin-Deficiency

The four saposins A–D arise from a common precursor protein, the prosaposin. The 70 kDa glycoprotein, prosaposin, is proteolytically processed to the four lipid-binding proteins in the late endosomes and lysosomes.[266,355] Intracellularly, prosaposin is targeted to the lysosomes via mannose-6-phosphate receptors and sortilin.[356]

Mutations in the start codon of the prosaposin gene lead to a complete deficiency of the protein and of all four mature saposins.[290,293] Prosaposin deficiency is a fatal perinatal disorder characterized by hepatosplenomegaly and severe neurological symptoms. Patients develop storage of many sphingolipids, including ceramide, glucosylceramide, LacCer, GG GM3, GalCer, sulfatides, digalactosylceramide, and globotriaosylceramide, accompanied by a massive accumulation of intralysosomal membranes.

In humans and in prosaposin deficient mice,[281,357] the disorder leads to the loss of the water permeability barrier in the skin due to a block in the extracellular processing of glucosylceramides containing ultra-long acyl chains.[283] The mice show an ichthyotic skin phenotype similar to β-glucocerebrosidase-deficient Gaucher mice.[281,351]

### 5.1.9 Sap A

Sap A mediates the degradation of GalCer by galactosylceramide-β-galactosidase. Inherited deficiency of galactosylceramide-β-galactosidase is the underlying cause of Krabbe disease. Likewise, human patients and mice with a mutation in the saposin A-domain of the saposin precursor suffer from a late-onset variant of the disease.[338,358]

### 5.1.10 Sap B

Sap B is required for the degradation of sulfatide by arylsulfatase A,[359] globotriaosylceramide and digalactosylceramide by α-galactosidase A. It can also mediate the degradation of other glycolipids, (e.g., GG GM2 can be degraded in the presence of Sap B together with the GM2-activator protein (GM2AP)).[324] In vitro, glycosylated saposins bind to lipid bilayers at acidic pH and are able to extract lipids.[313]

### 5.1.11 Sap C

Sap C mediates the lysosomal degradation of glucosylceramide by glucosylceramide-β-glucosidase.[309,319,360] Sap C deficiency leads to an abnormal juvenile form of Gaucher disease with an accumulation of glucosylceramide.[361]

### 5.1.12 Sap D

Sap D is required for lysosomal degradation of ceramide by acid ceramidase.[327] Saposin D deficient mice show accumulation of ceramides containing α-hydroxy fatty acids in the brain and kidney, renal tubular degeneration and cerebellar Purkinje cell loss.[362]

## 6. REGULATION OF GANGLIOSIDE CATABOLISM

So far, proteins essential for lysosomal degradation have been discussed, but furthermore, as already stated previously, GG catabolism is highly affected by lipid and protein composition of its surrounding.

### 6.1 Ganglioside GM2 Catabolism is Stimulated by Anionic Lipids and Inhibited by Sphingomyelin and High Cholesterol Levels

Fig. 6A shows a model of GM2 degradation by human β-hexosaminidase A, stimulated by GM2AP. In reconstitution experiments using liposomes, which mimic luminal vesicles of the late endosomal/lysosomal compartment, purified β-hexosaminidase A and GM2 activator protein showed that

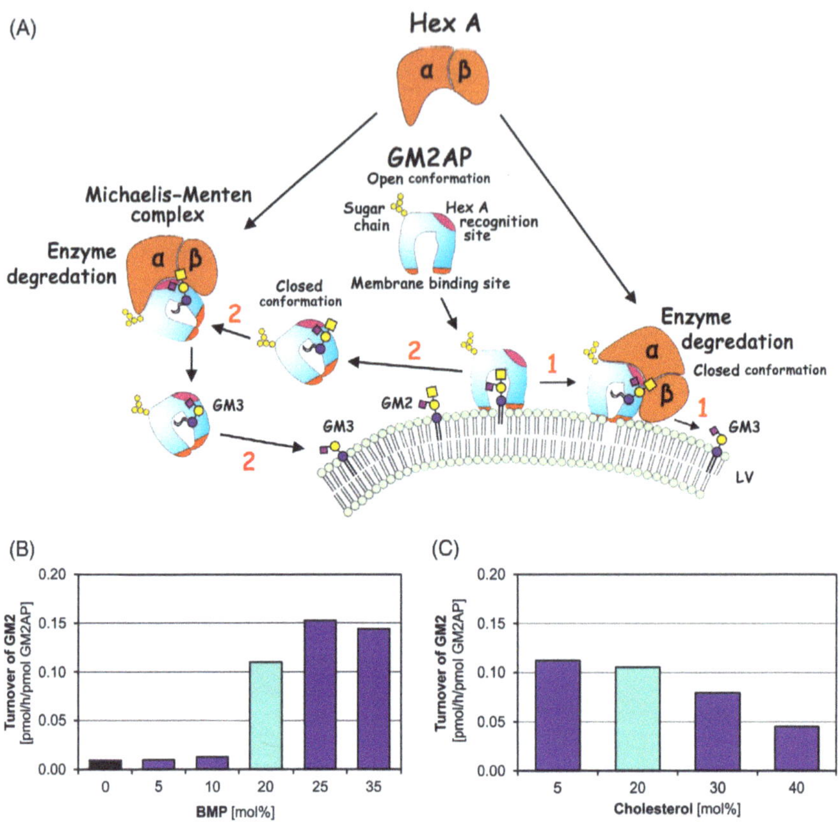


**Fig. 6** Model of GM2 degradation by human β-hexosaminidase A (Hex A), stimulated by GM2AP. (A) The GM2AP contains a hydrophobic cavity, suitable for the ceramide portion of GM2 and other lipids, lined by surface loops and a single short helix. The activator binds in its open conformation to the membrane using the hydrophobic loops (orange) and penetrates into the hydrophobic region of the bilayer. The lipid recognition site of the activator can interact with the substrate, and its ceramide portion can move inside the hydrophobic cavity. At this point, the conformation of the lipid-loaded activator may change to the closed one, thus the complex becomes more water soluble and can stay (1) or leave the membrane (2), exposing GM2 to the water-soluble Hex A to be degraded (modified after Breiden B, Sandhoff K. Ganglioside metabolism and its inherited diseases. *Methods Mol Biol.* 2018 [in press]). (B, C) GM2 hydrolysis is stimulated by anionic lipids like bis(monoacylglycero)phosphate (BMP) (B) and inhibited by cholesterol (Chol) (C) or sphingomyelin. Liposomes contain 5 mol% cholesterol, 0–35 mol% BMP or phosphatidylcholine as host lipid (B) and 5–40 mol% cholesterol, 20 mol% BMP and phosphatidylcholine as host lipid (C).[310] The *lighter blue columns* represent identical liposomal conditions.

anionic membrane lipids, like BMP, PI, and PG, are needed in the vesicular GM2-carrying membranes to achieve a physiologically relevant catabolic rate of GM2 degradation[310,363] (Fig. 6B). During endocytosis the lipid composition of luminal endo-lysosomal vesicles changes drastically due to digestion of membrane lipids and sorting out of cholesterol along the endocytotic pathway. Cholesterol is sorted out at the late endosomes mainly by two sterol-binding proteins, Niemann–Pick protein types C (NPC1 and NPC2).[364,365] In Niemann–Pick disease type C1, NPC1 and in type C2 NPC2 is defective. The disorder is characterized by lysosomal accumulation of cholesterol due to the defective efflux and by accumulation of SM and GSLs such as GlcCer, LacCer, and gangliosides GM3 and GM2.[311,312] Cholesterol transfer by NPC2 and thereby cholesterol efflux from the lysosome is blocked by high SM concentration,[272,273] which explains the secondary cholesterol accumulation triggered by the primary storage compound, SM, in ASM deficient Niemann–Pick disease types A and B. Apparently, SM degradation by ASM is required for physiological secretion of cholesterol from the late endosomal compartment, and is a prerequisite for endolysosomal lipid digestion. The resulting high cholesterol (Fig. 6C) and SM levels inhibit the BMP-enhanced hydrolysis of GM2 strongly.[310] Furthermore, the secondary storage of minor GSLs such as GlcCer, LacCer and GG GM3 as observed in NPC patients[311] might be explained by the strong SM-inhibition of GlcCer hydrolysis[309] and the cholesterol-inhibition of Sap A,[366] Sap B[313] and GM2 activator protein.[310] The clinical course of the disease seems to be negatively influenced by the secondary accumulation of GSLs. Inhibition of SL biosynthesis by Miglutat improves or stabilizes the neurological manifestation of the disease.[367,368]

## 6.2 Lysosomal Ganglioside and Glucosylceramide Degradation is Stimulated by SAPs and Negatively Charged Lipids and Inhibited by Sphingomyelin

In the absence of anionic lipids, such as bis(monoacylglycero)phosphate (BMP), phosphatidic acid, phosphatidylglycerol, PI, or PS, membrane-bound complex GSLs like GM1 (Fig. 7) and GM2 (Fig. 6) are hardly hydrolyzed by their respective hydrolases, and, even less so, in the absence of both, SAPs and anionic lipids.[310,363] In vitro addition of anionic lipids to liposomal membranes can stimulate the degradation immensely. Quantitative data on the stimulation of GG GM1 hydrolysis by β-galactosidase in the presence of SAPs (GM2AP and Sap B) and anionic lipids are presented in Fig. 7.

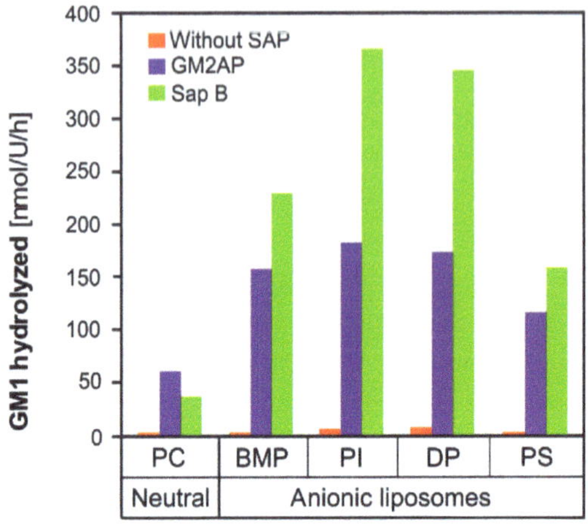


**Fig. 7** Lysosomal anionic lipids enhance the degradation of GM1 by β-galactosidase. In the absence of anionic lipids and SAPs, β-galactosidase hardly hydrolyzes any detectable amounts of ganglioside GM1 in phosphatidylcholine (PC) containing neutral liposomes (red column far left). Incorporation of anionic lipids (bis(monoacylglycero)phosphate (BMP), phosphatidylinositol (PI), dolicholphosphate (DP), phosphatidylserine (PS)) into the GM1-carrying liposomes and addition of a SAP, either GM2AP or Sap B, stimulates the GM1 hydrolysis up to 100-fold. As indicated, lysosomal anionic lipids (10 mol%) were incorporated in lipid unilamellar vesicles, composed of 10 mol% ganglioside GM1, 20 mol% cholesterol, and 60 mol% PC. Assays were carried out in the absence of an activator protein and in the presence of 5 mM of GM2AP or 5 mM of Sap B, both isolated from human spleen. Mean error was determined to be less than 10%.[151,324] *This research was originally published in the Journal of Biological Chemistry. Wilkening et al. Degradation of membrane-bound ganglioside GM1. J Biol Chem. 2000;275:35814–35819. © The American Society for Biochemistry and Molecular Biology.*

Also, the degradation of GSLs with short glycosylation chains is regulated by the lipid surrounding. Glucosylceramide is the primary storage lipid in the lysosomes of Gaucher patients and a secondary one in Niemann–Pick disease types A, B, and C. Glucosylceramide is already cleaved by β-glucosidase in the presence of anionic lipids and absence of SAPs.[299,309] However, anionic lipids can stimulate glucosylceramide hydrolysis at low pH by up to 1000-fold (Fig. 8) while cationic lipids inhibit the degradation, especially the minor cytotoxic storage compound glucosylsphingosine,[319] a biomarker for Gaucher disease.[369] Ceramide, fatty acids, monoacylglycerol, and diacylglycerol also stimulate glucosylceramide hydrolysis in the presence of anionic lipids while SM, sphingosine, and sphinganine play strong inhibitory roles (Fig. 8), thereby explaining the secondary storage of glucosylceramide in Niemann–Pick diseases.[309]

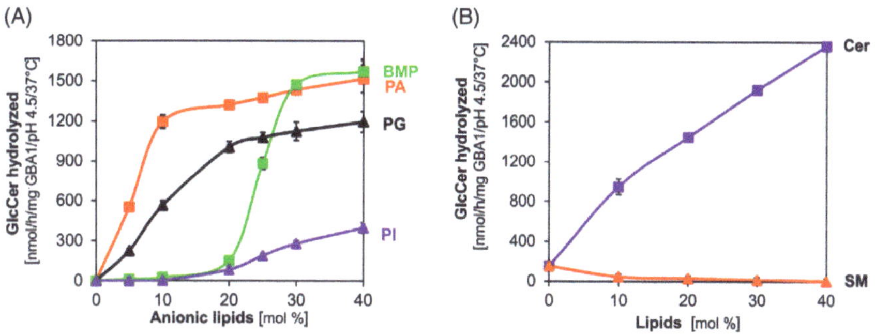


**Fig. 8** Anionic lipids stimulate membrane bound GlcCer hydrolysis by GBA1 in a concentration-dependent manner (A). Cer strongly stimulates membrane bound GlcCer hydrolysis by GBA1 in the presence of 20 mol% of BMP, while SM plays a strong inhibitory role (B). Liposomes in A contain 5 mol% cholesterol, 0–40 mol% anionic lipids and phosphatidylcholine as host lipid and in B 5 mol% cholesterol, 20 mol% anionic BMP, ceramide and sphingomyelin as indicated and phosphatidylcholine as host lipid.[151,309] *BMP*, Bis(monoacylglycero)phosphate; *Cer*, ceramide; *PA*, phosphatidic acid; *PG*, phosphatidylglycerol; *PI*, phosphatidylinositol; *PS*, phosphatidylserine; *SM*, sphingomyelin. *This research was originally published in The Journal of Lipid Research. Abdul-Hammed M, Breiden B, Schwarzmann G, Sandhoff K. Lipids regulate the hydrolysis of membrane bound glucosylceramide by lysosomal β-glucocerebrosidase. J Lipid Res. 2017;58:563–577. © The American Society for Biochemistry and Molecular Biology.*

All hydrolases and SAPs are N-glycosylated and glycosylation regulates its function, as has been shown for Sap B, where the unglycosylated mutant protein causes a fatal sphingolipid storage disease in patients with metachromatic leukodystrophy, despite the presence of the unglycosylated protein in the lysosomal compartment and its ability to stimulate the enzymatic hydrolysis of sulfatides, in vitro, even better than the glycosylated wild type Sap B.[313]

Functional assays of the membrane-related bioactivity of chemically synthesized variant Sap D glycoforms[370] revealed that the lipid-binding properties are highly affected by the single oligosaccharide chain of Sap D showing a dependency of the size and the type of the N-glycan.

Although all chemically synthesized Sap D glycoforms stimulated ceramide cleavage by acid ceramidase to a similar extent, their membrane-binding and fusion capacity increased substantially by shortening their N-glycan moiety. As expected, however, their ability to form soluble ceramide-Sap D complexes increased with increasing oligosaccharide side chains.

## ACKNOWLEDGMENT

We thank the very helpful library service at the German Cancer Research Center in retrieving all the literature. The authors also acknowledge the helpful discussion with Dr. Misbaudeen Abdul-Hammed. The work of the authors was supported by DFG Grant TRR83 and by the *Fonds der Chemischen Industrie*.

## ABBREVIATIONS

| | |
|---|---|
| ASM | acid sphingomyelinase |
| ASA | arylsulfatase A |
| BMP | bis(monoacylglycero)phosphate |
| Cer | ceramide |
| CERT | ceramide transfer protein |
| CerS | ceramide synthases |
| ER | endoplasmatic reticulum |
| Gal | galactose |
| GalCer | galactosylceramide |
| GalNAc | N-acetylgalactosamine |
| GG | ganglioside |
| Glc | glucose |
| GlcCer | glucosylceramide |
| GM2AP | GM2 activator protein |
| GSL | glycosphingolipid |
| LacCer | lactosylceramide |
| LV | intraendolysosomal luminal vesicles |
| NE | nuclear envelope |
| Neu | neuraminidase |
| Neu5Ac | N-acetylneuraminic acid |
| NPC | Niemann–Pick disease type C |
| NPC1/2 | Niemann–Pick disease type C protein1/2 |
| PA | phosphatidic acid |
| PC | phosphatidylcholine |
| PG | phosphatidylglycerol |
| PI | phosphatidylinositol |
| PS | phosphatidylserine |
| Sa | sphinganine/dihydrosphingosine |
| SAP | sphingolipid activator protein |
| Sap | saposin |
| SL | shingolipid |
| SM | sphingomyelin |
| So | sphingosine |
| TGN | trans Golgi network |

## REFERENCES

1. Klenk E. Die Fettstoffe des Gehirns bei Amaurotischer Idiotie und Niemann–Pick'sher Krankheit. *Ber Ges Physiol.* 1937;96:659–660.
2. Klenk E. Beiträge zur Chemie der Lipidosen – Niemann–Picksche Krankheit und amaurotische Idiotie. *Hoppe Seylers Z Physiol Chem.* 1939;262:128–143.
3. Klenk E. Über die Ganglioside, eine neue Gruppe von zuckerhaltigen Gehirnlipoiden. *Hoppe Seylers Z Physiol Chem.* 1942;273:76–86.
4. Wiegandt H. The chemical constitution of gangliosides of the vertebrate nervous system. *Behav Brain Res.* 1995;66(1–2):85–97.
5. Rahmann H. Functional meaning of neuronal gangliosides for the process of thermal adaptation in vertebrates. *J Therm Biol.* 1983;8(4):404–407.
6. Kuhn R, Wiegandt H. Die Konstitution der Ganglio-N-Tetraose und des Gangliosids G1. *Chem Ber.* 1963;96(3):866.
7. Schauer R. Sialic acids as link to Japanese scientists. *Proc Jpn Acad Ser B Phys Biol Sci.* 2016;92(4):109–120.
8. Schwarzkopf M, Knobeloch KP, Rohde E, et al. Sialylation is essential for early development in mice. *Proc Natl Acad Sci USA.* 2002;99(8):5267–5270.
9. Tanner ME. The enzymes of sialic acid biosynthesis. *Bioorg Chem.* 2005;33(3):216–228.
10. Inokuchi J. Membrane microdomains and insulin resistance. *FEBS Lett.* 2010;584(9):1864–1871.
11. Regina Todeschini A, Hakomori SI. Functional role of glycosphingolipids and gangliosides in control of cell adhesion, motility, and growth, through glycosynaptic microdomains. *Biochim Biophys Acta.* 2008;1780(3):421–433.
12. Ohmi Y, Ohkawa Y, Yamauchi Y, Tajima O, Furukawa K, Furukawa K. Essential roles of gangliosides in the formation and maintenance of membrane microdomains in brain tissues. *Neurochem Res.* 2012;37(6):1185–1191.
13. Furukawa K, Hamamura K, Ohkawa Y, Ohmi Y, Furukawa K. Disialyl gangliosides enhance tumor phenotypes with differential modalities. *Glycoconj J.* 2012;29(8–9):579–584.
14. Hakomori SI, Handa K. GM3 and cancer. *Glycoconj J.* 2015;32(1–2):1–8.
15. Ledeen RW, Wu G, Andre S, et al. Beyond glycoproteins as galectin counterreceptors: tumor-effector T cell growth control via ganglioside GM1 [corrected]. *Ann N Y Acad Sci.* 2012;1253:206–221.
16. Lipina C, Hundal HS. Ganglioside GM3 as a gatekeeper of obesity-associated insulin resistance: evidence and mechanisms. *FEBS Lett.* 2015;589(21):3221–3227.
17. Allende ML, Proia RL. Simplifying complexity: genetically resculpting glycosphingolipid synthesis pathways in mice to reveal function. *Glycoconj J.* 2014;31(9):613–622.
18. Nordstrom V, Willershauser M, Herzer S, et al. Neuronal expression of glucosylceramide synthase in central nervous system regulates body weight and energy homeostasis. *PLoS Biol.* 2013;11(3):e1001506.
19. Kishimoto Y, Radin NS. Determination of brain gangliosides by determination of ganglioside stearic acid. *J Lipid Res.* 1966;7(1):141–145.
20. Yang LJ, Zeller CB, Shaper NL, et al. Gangliosides are neuronal ligands for myelin-associated glycoprotein. *Proc Natl Acad Sci USA.* 1996;93(2):814–818.
21. Collins BE, Yang LJ, Mukhopadhyay G, et al. Sialic acid specificity of myelin-associated glycoprotein binding. *J Biol Chem.* 1997;272(2):1248–1255.
22. Schnaar RL. Gangliosides of the vertebrate nervous system. *J Mol Biol.* 2016;428(16):3325–3336.
23. Muthing J, Maurer U, Sostaric K, et al. Different distributions of glycosphingolipids in mouse and rabbit skeletal muscle demonstrated by biochemical and immunohistological analyses. *J Biochem.* 1994;115(2):248–256.

24. Levy M, Futerman AH. Mammalian ceramide synthases. *IUBMB Life.* 2010;62 (5):347–356.
25. Yamashita T, Hashiramoto A, Haluzik M, et al. Enhanced insulin sensitivity in mice lacking ganglioside GM3. *Proc Natl Acad Sci U S A.* 2003;100(6):3445–3449.
26. Jennemann R, Kaden S, Sandhoff R, et al. Glycosphingolipids are essential for intestinal endocytic function. *J Biol Chem.* 2012;287:32598–32616.
27. Jennemann R, Rothermel U, Wang S, et al. Hepatic glycosphingolipid deficiency and liver function in mice. *Hepatology.* 2010;51(5):1799–1809.
28. Muthing J, Duvar S, Heitmann D, et al. Isolation and structural characterization of glycosphingolipids of in vitro propagated human umbilical vein endothelial cells. *Glycobiology.* 1999;9(5):459–468.
29. Yohe HC, Wallace PK, Berenson CS, Ye S, Reinhold BB, Reinhold VN. The major gangliosides of human peripheral blood monocytes/macrophages: absence of ganglio series structures. *Glycobiology.* 2001;11(10):831–841.
30. Kirsch S, Muthing J, Peter-Katalinic J, Bindila L. On-line nano-HPLC/ESI QTOF MS monitoring of alpha 2-3 and alpha 2-6 sialylation in granulocyte glycosphingolipidome. *Biol Chem.* 2009;390(7):657–672.
31. Sandhoff R, Hepbildikler ST, Jennemann R, et al. Kidney sulfatides in mouse models of inherited glycosphingolipid disorders: determination by nano-electrospray ionization tandem mass spectrometry. *J Biol Chem.* 2002;277(23):20386–20398.
32. Chinnapen DJ, Hsieh WT, te Welscher YM, et al. Lipid sorting by ceramide structure from plasma membrane to ER for the cholera toxin receptor ganglioside GM1. *Dev Cell.* 2012;23(3):573–586.
33. Saslowsky DE, te Welscher YM, Chinnapen DJ, et al. Ganglioside GM1-mediated transcytosis of cholera toxin bypasses the retrograde pathway and depends on the structure of the ceramide domain. *J Biol Chem.* 2013;288(36):25804–25809.
34. Cuatrecasas P. Gangliosides and membrane receptors for cholera toxin. *Biochemistry.* 1973;12(18):3558–3566.
35. Dong M, Tepp WH, Liu H, Johnson EA, Chapman ER. Mechanism of botulinum neurotoxin B and G entry into hippocampal neurons. *J Cell Biol.* 2007;179 (7):1511–1522.
36. Eidels L, Proia RL, Hart DA. Membrane receptors for bacterial toxins. *Microbiol Rev.* 1983;47(4):596–620.
37. Hamark C, Berntsson RP, Masuyer G, et al. Glycans confer specificity to the recognition of ganglioside receptors by botulinum neurotoxin A. *J Am Chem Soc.* 2017;139 (1):218–230.
38. Lingwood CA. Glycolipid receptors for verotoxin and Helicobacter pylori: role in pathology. *Biochim Biophys Acta.* 1999;1455(2–3):375–386.
39. Sandvig K, Bergan J, Kavaliauskiene S, Skotland T. Lipid requirements for entry of protein toxins into cells. *Prog Lipid Res.* 2014;54:1–13.
40. Van Heyningen WE, Miller PA. The fixation of tetanus toxin by ganglioside. *J Gen Microbiol.* 1961;24:107–119.
41. Melton-Celsa AR. Shiga toxin (Stx) classification, structure, and function. *Microbiol Spectr.* 2014;2(4). EHEC-0024-2013.
42. van Echten G, Sandhoff K. Modulation of ganglioside biosynthesis in primary cultured neurons. *J Neurochem.* 1989;52(1):207–214.
43. Svennerholm L. Ganglioside designation. *Adv in Exp Med Biol.* 1980;125:11.
44. Mandon EC, Ehses I, Rother J, van Echten G, Sandhoff K. Subcellular localization and membrane topology of serine palmitoyltransferase, 3-dehydrosphinganine reductase, and sphinganine N-acyltransferase in mouse liver. *J Biol Chem.* 1992;267(16):11144–11148.

45. Braun PE, Snell EE. Biosynthesis of sphingolipid bases. II. Keto intermediates in synthesis of sphingosine and dihydrosphingosine by cell-free extracts of Hansenula ciferri. *J Biol Chem.* 1968;243(14):3775–3783.
46. Kolter T, Sandhoff K. Sphingolipids – their metabolic pathways and the pathobiochemistry of neurodegenerative diseases. *Angew Chem Int Ed.* 1999;38(11):1532–1568.
47. Raman MC, Johnson KA, Yard BA, et al. The external aldimine form of serine palmitoyltransferase: structural, kinetic, and spectroscopic analysis of the wild-type enzyme and HSAN1 mutant mimics. *J Biol Chem.* 2009;284 (25):17328–17339.
48. Stoffel W, LeKim D, Sticht G. Biosynthesis of dihydrosphingosine in vitro. *Hoppe Seylers Z Physiol Chem.* 1968;349(5):664–670.
49. Braun PE, Morell P, Radin NS. Synthesis of C18- and C20-dihydrosphingosines, ketodihydrosphingosines, and ceramides by microsomal preparations from mouse brain. *J Biol Chem.* 1970;245(2):335–341.
50. Williams RD, Wang E, Merrill Jr AH. Enzymology of long-chain base synthesis by liver: characterization of serine palmitoyltransferase in rat liver microsomes. *Arch Biochem Biophys.* 1984;228(1):282–291.
51. Breslow DK, Collins SR, Bodenmiller B, et al. Orm family proteins mediate sphingolipid homeostasis. *Nature.* 2010;463(7284):1048–1053.
52. Han S, Lone MA, Schneiter R, Chang A. Orm1 and Orm2 are conserved endoplasmic reticulum membrane proteins regulating lipid homeostasis and protein quality control. *Proc Natl Acad Sci U S A.* 2010;107(13):5851–5856.
53. Gupta SD, Gable K, Alexaki A, et al. Expression of the ORMDLS, modulators of serine palmitoyltransferase, is regulated by sphingolipids in mammalian cells. *J Biol Chem.* 2015;290(1):90–98.
54. Siow DL, Wattenberg BW. Mammalian ORMDL proteins mediate the feedback response in ceramide biosynthesis. *J Biol Chem.* 2012;287(48):40198–40204.
55. Hornemann T, Penno A, Rutti MF, et al. The SPTLC3 subunit of serine palmitoyltransferase generates short chain sphingoid bases. *J Biol Chem.* 2009;284 (39):26322–26330.
56. Harmon JM, Bacikova D, Gable K, et al. Topological and functional characterization of the ssSPTs, small activating subunits of serine palmitoyltransferase. *J Biol Chem.* 2013;288(14):10144–10153.
57. Kihara A, Igarashi Y. FVT-1 is a mammalian 3-ketodihydrosphingosine reductase with an active site that faces the cytosolic side of the endoplasmic reticulum membrane. *J Biol Chem.* 2004;279(47):49243–49250.
58. Merrill Jr AH, Wang E. Biosynthesis of long-chain (sphingoid) bases from serine by LM cells. Evidence for introduction of the 4-trans-double bond after de novo biosynthesis of N-acylsphinganine(s). *J Biol Chem.* 1986;261(8):3764–3769.
59. Rother J, van Echten G, Schwarzmann G, Sandhoff K. Biosynthesis of sphingolipids: dihydroceramide and not sphinganine is desaturated by cultured cells. *Biochem Biophys Res Commun.* 1992;189(1):14–20.
60. Morell P, Radin NS. Specificity in ceramide biosynthesis from long chain bases and various fatty acyl coenzyme A's by brain microsomes. *J Biol Chem.* 1970;245 (2):342–350.
61. Bauer R, Voelzmann A, Breiden B, et al. Schlank, a member of the ceramide synthase family controls growth and body fat in Drosophila. *Embo J.* 2009;28 (23):3706–3716.
62. Deng X, Yin X, Allan R, et al. Ceramide biogenesis is required for radiation-induced apoptosis in the germ line of *C. elegans*. *Science.* 2008;322(5898):110–115.

63. Pewzner-Jung Y, Ben-Dor S, Futerman AH. When do Lasses (longevity assurance genes) become CerS (ceramide synthases)?: insights into the regulation of ceramide synthesis. *J Biol Chem*. 2006;281(35):25001–25005.
64. Sandhoff R. Very long chain sphingolipids: tissue expression, function and synthesis. *FEBS Lett*. 2010;584(9):1907–1913.
65. Jiang JC, Kirchman PA, Zagulski M, Hunt J, Jazwinski SM. Homologs of the yeast longevity gene LAG1 in *Caenorhabditis elegans* and human. *Genome Res*. 1998;8 (12):1259–1272.
66. Sambasivarao K, McCluer RH. Lipid components of gangliosides. *J Lipid Res*. 1964;5:103–108.
67. Venkataraman K, Riebeling C, Bodennec J, et al. Upstream of growth and differentiation factor 1 (uog1), a mammalian homolog of the yeast longevity assurance gene 1 (LAG1), regulates N-stearoyl-sphinganine (C18-(dihydro)ceramide) synthesis in a fumonisin B1-independent manner in mammalian cells. *J Biol Chem*. 2002;277 (38):35642–35649.
68. Sassa T, Hirayama T, Kihara A. Enzyme activities of the ceramide synthases cers2-6 are regulated by phosphorylation in the c-terminal region. *J Biol Chem*. 2016;291 (14):7477–7487.
69. Venkataraman K, Futerman AH. Do longevity assurance genes containing Hox domains regulate cell development via ceramide synthesis? *FEBS Lett*. 2002;528 (1–3):3–4.
70. Bauer R. Towards understanding regulation of energy homeostasis by ceramide synthases. *Results Prob Cell Differ*. 2010;52:175–181.
71. Voelzmann A, Wulf AL, Eckardt F, et al. Nuclear Drosophila CerS Schlank regulates lipid homeostasis via the homeodomain, independent of the lag1p motif. *FEBS Lett*. 2016.
72. Sociale M, Wulf A-L, Breiden B, et al. Ceramide synthase Schlank is a Transriptional Regulator Adapting Gene Expression to Energy Requirements. *submitted*. 2017
73. Geeraert L, Mannaerts GP, van Veldhoven PP. Conversion of dihydroceramide into ceramide: involvement of a desaturase. *Biochem J*. 1997;327(Pt 1):125–132.
74. Michel C, van Echten-Deckert G, Rother J, Sandhoff K, Wang E, Merrill Jr AH. Characterization of ceramide synthesis. A dihydroceramide desaturase introduces the 4,5-trans-double bond of sphingosine at the level of dihydroceramide. *J Biol Chem*. 1997;272(36):22432–22437.
75. Breimer ME, Karlsson KA, Larson G, McKibbin JM. Chemical characterization of a blood group H type pentaglycosylceramide of human small intestine. *Chem Phys Lipids*. 1983;33(2):135–144.
76. Smith EL, McKibbin JM, Karlsson KA, Pascher I, Samuelsson BE. Characterization by mass spectrometry of blood group A active glycolipids from human and dog small intestins. *Biochemistry*. 1975;14(10):2120–2124.
77. Umesaki Y, Suzuki A, Kasama T, Tohyama K, Mutai M, Yamakawa T. Presence of asialo GM1 and glucosylceramide in the intestinal mucosa of mice and induction of fucosyl asialo GM1 by conventionalization of germ-free mice. *J Biochem*. 1981;90(6):1731–1738.
78. Enomoto A, Omae F, Miyazaki M, Kozutsumi Y, Yubisui T, Suzuki A. Dihydroceramide: sphinganine C-4-hydroxylation requires Des2 hydroxylase and the membrane form of cytochrome b5. *Biochem J*. 2006;397(2):289–295.
79. Omae F, Miyazaki M, Enomoto A, Suzuki M, Suzuki Y, Suzuki A. DES2 protein is responsible for phytoceramide biosynthesis in the mouse small intestine. *Biochem J*. 2004;379(Pt 3):687–695.
80. Ternes P, Franke S, Zahringer U, Sperling P, Heinz E. Identification and characterization of a sphingolipid delta 4-desaturase family. *J Biol Chem*. 2002;277 (28):25512–25518.

81. Sprong H, Kruithof B, Leijendekker R, Slot JW, van Meer G, van der Sluijs P. UDP-galactose: ceramide galactosyltransferase is a class I integral membrane protein of the endoplasmic reticulum. *J Biol Chem.* 1998;273(40):25880–25888.
82. Chisada S, Yoshimura Y, Sakaguchi K, et al. Zebrafish and mouse alpha2,3-sialyltransferases responsible for synthesizing GM4 ganglioside. *J Biol Chem.* 2009;284 (44):30534–30546.
83. Uemura S, Go S, Shishido F, Inokuchi J. Expression machinery of GM4: the excess amounts of GM3/GM4S synthase (ST3GAL5) are necessary for GM4 synthesis in mammalian cells. *Glycoconj J.* 2014;31(2):101–108.
84. Kuhn R, Wiegandt H. Further gangliosides from the human brain. *Z Naturforsch B.* 1964;19:256–257.
85. Tadano K, Ishizuka I. Isolation and partial characterization of a novel sulfoglycosphingolipid and ganglioside GM4 from rat kidney. *Biochem Biophys Res Commun.* 1980;97 (1):126–132.
86. Iwamori M, Shimomura J, Tsuyuhara S, Nagai Y. Gangliosides of various rat tissues: distribution of ganglio-N-tetraose-containing gangliosides and tissue-characteristic composition of gangliosides. *J Biochem.* 1984;95(3):761–770.
87. Yamaji T, Hanada K. Sphingolipid metabolism and interorganellar transport: localization of sphingolipid enzymes and lipid transfer proteins. *Traffic.* 2015;16(2):101–122.
88. Ichikawa S, Ozawa K, Hirabayashi Y. Molecular cloning and characterization of the mouse ceramide glucosyltransferase gene. *Biochem Biophys Res Commun.* 1998;253 (3):707–711.
89. Futerman AH, Pagano RE. Determination of the intracellular sites and topology of glucosylceramide synthesis in rat liver. *Biochem J.* 1991;280(Pt 2):295–302.
90. Jeckel D, Karrenbauer A, Burger KN, van Meer G, Wieland F. Glucosylceramide is synthesized at the cytosolic surface of various Golgi subfractions. *J Cell Biol.* 1992;117 (2):259–267.
91. Chalat M, Menon I, Turan Z, Menon AK. Reconstitution of glucosylceramide flip-flop across endoplasmic reticulum: implications for mechanism of glycosphingolipid biosynthesis. *J Biol Chem.* 2012;287(19):15523–15532.
92. Halter D, Neumann S, van Dijk SM, et al. Pre- and post-Golgi translocation of glucosylceramide in glycosphingolipid synthesis. *J Cell Biol.* 2007;179(1):101–115.
93. Nishie T, Hikimochi Y, Zama K, et al. Beta4-galactosyltransferase-5 is a lactosylceramide synthase essential for mouse extra-embryonic development. *Glycobiology.* 2010;20(10):1311–1322.
94. Tokuda N, Numata S, Li X, et al. beta4GalT6 is involved in the synthesis of lactosylceramide with less intensity than beta4GalT5. *Glycobiology.* 2013;23(10):1175–1183.
95. Kolter T. Ganglioside biochemistry. *ISRN Biochem.* 2012;2012:506160.
96. Symington FW, Murray WA, Bearman SI, Hakomori S. Intracellular localization of lactosylceramide, the major human neutrophil glycosphingolipid. *J Biol Chem.* 1987;262 (23):11356–11363.
97. Suzuki M, Cheung NK. Disialoganglioside GD2 as a therapeutic target for human diseases. *Expert Opin Ther Targets.* 2015;19(3):349–362.
98. Hitoshi S, Kusunoki S, Kon K, et al. A novel ganglioside, 9-O-acetyl GD1b, is recognized by serum antibodies in Guillain-Barre syndrome. *J Neuroimmunol.* 1996;66 (1–2):95–101.
99. Sonnino S, Ghidoni R, Chigorno V, Masserini M, Tettamanti G. Recognition by two-dimensional thin-layer chromatography and densitometric quantification of alkali-labile gangliosides from the brain of different animals. *Anal Biochem.* 1983;128(1):104–114.
100. Riboni L, Sonnino S, Acquotti D, et al. Natural occurrence of ganglioside lactones. Isolation and characterization of GD1b inner ester from adult human brain. *J Biol Chem.* 1986;261(18):8514–8519.

101. Svennerholm L, Rynmark BM, Vilbergsson G, et al. Gangliosides in human fetal brain. *J Neurochem.* 1991;56(5):1763–1768.
102. Ishii A, Ohta M, Watanabe Y, et al. Expression cloning and functional characterization of human cDNA for ganglioside GM3 synthase. *J Biol Chem.* 1998;273(48):31652–31655.
103. Kono M, Yoshida Y, Kojima N, Tsuji S. Molecular cloning and expression of a fifth type of alpha2,8-sialyltransferase (ST8Sia V). Its substrate specificity is similar to that of SAT-V/III, which synthesize GD1c, GT1a, GQ1b and GT3. *J Biol Chem.* 1996;271(46):29366–29371.
104. Yoshida Y, Kojima N, Kurosawa N, Hamamoto T, Tsuji S. Molecular cloning of Sia alpha 2,3 Gal beta 1,4GlcNAc alpha 2,8-sialyltransferase from mouse brain. *J Biol Chem.* 1995;270(24):14628–14633.
105. Roseman S. The synthesis of complex carbohydrates by multiglycosyltransferase systems and their potential function in intercellular adhesion. *Chem Phys Lipids.* 1970;5(1):270–297.
106. Yusuf HK, Pohlentz G, Schwarzmann G, Sandhoff K. Ganglioside biosynthesis in Golgi apparatus of rat liver. Stimulation by phosphatidylglycerol and inhibition by tunicamycin. *Eur J Biochem.* 1983;134(1):47–54.
107. Yusuf HK, Pohlentz G, Sandhoff K. Ganglioside biosynthesis in Golgi apparatus: new perspectives on its mechanism. *J Neurosci Res.* 1984;12(2–3):161–178.
108. Giraudo CG, Daniotti JL, Maccioni HJ. Physical and functional association of glycolipid N-acetyl-galactosaminyl and galactosyl transferases in the Golgi apparatus. *Proc Natl Acad Sci U S A.* 2001;98(4):1625–1630.
109. Chumpen Ramirez S, Ruggiero FM, Daniotti JL, Valdez Taubas J. Ganglioside glycosyltransferases are S-acylated at conserved cysteine residues involved in homodimerisation. *Biochem J.* 2017;474(16):2803–2816.
110. Daniotti JL, Pedro MP, Valdez Taubas J. The role of S-acylation in protein trafficking. *Traffic.* 2017.
111. Keenan TW, Morre DJ, Basu S. Ganglioside biosynthesis. Concentration of glycosphingolipid glycosyltransferases in Golgi apparatus from rat liver. *J Biol Chem.* 1974;249(1):310–315.
112. Kaufman B, Basu S, Roseman S. Enzymatic synthesis of disialogangliosides from monosialogangliosides by sialyltransferases from embryonic chicken brain. *J Biol Chem.* 1968;243(21):5804–5807.
113. Pohlentz G, Klein D, Schwarzmann G, Schmitz D, Sandhoff K. Both GA2, GM2, and GD2 synthases and GM1b, GD1a, and GT1b synthases are single enzymes in Golgi vesicles from rat liver. *Proc Natl Acad Sci U S A.* 1988;85(19):7044–7048.
114. Iber H, Zacharias C, Sandhoff K. The c-series gangliosides GT3, GT2 and GP1c are formed in rat liver Golgi by the same set of glycosyltransferases that catalyse the biosynthesis of asialo-, a- and b-series gangliosides. *Glycobiology.* 1992;2(2):137–142.
115. Iber H, Sandhoff K. Identity of GD1C, GT1a and GQ1b synthase in GOLGI vesicles from rat liver. *FEBS Lett.* 1989;254(1–2):124–128.
116. Iber H, van Echten G, Sandhoff K. Substrate specificity of alpha 2——3-sialyltransferases in ganglioside biosynthesis of rat liver golgi. *Eur J Biochem.* 1991;195(1):115–120.
117. Iber H, Kaufmann R, Pohlentz G, Schwarzmann G, Sandhoff K. Identity of GA1, GM1a and GD1b synthase in Golgi vesicles from rat liver. *FEBS Lett.* 1989;248(1–2):18–22.
118. Kolter T, Proia RL, Sandhoff K. Combinatorial ganglioside biosynthesis. *J Biol Chem.* 2002;277(29):25859–25862.
119. Inoue M, Fujii Y, Furukawa K, et al. Refractory skin injury in complex knock-out mice expressing only the GM3 ganglioside. *J Biol Chem.* 2002;277(33):29881–29888.
120. Kawai H, Allende ML, Wada R, et al. Mice expressing only monosialoganglioside GM3 exhibit lethal audiogenic seizures. *J Biol Chem.* 2001;276(10):6885–6888.

121. Sturgill ER, Aoki K, Lopez PHH, et al. Biosynthesis of the major brain gangliosides GD1a and GT1b. *Glycobiology*. 2012;22(10):1289–1301.
122. Takamiya K, Yamamoto A, Furukawa K, et al. Mice with disrupted GM2/GD2 synthase gene lack complex gangliosides but exhibit only subtle defects in their nervous system. *Proc Natl Acad Sci U S A*. 1996;93(20):10662–10667.
123. Yamashita T, Wu YP, Sandhoff R, et al. Interruption of ganglioside synthesis produces central nervous system degeneration and altered axon-glial interactions. *Proc Natl Acad Sci U S A*. 2005;102(8):2725–2730.
124. Pearce OM, Laubli H. Sialic acids in cancer biology and immunity. *Glycobiology*. 2016;26(2):111–128.
125. Varki A. Loss of N-glycolylneuraminic acid in humans: mechanisms, consequences, and implications for hominid evolution. *Am J Phys Anthropol*. 2001;suppl 33:54–69.
126. Krengel U, Bousquet PA. Molecular recognition of gangliosides and their potential for cancer immunotherapies. *Front Immunol*. 2014;5:325.
127. Davies LR, Pearce OM, Tessier MB, et al. Metabolism of vertebrate amino sugars with N-glycolyl groups: resistance of alpha2-8-linked N-glycolylneuraminic acid to enzymatic cleavage. *J Biol Chem*. 2012;287(34):28917–28931.
128. Naito-Matsui Y, Davies LR, Takematsu H, et al. Physiological exploration of the long term evolutionary selection against expression of N-Glycolylneuraminic acid in the brain. *J Biol Chem*. 2017;292(7):2557–2570.
129. Mlinac K, Fabris D, Vukelic Z, Rozman M, Heffer M, Bognar SK. Structural analysis of brain ganglioside acetylation patterns in mice with altered ganglioside biosynthesis. *Carbohydr Res*. 2013;382:1–8.
130. Manzi AE, Sjoberg ER, Diaz S, Varki A. Biosynthesis and turnover of O-acetyl and N-acetyl groups in the gangliosides of human melanoma cells. *J Biol Chem*. 1990;265(22):13091–13103.
131. Kohla G, Stockfleth E, Schauer R. Gangliosides with O-acetylated sialic acids in tumors of neuroectodermal origin. *Neurochem Res*. 2002;27(7–8):583–592.
132. Rogers GN, Herrler G, Paulson JC, Klenk HD. Influenza C virus uses 9-O-acetyl-N-acetylneuraminic acid as a high affinity receptor determinant for attachment to cells. *J Biol Chem*. 1986;261(13):5947–5951.
133. Matrosovich M, Herrler G, Klenk HD. Sialic acid receptors of viruses. *Top Curr Chem*. 2015;367:1–28.
134. Herrler G, Klenk HD. The surface receptor is a major determinant of the cell tropism of influenza C virus. *Virology*. 1987;159(1):102–108.
135. Tiralongo J, Schauer R. The enigma of enzymatic sialic acid O-acetylation. *Trends in glycoscience and glycotechnology*. 2004;16(87):1–15.
136. Wipfler D, Srinivasan GV, Sadick H, et al. Differentially regulated expression of 9-O-acetyl GD3 (CD60b) and 7-O-acetyl-GD3 (CD60c) during differentiation and maturation of human T and B lymphocytes. *Glycobiology*. 2011;21(9):1161–1172.
137. Corfield AP, Sander-Wewer M, Veh RW, Wember M, Schauer R. The action of sialidases on substrates containing O-acetylsialic acids. *Biol Chem Hoppe Seyler*. 1986;367(5):433–439.
138. Oehler C, Kopitz J, Cantz M. Substrate specificity and inhibitor studies of a membrane-bound ganglioside sialidase isolated from human brain tissue. *Biol Chem*. 2002;383(11):1735–1742.
139. Mandal C, Schwartz-Albiez R, Vlasak R. Functions and biosynthesis of O-acetylated sialic Acids. *Top Curr Chem*. 2015;366:1–30.
140. Ravindranaths MH, Paulson JC, Irie RF. Human melanoma antigen O-acetylated ganglioside GD3 is recognized by cancer antennarius lectin. *J Biol Chem*. 1988;263(4):2079–2086.

141. Mukherjee K, Chava AK, Mandal C, et al. O-acetylation of GD3 prevents its apoptotic effect and promotes survival of lymphoblasts in childhood acute lymphoblastic leukaemia. *J Cell Biochem.* 2008;105(3):724–734.
142. Birks SM, Danquah JO, King L, Vlasak R, Gorecki DC, Pilkington GJ. Targeting the GD3 acetylation pathway selectively induces apoptosis in glioblastoma. *Neuro Oncol.* 2011;13(9):950–960.
143. Fahr C, Schauer R. Detection of sialic acids and gangliosides with special reference to 9-O-acetylated species in basaliomas and normal human skin. *J Invest Dermatol.* 2001;116(2):254–260.
144. Kamerling JP, Schauer R, Shukla AK, Stoll S, Van Halbeek H, Vliegenthart JF. Migration of O-acetyl groups in N,O-acetylneuraminic acids. *Eur J Biochem.* 1987;162(3):601–607.
145. Schauer R. Sialic acids: fascinating sugars in higher animals and man. *Zoology (Jena).* 2004;107(1):49–64.
146. Baumann AM, Bakkers MJ, Buettner FF, et al. 9-O-Acetylation of sialic acids is catalysed by CASD1 via a covalent acetyl-enzyme intermediate. *Nat Commun.* 2015;6:7673.
147. Arming S, Wipfler D, Mayr J, et al. The human Cas1 protein: a sialic acid-specific O-acetyltransferase? *Glycobiology.* 2011;21(5):553–564.
148. Iwersen M, Vandamme-Feldhaus V, Schauer R. Enzymatic 4-O-acetylation of N-acetylneuraminic acid in guinea-pig liver. *Glycoconj J.* 1998;15(9):895–904.
149. Terabayashi T, Kawanishi Y. Naturally occurring ganglioside lactones in Minke whale brain. *Carbohydr Res.* 1998;307(3–4):281–290.
150. Sonnino S, Chigorno V, Aureli M, et al. Role of gangliosides and plasma membrane-associated sialidase in the process of cell membrane organization. *Adv Exp Med Biol.* 2011;705:297–316.
151. Breiden B, Sandhoff K. Ganglioside metabolism and its inherited diseases. *Methods Mol Biol* 2018 [in press].
152. Leskawa KC, Rosenberg A. The organization of gangliosides and other lipid components in synaptosomal plasma membranes and modifying effects of calcium ion. *Cell Mol Neurobiol.* 1981;1(4):373–388.
153. Hansson HA, Holmgren J, Svennerholm L. Ultrastructural localization of cell membrane GM1 ganglioside by cholera toxin. *Proc Natl Acad Sci U S A.* 1977;74(9):3782–3786.
154. Sonnino S, Prinetti A. The role of sphingolipids in neuronal plasticity of the brain. *J Neurochem.* 2016;137(4):485–488.
155. Prendergast J, Umanah GKE, Yoo SW, et al. Ganglioside regulation of AMPA receptor trafficking. *J Neurosci.* 2014;34(39):13246–13258.
156. Preti A, Fiorilli A, Lombardo A, Caimi L, Tettamanti G. Occurrence of sialyltransferase activity in the synaptosomal membranes prepared from calf brain cortex. *J Neurochem.* 1980;35(2):281–296.
157. Durrie R, Saito M, Rosenberg A. Endogenous glycosphingolipid acceptor specificity of sialosyltransferase systems in intact Golgi membranes, synaptosomes, and synaptic plasma membranes from rat brain. *Biochemistry.* 1988;27(10):3759–3764.
158. Ohman R. Subcellular fractionation of ganglioside sialidase from human brain. *J Neurochem.* 1971;18(1):89–95.
159. Tettamanti G, Preti A, Cestaro B, et al. Gangliosides, neuraminidase and sialyltransferase at the nerve endings. *Adv Exp Med Biol.* 1980;125:263–281.
160. Saito M, Tanaka Y, Tang CP, Yu RK, Ando S. Characterization of sialidase activity in mouse synaptic plasma membranes and its age-related changes. *J Neurosci Res.* 1995;40(3):401–406.

161. Scheel G, Acevedo E, Conzelmann E, Nehrkorn H, Sandhoff K. Model for the interaction of membrane-bound substrates and enzymes. Hydrolysis of ganglioside GD1a by sialidase of neuronal membranes isolated from calf brain. *Eur J Biochem.* 1982;127(2):245–253.
162. Sandhoff K, Pallmann B. Membrane-bound neuraminidase from calf brain: regulation of oligosialoganglioside degradation by membrane fluidity and membrane components. *Proc Natl Acad Sci U S A.* 1978;75(1):122–126.
163. Miyagi T, Yamaguchi K. Mammalian sialidases: physiological and pathological roles in cellular functions. *Glycobiology.* 2012;22(7):880–896.
164. Shiozaki K, Takahashi K, Hosono M, et al. Phosphatidic acid-mediated activation and translocation to the cell surface of sialidase NEU3, promoting signaling for cell migration. *FASEB J.* 2015;29(5):2099–2111.
165. Aureli M, Loberto N, Lanteri P, Chigorno V, Prinetti A, Sonnino S. Cell surface sphingolipid glycohydrolases in neuronal differentiation and aging in culture. *J Neurochem.* 2011;116(5):891–899.
166. Schwarzmann G, Hoffmann-Bleihauer P, Schubert J, Sandhoff K, Marsh D. Incorporation of ganglioside analogues into fibroblast cell membranes. A spin-label study. *Biochemistry.* 1983;22(21):5041–5048.
167. Sonderfeld S, Conzelmann E, Schwarzmann G, Burg J, Hinrichs U, Sandhoff K. Incorporation and metabolism of ganglioside GM2 in skin fibroblasts from normal and GM2 gangliosidosis subjects. *Eur J Biochem.* 1985;149(2):247–255.
168. Lucas M, Gershlick DC, Vidaurrazaga A, Rojas AL, Bonifacino JS, Hierro A. Structural mechanism for cargo recognition by the retromer complex. *Cell.* 2016;167(6). 1623–1635.e1614.
169. Ghidoni R, Trinchera M, Sonnino S, Chigorno V, Tettamanti G. The sialic acid residue of exogenous GM1 ganglioside is recycled for biosynthesis of sialoglycoconjugates in rat liver. *Biochem J.* 1987;247(1):157–164.
170. Ghidoni R, Trinchera M, Venerando B, Fiorilli A, Sonnino S, Tettamanti G. Incorporation and metabolism of exogenous GM1 ganglioside in rat liver. *Biochem J.* 1986;237(1):147–155.
171. Kok JW, Eskelinen S, Hoekstra K, Hoekstra D. Salvage of glucosylceramide by recycling after internalization along the pathway of receptor-mediated endocytosis. *Proc Natl Acad Sci U S A.* 1989;86(24):9896–9900.
172. Schwarzmann G, Hofmann P, Putz U, Albrecht B. Demonstration of direct glycosylation of nondegradable glucosylceramide analogs in cultured cells. *J Biol Chem.* 1995;270(36):21271–21276.
173. Gillard BK, Harrell RG, Marcus DM. Pathways of glycosphingolipid biosynthesis in SW13 cells in the presence and absence of vimentin intermediate filaments. *Glycobiology.* 1996;6(1):33–42.
174. Tettamanti G. Ganglioside/glycosphingolipid turnover: new concepts. *Glycoconj J.* 2004;20(5):301–317.
175. Tettamanti G, Bassi R, Viani P, Riboni L. Salvage pathways in glycosphingolipid metabolism. *Biochimie.* 2003;85(3–4):423–437.
176. Hanus C, Ehlers MD. Specialization of biosynthetic membrane trafficking for neuronal form and function. *Curr Opin Neurobiol.* 2016;39:8–16.
177. Ledeen R, Wu G. New findings on nuclear gangliosides: overview on metabolism and function. *J Neurochem.* 2011;116(5):714–720.
178. Fredman P. Gangliosides associated with primary brain tumors and their expression in cell lines established from these tumors. *Prog Brain Res.* 1994;101: 225–240.
179. Posse de Chaves E, Sipione S. Sphingolipids and gangliosides of the nervous system in membrane function and dysfunction. *FEBS Lett.* 2010;584(9):1748–1759.

180. Rosenberg A, Stern N. Changes in sphingosine and fatty acid components of the gangliosides in developing rat and human brain. *J Lipid Res.* 1966;7(1):122–131.
181. Naoi M, Klenk E. The sphingosine bases of the gangliosides from developing human brain and from brains of amaurotic idiots. *Hoppe Seylers Z Physiol Chem.* 1972;353 (11):1677–1683.
182. Mansson JE, Vanier MT, Svennerholm L. Changes in the fatty acid and sphingosine composition of the major gangliosides of human brain with age. *J Neurochem.* 1978;30 (1):273–275.
183. Sugiura Y, Shimma S, Konishi Y, Yamada MK, Setou M. Imaging mass spectrometry technology and application on ganglioside study; visualization of age-dependent accumulation of C20-ganglioside molecular species in the mouse hippocampus. *PLoS One.* 2008;3(9).
184. Zhao L, Spassieva S, Gable K, et al. Elevation of 20-carbon long chain bases due to a mutation in serine palmitoyltransferase small subunit b results in neurodegeneration. *Proc Natl Acad Sci U S A.* 2015;112(42):12962–12967.
185. Penno A, Reilly MM, Houlden H, et al. Hereditary sensory neuropathy type 1 is caused by the accumulation of two neurotoxic sphingolipids. *J Biol Chem.* 2010;285 (15):11178–11187.
186. Alecu I, Othman A, Penno A, et al. Cytotoxic 1-deoxysphingolipids are metabolized by a cytochrome P450-dependent pathway. *J Lipid Res.* 2017;58(1):60–71.
187. Zhao L, Spassieva SD, Jucius TJ, et al. A deficiency of ceramide biosynthesis causes cerebellar purkinje cell neurodegeneration and lipofuscin accumulation. *PLoS Genet.* 2011;7(5):e1002063.
188. Ginkel C, Hartmann D, Vom Dorp K, et al. Ablation of neuronal ceramide synthase 1 in mice decreases ganglioside levels and expression of Myelin associated glycoprotein in oligodendrocytes. *J Biol Chem.* 2012;287(50):41888–41902.
189. Vanni N, Fruscione F, Ferlazzo E, et al. Impairment of ceramide synthesis causes a novel progressive myoclonus epilepsy. *Ann Neurol.* 2014;76(2):206–212.
190. Imgrund S, Hartmann D, Farwanah H, et al. Adult ceramide synthase 2 (CERS2)-deficient mice exhibit myelin sheath defects, cerebellar degeneration, and hepatocarcinomas. *J Biol Chem.* 2009;284(48):33549–33560.
191. Pewzner-Jung Y, Park H, Laviad EL, et al. A critical role for ceramide synthase 2 in liver homeostasis: I. alterations in lipid metabolic pathways. *J Biol Chem.* 2010;285 (14):10902–10910.
192. Park JW, Park WJ, Kuperman Y, Boura-Halfon S, Pewzner-Jung Y, Futerman AH. Ablation of very long acyl chain sphingolipids causes hepatic insulin resistance in mice due to altered detergent-resistant membranes. *Hepatology.* 2013;57(2):525–532.
193. Sandhoff R, Geyer R, Jennemann R, et al. Novel class of glycosphingolipids involved in male fertility. *J Biol Chem.* 2005;280(29):27310–27318.
194. Rabionet M, van der Spoel AC, Chuang CC, et al. Male germ cells require polyenoic sphingolipids with complex glycosylation for completion of meiosis: a link to ceramide synthase-3. *J Biol Chem.* 2008;283(19):13357–13369.
195. Rabionet M, Bayerle A, Jennemann R, et al. Male meiotic cytokinesis requires ceramide synthase 3-dependent sphingolipids with unique membrane anchors. *Hum Mol Genet.* 2015;24(17):4792–4808.
196. Penalva DA, Furland NE, Lopez GH, Aveldano MI, Antollini SS. Unique thermal behavior of sphingomyelin species with nonhydroxy and 2-hydroxy very-long-chain (C28-C32) PUFAs. *J Lipid Res.* 2013;54(8):2225–2235.
197. Penalva DA, Oresti GM, Dupuy F, et al. Atypical surface behavior of ceramides with nonhydroxy and 2-hydroxy very long-chain (C28-C32) PUFAs. *Biochim Biophys Acta.* 2014;1838(3):731–738.

198. Yamashita T, Wada R, Sasaki T, et al. A vital role for glycosphingolipid synthesis during development and differentiation. *Proc Natl Acad Sci U S A*. 1999;96(16):9142–9147.
199. Jennemann R, Sandhoff R, Wang S, et al. Cell-specific deletion of glucosylceramide synthase in brain leads to severe neural defects after birth. *Proc Natl Acad Sci U S A*. 2005;102(35):12459–12464.
200. Wu G, Lu ZH, Andre S, Gabius HJ, Ledeen RW. Functional interplay between ganglioside GM1 and cross-linking galectin-1 induces axon-like neuritogenesis via integrin-based signaling and TRPC5-dependent $Ca^{2+}$ influx. *J Neurochem*. 2016;136(3):550–563.
201. Watanabe S, Endo S, Oshima E, et al. Glycosphingolipid synthesis in cerebellar Purkinje neurons: roles in myelin formation and axonal homeostasis. *Glia*. 2010;58(10):1197–1207.
202. Herzer S, Meldner S, Grone HJ, Nordstrom V. Fasting-induced lipolysis and hypothalamic insulin signaling are regulated by neuronal glucosylceramide synthase. *Diabetes*. 2015;64(10):3363–3376.
203. Herzer S, Meldner S, Rehder K, Grone HJ, Nordstrom V. Lipid microdomain modification sustains neuronal viability in models of Alzheimer's disease. *Acta Neuropathol Commun*. 2016;4(1):103.
204. Simpson MA, Cross H, Proukakis C, et al. Infantile-onset symptomatic epilepsy syndrome caused by a homozygous loss-of-function mutation of GM3 synthase. *Nat Genet*. 2004;36:1225–1229.
205. Farukhi F, Dakkouri C, Wang H, Wiztnitzer M, Traboulsi EI. Etiology of vision loss in ganglioside GM3 synthase deficiency. *Ophthalmic Genet*. 2006;27(3):89–91.
206. Wang H, Bright A, Xin B, Bockoven JR, Paller AS. Cutaneous dyspigmentation in patients with ganglioside GM3 synthase deficiency. *Am J Med Genet A*. 2013;161A(4):875–879.
207. Yoshikawa M, Go S, Suzuki S, et al. Ganglioside GM3 is essential for the structural integrity and function of cochlear hair cells. *Hum Mol Genet*. 2015;24(10):2796–2807.
208. Fragaki K, Ait-El-Mkadem S, Chaussenot A, et al. Refractory epilepsy and mitochondrial dysfunction due to GM3 synthase deficiency. *Eur J Hum Genet*. 2013;21(5):528–534.
209. Boccuto L, Aoki K, Flanagan-Steet H, et al. A mutation in a ganglioside biosynthetic enzyme, ST3GAL5, results in salt & pepper syndrome, a neurocutaneous disorder with altered glycolipid and glycoprotein glycosylation. *Hum Mol Genet*. 2014;23(2):418–433.
210. Wang H, Wang A, Wang D, et al. Early growth and development impairments in patients with ganglioside GM3 synthase deficiency. *Clin Genet*. 2016;89(5):625–629.
211. Collins BE, Ito H, Sawada N, Ishida H, Kiso M, Schnaar RL. Enhanced binding of the neural siglecs, myelin-associated glycoprotein and Schwann cell myelin protein, to Chol-1 (alpha-series) gangliosides and novel sulfated Chol-1 analogs. *J Biol Chem*. 1999;274(53):37637–37643.
212. Vyas AA, Schnaar RL. Brain gangliosides: functional ligands for myelin stability and the control of nerve regeneration. *Biochimie*. 2001;83(7):677–682.
213. Niimi K, Nishioka C, Miyamoto T, et al. Impairment of neuropsychological behaviors in ganglioside GM3-knockout mice. *Biochem Biophys Res Commun*. 2011;406(4):524–528.
214. Yoshikawa M, Go S, Takasaki K, et al. Mice lacking ganglioside GM3 synthase exhibit complete hearing loss due to selective degeneration of the organ of Corti. *Proc Natl Acad Sci U S A*. 2009.
215. Wang XQ, Lee S, Wilson H, et al. Ganglioside GM3 depletion reverses impaired wound healing in diabetic mice by activating IGF-1 and insulin receptors. *J Invest Dermatol*. 2014;134(5):1446–1455.

216. Menichella DM, Jayaraj ND, Wilson HM, et al. Ganglioside GM3 synthase depletion reverses neuropathic pain and small fiber neuropathy in diet-induced diabetic mice. *Mol Pain.* 2016;12.
217. Nagafuku M, Okuyama K, Onimaru Y, et al. CD4 and CD8 T cells require different membrane gangliosides for activation. *Proc Natl Acad Sci U S A.* 2012;109(6): E336–E342.
218. Inokuchi J, Nagafuku M, Ohno I, Suzuki A. Heterogeneity of gangliosides among T cell subsets. *Cell Mol Life Sci.* 2013;70(17):3067–3075.
219. Kiguchi K, Henning-Chubb CB, Huberman E. Glycosphingolipid patterns of peripheral blood lymphocytes, monocytes, and granulocytes are cell specific. *J Biochem (Tokyo).* 1990;107(1):8–14.
220. Stein KE, Marcus DM. Glycosphingolipids of purified human lymphocytes. *Biochemistry.* 1977;16(24):5285–5291.
221. Tsukuda Y, Iwasaki N, Seito N, et al. Ganglioside GM3 has an essential role in the pathogenesis and progression of rheumatoid arthritis. *PLoS One.* 2012;7(6):e40136.
222. David MJ, Portoukalian J, Rebbaa A, Vignon E, Carret JP, Richard M. Characterization of gangliosides from normal and osteoarthritic human articular cartilage. *Arthritis Rheum.* 1993;36(7):938–942.
223. Sasazawa F, Onodera T, Yamashita T, et al. Depletion of gangliosides enhances cartilage degradation in mice. *Osteoarthritis Cartilage.* 2014;22(2):313–322.
224. Matsuoka M, Onodera T, Homan K, et al. Depletion of gangliosides enhances articular cartilage repair in mice. *Sci Rep.* 2017;7:43729.
225. Nagahori N, Yamashita T, Amano M, Nishimura S. Effect of ganglioside GM3 synthase gene knockout on the glycoprotein N-glycan profile of mouse embryonic fibroblast. *Chembiochem.* 2013;14(1):73–82.
226. Natoli TA, Husson H, Rogers KA, et al. Loss of GM3 synthase gene, but not sphingosine kinase 1, is protective against murine nephronophthisis-related polycystic kidney disease. *Hum Mol Genet.* 2012;21(15):3397–3407.
227. Kawai H, Sango K, Mullin KA, Proia RL. Embryonic stem cells with a disrupted GD3 synthase gene undergo neuronal differentiation in the absence of b-series gangliosides. *J Biol Chem.* 1998;273(31):19634–19638.
228. Okada M, Itoh Mi M, Haraguchi M, et al. b-series Ganglioside deficiency exhibits no definite changes in the neurogenesis and the sensitivity to fas-mediated apoptosis but impairs regeneration of the lesioned hypoglossal nerve. *J Biol Chem.* 2002;277(3):1633–1636.
229. Bernardo A, Harrison FE, McCord M, et al. Elimination of GD3 synthase improves memory and reduces amyloid-beta plaque load in transgenic mice. *Neurobiol Aging.* 2009;30(11):1777–1791.
230. Ariga T, Itokazu Y, McDonald MP, Hirabayashi Y, Ando S, Yu RK. Brain gangliosides of a transgenic mouse model of Alzheimer's disease with deficiency in GD3-synthase: expression of elevated levels of a cholinergic-specific ganglioside, GT1aalpha. *ASN Neuro.* 2013;5(2):141–148.
231. Ribeiro-Resende VT, Araujo Gomes T, de Lima S, et al. Mice lacking GD3 synthase display morphological abnormalities in the sciatic nerve and neuronal disturbances during peripheral nerve regeneration. *PLoS One.* 2014;9(10):e108919.
232. Wang J, Yu RK. Interaction of ganglioside GD3 with an EGF receptor sustains the self-renewal ability of mouse neural stem cells in vitro. *Proc Natl Acad Sci U S A.* 2013;110(47):19137–19142.
233. Wang J, Cheng A, Wakade C, Yu RK. Ganglioside GD3 is required for neurogenesis and long-term maintenance of neural stem cells in the postnatal mouse brain. *J Neurosci.* 2014;34(41):13790–13800.

234. Akkhawattanangkul Y, Maiti P, Xue Y, et al. Targeted deletion of GD3 synthase protects against MPTP-induced neurodegeneration. *Genes Brain Behav.* 2017.
235. Handa Y, Ozaki N, Honda T, et al. GD3 synthase gene knockout mice exhibit thermal hyperalgesia and mechanical allodynia but decreased response to formalin-induced prolonged noxious stimulation. *Pain.* 2005;117(3):271–279.
236. Daniotti JL, Crespo PM, Yamashita T. In vivo modulation of epidermal growth factor receptor phosphorylation in mice expressing different gangliosides. *J Cell Biochem.* 2006;99(5):1442–1451.
237. Ji S, Tokizane K, Ohkawa Y, et al. Increased a-series gangliosides positively regulate leptin/Ob receptor-mediated signals in hypothalamus of GD3 synthase-deficient mice. *Biochem Biophys Res Commun.* 2016;479(3):453–460.
238. Ji S, Ohkawa Y, Tokizane K, et al. b-Series gangliosides crucially regulate leptin secretion in adipose tissues. *Biochem Biophys Res Commun.* 2015;459(2):189–195.
239. Harlalka GV, Lehman A, Chioza B, et al. Mutations in B4GALNT1 (GM2 synthase) underlie a new disorder of ganglioside biosynthesis. *Brain.* 2013;136(Pt 12):3618–3624.
240. Liu Y, Wada R, Kawai H, et al. A genetic model of substrate deprivation therapy for a glycosphingolipid storage disorder [see comments]. *J Clin Invest.* 1999;103(4):497–505.
241. Furukawa K, Aixinjueluo W, Kasama T, et al. Disruption of GM2/GD2 synthase gene resulted in overt expression of 9-O-acetyl GD3 irrespective of Tis21. *J Neurochem.* 2008;105(3):1057–1066.
242. Takamiya K, Yamamoto A, Furukawa K, et al. Complex gangliosides are essential in spermatogenesis of mice: possible roles in the transport of testosterone. *Proc Natl Acad Sci USA.* 1998;95(21):12147–12152.
243. Sheikh KA, Sun J, Liu Y, et al. Mice lacking complex gangliosides develop Wallerian degeneration and myelination defects. *Proc Natl Acad Sci USA.* 1999;96(13):7532–7537.
244. Chiavegatto S, Sun J, Nelson RJ, Schnaar RL. A functional role for complex gangliosides: motor deficits in GM2/GD2 synthase knockout mice. *Exp Neurol.* 2000;166(2):227–234.
245. Bullens RW, O'Hanlon GM, Wagner E, et al. Roles of complex gangliosides at the neuromuscular junction. *Ann N Y Acad Sci.* 2003;998:401–403.
246. Zitman FM, Todorov B, Verschuuren JJ, et al. Neuromuscular synaptic transmission in aged ganglioside-deficient mice. *Neurobiol Aging.* 2011;32(1):157–167.
247. Rupp A, Cunningham ME, Yao D, Furukawa K, Willison HJ. The effects of age and ganglioside composition on the rate of motor nerve terminal regeneration following antibody-mediated injury in mice. *Synapse.* 2013;67(7):382–389.
248. Wu G, Xie X, Lu ZH, Ledeen RW. Cerebellar neurons lacking complex gangliosides degenerate in the presence of depolarizing levels of potassium. *Proc Natl Acad Sci USA.* 2001;98(1):307–312.
249. Wu G, Lu ZH, Xie X, Ledeen RW. Susceptibility of cerebellar granule neurons from GM2/GD2 synthase-null mice to apoptosis induced by glutamate excitotoxicity and elevated KCl: rescue by GM1 and LIGA20. *Glycoconj J.* 2004;21(6):305–313.
250. Wu G, Lu ZH, Wang J, et al. Enhanced susceptibility to kainate-induced seizures, neuronal apoptosis, and death in mice lacking gangliotetraose gangliosides: protection with LIGA 20, a membrane-permeant analog of GM1. *J Neurosci.* 2005;25(47):11014–11022.
251. Kittaka D, Itoh M, Ohmi Y, et al. Impaired hypoglossal nerve regeneration in mutant mice lacking complex gangliosides: down-regulation of neurotrophic factors and receptors as possible mechanisms. *Glycobiology.* 2008;18(7):509–516.
252. Oikawa N, Yamaguchi H, Ogino K, et al. Gangliosides determine the amyloid pathology of Alzheimer's disease. *Neuroreport.* 2009;20(12):1043–1046.

253. Mohammed EE, Snella EM, Rutz-Mendicino MM, et al. Accelerated clinical disease and pathology in mucopolysaccharidosis type IIIB and GalNAc transferase double knockout mice. *Mol Genet Metab.* 2012;107(1–2):129–135.
254. Hadaczek P, Wu G, Sharma N, et al. GDNF signaling implemented by GM1 ganglioside; failure in Parkinson's disease and GM1-deficient murine model. *Exp Neurol.* 2015;263:177–189.
255. Zhao J, Furukawa K, Fukumoto S, et al. Attenuation of interleukin 2 signal in the spleen cells of complex ganglioside-lacking mice. *J Biol Chem.* 1999;274(20):13744–13747.
256. Tajima O, Egashira N, Ohmi Y, et al. Reduced motor and sensory functions and emotional response in GM3-only mice: emergence from early stage of life and exacerbation with aging. *Behav Brain Res.* 2009;198(1):74–82.
257. Tajima O, Egashira N, Ohmi Y, et al. Dysfunction of muscarinic acetylcholine receptors as a substantial basis for progressive neurological deterioration in GM3-only mice. *Behav Brain Res.* 2010;206(1):101–108.
258. Ohkawa Y, Ohmi Y, Tajima O, Yamauchi Y, Furukawa K, Furukawa K. Wisp2/CCN5 up-regulated in the central nervous system of GM3-only mice facilitates neurite formation in Neuro2a cells via integrin-Akt signaling. *Biochem Biophys Res Commun.* 2011;411(3):483–489.
259. Ohmi Y, Tajima O, Ohkawa Y, et al. Gangliosides play pivotal roles in the regulation of complement systems and in the maintenance of integrity in nerve tissues. *Proc Natl Acad Sci U S A.* 2009;106(52):22405–22410.
260. Ohmi Y, Tajima O, Ohkawa Y, et al. Gangliosides are essential in the protection of inflammation and neurodegeneration via maintenance of lipid rafts: elucidation by a series of ganglioside-deficient mutant mice. *J Neurochem.* 2011;116(5):926–935.
261. Ohmi Y, Ohkawa Y, Tajima O, Sugiura Y, Furukawa K, Furukawa K. Ganglioside deficiency causes inflammation and neurodegeneration via the activation of complement system in the spinal cord. *J Neuroinflammation.* 2014;11:61.
262. Zitman FM, Todorov B, Jacobs BC, et al. Neuromuscular synaptic function in mice lacking major subsets of gangliosides. *Neuroscience.* 2008;156(4):885–897.
263. Yoo SW, Motari MG, Susuki K, et al. Sialylation regulates brain structure and function. *FASEB J.* 2015;29(7):3040–3053.
264. Lopez PH, Aja S, Aoki K, et al. Mice lacking sialyltransferase ST3Gal-II develop late-onset obesity and insulin resistance. *Glycobiology.* 2017;27(2):129–139.
265. Sandhoff K. My journey into the world of sphingolipids and sphingolipidoses. *Proc Jpn Acad Ser B Phys Biol Sci.* 2012;88(10):554–582.
266. Kolter T, Sandhoff K. Principles of lysosomal membrane digestion: stimulation of sphingolipid degradation by sphingolipid activator proteins and anionic lysosomal lipids. *Annu Rev Cell Dev Biol.* 2005;21:81–103.
267. Fürst W, Sandhoff K. Activator proteins and topology of lysosomal sphingolipid catabolism. *Biochim Biophys Acta.* 1992;1126(1):1–16.
268. Wollert T, Hurley JH. Molecular mechanism of multivesicular body biogenesis by ESCRT complexes. *Nature.* 2010;464(7290):864–869.
269. Florey O, Overholtzer M. Autophagy proteins in macroendocytic engulfment. *Trends Cell Biol.* 2012;22(7):374–380.
270. Gallala HD, Breiden B, Sandhoff K. Regulation of the NPC2 protein-mediated cholesterol trafficking by membrane lipids. *J Neurochem.* 2011;116(5):702–707.
271. Quintern LE, Weitz G, Nehrkorn H, Tager JM, Schram AW, Sandhoff K. Acid sphingomyelinase from human urine: purification and characterization. *Biochim Biophys Acta.* 1987;922(3):323–336.
272. Oninla VO, Breiden B, Babalola JO, Sandhoff K. Acid sphingomyelinase activity is regulated by membrane lipids and facilitates cholesterol transfer by NPC2. *J Lipid Res.* 2014;55(12):2606–2619.

273. Abdul-Hammed M, Breiden B, Adebayo MA, Babalola JO, Schwarzmann G, Sandhoff K. Role of endosomal membrane lipids and NPC2 in cholesterol transfer and membrane fusion. *J Lipid Res.* 2010;51(7):1747–1760.
274. Gallala HD, Sandhoff K. Biological function of the cellular lipid BMP-BMP as a key activator for cholesterol sorting and membrane digestion. *Neurochem Res.* 2011;36 (9):1594–1600.
275. Appelqvist H, Sandin L, Bjornstrom K, et al. Sensitivity to lysosome-dependent cell death is directly regulated by lysosomal cholesterol content. *PLoS One.* 2012;7(11): e50262.
276. Eskelinen EL, Tanaka Y, Saftig P. At the acidic edge: emerging functions for lysosomal membrane proteins. *Trends Cell Biol.* 2003;13(3):137–145.
277. Kolter T, Winau F, Schaible UE, Leippe M, Sandhoff K. Lipid-binding proteins in membrane digestion, antigen presentation, and antimicrobial defense. *J Biol Chem.* 2005;280(50):41125–41128.
278. Kolter T, Sandhoff K. Sphingolipid metabolism diseases. *Biochim Biophys Acta.* 2006;1758(12):2057–2079.
279. Sandhoff K, Harzer K. Gangliosides and gangliosidoses: principles of molecular and metabolic pathogenesis. *J Neurosci.* 2013;33(25):10195–10208.
280. Schulze H, Sandhoff K. Sphingolipids and lysosomal pathologies. *Biochim Biophys Acta.* 2014;1841(5):799–810.
281. Doering T, Holleran WM, Potratz A, et al. Sphingolipid activator proteins are required for epidermal permeability barrier formation. *J Biol Chem.* 1999;274(16):11038–11045.
282. Hulkova H, Cervenkova M, Ledvinova J, et al. A novel mutation in the coding region of the prosaposin gene leads to a complete deficiency of prosaposin and saposins, and is associated with a complex sphingolipidosis dominated by lactosylceramide accumulation. *Hum Mol Genet.* 2001;10(9):927–940.
283. Breiden B, Sandhoff K. The role of sphingolipid metabolism in cutaneous permeability barrier formation. *Biochim Biophys Acta.* 2014;1841(3):441–452.
284. Monti E, Bonten E, D'Azzo A, et al. Sialidases in vertebrates: a family of enzymes tailored for several cell functions. *Adv Carbohydr Chem Biochem.* 2010;64:403–479.
285. Smutova V, Albohy A, Pan X, et al. Structural basis for substrate specificity of mammalian neuraminidases. *PLoS One.* 2014;9(9):e106320.
286. Timur ZK, Akyildiz Demir S, Marsching C, Sandhoff R, Seyrantepe V. Neuraminidase-1 contributes significantly to the degradation of neuronal B-series gangliosides but not to the bypass of the catabolic block in Tay–Sachs mouse models. *Mol Genet Metab Rep.* 2015;4:72–82.
287. Bonten E, van der Spoel A, Fornerod M, Grosveld G, d'Azzo A. Characterization of human lysosomal neuraminidase defines the molecular basis of the metabolic storage disorder sialidosis. *Genes Dev.* 1996;10(24):3156–3169.
288. d'Azzo A, Bonten E. Molecular mechanisms of pathogenesis in a glycosphingolipid and a glycoprotein storage disease. *Biochem Soc Trans.* 2010;38(6):1453–1457.
289. Mobius W, Herzog V, Sandhoff K, Schwarzmann G. Intracellular distribution of a biotin-labeled ganglioside, GM1, by immunoelectron microscopy after endocytosis in fibroblasts. *J Histochem Cytochem.* 1999;47(8):1005–1014.
290. Bradova V, Smid F, Ulrich-Bott B, Roggendorf W, Paton BC, Harzer K. Prosaposin deficiency: further characterization of the sphingolipid activator protein-deficient sibs. Multiple glycolipid elevations (including lactosylceramidosis), partial enzyme deficiencies and ultrastructure of the skin in this generalized sphingolipid storage disease. *Hum Genet.* 1993;92(2):143–152.
291. Burkhardt JK, Huttler S, Klein A, et al. Accumulation of sphingolipids in SAP-precursor (prosaposin)-deficient fibroblasts occurs as intralysosomal membrane structures and

can be completely reversed by treatment with human SAP-precursor. *Eur J Cell Biol.* 1997;73(1):10–18.
292. Harzer K, Paton BC, Poulos A, et al. Sphingolipid activator protein deficiency in a 16-week-old atypical Gaucher disease patient and his fetal sibling: biochemical signs of combined sphingolipidoses. *Eur J Pediatr.* 1989;149(1):31–39.
293. Schnabel D, Schroder M, Furst W, et al. Simultaneous deficiency of sphingolipid activator proteins 1 and 2 is caused by a mutation in the initiation codon of their common gene. *J Biol Chem.* 1992;267(5):3312–3315.
294. Mobius W, Herzog V, Sandhoff K, Schwarzmann G. Gangliosides are transported from the plasma membrane to intralysosomal membranes as revealed by immuno-electron microscopy. *Biosci Rep.* 1999;19(4):307–316.
295. Leinekugel P, Michel S, Conzelmann E, Sandhoff K. Quantitative correlation between the residual activity of beta-hexosaminidase A and arylsulfatase A and the severity of the resulting lysosomal storage disease. *Hum Genet.* 1992;88(5):513–523.
296. Conzelmann E, Sandhoff K. Partial enzyme deficiencies: residual activities and the development of neurological disorders. *Dev Neurosci.* 1983;6(1):58–71.
297. Mobius W, van Donselaar E, Ohno-Iwashita Y, et al. Recycling compartments and the internal vesicles of multivesicular bodies harbor most of the cholesterol found in the endocytic pathway. *Traffic.* 2003;4(4):222–231.
298. Kobayashi T, Beuchat MH, Lindsay M, et al. Late endosomal membranes rich in lysobisphosphatidic acid regulate cholesterol transport. *Nat Cell Biol.* 1999;1 (2):113–118.
299. Wilkening G, Linke T, Sandhoff K. Lysosomal degradation on vesicular membrane surfaces. Enhanced glucosylceramide degradation by lysosomal anionic lipids and activators. *J Biol Chem.* 1998;273(46):30271–30278.
300. Kolzer M, Werth N, Sandhoff K. Interactions of acid sphingomyelinase and lipid bilayers in the presence of the tricyclic antidepressant desipramine. *FEBS Lett.* 2004;559(1–3):96–98.
301. Hurwitz R, Ferlinz K, Sandhoff K. The tricyclic antidepressant desipramine causes proteolytic degradation of lysosomal sphingomyelinase in human fibroblasts. *Biol Chem Hoppe Seyler.* 1994;375(7):447–450.
302. Elojeimy S, Holman DH, Liu X, et al. New insights on the use of desipramine as an inhibitor for acid ceramidase. *FEBS Lett.* 2006;580(19):4751–4756.
303. Lullmann H, Lullmann-Rauch R, Wassermann O. Lipidosis induced by amphiphilic cationic drugs. *Biochem Pharmacol.* 1978;27(8):1103–1108.
304. Suzuki K, Chen GC. GM1-gangliosidosis (generalized gangliosidosis). morphology and chemical pathology. *Pathol Eur.* 1968;3(2):389–408.
305. Terry RD, Weiss M. Studies in Tay–Sachs disease. II. Ultrastructure of the cerebrum. *J Neuropathol Exp Neurol.* 1963;22:18–55.
306. Lee RE. The fine structure of the cerebroside occurring in Gaucher's disease. *Proc Natl Acad Sci U S A.* 1968;61(2):484–489.
307. Liu H, Chakravarty D, Maceyka M, Milstien S, Spiegel S. Sphingosine kinases: a novel family of lipid kinases. *Prog Nucleic Acid Res Mol Biol.* 2002;71:493–511.
308. van Meer G, Voelker DR, Feigenson GW. Membrane lipids: where they are and how they behave. *Nat Rev Mol Cell Biol.* 2008;9(2):112–124.
309. Abdul-Hammed M, Breiden B, Schwarzmann G, Sandhoff K. Lipids regulate the hydrolysis of membrane bound glucosylceramide by lysosomal beta-glucocerebrosidase. *J Lipid Res.* 2017;58(3):563–577.
310. Anheuser S, Breiden B, Schwarzmann G, Sandhoff K. Membrane lipids regulate ganglioside GM2 catabolism and GM2 activator protein activity. *J Lipid Res.* 2015;56 (9):1747–1761.
311. Vanier MT. Complex lipid trafficking in Niemann–Pick disease type C. *J Inherit Metab Dis.* 2015;38(1):187–199.

312. Vanier MT. Biochemical studies in Niemann–Pick disease. I. Major sphingolipids of liver and spleen. *Biochim Biophys Acta.* 1983;750(1):178–184.
313. Remmel N, Locatelli-Hoops S, Breiden B, Schwarzmann G, Sandhoff K. Saposin B mobilizes lipids from cholesterol-poor and bis(monoacylglycero)phosphate-rich membranes at acidic pH. Unglycosylated patient variant saposin B lacks lipid-extraction capacity. *FEBS J.* 2007;274(13):3405–3420.
314. Sandhoff K. Metabolic and cellular bases of sphingolipidoses. *Biochem Soc Trans.* 2013;41 (6):1562–1568.
315. Ballabio A, Gieselmann V. Lysosomal disorders: from storage to cellular damage. *Biochim Biophys Acta.* 2009;1793(4):684–696.
316. Suzuki K. My encounters with Krabbe disease: A personal recollection of a 40-Year journey with young colleagues. *J Neurosci Res.* 2016;94(11):965–972.
317. Nilsson O, Svennerholm L. Accumulation of glucosylceramide and glucosylsphingosine (psychosine) in cerebrum and cerebellum in infantile and juvenile Gaucher disease. *J Neurochem.* 1982;39(3):709–718.
318. Raghavan SS, Mumford RA, Kanfer JN. Isolation and characterization of glucosylsphingosine from Gaucher's spleen. *J Lipid Res.* 1974;15(5):484–490.
319. Sarmientos F, Schwarzmann G, Sandhoff K. Specificity of human glucosylceramide beta-glucosidase towards synthetic glucosylsphingolipids inserted into liposomes. Kinetic studies in a detergent-free assay system. *Eur J Biochem.* 1986;160(3):527–535.
320. Mizukami H, Mi Y, Wada R, et al. Systemic inflammation in glucocerebrosidase-deficient mice with minimal glucosylceramide storage. *J Clin Invest.* 2002;109 (9):1215–1221.
321. Jeyakumar M, Thomas R, Elliot-Smith E, et al. Central nervous system inflammation is a hallmark of pathogenesis in mouse models of GM1 and GM2 gangliosidosis. *Brain.* 2003;126(Pt 4):974–987.
322. Graber D, Salvayre R, Levade T. Accurate differentiation of neuronopathic and non-neuronopathic forms of Niemann–Pick disease by evaluation of the effective residual lysosomal sphingomyelinase activity in intact cells. *J Neurochem.* 1994;63(3):1060–1068.
323. Meivar-Levy I, Horowitz M, Futerman AH. Analysis of glucocerebrosidase activity using N-(1-[14C]hexanoyl)-D-erythroglucosylsphingosine demonstrates a correlation between levels of residual enzyme activity and the type of Gaucher disease. *Biochem J.* 1994;303(Pt 2):377–382.
324. Wilkening G, Linke T, Uhlhorn-Dierks G, Sandhoff K. Degradation of membrane-bound ganglioside GM1. Stimulation by bis(monoacylglycero)phosphate and the activator proteins SAP-B and GM2-AP. *J Biol Chem.* 2000;275(46):35814–35819.
325. Fingerhut R, van der Horst GT, Verheijen FW, Conzelmann E. Degradation of gangliosides by the lysosomal sialidase requires an activator protein. *Eur J Biochem.* 1992;208 (3):623–629.
326. Zschoche A, Furst W, Schwarzmann G, Sanhoff K. Hydrolysis of lactosylceramide by human galactosylceramidase and GM1-beta-galactosidase in a detergent-free system and its stimulation by sphingolipid activator proteins, sap-B and sap-C. Activator proteins stimulate lactosylceramide hydrolysis. *Eur J Biochem.* 1994;222(1):83–90.
327. Klein A, Henseler M, Klein C, Suzuki K, Harzer K, Sandhoff K. Sphingolipid activator protein D (sap-D) stimulates the lysosomal degradation of ceramide in vivo. *Biochem Biophys Res Commun.* 1994;200(3):1440–1448.
328. Valle D, (Editor-in-Chief), Beaudet AL, Vogelstein B, Kinzler KW, Antonarakis SE, Ballabio A, Michael Gibson, Mitchell G, eds. Lysosomal Disorders in the Online Metabolic and Molecular Bases of Inherited Disease. McGraw-Hill Education; 2001. https://ommbid.mhmedical.com/
329. Patterson MC, Vanier MT, Suzuki K, Morris JA, Carsteu E, Neufeld EB, Niemann–Pick disease type C. A lipid trafficking disorder. 8th ed. Scriver CR, ed.

et al. *The Metabolic and Molecular Basis Of Inherited Disease.* Vol II. New York: McGraw-Hill; 2001:3611–3633.
330. Okada S, O'Brien JS. Generalized gangliosidosis: beta-galactosidase deficiency. *Science.* 1968;160(3831):1002–1004.
331. Jatzkewitz H, Sandhoff K. On a biochemically special form of infantile amaturotic idiocy. *Biochim Biophys Acta.* 1963;70:354–356.
332. Suzuki Y, Oshima A, Nanba E, Beta-galactosidase deficiency (beta-galactosidosis): GM1 gangliosidosis and Morquio B disease. 8th ed. Scriver CR, Beaudet AL, Sly WS, Valle D, eds. *The Metabolic and Molecular Bases of Inherited Disease.* Vol II. New York: McGraw-Hill (pub); 2001:3775–3809.
333. Morreau H, Galjart NJ, Gillemans N, Willemsen R, van der Horst GT, d'Azzo A. Alternative splicing of beta-galactosidase mRNA generates the classic lysosomal enzyme and a beta-galactosidase-related protein. *J Biol Chem.* 1989;264(34):20655–20663.
334. Hinek A, Rabinovitch M, Keeley F, Okamura-Oho Y, Callahan J. The 67-kD elastin/laminin-binding protein is related to an enzymatically inactive, alternatively spliced form of beta-galactosidase. *J Clin Invest.* 1993;91(3):1198–1205.
335. Pshezhetsky AV, Ashmarina M. Lysosomal multienzyme complex: biochemistry, genetics, and molecular pathophysiology. *Prog Nucleic Acid Res Mol Biol.* 2001;69:81–114.
336. Caciotti A, Garman SC, Rivera-Colon Y, et al. GM1 gangliosidosis and Morquio B disease: an update on genetic alterations and clinical findings. *Biochim Biophys Acta.* 2011;1812(7):782–790.
337. Brunetti-Pierri N, Scaglia F. GM1 gangliosidosis: review of clinical, molecular, and therapeutic aspects. *Mol Genet Metab.* 2008;94(4):391–396.
338. Matsuda J, Suzuki O, Oshima A, et al. Beta-galactosidase-deficient mouse as an animal model for GM1-gangliosidosis. *Glycoconj J.* 1997;14(6):729–736.
339. Neufeld EF, d'Azzo A. Biosynthesis of normal and mutant beta-hexosaminidases. *Adv Genet.* 2001;44:165–171.
340. Okumiya T, Sakuraba H, Kase R, Sugiura T. Imbalanced substrate specificity of mutant beta-galactosidase in patients with Morquio B disease. *Mol Genet Metab.* 2003;78(1):51–58.
341. Sandhoff K, Harzer K, Wassle W, Jatzkewitz H. Enzyme alterations and lipid storage in three variants of Tay–Sachs disease. *J Neurochem.* 1971;18(12):2469–2489.
342. Sandhoff K, Jatzkewitz H, Peters G. [Infantile amaurotic idiocy and related forms as ganglioside storage diseases]. *Naturwissenschaften.* 1969;56(7):356–362.
343. Sango K, Yamanaka S, Hoffmann A, et al. Mouse models of Tay–Sachs and Sandhoff diseases differ in neurologic phenotype and ganglioside metabolism. *Nat Genet.* 1995;11(2):170–176.
344. Kytzia HJ, Sandhoff K. Evidence for two different active sites on human beta-hexosaminidase A. Interaction of GM2 activator protein with beta-hexosaminidase A. *J Biol Chem.* 1985;260(12):7568–7572.
345. Hepbildikler ST, Sandhoff R, Kolzer M, Proia RL, Sandhoff K. Physiological substrates for human lysosomal beta -hexosaminidase S. *J Biol Chem.* 2002;277(4):2562–2572.
346. Conzelmann E, Sandhoff K. AB variant of infantile GM2 gangliosidosis: deficiency of a factor necessary for stimulation of hexosaminidase A-catalyzed degradation of ganglioside GM2 and glycolipid GA2. *Proc Natl Acad Sci U S A.* 1978;75(8):3979–3983.
347. Brady RO, Kanfer JN, Shapiro D. Metabolism of glucocerebrosides II. Evidence of an enzymatic deficiency in Gaucher's disease. *Biochem Biophys Res Commun.* 1965;18:221–225.
348. Patrick A. A deficiency of glucocerebrosidase in Gauchers disease. *Biochem J.* 1965;97(2).C17.

349. Barton NW, Furbish FS, Murray GJ, Garfield M, Brady RO. Therapeutic response to intravenous infusions of glucocerebrosidase in a patient with Gaucher disease. *Proc Natl Acad Sci U S A.* 1990;87(5):1913–1916.
350. Brady RO. Enzyme replacement for lysosomal diseases. *Annu Rev Med.* 2006;57:283–296.
351. Doering T, Proia RL, Sandhoff K. Accumulation of protein-bound epidermal glucosylceramides in beta-glucocerebrosidase deficient type 2 Gaucher mice. *FEBS Lett.* 1999;447(2–3):167–170.
352. Bernardo K, Hurwitz R, Zenk T, et al. Purification, characterization, and biosynthesis of human acid ceramidase. *J Biol Chem.* 1995;270(19):11098–11102.
353. Koch J, Gartner S, Li CM, et al. Molecular cloning and characterization of a full-length complementary DNA encoding human acid ceramidase. Identification of the first molecular lesion causing Farber disease. *J Biol Chem.* 1996;271(51):33110–33115.
354. Shtraizent N, Eliyahu E, Park JH, He X, Shalgi R, Schuchman EH. Autoproteolytic cleavage and activation of human acid ceramidase. *J Biol Chem.* 2008;283 (17):11253–11259.
355. Furst W, Machleidt W, Sandhoff K. The precursor of sulfatide activator protein is processed to three different proteins. *Biol Chem Hoppe Seyler.* 1988;369(5):317–328.
356. Lefrancois S, Zeng J, Hassan AJ, Canuel M, Morales CR. The lysosomal trafficking of sphingolipid activator proteins (SAPs) is mediated by sortilin. *EMBO J.* 2003;22 (24):6430–6437.
357. Fujita N, Suzuki K, Vanier MT, et al. Targeted disruption of the mouse sphingolipid activator protein gene: a complex phenotype, including severe leukodystrophy and wide-spread storage of multiple sphingolipids. *Hum Mol Genet.* 1996;5(6):711–725.
358. Spiegel R, Bach G, Sury V, et al. A mutation in the saposin A coding region of the prosaposin gene in an infant presenting as Krabbe disease: first report of saposin A deficiency in humans. *Mol Genet Metab.* 2005;84(2):160–166.
359. Mehl E, Jatzkewitz H. On an enzyme from swine kidneys which cleaves cerebroside sulfuric acid esters. *Hoppe Seylers Z Physiol Chem.* 1963;331:292–294.
360. Alattia JR, Shaw JE, Yip CM, Prive GG. Molecular imaging of membrane interfaces reveals mode of beta-glucosidase activation by saposin C. *Proc Natl Acad Sci U S A.* 2007;104(44):17394–17399.
361. Christomanou H, Aignesberger A, Linke RP. Immunochemical characterization of two activator proteins stimulating enzymic sphingomyelin degradation in vitro. Absence of one of them in a human Gaucher disease variant. *Biol Chem Hoppe Seyler.* 1986;367 (9):879–890.
362. Matsuda J, Kido M, Tadano-Aritomi K, et al. Mutation in saposin D domain of sphingolipid activator protein gene causes urinary system defects and cerebellar Purkinje cell degeneration with accumulation of hydroxy fatty acid-containing ceramide in mouse. *Hum Mol Genet.* 2004;13(21):2709–2723.
363. Werth N, Schuette CG, Wilkening G, Lemm T, Sandhoff K. Degradation of membrane-bound ganglioside GM2 by beta-hexosaminidase A. Stimulation by GM2 activator protein and lysosomal lipids. *J Biol Chem.* 2001;276(16):12685–12690.
364. Cheruku SR, Xu Z, Dutia R, Lobel P, Storch J. Mechanism of cholesterol transfer from the Niemann–Pick type C2 protein to model membranes supports a role in lysosomal cholesterol transport. *J Biol Chem.* 2006;281(42):31594–31604.
365. Naureckiene S, Sleat DE, Lackland H, et al. Identification of HE1 as the second gene of Niemann–Pick C disease. *Science.* 2000;290(5500):2298–2301.
366. Locatelli-Hoops S, Remmel N, Klingenstein R, et al. Saposin A mobilizes lipids from low cholesterol and high bis(monoacylglycerol)phosphate-containing membranes: patient variant Saposin A lacks lipid extraction capacity. *J Biol Chem.* 2006;281 (43):32451–32460.

367. Bowman EA, Walterfang M, Abel L, Desmond P, Fahey M, Velakoulis D. Longitudinal changes in cerebellar and subcortical volumes in adult-onset Niemann–Pick disease type C patients treated with miglustat. *J Neurol.* 2015;262(9):2106–2114.
368. Patterson MC, Mengel E, Vanier MT, et al. Stable or improved neurological manifestations during miglustat therapy in patients from the international disease registry for Niemann–Pick disease type C: an observational cohort study. *Orphanet J Rare Dis.* 2015;10:65.
369. Murugesan V, Chuang WL, Liu J, et al. Glucosylsphingosine is a key biomarker of Gaucher disease. *Am J Hematol.* 2016;91(11):1082–1089.
370. Graf CGF, Schulz C, Schmalzlein M, et al. Synthetic glycoforms reveal carbohydrate-dependent bioactivity of human saposin D. *Angew Chem Int Ed Engl.* 2017;56 (19):5252–5257.

CHAPTER TWO

# Congenital Disorders of Ganglioside Biosynthesis

**T. August Li, Ronald L. Schnaar[1]**
Johns Hopkins University School of Medicine, Baltimore, MD, United States
[1] Corresponding author. E-mail address: schnaar@jhu.edu

## Contents



## Abstract

Gangliosides are expressed on all vertebrate cells and tissues, but are particularly abundant in the mammalian brain, where they constitute major cell-surface determinants on all nerve cells. The same four ganglioside structures, GM1, GD1a, GD1b, and GT1b, constitute the great majority of brain gangliosides in all mammals. Biosynthesis of these major brain gangliosides starts with addition of glucose to the ceramide lipid carrier followed by stepwise addition of up to six additional monosaccharides. This primarily involves the sequential action of seven glycosyltransferases, many of which appear to act specifically on glycolipid (rather than glycoprotein) acceptors. Congenital human disorders of ganglioside biosynthesis are exceedingly rare, but provide a glimpse into the functions of these major nerve cell surface components. To date, less than 100 individuals from 36 family pedigrees have been confirmed to

*Progress in Molecular Biology and Translational Science*, Volume 156
ISSN 1877-1173
http://dx.doi.org/10.1016/bs.pmbts.2018.01.001

carry deleterious mutations in ganglioside biosynthetic genes. Mutations in *ST3GAL5* (coding GM3 synthase) were discovered as the basis for severe congenital infantile seizures, whereas mutations in *B4GALNT1* (coding GM2/GD2 synthase) are the basis of hereditary spastic paraplegia accompanied by intellectual disability. In this review, we compile and summarize the findings of these studies, compare human disorders with mouse genetic models, and reflect on the implications for understanding ganglioside function and dysfunction in the nervous system.

# 1. INTRODUCTION

## 1.1 Congenital Disorders of Ganglioside Biosynthesis

Gangliosides are components of the cell surface of all mammalian cells, but are particularly abundant and typically more structurally complex in the mammalian brain.[1,2] The same four complex ganglioside structures, GM1, GD1a, GD1b, and GT1b (Fig. 1) comprise the vast majority (near or exceeding 90%) of gangliosides in the brains of all mammals tested including humans. Although found on all types of brain cells (neurons and glia), they appear

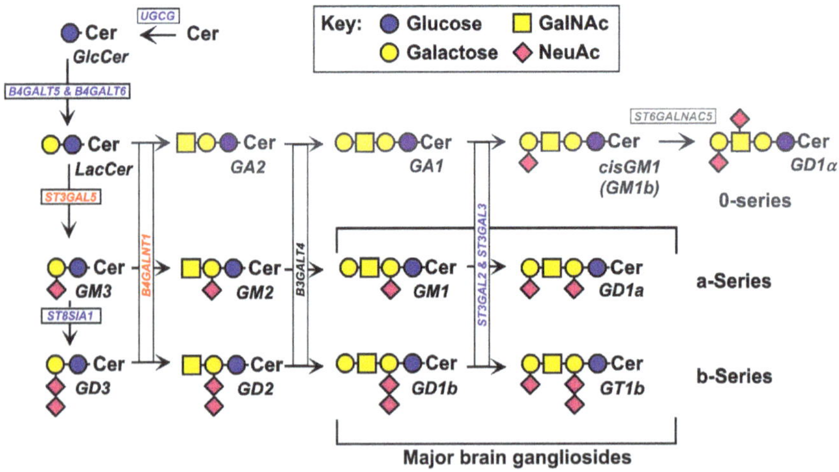


**Fig. 1** Biosynthetic pathways of major brain gangliosides in humans. The genes encoding the enzymes responsible for each step are boxed. Mutations in the genes in *red font* are responsible for human congenital disorders of ganglioside biosynthesis. Mouse genetic models have been generated for genes homologous to those in *red*, as wells as those in *blue font*. Minor brain gangliosides of the 0-series are faded in the image.

to be more abundant and complex on neurons.[3,4] As such, gangliosides comprise quantitatively significant molecular determinants on all nerve cell surfaces. Environmental or genetic modifications in normal ganglioside expression are expected to result in pathology.

Human diseases of ganglioside metabolism are rare, with congenital errors in biosynthesis reported in fewer than 100 individuals (Table 1) and errors in catabolism occurring in 1 of ~100,000 live births.[5,6] Both anabolic and catabolic (lysosomal storage) disorders are marked by neurological deficits. Major pathologies in catabolic disorders are believed to be primarily (although not exclusively) due to intracellular accumulation of gangliosides leading to cellular dysfunction and death.[7] In contrast, based on animal models,[1] genetic disruptions in ganglioside biosynthetic pathways may result in little change in total brain ganglioside concentration with equivalent buildup of ganglioside species behind the biosynthetic block. In congenital disorders of ganglioside biosynthesis (CDGg), pathology may be due to the disappearance of essential ganglioside species downstream of the biosynthetic block, an increase in precursor gangliosides, or both.

Genetic linkage analyses, made possible by enhanced capabilities in gene sequencing, revealed that mutations in genes encoding two enzymes involved in the biosynthesis of gangliosides (*B4GALNT1* or *ST3GAL5*) result in human disease.[8,10–14,16,17] Based on reports to date, CDGg are rare autosomal recessive disorders with reports of 96 affected individuals in 36 family pedigrees (Table 1). These disorders are part of a larger group of human congenital disorders of glycosylation (CDG) discovered using enhanced genetic and biochemical correlations in the field of glycobiology.[18] The true extent of CDGgs is unknown, as they have been reported to have been misdiagnosed as other neurological syndromes. Regardless of the number of individuals affected, consideration of the clinical outcomes of altered ganglioside expression in humans provides insights into ganglioside metabolism and function. The knowledge gained may enhance understanding of ganglioside function in the human brain, the extent to which gangliosides may be more broadly involved in human pathology, and potential avenues to treat disorders of altered ganglioside expression.

## 1.2 Biosynthesis of Major Brain Gangliosides

The quantitatively major gangliosides of the mammalian brain (GM1, GD1a, GD1b, GT1b) are synthesized stepwise (Fig. 1) starting with the transfer of glucose from UDP-Glc to ceramide on the cytosolic leaflet of the early Golgi

**Table 1** Human Mutations in Ganglioside Biosynthetic Genes.

| Mutation | No. of Affected | Refer-ences | Onset (Years) | Motor Deficits | Develop-mental Delay | Cognitive Deficits | Seizures Affected/Observed | Other features |
|---|---|---|---|---|---|---|---|---|
| ***B4GALNT1*** | | | | | | | | |
| c.852G>C (p.K284N)<br>c.1514G>A (p.R505H)<br>c.1458dupA (p.L487TfsX77) | 2<br>4<br>6 | 8,9 | 6–11 | Progressive weakness, spasticity | None reported | ID | 2/12 | Sensory polyneuro pathy, pes cavus, emotional lability, psychiatric illness |
| c. 395delC (p.P132QfsX7)<br>c.898C>T (p.R300C)<br>c.682C>T (p.R228X)<br>c.263dupG (p.L89PfsX13)<br>c.358C>T (p.Q120X)<br>c.1298A>C (p.D433A)<br>c.917_922dup (p.T307_Val308dup); c1315_1317delTTC (p.F439del) | 3<br>4<br>5<br>1<br>4<br>3<br>1 | 10 | 2–19 | Spasticity, hyperreflexia, dysarthria, extrapyramidal signs | Mild-moderate | Mild-moderate ID | 0/21 | Sensory polyneuropathy, pes cavus, strabismus |
| c.838-2A>G<br>c.1458_1459insA (p.L487fs) | 4<br>5 | 11 | <3–7 | Delayed milestones, spasticity, dysarthria | None reported | Mild-severe ID, low verbal skills | 0/9 | Autistic features, paranoia, social phobia, emotionally labile |

| | | | | | | | | |
|---|---|---|---|---|---|---|---|---|
| ***ST3GAL5*** | | | | | | | | |
| c.862C>T (p.R288X) | 49 | 12–15 | <1 | Severely delayed, choreoathetosis, nonambulatory | Profound stagnation, regression | ID, nonverbal | 46/46 | Microcephaly, blind, deaf |
| c.1063G>A (p.E355K) | 3 | 16 | 2 | Choreoathetosis, spasticity, nonambulatory | Severely delayed landmarks | ID, nonverbal | 1/3 | Microcephaly, normal vision |
| c.584G>C (p.C195S); c.601G>A (p.G201R) | 2 | 17 | <1 | Choreoathetosis, nonambulatory | Profound stagnation, regression | ID, nonverbal | 0/2 | Microcephaly, normal vision |

apparatus by the enzyme glucosylceramide synthase, encoded by the *UGCG* gene.[1] Glucosylceramide is then flipped into the lumen of the Golgi apparatus where synthesis continues by stepwise addition of monosaccharides to the growing glycan chain by up to six additional glycosyltransferases (Fig. 1). Some of the glycosyltransferases in this pathway predominantly or exclusively transfer sugars to glycosphingolipids rather than to glycoproteins based on in vitro acceptor specificity and mouse genetic studies. These include the products of the following genes (Fig. 1): *UGCG* (glucosylceramide synthase),[19] *ST3GAL5* (GM3 synthase),[20] *ST8SIA1* (GD3 synthase),[21] and *B4GALNT1* (GM2/GD2 synthase).[22] In mice, where quantitative studies have been performed, knockout of *St3gal5* (GM3 synthase), *St8sia1* (GD3 synthase), or *B4galnt1* (GM2/GD2 synthase) results in buildup of gangliosides behind the block such that the total brain ganglioside concentration remains approximately the same.[23–26] Human congenital disorders have thus far been associated with only two of these genes, *B4GALNT1* and *ST3GAL5*. The clinical and biochemical outcomes of human mutations in these genes are the focus of this review.

## 2. GM2/GD2 SYNTHASE DEFICIENCIES (*B4GALNT1* MUTATIONS) IN HUMAN DISEASE

### 2.1 Case Study: "GM3 (Hematoside) Sphingolipodystrophy"

In 1974, Max *et al.* reported a case study of a single infant suspected of lacking the ability to synthesize brain or liver gangliosides more complex than GM3 or GD3.[27] Evidence provided in four reports about this infant, published in the period 1974–76,[27–30] contain the most thorough analyses of postmortem frozen and fixed brain tissue from any of the CDGg's reported to date. In light of other cases with related enzyme deficits, this decades old case study is worthy of detailed review, even though molecular genetic confirmation of a gene mutation is lacking.

Within the first weeks after birth, an infant born to healthy unrelated parents of European Jewish descent was found to be unresponsive and hypotonic with frequent seizures.[27] An electroencephalogram was grossly abnormal with bilateral epileptiform foci. The infant died at 15 weeks of age from pneumonia, and both frozen and fixed brain tissues were collected for biochemical and histological analyses. Brain gangliosides were extracted and subjected to thin layer chromatography (TLC) analysis. At 15 weeks of age,

>90% of normal human brain gangliosides are represented by GM1, GD1a, GD1b, and GT1b, with minor amounts of GM3, GM2, and GD3 (Fig. 2). In contrast, the affected infant brain contained only GM3 and GD3 with no evidence of any of the major brain gangliosides. Notably, in the brain tissue sample tested, the total ganglioside expression level (per milligram protein) was reported to be 20% of that in a comparable normal human brain. A subsequent study[29] revealed no significant decrease in the catabolic enzyme that breaks down GM3 (sialidase) in patient brain, but a marked (88%) decrease in the enzyme activity that synthesizes GM2 from GM3, UDP-GalNAc:GM3 *N*-acetylgalactosaminyltransferase (GM2/GD2 synthase). Together with anecdotal evidence of a similar phenotypic disorder in a maternal uncle, these biochemical findings are consistent with a congenital mutation of the gene responsible for coding GM2/GD2 synthase, *B4GALNT1*.

More detailed clinical findings at 4 weeks of age[28] reported a limp infant with reduced spontaneous movement, infrequent crying, no head control,

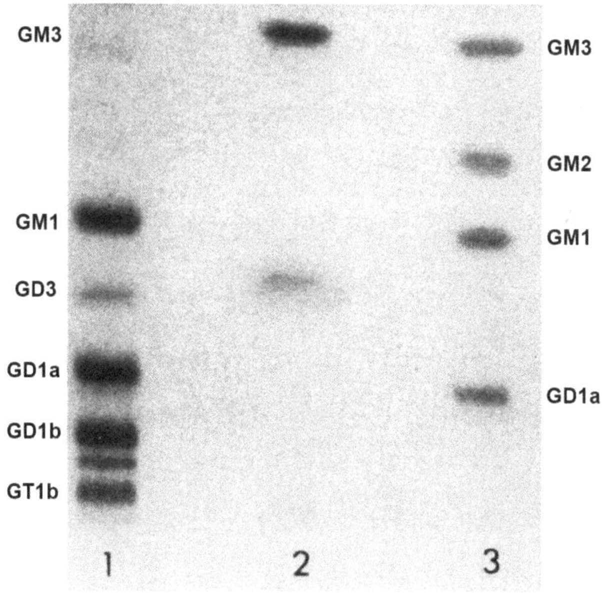


**Fig. 2** Thin layer chromatography (TLC) analysis of normal human infant brain gangliosides (lane 1), gangliosides from the brain of an infant lacking GM2/GD2 synthase (lane 2), and ganglioside standards (lane 3). The chromatogram is from Max *et al.*[27] with flanking ganglioside labels added to reflect the standards used (right) and human gangliosides identified by their relative TLC mobility (left). *Reprinted with permission from the Massachusetts Medical Society.*

and poor grasp reflex. Limb spasms were frequent with periodic progression to generalized seizures. Histological findings[28,30] noted vacuolation of white matter tracts (spaces between widely separated myelin lamellae) throughout the brain and a thin corpus callosum. Ultrastructure of central nervous system white matter revealed unmyelinated and thinly myelinated axons and structural deficits in myelin sheaths. Consideration of these findings in light of other subjects with mutations in *B4GALNT1*, as well as a mouse genetic model is addressed later.

## 2.2 Genetic Mutations in *B4GALNT1* Cause Hereditary Spastic Paraplegia

Nearly 40 years after the first reported case of a human deficit in GM2/GD2 synthase enzyme activity, advances in gene sequencing and linkage analysis led to identification of several mutations in the gene responsible, *B4GALNT1*, as the basis for complex hereditary spastic paraplegia (HSP) in 42 individuals associated with 12 family pedigrees with 12 different gene mutations (Table 1, Fig. 3).[8–11] HSP, also known as familial spastic paraparesis, is characterized by familial transmitted weakness and spasticity of the lower limbs, often with onset in childhood or adulthood and with progressing severity.[31] HSP's are highly heterogeneous; whereas "pure" HSP is associated only with lower limb dysfunction, complicated (or complex) HSP is accompanied by other neurological symptoms, including intellectual

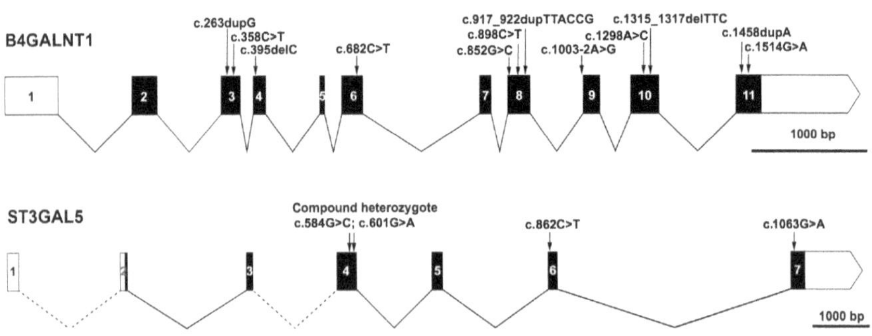


**Fig. 3** Mutations in human genes *B4GALNT1* and *ST3GAL5* responsible for CDGg mapped on exon-intron maps of each gene. Reference sequences for *B4GALNT1* are NM_001478.4 (nucleotide) and NG_033849.1 (gene). Reference sequences for *ST3GAL5* are NM_003896.3 (nucleotide) and NG_012807.1 (gene). In the *ST3GAL5* map, the very large intron 1 (25.3 kb) and intron 3 (13.0 kb) are not to scale *(dashed lines)* for presentation purposes. Due to transcript variants and/or alternative translational start sites, some studies provide different numbers for the mutations shown.[11,12,16]

disability (ID, see later). Different HSP's have been linked to nearly a hundred different gene mutations, many of which were first designated as spastic paraplegia genes (SPG's) and designated by number. Mutations in *B4GALNT1* were discovered by gene linkage analysis to be responsible for a form of HSP designated as SPG26.[9]

SPG26 was first described in a large consanguineous family (13 children) of Bedouin ancestry with 5 affected individuals.[9,32] A sixth affected family member of this pedigree, a niece to the other affected individuals, was later identified.[8] The onset of symptoms was between 6 and 11 years of age, progressing to spastic gait with other motor symptoms, including dysreflexia, muscle atrophy, and dysarthria (motor speech deficit). Subsequently, 11 additional family pedigrees from around the world, several including consanguineous families, presented, each with a distinct *B4GALNT1* mutation (Table 1, Fig. 3).[8,10,11] Although all affected individuals displayed spastic paraparesis, the course of the disease varied among pedigrees and individuals. In a representative study of 18 affected individuals from 7 pedigrees,[10] the age of onset of HSP, noted as gait difficulties, ranged from 2 to 19 years of age. Progression of motor dysfunction was typically slow, with 4 of 18 affected individuals either bedridden or confined to a wheelchair 13–47 years after the disease was first noted. In addition to uniform prominent spasticity of the lower limbs, about one-third of the affected individuals had upper limb spasticity and half had dysarthria. Most had muscle wasting and dysreflexia, with peripheral neuropathy predominantly of the axonal type (compare with *B4galnt1*-null mouse models, later). Brain magnetic resonance imaging revealed several affected individuals with cortical atrophy and/or white matter hyperintensities. In another study,[8] two affected siblings of one pedigree did not present with altered gait until they were >35 years old; whereas four affected individuals from another pedigree all displayed early motor deficits (from <2 years old).

SPG26-related gene mutations have been found in 7 of the 10 coding exons of *B4GALNT1* scattered across much of the gene (Fig. 3). As biochemical data correlating these mutations with altered ganglioside expression in human subjects is unavailable, relating severity and type of symptoms to biochemical alterations is not yet possible. Evidence of altered ganglioside biosynthesis in these patients comes from a study using cultured skin fibroblasts from two affected siblings in which decreased GM2 and increased GM3 were documented.[8] However, ganglioside expression in cultured human fibroblasts is very limited compared to that in nerve cells or tissues, in which gangliosides are much more abundant and complex. The single case study from which brain ganglioside expression is available (Fig. 2) indicates that the severely affected subject had a complete lack of major brain gangliosides.[27] Linking phenotype to brain ganglioside expression over the larger population of affected individuals will require additional biochemical data.

## 2.3 Intellectual Disability is Invariably Associated With *B4GALNT1* Mutations

In addition to uniform lower limb spasticity and gait disorders, all individuals with *B4GALNT1* mutations (Table 1) suffer mild to severe intellectual impairment, with reported IQ scores typically in the range 50–70 indicating a limitation in intellectual functioning (http://aaidd.org). Notably, ID did not appear to correlate with onset of motor deficits. The siblings noted above who were diagnosed with gait abnormalities in their 4th decade of life, displayed ID (IQ 59–64) in childhood.[8] In contrast, several affected individuals from other pedigrees showed early onset of gait disorders (at 2–3 years old) with only mild intellectual impairment.[10] Even within one pedigree, affected individuals ranged from severe cognitive impairment (inability to self-care or interact socially) to mild ID.[8] A minority of these patients also suffered from seizures, with mild to severe clonus (3/42) or *grand mal* seizures (2/42).

## 2.4 Potential Mechanisms Responsible for Clinical Outcomes Due to *B4GALNT1* Mutations: Comparison of Human Disorders With Mouse Genetic Models

Progressive motor neuropathy and ID are uniform clinical outcomes in individuals with congenital *B4GALNT1* mutations. Genetically engineered mice with an interrupted *B4galnt1* gene, which fail to synthesize complex gangliosides, phenocopy the human motor neuropathy,[33] and cognitive deficits.[34] Progressive motor neuropathy of the axonal type in these mutant mice has been experimentally correlated with dysmyelination attributed to disruption of the interaction of gangliosides GD1a and GT1b with a specific myelin protein, myelin-associated glycoprotein (MAG).[33] ID appears to be a separate phenotype and its mechanistic basis is as yet incompletely understood.

Two of the four major brain gangliosides, GD1a and GT1b, bind specifically to MAG, which is found on the innermost leaflet of myelin apposing the axon surface.[1,33,35,36] Structure-function-binding studies,[35,37,38] as well as crystallographic structure determinations,[39] confirm the specific interaction of MAG with GD1a and GT1b, which bear α2–3-linked sialic acids on their terminal galactose residue, but not to GM1 and GD1b, which do not (Fig. 1). Notably, the mouse and human forms of MAG share 94% identity at the amino acid level, and all four major nervous system gangliosides (GM1, GD1a, GD1b, and GT1b) are the same in mice and men.

Genetic disruption of the *Mag* gene in mice results in progressive axon degeneration downstream of dysmyelination in both the central and peripheral nervous systems.[33,40,41] Disruption of the *B4galnt1* gene results in the same degree of axon degeneration, and *Mag/B4galnt1* double-null mice essentially phenocopy the single-null mice.[33] Other outcomes of *Mag* gene disruption are also phenocopied in *B4galnt1*-null and *Mag/B4galnt1* double-null mice, including diminished axon caliber, diminished neurofilament spacing, and disruptions at nodes of Ranvier.[33,42] Behaviorally, *B4galnt1*-null mice begin to show altered gait in adulthood (3 months of age) and by 1 year old display hindlimb ataxia.[43] *B4galnt1*-null mice exhibit peripheral neuropathy predominantly of the axonal type,[44] a phenotype that matches a primary clinical characteristic of human subjects with *B4GALNT1* mutations. These observations implicate the loss of MAG-ganglioside interactions in the etiology of this clinical outcome.[10] In support of this hypothesis, a case study of two siblings diagnosed with progressive axonal sensorimotor polyneuropathy[45] traced the disease to a specific gene mutation in *MAG* that altered the key residue responsible for sialic acid binding (Arg118, identified by site directed mutagenesis[46] and confirmed by crystallography[39]). This finding is consistent with other studies that discovered other pathogenic mutations in the *MAG* gene that result in HSP.[47,48] Together, these data strongly support the conclusion that the pathology of axonal neuropathy in complex HSP, as a result of *B4GALNT1* mutations, can be attributed to alterations in ganglioside-MAG interactions. However, progressive neuropathy of the axonal type is not a cognitive disorder, suggesting that there are other pathological mechanism(s) responsible for ID in these subjects.

The cognitive dysfunction noted in all subjects with mutations in the *B4GALNT1* gene is at least partly phenocopied in *B4galnt1*-null mice.[34] Compared to their wild-type littermates, at 8 weeks of age these mice displayed modest deficits in memory tasks evidenced by Morris water maze, and more notably in short-term avoidance (fear-motivated) memory. Alterations in electrically-induced long-term potentiation (LTP), an electrophysiological correlate of learning and memory,[49] was also observed in field recordings of hippocampal slices from 8-week old *B4galnt1*-null mice compared to slices from wild type littermates.[34]

Although the molecular bases for altered cognition in these mice has not been established, biochemical findings have linked specific brain gangliosides to neurotransmitter receptors responsible for long-term potentiation, learning, and memory.[50] Glutamate is the major excitatory neurotransmitter in the brain, and AMPA-type glutamate receptors (AMPARs) are the most abundant receptors that translate synaptic glutamate release to neuronal excitation. AMPARs are heterotetramers comprised of pairs of dimers of four possible subunits, GluA1 through GluA4.[51] AMPARs are dynamically trafficked by their association with other molecules inside the cell and at the cell surface. Highly directed trafficking of AMPARs in and out of the postsynaptic membrane is thought to regulate learning and memory, and dysfunctions in AMPAR regulation have been linked to ID.[50] An unbiased affinity proteomics approach using immobilized gangliosides GT1b and GM1 as "bait" revealed that among proteins isolated from cultured neurons, GT1b selectively captured three proteins involved in neurotransmitter trafficking: Thorase, *N*-ethylmaleimide sensitive protein (NSF), and nicalin.[52] The first of these, Thorase, is involved in trafficking GluA2-containing AMPARs.[53] This is particularly relevant to learning and memory, as GluA2 subunits diminish calcium flux, and are thought to be enhanced regulators of glutamate excitability at the postsynaptic membrane.[54] Notably, GluA2 receptors were found to associate selectively with ganglioside GM1. Whether these molecular associations explain the deficits in LTP and cognition in animal models or human disorders of ganglioside expression has yet to be established.

## 3. GM3 SYNTHASE DEFICIENCIES (*ST3GAL5* MUTATIONS) IN HUMAN DISEASE

### 3.1 Genetic Mutations in *ST3GAL5* Cause Profound Developmental, Motor, and Cognitive Deficits

Ganglioside GM3 is the first ganglioside in the biosynthetic pathway to the major brain gangliosides (Fig. 1), and a common ganglioside in nonneural tissues and cells.[55] GM3 synthase, coded by the human *ST3GAL5* gene, transfers sialic acid from CMP-sialic acid to the 3-hydroxyl of the terminal galactose of lactosylceramide.[20] As an abundant nonneural ganglioside and the biosynthetic precursor to all major brain gangliosides, genetic deletion of GM3 synthase was predicted to result in profound deficits. Surprisingly, mice engineered to lack GM3 synthase

(*St3gal5*-null mice) are viable and appear without major abnormalities (see more on this later).[23] A year after the *St3gal5*-null mouse results were published, genetic linkage analysis revealed that a mutation in the same gene in humans (*ST3GAL5*) was responsible for a severe infantile-onset seizure disorder affecting nine individuals in a single family pedigree.[12] Since that time a total of 54 affected individuals associated with 24 family pedigrees displaying 3 different *ST3GAL5* gene mutations have been reported (Table 1, Fig. 3).[13,14,16,17]

All but five of the affected individuals were found to have the same mutation (c.862C>T, p.R288X) in *ST3GAL5* resulting in premature translational termination of the polypeptide between two major conserved sialylmotifs (large and small). Additional human missense mutations within the sialylmotifs were subsequently discovered to cause human disease, including compound heterozygous missense mutations in the large sialylmotif in one family (c.584G>C, p.C195S and c.601G>A, p.G201R) and a single missense mutation (c.1063G>A, p.E355K) in the small sialylmotif in another (Fig. 3). To date, nervous system gangliosides have not been analyzed in the affected subjects, although glycosphingolipids in plasma and cultured fibroblasts have been reported. Analysis of these samples from selected subjects with each *ST3GAL5* mutation revealed loss of GM3 and downstream gangliosides and increased globoseries glycosphingolipids.[12,14,16,17] Although these data imply profound alterations of brain gangliosides, plasma and fibroblast ganglioside patterns are simple compared to those in the nervous system and a more thorough biochemical analysis awaits availability of postmortem brain tissue.

The largest cohort of affected subjects (c.862C>T) experienced severe drug-resistant seizures starting in the first year of life with ongoing choreoathetosis.[12–14] Motor and cognitive development was markedly delayed with microcephaly, hypotonia, failure to thrive, and profound intellectual disability. Subjects were nonambulatory, nonverbal, blind,[15] and deaf.[56] Magnetic resonance imaging revealed delayed brain maturation, atrophy, and hypomyelination upon aging.[14,56] The five subjects with missense *ST3GAL5* mutations (Table 1) had serious motor and cognitive deficits, although only one of three subjects with the c.1063G>A mutation[16] and none of the compound heterozygotes[17] subjects experienced seizures and none were blind. Together, these data indicate that disruption of brain ganglioside biosynthesis restricts nervous system development resulting in severe motor and cognitive deficits in all subjects.

### 3.2 Potential Mechanisms Responsible for Clinical Outcomes Due to *ST3GAL5* Mutations: Comparison of Human Disorders With Mouse Genetic Models

Mice engineered to lack GM3 synthase (*St3gal5*-null mice) fail to phenocopy major neurological deficits uniformly reported in subjects with *ST3GAL5* mutations.[23] *St3gal5*-null mice were viable and appeared without major abnormalities.[23,57] Biochemical analyses of brain gangliosides in *St3gal5*-null mice revealed a bypass of major brain gangliosides with essentially quantitative replacement by the normally low-abundance "0-series" gangliosides *cis*GM1 (GM1b) and GD1α (Fig. 1). As GD1α is an excellent MAG-binding ganglioside,[58] it may support myelin–axon interactions in the absence of GD1a and GT1b and explain the lack of motor neuropathy in knockout mice.

A common phenotype in humans with *ST3GAL5* mutations and *St3gal5*-null mice is loss of hearing.[59] In the mouse model, hair cell deformities and selective degeneration of the organ of Corti were reported. Human subjects with *ST3GAL5* mutations have neonatal onset hearing impairment. Consistent with the mouse model, tympanometry of these subjects as young children revealed outer hair cell impairment, as well as significant dysfunction of inner hair cells, central auditory pathways, or both.[56]

Despite similar auditory loss, the severe phenotypes of loss of GM3 synthase in humans are absent in mice. The concept that 0-series gangliosides compensate for loss of major brain gangliosides in brain development in mice is supported by studies using mice engineered to have other ganglioside deficits (Table 2). In mice engineered with disrupted genes within the ganglioside biosynthetic pathway, total brain ganglioside expression remains approximately constant, with gangliosides building up behind the block(s) (Fig. 1).[1] In *B4galnt1*-null mice (described earlier) GM3 and GD3 build up in the absence of major brain gangliosides.[25,60] In *St8sia1*-null mice the absence of GD1b and GT1b is balanced by increases in GM1 and GD1a, and the mice do not display obvious abnormalities.[24] However, mice double null in *B4galnt1* and *St8sia1*, which lack all major brain ganglioside and in which GM3 builds up behind the blocks, had >80% mortality in the first 6 months of life, primarily due to lethal audiogenic seizures, in part phenocopying the seizure disorder of subjects with *ST3GAL5* mutations.[24] These mice, expressing only GM3, demonstrated reduced motor and sensory functions and degeneration of peripheral nerves.[61,62] Double mutant mice with disrupted *B4galnt1* and *St3gal5* genes, in which the entire ganglio-series pathway is

**Table 2** Outcomes of Ganglioside Biosynthetic Gene Deletion in Genetically Engineered Mice.

| Gene(s) Deleted | Major Brain Gangliosides | Major Nervous System Phenotype(s) | References |
|---|---|---|---|
| None (wild type) | GM1, GD1a, GD1b, GT1b | Normal | |
| *B4galnt1* | GM3, GD3 | Axon degeneration, dysmyelination, progressive motor behavioral deficits, hyperactivity, seizure susceptibility | 25,33,43,44,64 |
| *St8sia1* | GM1, GD1a | Mostly normal; selective reduction in axon regeneration | 24,65 |
| *St3gal5* | GM1b, GD1α | Mostly normal; hearing impaired | 23 |
| *B4galnt1/St8sia1* | GM3 (GM3 only) | Motor and sensory deficits, audiogenic seizure | 24,62 |
| *B4galnt1/St3gal5* | No glioseries: minor amount of neolactoseries | Short lived, severe early neurodegeneration, perturbed axon–myelin interactions | 63 |

blocked, had greatly reduced total brain gangliosides and profound neurodegenerative disease.[63] These mice express small amounts of sialylated glycosphingolipids in the brain on a neolactoseries glycosphingolipid core. They were born at the expected Mendelian frequency but demonstrated early axonal degeneration, perturbed axon-glial interactions, motor deficits (ataxia, tremors) within 2 weeks of birth, progressive microcephaly, and most died within 2 months.

Together, the above mouse models (Table 2) recapitulate much of the pathology of *ST3GAL5* mutations in humans but, for the most serious outcomes, only in the combined absence of both normal brain gangliosides (GM1, GD1a, GD1b, GT1b) and 0-series gangliosides (GM1b, GD1α). This raises the possibility that quantitative or qualitative brain ganglioside expression in human CDGg is unlike that in comparable mouse genetic models, even though the enzymes and ganglioside structures are well conserved across species. If so, this would be akin to the mouse model for the human ganglioside catabolic disorder, such as Tay–Sachs disease. Although human *HEXA* mutations result in severe brain lysosomal storage disease and early death, *Hexa*-null mice have only a modest neuronal phenotype. This is because mice, but not humans, circumvent the block in the catabolic pathway.[66] Similarly, mice may compensate for disruption of ganglioside biosynthetic enzymes in ways that humans do not. Access to postmortem brain from subjects with genetic disruptions in the ganglioside biosynthetic pathway may help resolve this apparent discrepancy with genetically equivalent mouse models.

## 4. PERSPECTIVES

Gangliosides are major nerve cell surface molecular determinants throughout the central and peripheral nervous systems, and major brain ganglioside structures have been well preserved throughout mammalian and avian evolution.[1] The discovery of human CDGg establish the importance of gangliosides in brain development, motor and sensory function, cognition and axon-myelin stability. The spectrum of clinical outcomes due to altered ganglioside expression is likely to involve the multiple ways in which gangliosides impact cellular functions. As recognition molecules at the nerve cell surface certain gangliosides can engage complementary receptors on apposing cells, such as MAG; whereas other gangliosides may interact laterally in the plane of the plasma membrane to regulate cell signaling receptors and ion channels.[2,52,67,68] As gene deletion in the ganglioside biosynthetic pathway in mice results in buildup of gangliosides behind the block, some clinical outcomes in human CDGg may be due to increased concentrations of simpler structures, as well as decreased concentrations of the major brain gangliosides. As experimental insights into ganglioside functions in the nervous system continue to emerge, a better understanding of the downstream results of specific increases and decreases in brain ganglioside expression will follow.

Correlating ganglioside expression and nervous system function in human disease is currently limited by the lack of postmortem brain tissue from subjects with different *B4GALNT1* and *ST3GAL5* mutations. Correlating specific changes in brain ganglioside expression with variations in the clinical outcomes is currently not possible. Likewise, the results of specific mutations on the enzymatic activities of GM2/GD2 synthase and

GM3 synthase are insufficient to be predictive of clinical outcomes. In the absence of clinical samples that would shed additional light on ganglioside structure–function relationships, the availability of patient fibroblasts provides valuable but limited information, as the ganglioside expression profile of human fibroblasts is very simple compared to that of nerve cells.[69] With technologies to revert fibroblasts to pluripotency and induce neuronal differentiation,[70] it is hoped that additional cellular (if not tissue) materials are within reach for future studies.

As knowledge about human CDGg progresses, a better understanding of the physiological and pathological roles of gangliosides will emerge. One can hope that this knowledge will provide clinical insights beyond these rare congenital disorders, and provide information that can lead to therapeutic intervention in these and other neurological diseases and disorders.

## REFERENCES

1. Schnaar RL, Gerardy-Schahn R, Hildebrandt H. Sialic acids in the brain: gangliosides and polysialic acid in nervous system development, stability, disease and regeneration. *Physiol Rev.* 2014;94:461–518.
2. Schnaar RL. Gangliosides of the vertebrate nervous system. *J Mol Biol.* 2016;428:3325–3336.
3. Svennerholm L, Rynmark BM, Vilbergsson G, et al. Gangliosides in human fetal brain. *J Neurochem.* 1991;56:1763–1768.
4. Satoh JI, Tai T, Kim SU. Differential expression of gangliosides and galactolipids in fetal human oligodendrocytes and astrocytes in culture. *Brain Res Dev Brain Res.* 1996;93:172–181.
5. Caciotti A, Garman SC, Rivera-Colon Y, et al. GM1 gangliosidosis and Morquio B disease: an update on genetic alterations and clinical findings. *Biochim Biophys Acta.* 2011;1812:782–790.
6. Bley AE, Giannikopoulos OA, Hayden D, Kubilus K, Tifft CJ, Eichler FS. Natural history of infantile G(M2) gangliosidosis. *Pediatrics.* 2011;128:e1233–e1241.
7. Schultz ML, Tecedor L, Chang M, Davidson BL. Clarifying lysosomal storage diseases. *Trends Neurosci.* 2011;34:401–410.
8. Harlalka G, Lehman A, Chioza B, et al. Mutations in *B4GALNT1* (GM2 synthase) underlie a new disorder of ganglioside biosynthesis. *Brain.* 2013;136:3618–3624.
9. Wilkinson PA, Simpson MA, Bastaki L, et al. A new locus for autosomal recessive complicated hereditary spastic paraplegia (SPG26) maps to chromosome 12p11.1-12q14. *J Med Genet.* 2005;42:80–82.
10. Boukhris A, Schule R, Loureiro JL, et al. Alteration of ganglioside biosynthesis responsible for complex hereditary spastic paraplegia. *Am J Hum Genet.* 2013;93:118–123.
11. Wakil SM, Monies DM, Ramzan K, et al. Novel B4GALNT1 mutations in a complicated form of hereditary spastic paraplegia. *Clin Genet.* 2014;86:500–501.
12. Simpson MA, Cross H, Gurtz K, et al. Infantile onset symptomatic epilepsy syndrome caused by homozygous loss of function mutations in GM3 synthase. *Nat Genet.* 2004;36:1225–1229.
13. Wang H, Bright A, Xin B, Bockoven JR, Paller AS. Cutaneous dyspigmentation in patients with ganglioside GM3 synthase deficiency. *Am J Med Genet A.* 2013;161A:875–879.

14. Fragaki K, Ait-El-Mkadem S, Chaussenot A, et al. Refractory epilepsy and mitochondrial dysfunction due to GM3 synthase deficiency. *Eur J Hum Genet.* 2013;21:528–534.
15. Farukhi F, Dakkouri C, Wang H, Wiztnitzer M, Traboulsi EI. Etiology of vision loss in ganglioside GM3 synthase deficiency. *Ophthalmic Genet.* 2006;27:89–91.
16. Boccuto L, Aoki K, Flanagan-Steet H, et al. A mutation in a ganglioside biosynthetic enzyme, ST3GAL5, results in salt & pepper syndrome, a neurocutaneous disorder with altered glycolipid and glycoprotein glycosylation. *Hum Mol Genet.* 2014;23:418–433.
17. Lee JS, Yoo Y, Lim BC, et al. GM3 synthase deficiency due to ST3GAL5 variants in two Korean female siblings: masquerading as Rett syndrome-like phenotype. *Am J Med Genet A.* 2016;170:2200–2205.
18. Freeze HH, Chong JX, Bamshad MJ, Ng BG. Solving glycosylation disorders: fundamental approaches reveal complicated pathways. *Am J Hum Genet.* 2014;94:161–175.
19. Hirabayashi Y, Ishibashi Y. UDP-glucose: ceramide glucosyltransferase (UGCG). In: Taniguchi N, Honke K, Fukuda M, Narimatsu H, Yamaguchi K, Angata T, eds. *Handbook of Glycosyltransferases and Related Genes 2nd ed. Tokyo: Springer; 2014:3–13.*
20. Inokuchi J, Uemura S. ST3 beta-galactoside alpha-2,3-sialyltransferase 5 (ST3GAL5). In: Taniguchi N, Honke K, Fukuda M, Narimatsu H, Yamaguchi Y, Angata K, eds. *Handbook of Glycosyltransferases and Related Genes. Tokyo: Springer; 2014:675–686.*
21. Bobowski M, Harduin-Lepers A, Delannoy P. ST8 alpha-*N*-acetyl-neuraminide alpha-2,8-sialyltransferase 1 (ST8SIA1). In: Taniguchi N, Honke K, Fukuda M, Narimatsu H, Yamaguchi K, Angata T, eds. *Handbook of Glycosyltransferases and Related Genes 2nd ed. Tokyo: Springer; 2014:767–780.*
22. Furukawa K, Furukawa K, Ohmi Y, et al. Beta-1,4 *N*-acetylgalactosaminyltransferase 1,2 (B4GALNT1,2). In: Taniguchi N, Honke K, Fukuda M, Narimatsu H, Yamaguchi K, Angata T, eds. *Handbook of Glycosyltransferases and Related Genes 2nd ed. Tokyo: Springer; 2014:417–428.*
23. Yamashita T, Hashiramoto A, Haluzik M, et al. Enhanced insulin sensitivity in mice lacking ganglioside GM3. *Proc Natl Acad Sci USA.* 2003;100:3445–3449.
24. Kawai H, Allende ML, Wada R, et al. Mice expressing only monosialoganglioside GM3 exhibit lethal audiogenic seizures. *J Biol Chem.* 2001;276:6885–6888.
25. Takamiya K, Yamamoto A, Furukawa K, et al. Mice with disrupted GM2/GD2 synthase gene lack complex gangliosides but exhibit only subtle defects in their nervous system. *Proc Natl Acad Sci USA.* 1996;93:10662–10667.
26. Sun J, Shaper NL, Itonori S, Heffer-Lauc M, Sheikh KA, Schnaar RL. Myelin-associated glycoprotein (Siglec-4) expression is progressively and selectively decreased in the brains of mice lacking complex gangliosides. *Glycobiology.* 2004;14:851–857.
27. Max SR, Maclaren NK, Brady RO, et al. GM3 (hematoside) sphingolipodystrophy. *N Engl J Med.* 1974;291:929–931.
28. Maclaren NK, Max SR, Cornblath M, et al. GM3 gangliosidosis: a novel human sphingolipodystrophy. *Pediatrics.* 1976;57:106–110.
29. Fishman PH, Max SR, Tallman JF, Brady RO, Maclaren NK, Cornblath M. Deficient ganglioside biosynthesis: a novel human sphingolipidosis. *Science.* 1975;187:68–70.
30. Tanaka J, Garcia JH, Max SR, et al. Cerebral sponginess and GM3 gangliosidosis; ultrastructure and probable pathogenesis. *J Neuropathol Exp Neurol.* 1975;34:249–262.
31. de Souza PV, de Rezende Pinto WB, de Rezende Batistella GN, Bortholin T, Oliveira AS. Hereditary spastic paraplegia: clinical and genetic hallmarks. *Cerebellum.* 2017;16:525–551.
32. Farag TI, El-Badramany MH, Al-Sharkawy S. Troyer syndrome: report of the first "non-Amish" sibship and review. *Am J Med Genet.* 1994;53:383–385.
33. Pan B, Fromholt SE, Hess EJ, et al. Myelin-associated glycoprotein and complementary axonal ligands, gangliosides, mediate axon stability in the CNS and PNS: neuropathology and behavioral deficits in single- and double-null mice. *Exp Neurol.* 2005;195:208–217.

34. Sha S, Zhou L, Yin J, et al. Deficits in cognitive function and hippocampal plasticity in GM2/GD2 synthase knockout mice. *Hippocampus.* 2014;24:369–382.
35. Schnaar RL, Collins BE, Wright LP, et al. Myelin-associated glycoprotein binding to gangliosides. Structural specificity and functional implications. *Ann N Y Acad Sci.* 1998;845:92–105.
36. Yang LJ., Zeller CB, Shaper NL, et al. Gangliosides are neuronal ligands for myelin-associated glycoprotein. *Proc Natl Acad Sci USA.* 1996;93:814–818.
37. Collins BE, Ito H, Sawada N, Ishida H, Kiso M, Schnaar RL. Enhanced binding of the neural siglecs, myelin-associated glycoprotein and Schwann cell myelin protein, to Chol-1 (α-series) gangliosides and novel sulfated Chol-1 analogs. *J Biol Chem.* 1999;274:37637–37643.
38. Collins BE, Yang LJ., Mukhopadhyay G, et al. Sialic acid specificity of myelin-associated glycoprotein binding. *J Biol Chem.* 1997;272:1248–1255.
39. Pronker MF, Lemstra S, Snijder J, et al. Structural basis of myelin-associated glycoprotein adhesion and signalling. *Nat Commun.* 2016;7:13584.
40. Nguyen T, Mehta NR, Conant K, et al. Axonal protective effects of the myelin associated glycoprotein. *J Neurosci.* 2009;29:630–637.
41. Nave KA, Trapp BD. Axon-glial signaling and the glial support of axon function. *Annu Rev Neurosci.* 2008;31:535–561.
42. Susuki K, Baba H, Tohyama K, et al. Gangliosides contribute to stability of paranodal junctions and ion channel clusters in myelinated nerve fibers. *Glia.* 2007;55:746–757.
43. Chiavegatto S, Sun J, Nelson RJ, Schnaar RL. A functional role for complex gangliosides: motor deficits in GM2/GD2 synthase knockout mice. *Exp Neurol.* 2000;166:227–234.
44. Sheikh KA, Sun J, Liu Y, et al. Mice lacking complex gangliosides develop Wallerian degeneration and myelination defects. *Proc Natl Acad Sci USA.* 1999;96:7532–7537.
45. Roda RH, FitzGibbon EJ, Boucekkine H, Schindler AB, Blackstone C. Neurologic syndrome associated with homozygous mutation at MAG sialic acid binding site. *Ann Clin Transl Neurol.* 2016;3:650–654.
46. Tang S, Shen YJ, DeBellard ME, et al. Myelin-associated glycoprotein interacts with neurons via a sialic acid binding site at ARG118 and a distinct neurite inhibition site. *J Cell Biol.* 1997;138:1355–1366.
47. Novarino G, Fenstermaker AG, Zaki MS, et al. Exome sequencing links corticospinal motor neuron disease to common neurodegenerative disorders. *Science.* 2014;343:506–511.
48. Lossos A, Elazar N, Lerer I, et al. Myelin-associated glycoprotein gene mutation causes Pelizaeus–Merzbacher disease-like disorder. *Brain.* 2015;138:2521–2536.
49. Scannevin RH, Huganir RL. Postsynaptic organization and regulation of excitatory synapses. *Nat Rev Neurosci.* 2000;1:133–141.
50. Volk L, Chiu SL, Sharma K, Huganir RL. Glutamate synapses in human cognitive disorders. *Annu Rev Neurosci.* 2015;38:127–149.
51. Greger IH, Watson JF, Cull-Candy SG. Structural and functional architecture of AMPA-type glutamate receptors and their auxiliary proteins. *Neuron.* 2017;94:713–730.
52. Prendergast J, Umanah GK, Yoo SW, et al. Ganglioside regulation of AMPA receptor trafficking. *J Neurosci.* 2014;34:13246–13258.
53. Zhang J, Wang Y, Chi Z, et al. The AAA+ ATPase Thorase regulates AMPA receptor-dependent synaptic plasticity and behavior. *Cell.* 2011;145:284–299.
54. Isaac JT, Ashby MC, McBain CJ. The role of the GluR2 subunit in AMPA receptor function and synaptic plasticity. *Neuron.* 2007;54:859–871.
55. Inokuchi J. Physiopathological function of hematoside (GM3 ganglioside). *Proc Jpn Acad Ser B Phys Biol Sci.* 2011;87:179–198.
56. Yoshikawa M, Go S, Suzuki S, et al. Ganglioside GM3 is essential for the structural integrity and function of cochlear hair cells. *Hum Mol Genet.* 2015;24:2796–2807.

57. Tsukamoto K, Kohda T, Mukamoto M, et al. Binding of Clostridium botulinum type C and D neurotoxins to ganglioside and phospholipid. Novel insights into the receptor for clostridial neurotoxins. *J Biol Chem.* 2005;280:35164–35171.
58. Sawada N, Ishida H, Collins BE, Schnaar RL, Kiso M. Ganglioside GD1α analogs as high affinity ligands for myelin-associated glycoprotein. *Carbohydr Res.* 1999;316:1–5.
59. Yoshikawa M, Go S, Takasaki K, et al. Mice lacking ganglioside GM3 synthase exhibit complete hearing loss due to selective degeneration of the organ of Corti. *Proc Natl Acad Sci USA.* 2009;106:9483–9488.
60. Liu Y, Wada R, Kawai H, et al. A genetic model of substrate deprivation therapy for a glycosphingolipid storage disorder. *J Clin Invest.* 1999;103:497–505.
61. Inoue M, Fujii Y, Furukawa K, et al. Refractory skin injury in complex knock-out mice expressing only the GM3 ganglioside. *J Biol Chem.* 2002;277:29881–29888.
62. Tajima O, Egashira N, Ohmi Y, et al. Reduced motor and sensory functions and emotional response in GM3-only mice: emergence from early stage of life and exacerbation with aging. *Behav Brain Res.* 2009;198:74–82.
63. Yamashita T, Wu YP, Sandhoff R, et al. Interruption of ganglioside synthesis produces central nervous system degeneration and altered axon-glial interactions. *Proc Natl Acad Sci USA.* 2005;102:2725–2730.
64. Wu G, Lu ZH, Wang J, et al. Enhanced susceptibility to kainate-induced seizures, neuronal apoptosis, and death in mice lacking gangliotetraose gangliosides: protection with LIGA 20, a membrane-permeant analog of GM1. *J Neurosci.* 2005;25:11014–11022.
65. Okada M, Itoh MM, Haraguchi M, et al. b-Series ganglioside deficiency exhibits no definite changes in the neurogenesis and the sensitivity to Fas-mediated apoptosis but impairs regeneration of the lesioned hypoglossal nerve. *J Biol Chem.* 2002;277:1633–1636.
66. Seyrantepe V, Demir SA, Timur ZK, et al. Murine Sialidase Neu3 facilitates GM2 degradation and bypass in mouse model of Tay–Sachs disease. *Exp Neurol.* 2017;299:26–41.
67. Todeschini AR, Hakomori SI. Functional role of glycosphingolipids and gangliosides in control of cell adhesion, motility, and growth, through glycosynaptic microdomains. *Biochim Biophys Acta.* 2008;1780:421–433.
68. Hakomori S. The glycosynapse. *Proc Natl Acad Sci USA.* 2002;99:225–232.
69. Sciannamblo M, Chigorno V, Passi A, Valaperta R, Zucchi I, Sonnino S. Changes of the ganglioside pattern and content in human fibroblasts by high density cell population subculture progression. *Glycoconj J.* 2002;19:181–186.
70. Velasco I, Salazar P, Giorgetti A, et al. Concise review: generation of neurons from somatic cells of healthy individuals and neurological patients through induced pluripotency or direct conversion. *Stem Cells.* 2014;32:2811–2817.

CHAPTER THREE

# Gangliosides in Membrane Organization

**Sandro Sonnino, Elena Chiricozzi, Sara Grassi, Laura Mauri, Simona Prioni, Alessandro Prinetti[1]**
University of Milano, Milano, Italy
[1] Corresponding author. E-mail address: alessandro.prinetti@unimi.it

## Contents



## Abstract

Since the structure of GM1 was elucidated 55 years ago, researchers have been attracted by the sialylated glycans of gangliosides. Gangliosides head groups, protruding toward the extracellular space, significantly contribute to the cell glycocalyx; and in certain cells, such as neurons, are major determinants of the features of the cell surface. Expression of glycosyltransferases involved in the de novo biosynthesis of gangliosides is tightly regulated along cell differentiation and activation, and is regarded as the main metabolic mechanism responsible for the acquisition of cell-specific ganglioside patterns. The resulting sialooligosaccharides are characterized by a high degree of geometrical complexity and by highly dynamic properties, which seem to be functional for complex interactions with other molecules sitting on the same cellular membrane (*cis*-interactions) or soluble molecules present in the extracellular environment, or molecules associated with the surface of other cells (*trans*-interactions). There is no doubt that the multifaceted biological functions of gangliosides are largely dependent on oligosaccharide-mediated molecular interactions. However, gangliosides are amphipathic membrane lipids, and their chemicophysical, aggregational, and, consequently, biological properties are dictated by the properties of the monomers as a whole, which are not merely dependent on the structures of

*Progress in Molecular Biology and Translational Science*, Volume 156
ISSN 1877-1173
http://dx.doi.org/10.1016/bs.pmbts.2017.12.007

their polar head groups. In this chapter, we would like to focus on the peculiar chemicophysical features of gangliosides (in particular, those of the nervous system), that represent an important driving force determining the organization and properties of cellular membranes, and to emphasize the causal connections between altered ganglioside-dependent membrane organization and relevant pathological conditions.

## 1. CHEMICAL STRUCTURE OF GANGLIOSIDES

Gangliosides are glycosphingolipids (GSLs), that is, amphipathic membrane lipids with oligosaccharide chains linked to the hydrophobic backbone, ceramide, responsible for the insertion in cellular membranes, via a β-glycosidic bond. Their nomenclature, as proposed by Svennerholm,[1] is based on the structure of the oligosaccharide head groups. The most abundant gangliosides in mammals belong to the ganglio-series, and the ones from brain (the tissue with the highest ganglioside content) are mostly represented by structures belonging to the ganglio-4 series, characterized by the tetrasaccharide-β-Gal-(1-3)-β-GalNAc-(1-4)-β-Gal-(1-4)-β-Glc-(1-1)-Cer. The signature sugar residue present in gangliosides, is sialic acid, which contains a carboxyl group. Ganglio-4 structures can bear one (a-series), two (b-series) or three (c-series) sialic acid residues linked to the internal galactose, and additional sialic acid residues on the external galactose (and rarely the *N*-acetylgalactosamine). Outside of the nervous system, the most abundant ganglioside in mammals is GM3, characterized by the trisaccharide-α-Neu5Ac-(2-3)-β-Gal-(1-4)-β-Glc(1-1)-Cer.

Sialic acids[2] are derivatives of the 5-amino-3,5-dideoxy-D-*glycero*-D-*galacto*-non-2-ulopyranosonic acid, or neuraminic acid. Two main *N*-acylated forms of sialic acid are known to be present in natural gangliosides: the 5-*N*-acetyl- and 5-*N*-glycolyl-derivative. Of the two derivatives, the former is the most abundant sialic acid inhealthy mammals, including humans. About 10% of the total 5-*N*-acetyl-neuraminic acid (Neu5Ac) in gangliosides is O-acetylated (i.e., 5-*N*-acetyl-9-O-acetyl-neuraminic acid), and polysialogangliosides containing this sugar structure have been characterized in mice brains.[3–5] Most mammals, including primates, are able to synthesize 5-*N*-glycolyl-neuraminic acid (Neu5Gc). Humans, however, are not able to do so due to the ancestral loss of the *CMAH* gene, encoding for the enzyme responsible for the conversion of CMP-5-*N*-acetyl-neuraminic acid to CMP-5-*N*-glycolyl-neuraminic acid, in the hominid lineage. Minor amounts of gangliosides containing 5-*N*-glycolyl-neuraminic acid, however, are detectable in human tissues, likely due to the recycling of food-derived Neu5Gc.[6] This acid is also usually present at relatively high levels in gangliosides from certain tumors.[7] Nevertheless, 5-*N*-glycolyl-neuraminic acid is almost absent in brain gangliosides.[8] Polysialogangliosides in the form of lactones, with the sialic acids linked together with ketosidic and ester linkages, have been detected in fish, whale, and human brain.[9–11]

Gangliosides share the same hydrophobic moiety, ceramide, with all other sphingolipids (SLs). This moiety is composed of a long-chain amino alcohol[12] (2*S*, 3*R*, 4*E*) 2-amino-1,3-dihydroxy-octadec-4-ene, commonly known as sphingosine, linked to a fatty acyl chain via an amide bond. In mammalian gangliosides, the C18 molecular species of sphingosine is the most common, however, less frequent structures with either shorter or longer alkyl chain have been identified, and gangliosides containing C20 sphingosine are relatively abundant in mammalian nervous tissue and cultured neuronal cells.[13] Further heterogeneity in the ceramide moiety is due to the presence of fatty acyl chains with different length and degree of unsaturation. The preference of the different known (dihydro)ceramide synthases for different acyl-CoA explains, at least in part, the different fatty acid composition of gangliosides and their tissue-specific composition from this point of view.[14] Brain gangliosides are highly enriched in saturated acyl chains, such as palmitic and stearic acid,[15] and ceramide synthase 1 is consistently expressed at high level in human brain.[16] Hydroxylated fatty acids, abundant in neutral glycolipids and sulfatides, are less common in gangliosides, however, GM3 ganglioside from human liver contains a significant amount of α-hydroxylated C16–C24 chains, which remarkably increase with aging.[17]

## 2. PHYSICOCHEMICAL PROPERTIES OF GANGLIOSIDES: A DRIVING FORCE FOR MEMBRANE ORGANIZATION

Life, as we conceive it, requires biological membranes, which provide the basic frame for the organization of every type of eukaryotic cell. Biological membranes separate the cellular and the extracellular environments and organize different compartments within cells, allowing the creation of gradients of water-soluble molecules (ions, enzymes, reaction substrates, and products) and restricting the occurrence of specific biochemical reactions into specialized subcellular environments. The finding that the basic molecular organization of biological membranes is represented by a bilayer of amphipathic phospholipids is undoubtedly one of the milestones of modern cellular and molecular biology.[18–20] The intrinsic properties of membrane lipids determine the fundamental, structural, and functional properties of biological membranes. In particular, the creation of the lipid bilayer, (which represents the first level of ordered organization of biological membranes) is lipid-driven, that is, it is the consequence of the aggregational properties of complex amphipathic membrane lipids. Glycerophospholipids (GPL) are by far the major structural lipid components of eukaryote cellular membranes, and phosphatidylcholine (PC), that accounts in "average" membranes for more than 50% of all cell membrane phospholipids, is the main bilayer-forming lipid. A GPL bilayer, as a whole, is a very stable structure, in agreement with the primary function of cellular membranes as physical barriers. However, a bulk membrane GPL is present in a large number of different molecular species. A large proportion of membrane phospholipids is rich in unsaturated fatty acids, ensuring that at physiological temperature the membrane bilayer is on a fluid, "liquid-disordered" phase, allowing a certain degree of lateral motility and conformational freedom to membrane components, including membrane-associated proteins. The behavior of biological lipid bilayers as two-dimensional fluids under physiological conditions is the tenet of the "fluid mosaic" model proposed by Singer and Nicolson in 1972.[21] However, different kinds of lateral interactions among membrane components can take place, leading to the establishment of multiple levels of lateral order (including the organization of "liquid-ordered" domains),[22] and early observations suggested that the membrane lipid components are not merely the solvent for membrane-associated proteins, but rather are active players in the creation of these levels of order. In fact, shortly after the fluid mosaic model had been formulated, studies of thermal effects in the behavior of membrane lipids suggested that phase behavior of lipid mixtures could be responsible for a certain degree of lateral organization in biological membranes.[23–25] In 1982, the concept that the existence of multiple phases in the membrane lipid environment can drive the "organization of the lipid components of membranes into domains"[26] was clearly formulated. Heterogeneity of membrane lipids is the driving force for the lateral organization of biological membranes. If we focus on GPL, a significant amount of molecular species with less fluid, saturated chains is present. Lipids with a high content of saturated acyl chains (that possess a high melting point and are tightly packed with a high degree of order in the hydrophobic core of a bilayer) are characterized usually by a higher transition temperature, within or above the physiological range. The difference of transition temperature due to the different acyl chain composition probably represents one of the main forces leading to phase separation in lipid bilayers,[27–29] as demonstrated by the observation that phase separation occurs in binary mixtures of PC molecular species differing in chain length and/or saturation.[30–37]

Sphingolipids (SLs) are ubiquitous but minor components of eukaryotic cell membranes, and in general complex GSLs, including gangliosides, are not bilayer-forming lipids. Gangliosides in water solution tend to form micellar aggregates,[38,39] with the notable exception of GM3, which spontaneously aggregates as stable unilamellar vesicles.[40] However, SL can be inserted in the GPL bilayer through their hydrophobic ceramide moiety. Even if they are minor membrane components, SLs and gangliosides are relatively abundant in some tissues or cell types. By far the most notable example is represented by the brain,[41] where gangliosides are present at high levels in neurons (e.g., in cultured cerebellar neurons they represent about 5% of total amphipathic lipids, i.e., about 10% of total lipids of outer membrane layer)[42] while sphingomyelin, galactosylceramide, and its sulfated derivative, 3-O-sulfogalactosylceramide or sulfatide, are enriched in myelin-forming oligodendrocytes.[43] In addition, there are striking differences in GSL content among different subcellular membranes, with the plasma membrane having the highest enrichment.[44,45] Finally, within plasma membranes, GSL are asymmetrically enriched in the external leaflet, with their oligosaccharide chain protruding toward the extracellular space, where its sugar residues can engage *cis*- and *trans*-interactions with a wide variety of cell surface and extracellular molecules.[46] Thus, the local concentration of GSL in specific cellular membranes can be relatively high, and their influence on the membrane organization can represent an important driving force for the creation of lateral order. It is worth to note that, even if the highest concentration of gangliosides is associated with the outer leaflet of the plasma membrane, consistent with the biological roles of oligosaccharide-dependent interactions mediated by gangliosides in the regulation of several aspects of the "social life" of the cell, ganglioside pools in other subcellular locations, for example, the nuclear membrane, seem to be relevant.[47]

SLs, and more specifically gangliosides, have several peculiar structural features that have important consequences in terms of the lateral organization of biological membranes. As discussed in the first part of this chapter, their molecular complexity is very pronounced: (1) the oligosaccharide chain of gangliosides is extremely variable due to the sugar structure, content, sequence, and linkages, and is characterized by the presence of one or more negative charges due to the sialic acid residues; (2) ceramide, the double tailed hydrophobic moiety of gangliosides, is also heterogeneous in terms of chain length and presence of unsaturations.

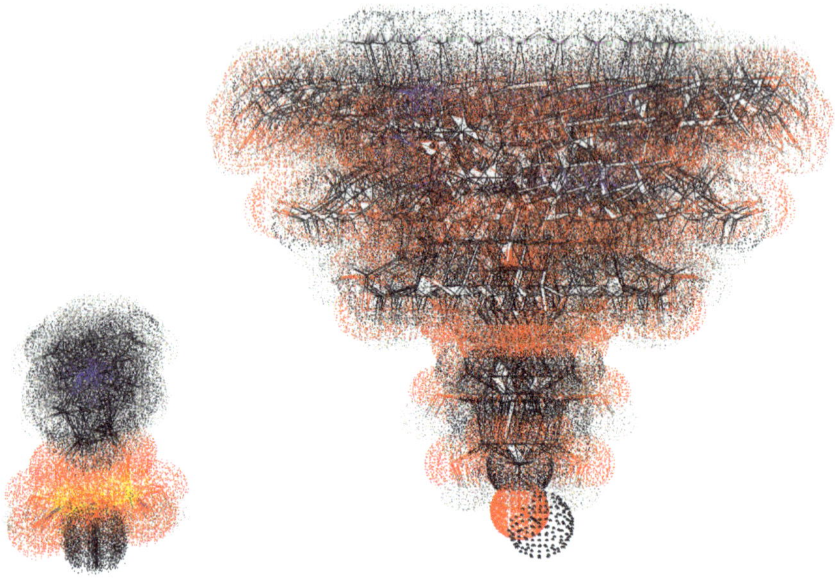

**Fig. 1** Schematic representation of the volumes of phosphocholine and GM1 oligosaccharide, based on the possible minimum energy conformations estimated by MM2 force field analysis.

The most remarkable feature of all gangliosides is the bulkiness of their hydrophilic head groups, as compared to that of phospholipids (Fig. 1). The volume mismatch of the head groups of amphipathic lipids present in the same membrane represents the main driving force for clustering and segregation of different membrane components, allowing minimization of the interfacial free energy. The area at the hydrophilic/hydrophobic surface of the membrane required to allow the insertion of the ganglioside molecule in the bilayer increases with the volume occupied by the head group. The larger the interfacial area required by ganglioside oligosaccharide structures, the more pronounced the segregation of ganglioside molecules within the phospholipid bilayer, and the more positive the curvature of the resulting ganglioside-rich membrane microenvironment. Thus, the volume of the sugar head groups directly determines the level of phase separation and the geometry in a specific membrane microenvironment. Published evidence clearly showed that gangliosides undergo phase separation in two-components, two-phases, as well as in one-component PC bilayers,[30–36] and in sphingomyelin bilayers.[48] Moreover, phase separation has been observed in mixed micelles of two different gangliosides (GM2 and GT1b,[49] GD1b and

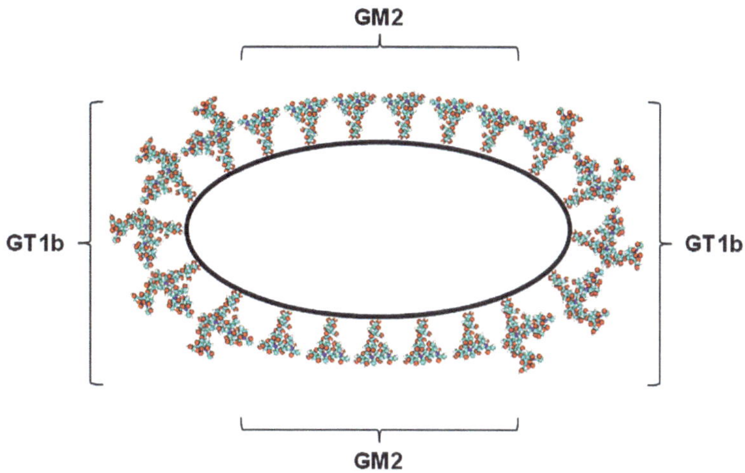


**Fig. 2** Schematic representation of phase separation driven by the differences in the oligosaccharide chains in a GM2/GT1b micelle. *Reproduced from Sonnino S, Prinetti A. Lipids and membrane lateral organization. Front Physiol. 2010;1:153.*[27]

GD1b-lactone,[50] or GM1 and GD1a)[51] with identical composition of the hydrophobic moiety (Fig. 2).

A number of studies on semisynthetic gangliosides with defined composition of the hydrophobic moiety, in monomeric form or inserted in model membranes (mainly phospholipid liposomes and single-component or mixed micelles) allowed elucidation of the contribution of different factors to the volume of the head groups (reviewed in Ref. 29). Obviously, the volume occupied by the head group increases with the complexity of the oligosaccharide chain, and the extent of ganglioside phase separation depends on the number of sugars in the oligosaccharide head group.[31,32] However, at least three additional factors are very relevant for the saccharide head groups: (1) the geometry and dynamics of the different glycosidic linkages present in the oligosaccharide chain; (2) the presence of the negative charge(s) of the sialic acid residue(s); and (3) the amount of hydration water associated with the oligosaccharide chains.

Molecular dynamics and high resolution NMR studies[52–60] [in particular carried out for the ganglio-series gangliosides, analysing the effect of the transition in the oligosaccharide chains (LacCer → GM3 → GM2 → GM1 → GD1a/b → GT1b)] showed that the addition of a single sugar to a given oligosaccharide chain has very dramatic consequences, giving rise to very different interresidual interactions within the oligosaccharide chains and generating very different dynamics for the glycosidic linkages even in very similar oligosaccharide structures. The result is that the number of possible conformations for very similar oligosaccharide chains differs substantially. The number of possible conformers of a given ganglioside oligosaccharide moiety directly influences the volume occupied by the ganglioside monomers at the membrane surface. The dynamic properties of the different glycosidic linkages in the oligosaccharides of GM2, GM1, and GD1a allow the existence of one, two. and four preferred conformations, respectively. Thus, the volume required to accommodate the GM1 and GD1a oligosaccharides (characterized by a higher number of preferred conformations) is much higher than the volume that could be predicted based on the features of the GM2 oligosaccharide, simply considering the addition of one or two more sugars, respectively.

In the GD1b tetrasaccharide-β-GalNAc-(1-4)[α-Neu5Ac-(2-8)-α-Neu5Ac-(2-3)-]-β-Gal-, geometrical constraints led by interactions between GalNAc and the external Neu5Ac determine the arrangement of the tetrasaccharide chain in a circle with a diameter of 3Å, with a central hollow characterized by a highly hydrophobic internal surface, apparently wide enough to accommodate cations.[61] This peculiar conformation confers a bulkier character to the oligosaccharide chain portion in the proximity of the hydrophobic-hydrophilic interface. Consequently, GD1b bears a significantly larger surface area than its isomer GD1a.

In ganglioside lactones, for example, GD1b-lactone,[9] the presence of the inner ester between the external sialic acid carboxyl group and the inner sialic acid residue, confers a rigid conformation to the trisaccharide-β-GalNAc-(1-4)[α-Neu5Ac-(2-3)-]-β-Gal-,which leads to a better lining up of the disialosyl chain with the neutral oligosaccharide chain, reducing the angle between the neutral chain and the inner sialic acid axis.[61] Consequently, the geometry of GD1b-lactone is closer to that of GD1a than to that of GD1b. This implies that the lactonization-delactonization process (occurring, e.g., due to local pH changes in the proximity of the cell surface) might have very deep influence on the aggregational properties of gangliosides.

As mentioned earlier, the volume of the sugar head group is also influenced by the presence of hydrating water associated with the oligosaccharide chain. The amount of water associated with the polar head groups, due to the strong hydrophilic character of the oligosaccharide chains, is remarkably high.[62] The number of water molecules tightly associated with a given oligosaccharide chain ranges from 22 to 60, depending on the structure of the oligosaccharide (while only 6–7 water molecules are associated with the phosphocholine head group of dipalmitoylPC).[63] NMR studies showed that the interactions between water molecules, the sugar residues sialic acid and inner galactose in the oligosaccharide of GM1[58] are very strong. The contribution of the associated water to the volume of the sugar head group is confirmed by calculations performed on GM2 molecules, indicating a difference of about 5Å between the dry and hydrated micellar radius.[49] Moreover, stable removal of water from the hydrophilic head of GSL in micellar dispersion lead to changes of the surface area occupied by the monomer inserted into the aggregate with consequent changes in surface curvature and in aggregate size.[64–66] The presence of water has two other consequences on the properties of the ganglioside oligosaccharide chains, which reflect on their aggregational behavior, as well as on their ability to engage *cis*- and *trans*-interactions with other molecules: (1) the associated water is likely relevant in avoiding repulsive effects between the negatively charged oligosaccharide chains, for example, in ganglioside-rich membrane microdomains; moreover, a network of water-mediated hydrogen bridges could significantly contribute to the stabilization of a glycolipid membrane clusters. (2) On the other hand, the presence of a thick and strongly associated water layer suggests that a direct role of carbohydrate–carbohydrate interactions in mediating ganglioside-dependent biological events is very unlikely.[67]

In addition to the features of the hydrophilic portions, we have recalled the importance of the differences in the transition temperatures due to the different acyl chain composition of membrane lipids as one of the main forces leading to phase separation in lipid bilayers, and of the high transition temperature of the hydrophobic chains of gangliosides in the segregation process. Lipids with a high content of saturated acyl chains (that possess a high melting point and can be tightly packed with a high degree of order in the hydrophobic core of a bilayer) are characterized usually by a higher transition temperature, within or above the physiological range.[29,68] Indeed, phase separation of sphingomyelin in dimyristoylPC bilayers depends on the degree of sphingomyelin chain mismatch,[69] and the lateral phase separation of GM1 in PC bilayers[34] is directly correlated with the acyl chain length and inversely correlated with the degree of unsaturation. In general, for a given oligosaccharide composition, the extent of ganglioside lateral phase separation depends upon the length and unsaturation difference between the ganglioside long-chain base and PC acyl chains.[30–32] It is important to recall that gangliosides containing saturated acyl chains, such as palmitic and stearic acid, are very abundant in some tissues, especially in the nervous system. Brain gangliosides are typical liquid-ordered phase lipids. In the liquid-ordered phase, a further stabilization element is represented by the enrichment in cholesterol, which can tightly intercalate between the ordered acyl chains of acyl-lipids.[29,70] The role of cholesterol in driving/stabilizing liquid-ordered phase formation in biological membranes has been very much emphasized.[71] Some studies in monolayer membrane models suggested the existence of a strong, somewhat specific lateral interaction between cholesterol and sphingomyelin, leading to the formation of liquid-condensed cholesterol- and sphingomyelin-rich domains.[72–74] However, studies in phospholipid bilayers seem to confute this evidence.[75] On the other hand, crosslinking of GM1 ganglioside using pentavalent cholera toxin at 37°C in a relatively physiological model system (plasma membrane spheres obtained by a cell-swelling procedure from A431 cells)[76] resulted in the cholesterol-dependent coalescence of GM1-rich domains with the formation of micrometer-scale domains, characterized by a lower translational diffusion, and higher degree of lateral order.[77]

In addition to the effect of the fatty acyl chains of GSL on their transition temperatures, the contribution of the part of the ceramide molecule sitting at the water lipid interface seems to be very relevant to the aggregation/segregation of these lipids within membranes. This is suggested by the increased order of C18-sphingomyelin molecules inserted into a dioleylPC bilayer due to the formation of sphingomyelin nanoclusters stabilized by hydrogen bonds.[78,79] Moreover, SL-rich domains can be formed in cells that contain sphingomyelin but not GSL[80]; and in subdomains enriched in sphingomyelin but with no gangliosides and a very low content of neutral GSLs.[81,82] The player here is the amide group of ceramide at the water/lipid interface, which behaves as a rigid system in a planar conformation, together with the hydroxyl group in position 3.[83,84] The availability of an amide nitrogen, carbonyl oxygen, and hydroxyl group, enables ceramide to act as hydrogen bond donor and acceptor at the same time. Therefore, this feature allows SL to form a stable net of interactions, which, in the case of gangliosides, further stabilizes the clustering driven by the properties of the hydrophilic head groups.

## 3. LOCAL MODULATION OF MEMBRANE GANGLIOSIDE COMPOSITION: A MECHANISM FOR THE FINE TUNING OF MEMBRANE LATERAL ORGANIZATION

The pattern of gangliosides expressed at the plasma membrane of a given cell type, and thus the resulting ganglioside-driven membrane organization, is specific and tightly regulated along development and dependent on the functional status of the cell. The genetic and epigenetic mechanism underlying this regulation have been particularly well studied in the nervous system,[85–87] not surprisingly considering the high enrichment of gangliosides in neurons. However, it is clear that, even for cell populations expressing relatively low amounts of gangliosides (e.g., immune system cells, such as T cells), the activation of different cellular subpopulations requires the expression of distinct ganglioside patterns.[88] In addition, even in the well-studied nervous system, the availability of analytical techniques with high spatial resolution, such as imaging mass spectrometry are revealing an amazing heterogeneity in ganglioside pattern among different cellular populations and substructures. For example, it has been reported that ganglioside distribution in the hippocampus is different in the three layers that constitute the molecular layer of the dentate gyrus, indicating a strikingly different expression of gangliosides in adjacent neuronal subpopulations and in neuronal axons versus dendrites.[89]

The regulation of plasma membrane ganglioside composition is mainly dependent on the biosynthetic and catabolic processes occurring in intracellular districts,[90] thus the turnover of plasma membrane gangliosides is intimately connected with a bidirectional traffic, from and to the plasma membrane, that mainly occurs via vesicular transport, even if nonvesicular transport via sphingolipid binding proteins plays an important role in specific steps.[91] On the other hand, several enzyme activities potentially involved in the local remodeling of SL composition are associated with the plasma membrane, providing the machinery for fast alterations of membrane organization in response to different stimuli. This concept was originally proposed for the interconversion of sphingomyelin into ceramide. Different sphingomyelinases are resident at the plasma membrane level or can be translocated to the plasma membrane from intracellular sites. In addition, at least one isoform of sphingomyelin synthase is enriched in the plasma membrane. Thus, the plasma membrane ratio between ceramide and sphingomyelin can be effectively regulated at the local level. Ceramide has important roles as a bioactive molecule, acting on plasmamembrane-associated and intracellular targets. In addition, ceramidases and sphingosine kinases are also associated with the plasma membrane, thus hydrolysis of sphingomyelin has been regarded by many authors mainly as a mechanism to generate bioactive ceramide and/or sphingosine 1-phosphate.[92] However, the sphingomyelin/ceramide interconversion at the plasma membrane has potentially important consequences on the membrane organization, strongly influencing membrane curvature and lateral organization. In particular, reorganization of lipid membrane domains into ceramide-rich signaling platforms has been reported to occur upon different receptor-dependent and -independent stimuli.[93,94] An increasing body of evidence indicates that active plasma membrane metabolic remodeling also occurs for GSL (Fig. 3). It has long been shown that synaptosomal membranes, a subset of neuronal membranes highly enriched in gangliosides, carry both a sialidase,[95–98] and a sialyltransferase[99] activity. The existence of a ganglioside-sialidase activity at the plasma membrane was further suggested by enzymatic and immunological studies,[100–103] and by metabolic studies in intact cells as well. Cultured rat cerebellar neurons and human neuroblastoma cells are able to desialylate exogenously added gangliosides under experimental conditions, preventing ganglioside internalization and/or blocking lysosomal function,[104,105] and a membrane-bound sialidase was purified from human[106] and bovine brain.[107] Finally, the existence of a specific membrane-linked sialidase, distinct from

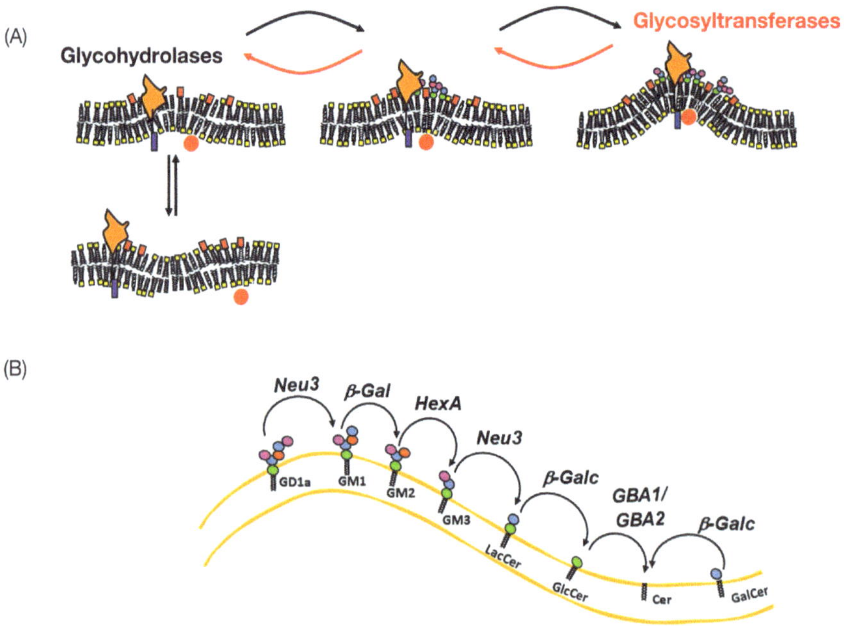


**Fig. 3** Remodeling of the plasma membrane ganglioside pattern by the local activity of hydrolytic enzymes or glycosyltransferases. Panel A: Changes of ganglioside glycosylation can affect either the clustering of signaling proteins within ganglioside-enriched membrane domains, or the membrane curvature. Panel B: schematic representation of the stepwise removal of saccharide units from the ganglioside oligosaccharide head group by the action of different hydrolytic enzymes, whose activity has been reported at the plasma membrane level.

other known sialidases, has been proven by the group of Miyagi, who cloned the cDNA sequence for human,[108] bovine,[109] and mouse[110] plasma membrane-associated sialidase, later named Neu3.[111] Following studies suggested that this enzyme plays a crucial role in modifying the cell surface ganglioside composition, in particular causing a shift from polysialylated species to GM1, a decrease of GM3 and a parallel increase in lactosylceramide, with deep consequences on very important cellular events, such as neuronal differentiation and apoptosis in colon cancer.[112] Neu3 is not randomly distributed at the cellular surface, but rather it is targeted to and dynamically associated with detergent-insoluble, GSL-enriched microdomains,[113–116] suggesting that membrane topology is highly favoring the accessibility of the putative substrates to the enzyme activity. On the other hand, several experimental data show that the local activation of Neu3 might represent an efficient mechanism to regulate the local ganglioside composition of specific membrane subdomains, thus affecting the biological activity of signaling proteins resident in or recruited to these membrane domains. In other words, the most relevant biological effects of Neu3 seem to be linked to its ability to modulate the composition of ganglioside-driven signaling units. From this point of view, the most striking example is represented by the ability of Neu3 to locally increase the surface concentration of GM1 in neurons, thus affecting the biological activity of the TrkA neurotrophin receptor. The physiological mechanisms responsible for Neu3 activation remain to be elucidated, however, it has been recently shown that phosphatidic acid, produced via epidermal growth factor (EGF)-induced activation of phospholipase D, promotes the activation and surface localization of Neu3.[117] It has long been known that ganglio-series gangliosides, and in particular GM1, possess neuritogenic and neuronotrophic properties,[118,119] exerting positive effects on neuronal growth, differentiation, and survival both in vivo and in cultured neurons, and a protective effect against neuronal injury.[120–123] Thus, GM1 seems to potentiate or replace neurotrophins in their action on neuronal cells,[124] and the underlying molecular mechanism is likely an effect on the activation of neurotrophin receptors. GM1, which is able to stimulate Trk receptor kinase activity, receptor autophosphorylation and dimerization, specifically and tightly binds Trk family receptors in cultured cells, brain tissue, and in vivo (recently reviewed in Ref. 125). A significant pool of Trk receptors is detergent-insoluble and associated with lipid rafts under basal conditions,[126–131] moreover, Trk receptors can be recruited to detergent-insoluble membranes from other cellular compartments upon stimulation.[126,127] In addition, the glycosylation of Trk is required for the formation of Trk-GM1 complexes and for the targeting of Trk into GM1-enriched domains.[132] This suggests that lateral interactions between the oligosaccharide chains of the receptor and of GM1, mediating the receptor localization within ganglioside-rich membrane areas, could be a relevant mechanism for the regulation of its function. From this point of view, the activity of Neu3 in modulating the local membrane ganglioside composition seems to be crucial. The artificial modulation of Neu3 expression, in a positive or negative sense, deeply affects neuritogenesis.[133,134] In addition, in cultured hippocampal neurons, the activity of the plasma membrane-associated ganglioside sialidase affects the local GM1 surface levels, and is essential for axonal growth and regeneration after axotomy.[135] In these cells, Neu3 activity is asymmetrically concentrated at the end of one single neurite, and determines the neurite's axonal fate by the rearrangement of GM1 TrkA microdomain with a consequent local increase in TrkA activity.[136]

Interestingly, Neu3 overexpression in cultured skin fibroblasts was able to modulate the production of bioactive ceramide at the cell surface, providing the first direct evidence of a link between glycosphingolipid metabolism and ceramide-mediated signaling, and suggesting that ganglioside hydrolysis, in addition to sphingomyelin hydrolysis, might contribute to the formation of ceramide-rich membrane platforms.[137] Neu3-assisted cell surface ceramide generation from ganglioside GM3 indirectly demonstrate the presence of the other two active glycosyl hydrolases, β-glucosidase and β-galactosidase, in the same plasma membrane district. Subsequently, several enzymes involved in GSL catabolism have been found associated with the external leaflet of the cell plasma membrane: β-hexosaminidase, β-galactosidase, β-glucosidase GBA1, and GBA2.[138–141]

These observations open the possibility that extensive hydrolytic remodeling (not onlydesialylation) of membrane gangliosides can occur and contribute to the local membrane glycolipid composition (Fig. 3). In addition, in situ sialylation of gangliosides also occurs at the cell surface. The existence of a synaptosomal membrane sialyltransferase in calf brain[99] has been confirmed by metabolic studies in chicken embryos[142] and rat brain.[143,144] Moreover, dexamethasone treatment in mouse thymus markedly increased GM3 synthesis, due to increased enzyme activity of GM3 synthase at the plasma membrane[145]; thus confirming that glycolipid sialylation might occur outside the Golgi compartment, contributing to the local modulation of cell surface glycolipid patterns. On the other hand, glycosylation/deglycosylation are possible mechanisms for the local regulation of the plasma membrane GSL composition and lateral organization. As mentioned earlier, some gangliosides do contain ester linkages, which are unstable under alkaline conditions (while the majority of gangliosides, lacking this kind of linkage, are stable under alkaline conditions). The so-called "alkali-labile" gangliosides have been detected in mammals, birds, reptiles, amphibians, teleost, elasmobranches, and cyclostomes,[146] and in particular they are present in the ganglioside mixtures extracted from the animal brains. In some cases, the amount of alkali-labile gangliosides can be around 10% of total brain gangliosides, thus they are not negligible membrane components at least in the nervous tissue. There are two main types of alkali labile gangliosides: ganglioside lactones and ganglioside containing O-acetylated-sialic acid. In both cases, the alkali-labile species are more hydrophobic than the corresponding alkali-stable parent structure, thus, their conformational/aggregational properties are significantly different (see earlier for the detailed discussion on the conformation of GD1b-lactone).

Ganglioside lactones are formed by a reaction between the carboxyl group of sialic acid and one of the hydroxyl groups available in the oligosaccharide structure. The most stable reported lactonic linkage is inside the disialyl residue, Neu5Ac-(2-8,1-9)-Neu5Ac-. Ganglioside lactones remained elusive for long time, due their possible artifactual formation at acidic pH during the extraction procedures. Under acidic pH, the disialyl residue of GD1b ganglioside spontaneously undergoes an equilibrium reaction between the two molecules that contain the Neu5Ac-(2-8)-Neu5Ac (GD1b) and the Neu5Ac-(2-8, 1-9)-Neu5Ac (GD1b-lactone) disialyl residue, respectively.[147] The existence of the ganglioside lactones in mammalian brains was originally established by two-dimensional TLC analyses by treating the plate with ammonia vapors before the second chromatographic run.[146,148,149] Under these conditions, the lactone is converted into the corresponding amide derivative that can be clearly separated in the second TLC run. The structure of GD1b-lactone extracted from human brain was subsequently characterized by mass spectrometry. Remarkably, the content of GD1b-lactone in the human brain progressively increased with age.[9] In vivo, GD1b seems to be the natural precursor of GD1b-lactone in brain. Intracisternal injection of tritium-labeled GD1b in rats led to a progressive increase of tritiated GD1b-lactone in the following 3 days. Then tritiated GD1b-lactone levels decreased together with tritiated GD1b due the membrane turnover and ganglioside catabolic pathway. This suggests that the lactonization process occurs at the cell surface.[150]

Ganglioside lactones are extremely resistant to the action of bacterial sialidase.[150] Based on the knowledge about the sensitivity of gangliosides to sialidase hydrolysis, this suggests that lactones should be resistant to the action of cell surface sialidase Neu3. This would imply that the lactonization process might be important in regulating the ability of Neu3 to modulate the production of GM1 from GD1b. GD1b and GD1b-lactone have opposite effects on phospholipase C (PKC)-mediated protein phosphorylation.[151] PKC is present in a ganglioside-rich membrane microenvironment, and it has been reported that its activity is modulated by changes in ganglioside membrane organization.[152] Thus, the equilibrium between ganglioside GD1b and its lactonic derivative could be instrumental for membrane reorganization and the concomitant regulation of protein phosphorylation by PKC.

O-acetylated sialic acid is component of both glycoproteins and glycolipids. Sialic acid can be O-acetylated in position 4, 7, 8, and 9. Three different sialate-O-acetyl transferase (SOAT) showing different position specificity have been identified.[153] Sialic acid acetylation occurs on the CMP-Neu5Ac before the sialic acid is transferred to the glycan acceptor by a sialyltransferase.[153]

Human brain gangliosides contain about 10% of O-acetylated Neu5Ac, with respect to the total sialic acid.[154] O-acetylation introduces a further element of great heterogeneity in the structure of ganglioside oligosaccharide chains, whose consequences are still poorly understood but potentially of great biological relevance. Two polysialylated O-acetylated gangliosides were purified and characterized from mouse brain, GT1b and GQ1b with a Neu5,9$Ac_2$ as external sialic acid of the disialyl residue linked to the internal galactose.[4,5] Other O-acetylated gangliosides were extracted from cells and characterized: O-acetylated-GD3, with Neu5,9$Ac_2$ as external sialic acid, from melanoma cells,[155] and two different O-acetylated-GM3, containing Neu5,4$Ac_2$[156] and Neu5,4,9$Ac_3$,[157] from erythrocytes. In addition, O-acetylated GD2 and GT3 gangliosides were reported as components of a plethora of normal and pathological tissues and cells.[153,158,159] The levels of O-acetylated gangliosides are relatively high in tissues characterized by a marked cell proliferation, suggesting that they might be markers for this biological condition. One of the most reported O-acetylated gangliosides is the O-acetylated-GD3, which is regarded as an oncofetal marker in many animal and human tumors (see as example Ref. 160).

As mentioned earlier, information on the functional role of nervous system O-acetylated gangliosides is scant and fragmentary. The nervous system of normothermic dormice (*Glisglis*) contains alkali-labile gangliosides many of which are O-acetylated gangliosides. They were present in all examined area ranging from about 10% in olfactory bulb to 30% in spinal cord. However, in hibernating dormice animals O-acetylated gangliosides were found only in spinal cord, where they covered only 3% of total gangliosides.[161] For some ganglioside species, the disappearance of O-acetylation in hibernating animals was paralleled by an increase of the corresponding alkali-stable counterparts, suggesting a different balance between SOAT and O-acetylhydrolases activity/expression in the two life conditions. However, this was not true for all alkali-labile species, suggesting that different Golgi biosynthetic processes also occur. In addition, an increase of GM1 ganglioside, which was never detected as the O-acetylated species, was observed in hibernating animals. This suggests that the adaptation of the animals implies a deep rearrangement and organization of the plasma membrane mediated by gangliosides, whose biological significance remains to be elucidated.

## 4. CHANGES OF GANGLIOSIDE-DRIVEN MEMBRANE ORGANIZATION AS A PATHOGENETIC MECHANISM

If changes in the plasma membrane ganglioside composition could represent a physiological mechanism for the regulation of cellular functions, which are dependent on the organization of the membrane microenvironment of signaling proteins, the obvious implication is that alterations of ganglioside expression associated with diseases might lead to abnormal membrane organization, which could contribute to the pathogenesis of the disease. Anomalies in gangliosides metabolism and/or traffic, leading to altered gangliosides patterns, have been reported in an incredible number of different pathological conditions. However, cancer, neurodegenerative diseases, and, more recently, diabetes are the diseases where a causal connection between altered ganglioside expression and consequent alterations in ganglioside-driven membrane organization as a relevant pathogenic mechanism seems to be convincingly proven. In this section, we will provide some examples with the aim to elucidate the molecular basis of this connection.

"Aberrant glycosylation" has been described for over 4 decades in almost all types of experimental and human tumors.[162–164] Even if altered expression of glycosylated epitopes in cancer is not restricted to GSL, ganglioside-dependent changes of membrane organization play a major role as determinants for aggressiveness and invasiveness of human cancer. Particular emphasis in this regard has been posed on the role of altered sialylation of glycolipids, and the role of gangliosides in regulating the function of membrane-associated signaling complexes, thus affecting tumor cell growth, survival, motility, and invasiveness, has been described in detail in the last 15 years. The clinical relevance of these observations has been highlighted by the finding that the expression of two of the main metabolic enzymes whose activity controls the sialylation levels of gangliosides, sialidase Neu3[165] and GM3 synthase (*St3gal5*),[166] is deeply altered in tumor tissues, suggesting that these metabolic enzymes could represent potentially druggable targets for novel therapeutic opportunities at least in some cancer types.

Growth factor receptor-dependent cell proliferation is controlled and modulated by gangliosides. In particular, EGF receptor (EGFR) autophosphorylation is subjected to inhibition by GM3,[167] due to a highly specific ganglioside-receptor interaction[167,168] involving side-by-side carbohydrate–carbohydrate interactions between the sialyllactose chain of GM3[168] and a N-linked glycan bearing multiple GlcNAc terminal residues on the receptor.[169,170] On the other hand, in addition to this specific GM3/EGFR interaction, GM3-driven membrane organization is an important additional factor in the regulation of EGFR activity. GM3/EGFR interaction is facilitated by the enrichment of EGFR in ganglioside-enriched, Triton X-100 insoluble membrane domains.[171,172] On the other hand, caveolae and caveolin-1 are involved in the modulation of EGFR signaling,[173,174] and caveolin membrane dynamics are deeply affected by gangliosides.[175] EGFR is localized within a caveolin-rich fraction in A431 cells, however, EGFR-containing membrane domains can be separated from caveolae,[176,177] suggesting a dynamic equilibrium between different membrane-associated EGFR pools. In a keratinocyte-derived cell line, GM3 overexpression induced a shift of caveolin-1 to EGFR-rich membrane regions, allowing its functional interaction with the EGFR receptor, that caused inhibition of EGFR tyrosine phosphorylation and dimerization.[178] Thus, increased membrane GM3 levels influence EGFR signaling by a second distinct molecular mechanism, modulating EGFR/caveolin-1 association. Moreover, GM3-rich membrane domains regulate the crosstalk of EGFR with integrin receptor signaling: in the presence of high membrane levels, the interaction of integrin β1 subunit with EGFR, and the consequent signaling crosstalk is perturbed.[179]

Independent of its crosstalk with EGFR signaling, integrin receptor function is heavily regulated by the association with ganglioside-rich membrane domains, and this regulation is likely the most relevant molecular mechanism underlying the effect of gangliosides on tumor cell motility and invasiveness. The term "glycosynapse" has been introduced by

Hakomori and coworkers[180–182] to describe a ganglioside-rich membrane domain involved in the regulation of carbohydrate-dependent cell adhesion and motility. Glycosynapses are noncaveolar, cholesterol-independent ganglioside-rich membrane domains,[183] characterized by the presence of tetraspanins (CD9, CD81, CD82) highly hydrophobic integral membrane proteins with four transmembrane stretches. Tetraspanins strongly interact with gangliosides in glycosynapses[184] and serve as the molecular pivot for the organization of membrane signaling complexes, whose function is in turn regulated by the levels of gangliosides in the membrane microenvironment. In particular, tetraspanins are associated with and regulate the function of integrin receptors,[185] involved in the regulation of tumor cell motility/invasiveness. Tetraspanin CD9 and integrin α3 or α5 receptor subunits are colocalized in glycosynapses. In a noninvasive bladder tumor cell line (originated by superficial, nonmetastatic human bladder cancer), characterized by high levels of GM3 ganglioside, CD9/integrin association is positively modulated, and a multimolecular complex between GM3, tetraspanin, and integrin α3β1 or α5β1 is stabilized.[186–188] In the presence of the GM3-dependent integrin/CD9 complex, the c-Src kinase Csk, responsible for the phosphorylation of c-Src at Tyr527, is translocated to the GM3-rich membrane domain with consequent c-Src inactivation.[188] Thus, the formation of integrin/CD9/GM3 complexes leads to the negative regulation of integrin-mediated cell adhesion and tumor cell motility. Conversely, in highly invasive bladder cancer cells, characterized by lower levels of GM3, the integrin/CD9 association is reduced and c-Src is highly active, leading to reduced adhesion and increased motility. A similar role in the stabilization of an integrin/tetraspanin complex has been suggested for GM2, with consequent regulation of Met tyrosine kinase activity.[189]

GM3-rich glycosynapses are also involved in the control of other forms of tumor cell adhesion, for example, in adhesion mediated by GSL–GSL interactions. In the case of GM3-dependent adhesion of melanoma cells to the neutral glycolipid Gg3, it has been shown that GM3 is closely associated with c-Src, Rho, and Ras within glycosphingolipid-enriched membrane domains and binding with Gg3 or anti-GM3 antibody stimulates focal adhesion kinase phosphorylation and c-Src activity.[81] A similar GM3-dependent association between c-Src and other related signaling molecules was observed also in neuroblastoma cells.[190]

In other tumors, regulation of Src-dependent tumor cell motility by ganglioside-rich signaling complexes might occur thanks to molecular mechanisms not involving tetraspanins. We showed that human ovarian carcinoma cells, characterized by low *St3gal5* expression, GM3 synthase activity and ganglioside content, are highly motile. In clones with increased GM3 synthase expression, obtained by *St3gal5* gene transfection (associated with an overall increase in cellular ganglioside levels), the in vitro motility was strongly reduced. GM3 synthase overexpression was paralleled by a marked overexpression of another hydrophobic membrane adapter protein, caveolin-1.[191] Caveolin-1 is supposed to concentrate signaling molecules within cholesterol- and sphingolipid-rich membrane domains, and the effect of caveolin-1 on tumor phenotype seems to be very heterogeneous and strongly dependent on the molecular partners interacting with this protein.[192] In high GM3 synthase, high caveolin-1 cells, caveolin and gangliosides are tightly interacting together forming a detergent-insoluble membrane complex.[193] Integrin receptor subunits and the nonreceptor tyrosine kinase c-Src are associated with the ganglioside/caveolin-1 complex. Caveolin-1[175,194] and Src kinases[195,196] are associated with sphingolipid-enriched membrane domains also in other cell types, and it has been suggested that caveolin-1 might act as a membrane adapter coupling integrin receptors to Src kinases,[197] and that caveolin-1-mediated inactivation of the integrin/Src/FAK pathway might be responsible for the inhibition of metastatic potential in melanoma.[198] Our results strongly support the hypothesis that the inactivation of c-Src by a ganglioside/caveolin-1 complex might be linked to the downregulation of ovarian carcinoma cell motility.

Complex alterations in brain lipid composition have been reported along with physiological ageing, including reduction of total lipid content, decreased ganglioside content, and altered sphingoid base compositions in SL.[199] These changes have deep consequences on the physicochemical properties of lipid rafts (e.g., on local membrane microviscosity), which seems to change with the decline of physiological performance of the brain along ageing. In other words, lipid-dependent membrane organization naturally undergoes an "ageing" process. In addition, alterations in membrane thermodynamics are more pronounced in mouse models of dementia.[200,201] Congenital disorders of sphingolipid metabolism, such as gangliosidosis, due to defects in the lysosomal degradation of gangliosides, some forms of Gaucher disease, due to the defective activity of lysosomal glucosylceramidase, and Niemann–Pick disease type A, due to the lack of acidic sphingomyelinase, are progressive and severe neurodegenerative diseases (reviewed in Refs. 202 and 203). These diseases are hallmarked by accumulation of the undegraded SL at the lysosomal level, as the consequence of the enzymatic defect. The accumulated SL are different depending on the specific defect, however, in all of the diseases traffic of lysosomal membranes with anomalous sphingolipid composition toward the plasma membrane[204] seems to have important consequences on the plasma membrane lipid composition and organization.[202,205–207] In addition, secondary alterations in ganglioside metabolism, likely due to extralysosomal events, have been reported in many of these diseases.[208]

However, alterations in the metabolism of gangliosides possibly impacting on membrane organization have been reported for several important nervous system diseases, including most of the neurodegenerative diseases and the major forms of dementing disease, even in the absence of genetic defects directly related to sphingolipid traffic and/or metabolism.[202,209] We will limit our discussion to Alzheimer's disease (AD), which in our opinion represents the most complex and intriguing example of the consequences of altered ganglioside-driven membrane organization in nervous system pathology.

Dysregulated brain ganglioside metabolism has been reported in brain of AD patients and in transgenic mice models of the disease (reviewed in Ref. 210). The patterns of ganglioside alterations in AD are very complex and differ depending on age of onset of the disease and on the type of mutation, suggesting that different ganglioside-regulated events contribute to the onset of different AD forms. However, reduced ganglioside with altered ratios between a-series and b-series structures, and elevated levels of simpler gangliosides have been observed in different brain regions of AD and dementia of the Alzheimer type (DAT) affected patients[211–216] with respect to age-matched healthy controls. Consistent with these findings, downregulation of several glycosyltransferases was observed in dementia and AD patient brains.[217] On the other hand, lipid rafts from the frontal and temporal cortices of AD patients contain a higher concentration of gangliosides GM1 and GM2 respect to age-matched control brains.[218]

Anomalous organization of ganglioside-enriched membrane domains seems to be relevant to the functions, traffic and proteolytic cleavage of the main player in AD, the amyloid precursor protein (APP) (Fig. 4). The exact physiological functions of APP are still to be fully elucidated,[219] however, it is apparently involved in the transduction across the membrane of signals relevant to neuronal adhesion, survival and synaptic functions.[220] Alterations in the membrane microenvironment of APP could alter the functions of APP as a signaling molecule.[221] On the other hand, great emphasis in AD pathogenesis has been given to overproduction of Aβ peptides (that accumulate in the brain lesions that are commonly thought to

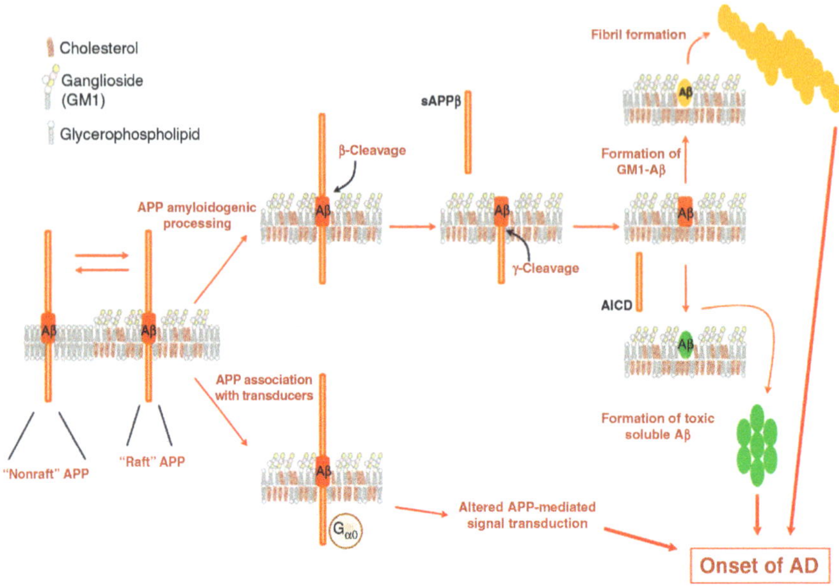


**Fig. 4** Targeting of APP to lipid raft with anomalous lipid composition can direct APP toward its amyloidogenic proteolytic processing, leading to the generation of amyloid Aβ peptide. Membrane bound Aβ can interact with ganglioside-rich domains at the cell surface with several different possible consequences: the trigger of the formation of insoluble amyloid fibrils, or the release of toxic soluble Aβ aggregates. In addition, association with anomalous lipid rafts can affect APP-mediated signal transduction. sAPPβ, soluble form of APPβ, released by β-cleavage of APP. AICD, APP intracellular C-terminus domain, released by subsequent γ-cleavage of APP. Aβ, amyloid Aβ peptide, generated by amyloidogenic processing of APP. *Reproduced from Hartmann T, Prinetti A. Going the wrong road: fyn and targeting of amyloid precursor protein to lipid rafts. J Neurochem. 2011;118:677–679.*[229]

cause AD).[222] Ganglioside-enriched lipid rafts contain not only APP, but also APP-derived proteolytic fragments, including Aβ, and the proteolytic enzymes involved in APP amyloidogenic processing. In particular they are enriched in active β- and γ-secretases, and the production of Aβ amyloid is preferentially, even if not exclusively localized within lipid rafts.[223] In other words, association with lipid rafts is a prerequisite for the amyloidogenic processing of APP, which could be favored by altered lipid raft organization driven by changes in their ganglioside composition. This notion has been recently supported by the observation that the introduction of *B4galnt1* in cells lacking the expression of this enzyme (with concomitant increase in the cellular levels of GD2, GM2, and GM1), positively regulates APP β-cleavage.[224] In turn, cleavage of APP also has a direct influence on the cellular lipid composition, as APP processing alters the synthesis of several lipids enriched in lipid rafts.[225] Moreover, γ-cleavage of APP, also taking place in lipid rafts, leads to the release of the APP intracellular C-terminus domain (AICD), which has distinct physiological functions, regulating the expression of several genes,[219] including serine-palmitoyltransferase,[226] the key enzyme in ganglioside de novo biosynthesis. On the other hand, targeting of APP to lipid rafts is regulated via by the Src family kinase Fyn, associated with lipid rafts in a ganglioside-dependent way.[227,228]

After the formation of soluble Aβ is formed by the amyloidogenic proteolytic cleavage the next step in the molecular mechanism underlying the onset of AD is Aβ conversion into aggregates, which are the seeds for the formation of insoluble amyloid fibrils. Soluble Aβ directly binds cholesterol and GM1 ganglioside (interacting with the ganglioside oligosaccharide chains by sugar-specific interactions),[230–233] and the binding of Aβ with membrane GM1,[234,235] which is modulated by cholesterol, drives its conformational transition by favoring the conversion of the aggregated form[236] leading to Aβfibrillogenesis[237] (Fig. 4). GM1-drivenAβ aggregation requires the interaction with a GM1-rich membrane microdomain rather than a simple Aβ-GM1 interaction.[238] On the other hand, interaction of GM1 with Aβ can lead to the formation of toxic soluble oligomers, that exert their detrimental effects through the high affinity binding with another protein enriched in lipid rafts, cellular PrP.[239]

More recently, a role for GM3-dependent membrane organization has been proposed in relation to type 2 diabetes. In 3T3-L1 adipocytes, the induction of insulin resistance by prolonged treatment with low concentrations of TNFα (an experimental condition mimicking the chronic low-grade inflammation status typical of the obese adipose tissue) was accompanied by the upregulation of GM3 synthase, leading to an increase of cellular GM3,[240,241] that accumulated in detergent-resistant membranes. This alteration of membrane organization had profound consequences on the activity of insulin receptor in adipocytes (Fig. 5). Insulin receptors (IR) are associated with detergent-resistant membranes from normal adipocytes[242] and localized in caveolae in intact cells.[241] In caveolar membrane domains, the β-subunit of IR interacts with caveolin-1 through a binding motif recognizing the scaffold domain of caveolin-1.[243] On the other hand, IR can be engaged in a lateral interaction by GM3. The interaction between GM3 and IR is direct, as the receptor can be crosslinked to a photoactivable GM3 derivative; it is specific, as it is abolished by the addition of exogenous GM3

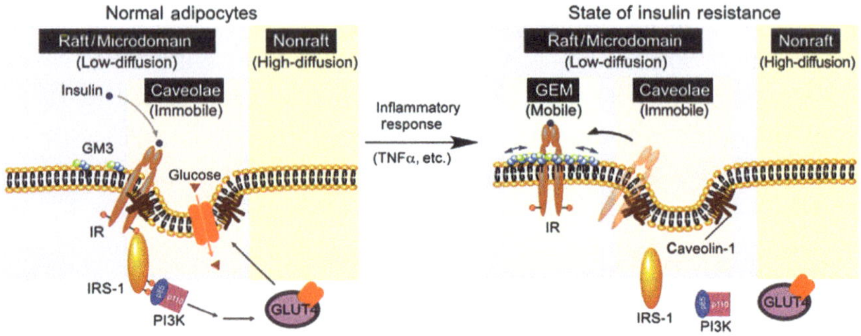


**Fig. 5** Schematic representation of the mechanism behind the shift of IR from the caveolae to glycolipid-enriched membrane domains (GEM) in insulin-resistant adipocytes. Binding of IR and caveolin-1 in caveolar domains is required for effective signal transduction downstream to IR. In insulin-resistant adipocytes, IR is sequestered in GM3-rich membrane domains and thus displaced from its association with caveolin-1. *Reproduced from Kabayama K, Sato T, Saito K, Loberto N, Prinetti A, Sonnino S, Kinjo M, Igarashi Y, Inokuchi J. Dissociation of the insulin receptor and caveolin-1 complex by ganglioside GM3 in the state of insulin resistance. Proc Natl Acad Sci USA. 2007;104:13678–13683.*[244]

in coimmunoprecipitation experiments and involves the lysine residue at 944 in the aminoacid sequence of the receptor, suggesting that an electrostatic interaction between the negatively charged sialyllactose chain of GM3 and the positively charged amino group of lysine 944 is essential for the formation of the GM3/IR complex. Coimmunoprecipitation, crosslinking, fluorescence microscopy, and FRAP experiments showed that IR can form distinct complexes with caveolin-1 and GM3 within lipid membrane domains.[244] Experimental increase of GM3 levels in normal adipocytes abolished insulin signaling and glucose uptake. Remarkably, in insulin-resistant adipocytes, the association of IR with GM3 was increased, while its association with caveolin-1 was decreased, indicating that the excess amount of GM3 in lipid membrane domains leads to the displacement of IR from the complex with caveolin-1. This suggests that the regulation of IR/caveolin-1 interaction by GM3 could be responsible for the changes in insulin response in adipocytes.[244] This is a very nice example showing that the function of a plasma membrane-associated protein can be modulated by its ganglioside-driven dynamic partitioning between distinct membrane microdomains. Interestingly enough, the possible clinical relevance of this phenomenon has been recently highlighted by the finding that the increased serum levels of specific GM3 molecular species are associated with risk factors of metabolic syndrome.[245]

## 5. CONCLUSIONS

The structural complexity of gangliosides oligosaccharide chains has attracted the attention of the researchers in this field for decades. Gangliosides head groups, protruding toward the extracellular space, are clearly designed by evolution to be engaged in interactions with molecules sitting on the same cellular membrane (*cis*-interactions), or present in the extracellular environment or on the surface of other cells (*trans*-interactions). Oligosaccharide-dependent recognition of gangliosides by interacting molecules is undoubtedly very important for the multifaceted biological functions of gangliosides. On the other hand, as discussed in this chapter, the chemicophysical properties of gangliosides do represent a major driving force for the organization of lateral order within biological membranes, and several biological functions of gangliosides are determined by this role, rather or in addition to the establishment of specific molecular interactions. Indeed, in some cases ganglioside-driven clustering of membrane molecules seems to facilitate their interaction with gangliosides themselves at the membrane level. This leads to an interesting speculation. The possible significance of alterations in gangliosides levels associated with several pathological conditions might be linked to the consequent alterations in membrane organization. Targeting aberrant ganglioside-driven membrane organization, rather than simply targeting ganglioside metabolism, might represent a valuable therapeutic strategy for some diseases, as proposed by the group of Inokuchi for type 2 diabetes.[246]

## REFERENCES

1. Svennerholm L. Designation and schematic structure of gangliosides and allied glycosphingolipids. *Prog Brain Res.* 1994;101:XI–XIV.
2. Schauer R. Chemistry, metabolism, and biological functions of sialic acids. *Adv Carbohydr Chem Biochem.* 1982;40:131–234.
3. Yu RK, Ledeen RW. Gangliosides of human, bovine, and rabbit plasma. *J Lipid Res.* 1972;13:680–686.
4. Ghidoni R, Sonnino S, Tettamanti G, Baumann N, Reuter G, Schauer R. Isolation and characterization of a trisialoganglioside from mouse brain, containing 9-O-acetyl-*N*-acetyl neuraminic acid. *J Biol Chem.* 1980;255:6990–6995.
5. Chigorno V, Sonnino S, Ghidoni R, Tettamanti G. Isolation and characterization of a tetrasialoganglioside from mouse brain, containing 9-O-acetyl, *N*-acetyl neuraminic acid. *Neurochem Int.* 1982;4:531–539.

6. Okerblom J, Varki A. Biochemical, cellular, physiological and pathological consequences of human loss of *N*-glycolylneuraminic acid. *Chembiochem.* 2017;18(13):1155–1171.
7. Marquina G, Waki H, Fernandez LE, Kon K, Carr A, Valiente O, Perez R, Ando S. Gangliosides expressed in human breast cancer. *Cancer Res.* 1996;56:5165–5171.
8. Davies LR, Varki A. Why Is *N*-glycolylneuraminic acid rare in the vertebrate brain? *Top Curr Chem.* 2015;366:31–54.
9. Riboni L, Sonnino S, Acquotti D, Malesci A, Ghidoni R, Egge H, Mingrino S, Tettamanti G. Natural occurrence of ganglioside lactones. Isolation and characterization of GD1b inner ester from adult human brain. *J Biol Chem.* 1986;261:8514–8519.
10. Avrova NF, Ghidoni R, Karpova OB, Nalivayeva NN, Malesci A, Tettamanti G. Systematic position of fish species and ganglioside composition and content. *Comp Biochem Physiol B.* 1986;83:669–676.
11. Terabayashi T, Kawanishi Y. Naturally occurring ganglioside lactones in Minke whale brain. *Carbohydr Res.* 1998;307:281–290.
12. Roisen FJ, Bartfeld H, Nagele R, Yorke G. Ganglioside stimulation of axonal sprouting in vitro. *Science.* 1981;214:577–578.
13. Sonnino S, Chigorno V. Ganglioside molecular species containing C18- and C20-sphingosine in mammalian nervous tissues and neuronal cell cultures. *Biochim Biophys Acta.* 2000;1469:63–77.
14. Mullen TD, Hannun YA, Obeid LM. Ceramide synthases at the centre of sphingolipid metabolism and biology. *Biochem J.* 2012;441:789–802.
15. Sastry PS. Lipids of nervous tissue: composition and metabolism. *Prog Lipid Res.* 1985;24:69–176.
16. Jiang JC, Kirchman PA, Zagulski M, Hunt J, Jazwinski SM. Homologs of the yeast longevity gene LAG1 in *Caenorhabditis elegans* and human. *Genome Res.* 1998;8:1259–1272.
17. Riboni L, Acquotti D, Casellato R, Ghidoni R, Montagnolo G, Benevento A, Zecca L, Rubino F, Sonnino S. Changes of the human liver GM3 ganglioside molecular species during aging. *Eur J Biochem.* 1992;203:107–113.
18. Tourtellotte ME, Branton D, Keith A. Membrane structure: spin labeling and freeze etching of Mycoplasma laidawii. *Proc Natl Acad Sci USA.* 1970;66:909–916.
19. Steim JM, Tourtellotte ME, Reinert JC, McElhaney RN, Rader RL. Calorimetric evidence for the liquid-crystalline state of lipids in a biomembrane. *Proc Natl Acad Sci USA.* 1969;63:104–109.
20. Engelman DM. X-ray diffraction studies of phase transitions in the membrane of Mycoplasma laidlawii. *J Mol Biol.* 1970;47:115–117.
21. Singer SJ, Nicolson GL. The fluid mosaic model of the structure of cell membranes. *Science.* 1972;175:720–731.
22. Sprong H, van der Sluijs P, van Meer G. How proteins move lipids and lipids move proteins. *Nat Rev Mol Cell Biol.* 2001;2:504–513.
23. Lee AG, Birdsall NJ, Metcalfe JC, Toon PA, Warren GB. Clusters in lipid bilayers and the interpretation of thermal effects in biological membranes. *Biochemistry.* 1974;13:3699–3705.
24. Wunderlich F, Ronai A, Speth V, Seelig J, Blume A. Thermotropic lipid clustering in tetrahymena membranes. *Biochemistry.* 1975;14:3730–3735.
25. Wunderlich F, Kreutz W, Mahler P, Ronai A, Heppeler G. Thermotropic fluid goes to ordered "discontinuous" phase separation in microsomal lipids of Tetrahymena. An X-ray diffraction study. *Biochemistry.* 1978;17:2005–2010.
26. Karnovsky MJ, Kleinfeld AM, Hoover RL, Klausner RD. The concept of lipid domains in membranes. *J Cell Biol.* 1982;94:1–6.
27. Sonnino S, Prinetti A. Lipids and membrane lateral organization. *Front Physiol.* 2010;1:153.
28. Sonnino S, Prinetti A. Membrane domains and the "lipid raft" concept. *Curr Med Chem.* 2013;20:4–21.

29. Sonnino S, Prinetti A, Mauri L, Chigorno V, Tettamanti G. Dynamic and structural properties of sphingolipids as driving forces for the formation of membrane domains. *Chem Rev.* 2006;106:2111–2125.
30. Masserini M, Freire E. Thermotropic characterization of phosphatidylcholine vesicles containing ganglioside GM1 with homogeneous ceramide chain length. *Biochemistry.* 1986;25:1043–1049.
31. Masserini M, Palestini P, Venerando B, Fiorilli A, Acquotti D, Tettamanti G. Interactions of proteins with ganglioside-enriched microdomains on the membrane: the lateral phase separation of molecular species of GD1a ganglioside, having homogeneous long-chain base composition, is recognized by Vibrio cholerae sialidase. *Biochemistry.* 1988;27:7973–7978.
32. Masserini M, Palestini P, Freire E. Influence of glycolipid oligosaccharide and long-chain base composition on the thermotropic properties of dipalmitoylphosphatidylcholine large unilamellar vesicles containing gangliosides. *Biochemistry.* 1989;28:5029–5034.
33. Terzaghi A, Tettamanti G, Masserini M. Interaction of glycosphingolipids and glycoproteins: thermotropic properties of model membranes containing GM1 ganglioside and glycophorin. *Biochemistry.* 1993;32:9722–9725.
34. Palestini P, Allietta M, Sonnino S, Tettamanti G, Thompson TE, Tillack TW. Gel phase preference of ganglioside GM1 at low concentration in two-component, two-phase phosphatidylcholine bilayers depends upon the ceramide moiety. *Biochim Biophys Acta.* 1995;1235:221–230.
35. Goins B, Masserini M, Barisas BG, Freire E. Lateral diffusion of ganglioside GM1 in phospholipid bilayer membranes. *Biophys J.* 1986;49:849–856.
36. Palestini P, Masserini M, Tettamanti G. Exposure to galactose oxidase of GM1 ganglioside molecular species embedded into phospholipid vesicles. *FEBS Lett.* 1994;350:219–222.
37. Rock P, Allietta M, Young Jr WW, Thompson TE, Tillack TW. Ganglioside GM1 and asialo-GM1 at low concentration are preferentially incorporated into the gel phase in two-component, two-phase phosphatidylcholine bilayers. *Biochemistry.* 1991;30:19–25.
38. Sonnino S, Cantu L, Corti M, Acquotti D, Venerando B. Aggregative properties of gangliosides in solution. *Chem Phys Lipids.* 1994;71:21–45.
39. Sonnino S, Cantu L, Corti M, Acquotti D, Kirschner G, Tettamanti G. Aggregation properties of semisynthetic GM1 ganglioside (II3Neu5AcGgOse4Cer) containing an acetyl group as acyl moiety. *Chem Phys Lipids.* 1990;56:49–57.
40. Sonnino S, Cantu L, Acquotti D, Corti M, Tettamanti G. Aggregation properties of GM3 ganglioside (II3Neu5AcLacCer) in aqueous solutions. *Chem Phys Lipids.* 1990;52:231–241.
41. Svennerholm L, Bostrom K, Jungbjer B, Olsson L. Membrane lipids of adult human brain: lipid composition of frontal and temporal lobe in subjects of age 20 to 100 years. *J Neurochem.* 1994;63:1802–1811.
42. Prinetti A, Chigorno V, Tettamanti G, Sonnino S. Sphingolipid-enriched membrane domains from rat cerebellar granule cells differentiated in culture. A compositional study. *J Biol Chem.* 2000;275:11658–11665.
43. Grassi S, Prioni S, Cabitta L, Aureli M, Sonnino S, Prinetti A. The role of 3-O-sulfogalactosylceramide, sulfatide, in the lateral organization of myelin membrane. *Neurochem Res.* 2015;41:130–143.
44. van Meer G, de Kroon AI. Lipid map of the mammalian cell. *J Cell Sci.* 2011;124:5–8.
45. van Meer G, Hoetzl S. Sphingolipid topology and the dynamic organization and function of membrane proteins. *FEBS Lett.* 2010;584:1800–1805.
46. Prinetti A, Loberto N, Chigorno V, Sonnino S. Glycosphingolipid behaviour in complex membranes. *Biochim Biophys Acta.* 2009;1788:184–193.

47. Ledeen R, Wu G. New findings on nuclear gangliosides: overview on metabolism and function. *J Neurochem.* 2011;116:714–720.
48. Ferraretto A, Pitto M, Palestini P, Masserini M. Lipid domains in the membrane: thermotropic properties of sphingomyelin vesicles containing GM1 ganglioside and cholesterol. *Biochemistry.* 1997;36:9232–9236.
49. Cantu L, Corti M, Sonnino S, Tettamanti G. Evidence for spontaneous segregation phenomena in mixed micelles of gangliosides. *Chem Phys Lipids.* 1990;55:223–229.
50. Cantù L, Corti M, Casellato R, Acquotti D, Sonnino S. Aggregation properties of GD1b, II3Neu5Ac2GgOse4Cer, and of GD1b-lactone, II3[alpha-Neu5Ac-(2——8, 1——9)-alpha-Neu5Ac]GgOse4Cer, in aqueous solution. *Chem Phys Lipids.* 1991;60:111–118.
51. Del Favero E, Brocca P, Motta S, Rondelli V, Sonnino S, Cantù L. Nanoscale structural response of ganglioside-containing aggregates to the interaction with sialidase. *J Neurochem.* 2010;116:833–839.
52. Acquotti D, Poppe L, Dabrowski J, Vonderlieth CW, Sonnino S, Tettamanti G. 3-Dimensional structure of the oligosaccharide chain of Gm1 ganglioside revealed by a distance-mapping procedure—a rotating and laboratory Ffame nuclear overhauser enhancement investigation of native glycolipid in dimethyl-sulfoxide and in water dodecylphosphocholine solutions. *J Am Chem Soc.* 1990;112:7772–7778.
53. Siebert HC, Reuter G, Schauer R, von der Lieth CW, Dabrowski J. Solution conformations of GM3 gangliosides containing different sialic acid residues as revealed by NOE-based distance mapping, molecular mechanics, and molecular dynamics calculations. *Biochemistry.* 1992;31:6962–6971.
54. Poppe L, van Halbeek H, Acquotti D, Sonnino S. Carbohydrate dynamics at a micellar surface: GD1a headgroup transformations revealed by NMR spectroscopy. *Biophys J.* 1994;66:1642–1652.
55. Jarrell HC, Jovall PA, Giziewicz JB, Turner LA, Smith IC. Determination of conformational properties of glycolipid head groups by 2H NMR of oriented multibilayers. *Biochemistry.* 1987;26:1805–1811.
56. Skarjune R, Oldfield E. Physical studies of cell surface and cell membrane structure. Deuterium nuclear magnetic resonance studies of *N*-palmitoylglucosylceramide (cerebroside) head group structure. *Biochemistry.* 1982;21:3154–3160.
57. Nyholm PG, Pascher I. Orientation of the saccharide chains of glycolipids at the membrane surface: conformational analysis of the glucose-ceramide and the glucose-glyceride linkages using molecular mechanics (MM3). *Biochemistry.* 1993;32:1225–1234.
58. Brocca P, Berthault P, Sonnino S. Conformation of the oligosaccharide chain of G(M1) ganglioside in a carbohydrate-enriched surface. *Biophys J.* 1998;74:309–318.
59. Brocca P, Acquotti D, Sonnino S. Nuclear overhauser effect investigation on GM1 ganglioside containing *N*-glycolyl-neuraminic acid (II3Neu5GcGgOse4Cer). *Glycoconj J.* 1996;13:57–62.
60. Brocca P, Acquotti D, Sonnino S. 1H-NMR study on ganglioside amide protons: evidence that the deuterium exchange kinetics are affected by the preparation of samples. *Glycoconj J.* 1993;10:441–446.
61. Acquotti D, Fronza G, Ragg E, Sonnino S. Three dimensional structure of GD1b and GD1b-monolactone gangliosides in dimethylsulphoxide: a nuclear overhauser effect investigation supported by molecular dynamics calculations. *Chem Phys Lipids.* 1991;59:107–125.
62. Ha JH, Spolar RS, Record Jr MT. Role of the hydrophobic effect in stability of site-specific protein-DNA complexes. *J Mol Biol.* 1989;209:801–816.
63. Bach D, Sela B, Miller IR. Compositional aspects of lipid hydration. *Chem Phys Lipids.* 1982;31:381–394.

64. Maggio B, Ariga T, Sturtevant JM, Yu RK. Thermotropic behavior of glycosphingolipids in aqueous dispersions. *Biochemistry.* 1985;24:1084–1092.
65. Cantu L, Corti M, Del Favero E, Muller E, Raudino A, Sonnino S. Thermal hysteresis in ganglioside micelles investigated by differential scanning calorimetry and light-scattering. *Langmuir.* 1999;15:4975–4980.
66. Koynova R, Caffrey M. Phases and phase transitions of the sphingolipids. *Biochim Biophys Acta.* 1995;1255:213–236.
67. Regina Todeschini A, Hakomori SI. Functional role of glycosphingolipids and gangliosides in control of cell adhesion, motility, and growth, through glycosynaptic microdomains. *Biochim Biophys Acta.* 2008;1780:421–433.
68. Prinetti A, Chigorno V, Prioni S, Loberto N, Marano N, Tettamanti G, Sonnino S. Changes in the lipid turnover, composition, and organization, as sphingolipid-enriched membrane domains, in rat cerebellar granule cells developing in vitro. *J Biol Chem.* 2001;276:21136–21145.
69. Kahya N, Scherfeld D, Schwille P. Differential lipid packing abilities and dynamics in giant unilamellar vesicles composed of short-chain saturated glycerol-phospholipids, sphingomyelin and cholesterol. *Chem Phys Lipids.* 2005;135:169–180.
70. Brown DA, London E. Structure and function of sphingolipid- and cholesterol-rich membrane rafts. *J Biol Chem.* 2000;275:17221–17224.
71. Simons K, Ikonen E. Functional rafts in cell membranes. *Nature.* 1997;387:569–572.
72. Li XM, Momsen MM, Brockman HL, Brown RE. Sterol structure and sphingomyelin acyl chain length modulate lateral packing elasticity and detergent solubility in model membranes. *Biophys J.* 2003;85:3788–3801.
73. Mattjus P, Slotte JP. Does cholesterol discriminate between sphingomyelin and phosphatidylcholine in mixed monolayers containing both phospholipids? *Chem Phys Lipids.* 1996;81:69–80.
74. Ohvo-Rekila H, Ramstedt B, Leppimaki P, Slotte JP. Cholesterol interactions with phospholipids in membranes. *Prog Lipid Res.* 2002;41:66–97.
75. Holopainen JM, Metso AJ, Mattila JP, Jutila A, Kinnunen PK. Evidence for the lack of a specific interaction between cholesterol and sphingomyelin. *Biophys J.* 2004;86:1510–1520.
76. Lingwood D, Ries J, Schwille P, Simons K. Plasma membranes are poised for activation of raft phase coalescence at physiological temperature. *Proc Natl Acad Sci USA.* 2008;105:10005–10010.
77. Kaiser HJ, Lingwood D, Levental I, Sampaio JL, Kalvodova L, Rajendran L, Simons K. Order of lipid phases in model and plasma membranes. *Proc Natl Acad Sci USA.* 2009;106:16645–16650.
78. Pandit SA, Jakobsson E, Scott HL. Simulation of the early stages of nano-domain formation in mixed bilayers of sphingomyelin, cholesterol, and dioleylphosphatidylcholine. *Biophys J.* 2004;87:3312–3322.
79. Mombelli E, Morris R, Taylor W, Fraternali F. Hydrogen-bonding propensities of sphingomyelin in solution and in a bilayer assembly: a molecular dynamics study. *Biophys J.* 2003;84:1507–1517.
80. Ostermeyer AG, Beckrich BT, Ivarson KA, Grove KE, Brown DA. Glycosphingolipids are not essential for formation of detergent-resistant membrane rafts in melanoma cells. methyl-beta-cyclodextrin does not affect cell surface transport of a GPI-anchored protein. *J Biol Chem.* 1999;274:34459–34466.
81. Iwabuchi K, Yamamura S, Prinetti A, Handa K, Hakomori S. GM3-enriched microdomain involved in cell adhesion and signal transduction through carbohydrate-carbohydrate interaction in mouse melanoma B16 cells. *J Biol Chem.* 1998;273:9130–9138.
82. Chigorno V, Palestini P, Sciannamblo M, Dolo V, Pavan A, Tettamanti G, Sonnino S. Evidence that ganglioside enriched domains are distinct from caveolae in MDCK II and human fibroblast cells in culture. *Eur J Biochem.* 2000;267:4187–4197.

83. Pascher I. Molecular arrangements in sphingolipids. Conformation and hydrogen bonding of ceramide and their implication on membrane stability and permeability. *Biochim Biophys Acta.* 1976;455:433–451.
84. Hakomori S, Handa K, Iwabuchi K, Yamamura S, Prinetti A. New insights in glycosphingolipid function: "glycosignaling domain," a cell surface assembly of glycosphingolipids with signal transducer molecules, involved in cell adhesion coupled with signaling. *Glycobiology.* 1998;8:xi–xix.
85. Tsai YT, Yu RK. Epigenetic activation of mouse ganglioside synthase genes: implications for neurogenesis. *J Neurochem.* 2014;128:101–110.
86. Yu RK, Tsai YT, Ariga T. Functional roles of gangliosides in neurodevelopment: an overview of recent advances. *Neurochem Res.* 2012;37:1230–1244.
87. Itokazu Y, Tsai YT, Yu RK. Epigenetic regulation of ganglioside expression in neural stem cells and neuronal cells. *Glycoconj J.* 2016;34:749–756.
88. Inokuchi J, Nagafuku M, Ohno I, Suzuki A. Distinct selectivity of gangliosides required for CD4(+) T and CD8(+) T cell activation. *Biochim Biophys Acta.* 2015;1851:98–106.
89. Colsch B, Jackson SN, Dutta S, Woods AS. Molecular microscopy of brain gangliosides: illustrating their distribution in hippocampal cell layers. *ACS Chem Neurosci.* 2011;2:213–222.
90. Sandhoff K, Kolter T. Biosynthesis and degradation of mammalian glycosphingolipids. *Philos Trans R Soc Lond B Biol Sci.* 2003;358:847–861.
91. D'Angelo G, Capasso S, Sticco L, Russo D. Glycosphingolipids: synthesis and functions. *FEBS J.* 2013;280:6338–6353.
92. Milhas D, Clarke CJ, Hannun YA. Sphingomyelin metabolism at the plasma membrane: implications for bioactive sphingolipids. *FEBS Lett.* 2010;584:1887–1894.
93. Grassme H, Riethmuller J, Gulbins E. Biological aspects of ceramide-enriched membrane domains. *Prog Lipid Res.* 2007;46:161–170.
94. Stancevic B, Kolesnick R. Ceramide-rich platforms in transmembrane signaling. *FEBS Lett.* 2010;584:1728–1740.
95. Schengrund CL, Rosenberg A. Intracellular location and properties of bovine brain sialidase. *J Biol Chem.* 1970;245:6196–6200.
96. Tettamanti G, Morgan IG, Gombos G, Vincendon G, Mandel P. Sub-synaptosomal localization of brain particulate neuraminidose. *Brain Res.* 1972;47:515–518.
97. Tettamanti G, Preti A, Lombardo A, Suman T, Zambotti V. Membrane-bound neuraminidase in the brain of different animals: behaviour of the enzyme on endogenous sialo derivatives and rationale for its assay. *J Neurochem.* 1975;25:451–456.
98. Tettamanti G, Preti A, Lombardo A, Bonali F, Zambotti V. Parallelism of subcellular location of major particulate neuraminidase and gangliosides in rabbit brain cortex. *Biochim Biophys Acta.* 1973;306:466–477.
99. Preti A, Fiorilli A, Lombardo A, Caimi L, Tettamanti G. Occurrence of sialyltransferase activity in the synaptosomal membranes prepared from calf brain cortex. *J Neurochem.* 1980;35:281–296.
100. Miyagi T, Sagawa J, Konno K, Handa S, Tsuiki S. Biochemical and immunological studies on two distinct ganglioside-hydrolyzing sialidases from the particulate fraction of rat brain. *J Biochem.* 1990;107:787–793.
101. Miyagi T, Sagawa J, Konno K, Tsuiki S. Immunological discrimination of intralysosomal, cytosolic, and two membrane sialidases present in rat tissues. *J Biochem.* 1990;107:794–798.
102. Schneider-Jakob HR, Cantz M. Lysosomal and plasma membrane ganglioside GM3 sialidases of cultured human fibroblasts. Differentiation by detergents and inhibitors. *Biol Chem Hoppe Seyler.* 1991;372:443–450.
103. Kopitz J, von Reitzenstein C, Muhl C, Cantz M. Role of plasma membrane ganglioside sialidase of human neuroblastoma cells in growth control and differentiation. *Biochem Biophys Res Commun.* 1994;199:1188–1193.

104. Riboni L, Prinetti A, Bassi R, Tettamanti G. Cerebellar granule cells in culture exhibit a ganglioside-sialidase presumably linked to the plasma membrane. *FEBS Lett.* 1991;287:42–46.
105. Kopitz J, Muhl C, Ehemann V, Lehmann C, Cantz M. Effects of cell surface ganglioside sialidase inhibition on growth control and differentiation of human neuroblastoma cells. *Eur J Cell Biol.* 1997;73:1–9.
106. Kopitz J, Sinz K, Brossmer R, Cantz M. Partial characterization and enrichment of a membrane-bound sialidase specific for gangliosides from human brain tissue. *Eur J Biochem.* 1997;248:527–534.
107. Hata K, Wada T, Hasegawa A, Kiso M, Miyagi T. Purification and characterization of a membrane-associated ganglioside sialidase from bovine brain. *J Biochem.* 1998;123:899–905.
108. Wada T, Yoshikawa Y, Tokuyama S, Kuwabara M, Akita H, Miyagi T. Cloning, expression, and chromosomal mapping of a human ganglioside sialidase. *Biochem Biophys Res Commun.* 1999;261:21–27.
109. Miyagi T, Wada T, Iwamatsu A, Hata K, Yoshikawa Y, Tokuyama S, Sawada M. Molecular cloning and characterization of a plasma membrane-associated sialidase specific for gangliosides. *J Biol Chem.* 1999;274:5004–5011.
110. Hasegawa T, Yamaguchi K, Wada T, Takeda A, Itoyama Y, Miyagi T. Molecular cloning of mouse ganglioside sialidase and its increased expression in neuro2a cell differentiation. *J Biol Chem.* 2000;275:14778.
111. Monti E, Bassi MT, Papini N, Riboni M, Manzoni M, Venerando B, Croci G, Preti A, Ballabio A, Tettamanti G, Borsani G. Identification and expression of NEU3, a novel human sialidase associated to the plasma membrane. *Biochem J.* 2000;349:343–351.
112. Miyagi T, Yamaguchi K. Mammalian sialidases: physiological and pathological roles in cellular functions. *Glycobiology.* 2012;22:880–896.
113. Kalka D, von Reitzenstein C, Kopitz J, Cantz M. The plasma membrane ganglioside sialidase cofractionates with markers of lipid rafts. *Biochem Biophys Res Commun.* 2001;283:989–993.
114. Wang Y, Yamaguchi K, Wada T, Hata K, Zhao X, Fujimoto T, Miyagi T. A close association of the ganglioside-specific sialidase Neu3 with caveolin in membrane microdomains. *J Biol Chem.* 2002;277:26252–26259.
115. Bonardi D, Papini N, Pasini M, Dileo L, Orizio F, Monti E, Caimi L, Venerando B, Bresciani R. Sialidase NEU3 dynamically associates to different membrane domains specifically modifying their ganglioside pattern and triggering Akt phosphorylation. *PLoS One.* 2014;9:e99405.
116. Cirillo F, Ghiroldi A, Fania C, Piccoli M, Torretta E, Tettamanti G, Gelfi C, Anastasia L. NEU3 sialidase protein interactors in the plasma membrane and in the endosomes. *J Biol Chem.* 2016;291:10615–10624.
117. Shiozaki K, Takahashi K, Hosono M, Yamaguchi K, Hata K, Shiozaki M, Bassi R, Prinetti A, Sonnino S, Nitta K, Miyagi T. Phosphatidic acid-mediated activation and translocation to the cell surface of sialidase NEU3, promoting signaling for cell migration. *FASEB J.* 2015;29:2099–2111.
118. Ledeen RW. Biology of gangliosides: neuritogenic and neuronotrophic properties. *J Neurosci Res.* 1984;12:147–159.
119. Ferrari G, Fabris M, Gorio A. Gangliosides enhance neurite outgrowth in PC12 cells. *Brain Res.* 1983;284:215–221.
120. Favaron M, Manev H, Alho H, Bertolino M, Ferret B, Guidotti A, Costa E. Gangliosides prevent glutamate and kainate neurotoxicity in primary neuronal cultures of neonatal rat cerebellum and cortex. *Proc Natl Acad Sci USA.* 1988;85:7351–7355.

121. Skaper SD, Facci L, Milani D, Leon A. Monosialoganglioside GM1 protects against anoxia-induced neuronal death in vitro. *Exp Neurol*. 1989;106:297–305.
122. Ferrari G, Anderson BL, Stephens RM, Kaplan DR, Greene LA. Prevention of apoptotic neuronal death by GM1 ganglioside. Involvement of Trk neurotrophin receptors. *J Biol Chem*. 1995;270:3074–3080.
123. Ferrari G, Batistatou A, Greene LA. Gangliosides rescue neuronal cells from death after trophic factor deprivation. *J Neurosci*. 1993;13:1879–1887.
124. Hadjiconstantinou M, Neff NH. GM1 ganglioside: in vivo and in vitro trophic actions on central neurotransmitter systems. *J Neurochem*. 1998;70:1335–1345.
125. Aureli M, Mauri L, Ciampa MG, Prinetti A, Toffano G, Secchieri C, Sonnino S. GM1 ganglioside: past studies and future potential. *Mol Neurobiol*. 2016;53:1824–1842.
126. Guirland C, Suzuki S, Kojima M, Lu B, Zheng JQ. Lipid rafts mediate chemotropic guidance of nerve growth cones. *Neuron*. 2004;42:51–62.
127. Suzuki S, Numakawa T, Shimazu K, Koshimizu H, Hara T, Hatanaka H, Mei L, Lu B, Kojima M. BDNF-induced recruitment of TrkB receptor into neuronal lipid rafts: roles in synaptic modulation. *J Cell Biol*. 2004;167:1205–1215.
128. Hibbert AP, Kramer BM, Miller FD, Kaplan DR. The localization, trafficking and retrograde transport of BDNF bound to p75NTR in sympathetic neurons. *Mol Cell Neurosci*. 2006;32:387–402.
129. Mihara T, Ueda A, Hirayama M, Takeuchi T, Yoshida S, Naito K, Yamamoto H, Mutoh T. Detection of new anti-neutral glycosphingolipids antibodies and their effects on Trk neurotrophin receptors. *FEBS Lett*. 2006;580:4991–4995.
130. Mojsilovic-Petrovic J, Jeong GB, Crocker A, Arneja A, David S, Russell DS, Kalb RG. Protecting motor neurons from toxic insult by antagonism of adenosine A2a and Trk receptors. *J Neurosci*. 2006;26:9250–9263.
131. Pereira DB, Chao MV. The tyrosine kinase Fyn determines the localization of TrkB receptors in lipid rafts. *J Neurosci*. 2007;27:4859–4869.
132. Mutoh T, Hamano T, Tokuda A, Kuriyama M. Unglycosylated Trk protein does not co-localize nor associate with ganglioside GM1 in stable clone of PC12 cells overexpressing Trk (PCtrk cells). *Glycoconj J*. 2000;17:233–237.
133. Proshin S, Yamaguchi K, Wada T, Miyagi T. Modulation of neuritogenesis by ganglioside-specific sialidase (Neu 3) in human neuroblastoma NB-1 cells. *Neurochem Res*. 2002;27:841–846.
134. Valaperta R, Valsecchi M, Rocchetta F, Aureli M, Prioni S, Prinetti A, Chigorno V, Sonnino S. Induction of axonal differentiation by silencing plasma membrane-associated sialidase Neu3 in neuroblastoma cells. *J Neurochem*. 2007;100:708–719.
135. Rodriguez JA, Piddini E, Hasegawa T, Miyagi T, Dotti CG. Plasma membrane ganglioside sialidase regulates axonal growth and regeneration in hippocampal neurons in culture. *J Neurosci*. 2001;21:8387–8395.
136. Da Silva JS, Hasegawa T, Miyagi T, Dotti CG, Abad-Rodriguez J. Asymmetric membrane ganglioside sialidase activity specifies axonal fate. *Nat Neurosci*. 2005;8:606–615.
137. Valaperta R, Chigorno V, Basso L, Prinetti A, Bresciani R, Preti A, Miyagi T, Sonnino S. Plasma membrane production of ceramide from ganglioside GM3 in human fibroblasts. *FASEB J*. 2006;20:1227–1229.
138. Sonnino S, Aureli M, Loberto N, Chigorno V, Prinetti A. Fine tuning of cell functions through remodeling of glycosphingolipids by plasma membrane-associated glycohydrolases. *FEBS Lett*. 2010;584:1914–1922.
139. Aureli M, Masilamani AP, Illuzzi G, Loberto N, Scandroglio F, Prinetti A, Chigorno V, Sonnino S. Activity of plasma membrane beta-galactosidase and beta-glucosidase. *FEBS Lett*. 2009;583:2469–2473.
140. Mencarelli S, Cavalieri C, Magini A, Tancini B, Basso L, Lemansky P, Hasilik A, Li YT, Chigorno V, Orlacchio A, Emiliani C, Sonnino S. Identification of plasma membrane

associated mature beta-hexosaminidase A, active towards GM2 ganglioside, in human fibroblasts. *FEBS Lett.* 2005;579:5501–5506.
141. Boot RG, Verhoek M, Donker-Koopman W, Strijland A, van Marle J, Overkleeft HS, Wennekes T, Aerts JM. Identification of the non-lysosomal glucosylceramidase as beta-glucosidase 2. *J Biol Chem.* 2007;282:1305–1312.
142. Matsui Y, Lombard D, Massarelli R, Mandel P, Dreyfus H. Surface glycosyltransferase activities during development of neuronal cell cultures. *J Neurochem.* 1986;46:144–150.
143. Durrie R, Saito M, Rosenberg A. Endogenous glycosphingolipid acceptor specificity of sialosyltransferase systems in intact Golgi membranes, synaptosomes, and synaptic plasma membranes from rat brain. *Biochemistry.* 1988;27:3759–3764.
144. Durrie R, Rosenberg A. Anabolic sialosylation of gangliosides in situ in rat brain cortical slices. *J Lipid Res.* 1989;30:1259–1266.
145. Iwamori M, Iwamori Y. Changes in the glycolipid composition and characteristic activation of GM3 synthase in the thymus of mouse after administration of dexamethasone. *Glycoconj J.* 2005;22:119–126.
146. Ghidoni R, Sonnino S, Chigorno V, Malesci A, Tettamanti G. Comparative and developmental behavior of alkali labile gangliosides in the brain. *Adv Exp Med Biol.* 1984;174:307–318.
147. Bassi R, Riboni L, Sonnino S, Tettamanti G. Lactonization of GD1b ganglioside under acidic conditions. *Carbohydr Res.* 1989;193:141–146.
148. Sonnino S, Ghidoni R, Chigorno V, Masserini M, Tettamanti G. Recognition by two-dimensional thin-layer chromatography and densitometric quantification of alkali-labile gangliosides from the brain of different animals. *Anal Biochem.* 1983;128:104–114.
149. Riboni L, Malesci A, Gaini SM, Sonnino S, Ghidoni R, Tettamanti G. Ganglioside pattern of normal human brain, from samples obtained at surgery. A study especially referred to alkali labile species. *J Biochem.* 1984;96:1943–1946.
150. Sonnino S, Acquotti D, Fronza G, Kirschner G, Ghidoni R, Tettamanti G, eds. In: *Biochemistry of gangliosides lactones.* Berlin Heidelberg, New York, Tokyo: Liviana Press Springer Verlag; 1988.
151. Bassi R, Chigorno V, Fiorilli A, Sonnino S, Tettamanti G. Exogenous gangliosides GD1b and GD1b-lactone, stably associated to rat brain P2 subcellular fraction, modulate differently the process of protein phosphorylation. *J Neurochem.* 1991;57:1207–1211.
152. Palestini P, Pitto M, Ferraretto A, Tettamanti G, Masserini M. Change of ganglioside accessibility at the plasma membrane surface of cultured neurons, following protein kinase C activation. *Biochemistry.* 1998;37:3143–3148.
153. Schauer R, Srinivasan GV, Wipfler D, Kniep B, Schwartz-Albiez R. O-Acetylated sialic acids and their role in immune defense. *Adv Exp Med Biol.* 2011;705:525–548.
154. Kamerling JP, Vliegenthart JF. Identification of O-acetylated *N*-acylneuraminic acids by mass spectrometry. *Carbohydr Res.* 1975;41:7–17.
155. Cheresh DA, Reisfeld RA, Varki AP. O-acetylation of disialoganglioside GD3 by human melanoma cells creates a unique antigenic determinant. *Science.* 1984;225:844–846.
156. Hakomori S, Saito T. Isolation and characterization of a glycosphingolipid having a new sialic acid. *Biochemistry.* 1969;8:5082–5088.
157. Yachida Y, Tsuchihashi K, Gasa S. Novel di-O-acetylated GM3s from equine erythrocytes, one containing 4,9-di-O-acetyl-*N*-glycolylneuraminic acid and another containing 4-O-acetyl-*N*-glycolylneuraminic acid and 6-O-acetyl-D-galactose. *Carbohydr Res.* 1997;298:201–212.
158. Kohla G, Stockfleth E, Schauer R. Gangliosides with O-acetylated sialic acids in tumors of neuroectodermal origin. *Neurochem Res.* 2002;27:583–592.
159. Mandal C, Schwartz-Albiez R, Vlasak R. Functions and biosynthesis of O-acetylated sialic acids. *Top Curr Chem.* 2015;366:1–30.

160. Shi WX, Chammas R, Varki A. Regulation of sialic acid 9-O-acetylation during the growth and differentiation of murine erythroleukemia cells. *J Biol Chem.* 1996;271:31517–31525.
161. Sonnino S, Ghidoni R, Malesci A, Tettamanti G, Marx J, Hilbig R, Rahmann H. Nervous system ganglioside composition of normothermic and hibernating dormice (*Glis glis*). *Neurochem Int.* 1984;6:677–683.
162. Hakomori S. Aberrant glycosylation in cancer cell membranes as focused on glycolipids: overview and perspectives. *Cancer Res.* 1985;45:2405–2414.
163. Hakomori S. Glycosylation defining cancer malignancy: new wine in an old bottle. *Proc Natl Acad Sci USA.* 2002;99:10231–10233.
164. Hakomori SI, Handa K. GM3 and cancer. *Glycoconj J.* 2015;32:1–8.
165. Kakugawa Y, Wada T, Yamaguchi K, Yamanami H, Ouchi K, Sato I, Miyagi T. Up-regulation of plasma membrane-associated ganglioside sialidase (Neu3) in human colon cancer and its involvement in apoptosis suppression. *Proc Natl Acad Sci USA.* 2002;99:10718–10723.
166. Noguchi M, Suzuki T, Kabayama K, Takahashi H, Chiba H, Shiratori M, Abe S, Watanabe A, Satoh M, Hasegawa T, Tagami S, Ishii A, Saitoh M, Kaneko M, Iseki K, Igarashi Y, Inokuchi J. GM3 synthase gene is a novel biomarker for histological classification and drug sensitivity against epidermal growth factor receptor tyrosine kinase inhibitors in non-small cell lung cancer. *Cancer Sci.* 2007;98:1625–1632.
167. Bremer EG, Schlessinger J, Hakomori S. Ganglioside-mediated modulation of cell growth. Specific effects of GM3 on tyrosine phosphorylation of the epidermal growth factor receptor. *J Biol Chem.* 1986;261:2434–2440.
168. Miljan EA, Meuillet EJ, Mania-Farnell B, George D, Yamamoto H, Simon HG, Bremer EG. Interaction of the extracellular domain of the epidermal growth factor receptor with gangliosides. *J Biol Chem.* 2002;277:10108–10113.
169. Yoon SJ, Nakayama K, Hikita T, Handa K, Hakomori SI. Epidermal growth factor receptor tyrosine kinase is modulated by GM3 interaction with N-linked GlcNAc termini of the receptor. *Proc Natl Acad Sci USA.* 2006;103:18987–18991.
170. Yoon SJ, Nakayama K, Takahashi N, Yagi H, Utkina N, Wang HY, Kato K, Sadilek M, Hakomori SI. Interaction of N-linked glycans, having multivalent GlcNAc termini, with GM3 ganglioside. *Glycoconj J.* 2006;23:639–649.
171. Ringerike T, Blystad FD, Levy FO, Madshus IH, Stang E. Cholesterol is important in control of EGF receptor kinase activity but EGF receptors are not concentrated in caveolae. *J Cell Sci.* 2002;115:1331–1340.
172. Roepstorff K, Thomsen P, Sandvig K, van Deurs B. Sequestration of epidermal growth factor receptors in non-caveolar lipid rafts inhibits ligand binding. *J Biol Chem.* 2002;277:18954–18960.
173. Pike LJ. Growth factor receptors, lipid rafts and caveolae: an evolving story. *Biochim Biophys Acta.* 2005;1746:260–273.
174. Pike LJ, Han X, Gross RW. Epidermal growth factor receptors are localized to lipid rafts that contain a balance of inner and outer leaflet lipids: a shotgun lipidomics study. *J Biol Chem.* 2005;280:26796–26804.
175. Prinetti A, Prioni S, Loberto N, Aureli M, Chigorno V, Sonnino S. Regulation of tumor phenotypes by caveolin-1 and sphingolipid-controlled membrane signaling complexes. *Biochim Biophys Acta.* 2008;1780:585–596.
176. Waugh MG, Lawson D, Hsuan JJ. Epidermal growth factor receptor activation is localized within low-buoyant density, non-caveolar membrane domains. *Biochem J.* 1999;337(pt 3):591–597.
177. Waugh MG, Minogue S, Anderson JS, dos Santos M, Hsuan JJ. Signalling and non-caveolar rafts. *Biochem Soc Trans.* 2001;29:509–511.

178. Wang X-Q, Sun P, Paller AS. Ganglioside induces Ccveolin-1 redistribution and interaction with the epidermal growth factor receptor. *J Biol Chem.* 2002;277:47028–47034.
179. Wang XQ, Sun P, Paller AS. Ganglioside GM3 blocks the activation of epidermal growth factor receptor induced by integrin at specific tyrosine sites. *J Biol Chem.* 2003;278:48770–48778.
180. Hakomori S. Glycosylation defining cancer malignancy: new wine in an old bottle. *PNAS.* 2002;99:10231–10233.
181. Hakomori S. Glycosynapses: microdomains controlling carbohydrate-dependent cell adhesion and signaling. *An Acad Bras Cienc.* 2004;76:553–572.
182. Hakomori S, Handa K. Glycosphingolipid-dependent cross-talk between glycosynapses interfacing tumor cells with their host cells: essential basis to define tumor malignancy. *FEBS Letters.* 2002;531:88–92.
183. Ono M, Handa K, Withers DA, Hakomori S-i. Glycosylation effect on membrane domain (GEM) involved in cell adhesion and motility: a preliminary note on functional [alpha]3, [alpha]5-CD82 glycosylation complex in ldlD 14 cells. *Biochem Biophys Res Commun.* 2000;279:744–750.
184. Kawakami Y, Kawakami K, Steelant WF, Ono M, Baek RC, Handa K, Withers DA, Hakomori S. Tetraspanin CD9 is a "proteolipid," and its interaction with alpha 3 integrin in microdomain is promoted by GM3 ganglioside, leading to inhibition of laminin-5-dependent cell motility. *J Biol Chem.* 2002;277:34349–34358.
185. Hemler ME. Integrin associated proteins. *Curr Opin Cell Biol.* 1998;10:578–585.
186. Kawakami Y, Kawakami K, Steelant WFA, Ono M, Baek RC, Handa K, Withers DA, Hakomori S. Tetraspanin CD9 is a "proteolipid," and its interaction with alpha 3 integrin in microdomain is promoted by GM3 ganglioside, leading to inhibition of laminin-5-dependent cell motility. *J Biol Chem.* 2002;277:34349–34358.
187. Miura Y, Kainuma M, Jiang H, Velasco H, Vogt PK, Hakomori S. Reversion of the Jun-induced oncogenic phenotype by enhanced synthesis of sialosyllactosylceramide (GM3 ganglioside). *Proc Natl Acad Sci USA.* 2004;101:16204–16209.
188. Mitsuzuka K, Handa K, Satoh M, Arai Y, Hakomori S. A specific microdomain ("glycosynapse 3") controls phenotypic conversion and reversion of bladder cancer cells through GM3-mediated interaction of alpha3beta1 integrin with CD9. *J Biol Chem.* 2005;280:35545–35553.
189. Todeschini AR, Dos Santos JN, Handa K, Hakomori S-i. Ganglioside GM2-tetraspanin CD82 complex inhibits met and its cross-talk with integrins, providing a basis for control of cell motility through glycosynapse. *J Biol Chem.* 2007;282:8123–8133.
190. Prinetti A, Iwabuchi K, Hakomori S. Glycosphingolipid-enriched signaling domain in mouse neuroblastoma Neuro2a cells. Mechanism of ganglioside-dependent neuritogenesis. *J Biol Chem.* 1999;274:20916–20924.
191. Prinetti A, Aureli M, Illuzzi G, Prioni S, Nocco V, Scandroglio F, Gagliano N, Tredici G, Rodriguez-Menendez V, Chigorno V, Sonnino S. GM3 synthase overexpression results in reduced cell motility and in caveolin-1 upregulation in human ovarian carcinoma cells. *Glycobiology.* 2010;20:62–77.
192. Prinetti A, Prioni S, Loberto N, Aureli M, Chigorno V, Sonnino S. Caveolin-1 and sphingolipid-controlled membrane signaling complexes regulating the tumor phenotype. *Biochim Biophys Acta.* 2007.
193. Prinetti A, Cao T, Illuzzi G, Prioni S, Aureli M, Gagliano N, Tredici G, Rodriguez-Menendez V, Chigorno V, Sonnino S. A glycosphingolipid/caveolin-1 signaling complex inhibits motility of human ovarian carcinoma cells. *J Biol Chem.* 2011;286:40900–40910.

194. Sonnino S, Prinetti A. Sphingolipids and membrane environments for caveolin. *FEBS Lett.* 2009;583:597–606.
195. Benistant C, Chapuis H, Mottet N, Noletti J, Crapez E, Bali JP, Roche S. Deregulation of the cytoplasmic tyrosine kinase cSrc in the absence of a truncating mutation at codon 531 in human bladder carcinoma. *Biochem Biophys Res Commun.* 2000;273:425–430.
196. Benistant C, Bourgaux J-F, Chapuis H, Mottet N, Roche S, Bali J-P. The COOH-terminal Src kinase Csk is a tumor antigen in human carcinoma. *Cancer Res.* 2001;61:1415–1420.
197. Wary KK, Mariotti A, Zurzolo C, Giancotti FG. A requirement for caveolin-1 and associated kinase Fyn in integrin signaling and anchorage-dependent cell growth. *Cell.* 1998;94:625–634.
198. Trimmer C, Whitaker-Menezes D, Bonuccelli G, Milliman JN, Daumer KM, Aplin AE, Pestell RG, Sotgia F, Lisanti MP, Capozza F. CAV1 inhibits metastatic potential in melanomas through suppression of the integrin/Src/FAK signaling pathway. *Cancer Res.* 2010;70:7489–7499.
199. Ledesma MD, Martin MG, Dotti CG. Lipid changes in the aged brain: effect on synaptic function and neuronal survival. *Prog Lipid Res.* 2012;51:23–35.
200. Diaz ML, Fabelo N, Marin R. Genotype-induced changes in biophysical properties of frontal cortex lipid raft from APP/PS1 transgenic mice. *Front Physiol.* 2012;3:454.
201. Fabelo N, Martin V, Marin R, Santpere G, Aso E, Ferrer I, Diaz M. Evidence for premature lipid raft aging in APP/PS1 double-transgenic mice, a model of familial Alzheimer disease. *J Neuropathol Exp Neurol.* 2012;71:868–881.
202. Piccinini M, Scandroglio F, Prioni S, Buccinna B, Loberto N, Aureli M, Chigorno V, Lupino E, DeMarco G, Lomartire A, Rinaudo MT, Sonnino S, Prinetti A. Deregulated sphingolipid metabolism and membrane organization in neurodegenerative disorders. *Mol Neurobiol.* 2010;41:314–340.
203. Sonnino S, Aureli M, Grassi S, Mauri L, Prioni S, Prinetti A. Lipid rafts in neurodegeneration and neuroprotection. *Mol Neurobiol.* 2014;50:130–148.
204. Reddy A, Caler EV, Andrews NW. Plasma membrane repair is mediated by Ca(2 +)-regulated exocytosis of lysosomes. *Cell.* 2001;106:157–169.
205. Scandroglio F, Venkata JK, Loberto N, Prioni S, Schuchman EH, Chigorno V, Prinetti A, Sonnino S. Lipid content of brain, brain membrane lipid domains, and neurons from acid sphingomyelinase deficient mice. *J Neurochem.* 2008;107:329–338.
206. Buccinna B, Piccinini M, Prinetti A, Scandroglio F, Prioni S, Valsecchi M, Votta B, Grifoni S, Lupino E, Ramondetti C, Schuchman EH, Giordana MT, Sonnino S, Rinaudo MT. Alterations of myelin-specific proteins and sphingolipids characterize the brains of acid sphingomyelinase-deficient mice, an animal model of Niemann-Pick disease type A. *J Neurochem.* 2009;109:105–115.
207. Ledesma MD, Prinetti A, Sonnino S, Schuchman EH. Brain pathology in Niemann Pick disease type A: insights from the acid sphingomyelinase knockout mice. *J Neurochem.* 2011;116:779–788.
208. Prinetti A, Prioni S, Chiricozzi E, Schuchman EH, Chigorno V, Sonnino S. Secondary alterations of sphingolipid metabolism in lysosomal storage diseases. *Neurochem Res.* 2011;36:1654–1668.
209. Vance JE. Dysregulation of cholesterol balance in the brain: contribution to neurodegenerative diseases. *Dis Model Mech.* 2012;5:746–755.
210. Ariga T, McDonald MP, Yu RK. Role of ganglioside metabolism in the pathogenesis of Alzheimer's disease: a review. *J Lipid Res.* 2008;49:1157–1175.
211. Svennerholm L, Gottfries CG. Membrane lipids, selectively diminished in Alzheimer brains, suggest synapse loss as a primary event in early-onset form (type I) and demyelination in late-onset form (type II). *J Neurochem.* 1994;62:1039–1047.

212. Kalanj S, Kracun I, Rosner H, Cosovic C. Regional distribution of brain gangliosides in Alzheimer's disease. *Neurol Croat.* 1991;40:269–281.
213. Brooksbank BW, McGovern J. Gangliosides in the brain in adult Down's syndrome and Alzheimer's disease. *Mol Chem Neuropathol.* 1989;11:143–156.
214. Kracun I, Rosner H, Drnovsek V, Heffer-Lauc M, Cosovic C, Lauc G. Human brain gangliosides in development, aging and disease. *Int J Dev Biol.* 1991;35:289–295.
215. Kracun I, Kalanj S, Talan-Hranilovic J, Cosovic C. Cortical distribution of gangliosides in Alzheimer's disease. *Neurochem Int.* 1992;20:433–438.
216. Crino PB, Ullman MD, Vogt BA, Bird ED, Volicer L. Brain gangliosides in dementia of the Alzheimer type. *Arch Neurol.* 1989;46:398–401.
217. Katsel P, Li C, Haroutunian V. Gene expression alterations in the sphingolipid metabolism pathways during progression of dementia and Alzheimer's disease: a shift toward ceramide accumulation at the earliest recognizable stages of Alzheimer's disease? *Neurochem Res.* 2007;32:845–856.
218. Molander-Melin M, Blennow K, Bogdanovic N, Dellheden B, Mansson JE, Fredman P. Structural membrane alterations in Alzheimer brains found to be associated with regional disease development; increased density of gangliosides GM1 and GM2 and loss of cholesterol in detergent-resistant membrane domains. *J Neurochem.* 2005;92:171–182.
219. Nalivaeva NN, Turner AJ. The amyloid precursor protein: a biochemical enigma in brain development, function and disease. *FEBS Lett.* 2013;587:2046–2054.
220. Muller UC, Zheng H. Physiological functions of APP family proteins. *Cold Spring Harb Perspect Med.* 2012;2:a006288.
221. Brouillet E, Trembleau A, Galanaud D, Volovitch M, Bouillot C, Valenza C, Prochiantz A, Allinquant B. The amyloid precursor protein interacts with Go heterotrimeric protein within a cell compartment specialized in signal transduction. *J Neurosci.* 1999;19:1717–1727.
222. Ashall F, Goate AM. Role of the beta-amyloid precursor protein in Alzheimer's disease. *Trends Biochem Sci.* 1994;19:42–46.
223. Hartmann T, Kuchenbecker J, Grimm MO. Alzheimer's disease: the lipid connection. *J Neurochem.* 2007;103(Suppl 1):159–170.
224. Yamaguchi T, Yamauchi Y, Furukawa K, Ohmi Y, Ohkawa Y, Zhang Q, Okajima T, Furukawa K. Expression of B4GALNT1, an essential glycosyltransferase for the synthesis of complex gangliosides, suppresses BACE1 degradation and modulates APP processing. *Sci Rep.* 2016;6:34505.
225. Grimm MO, Grimm HS, Patzold AJ, Zinser EG, Halonen R, Duering M, Tschape JA, De Strooper B, Muller U, Shen J, Hartmann T. Regulation of cholesterol and sphingomyelin metabolism by amyloid-beta and presenilin. *Nat Cell Biol.* 2005;7:1118–1123.
226. Grimm MO, Grosgen S, Rothhaar TL, Burg VK, Hundsdorfer B, Haupenthal VJ, Friess P, Muller U, Fassbender K, Riemenschneider M, Grimm HS, Hartmann T. Intracellular APP domain regulates serine-palmitoyl-CoA transferase expression and is affected in Alzheimer's disease. *Int J Alzheimers Dis.* 2011;2011:695413.
227. Minami SS, Hoe HS, Rebeck GW. Fyn kinase regulates the association between amyloid precursor protein and Dab1 by promoting their localization to detergent-resistant membranes. *J Neurochem.* 2011;118:879–890.
228. Kasahara K, Watanabe Y, Yamamoto T, Sanai Y. Association of Src family tyrosine kinase Lyn with ganglioside GD3 in rat brain. *Possible regulation of Lyn by glycosphingolipid in caveolae-like domains J Biol Chem.* 1997;272:29947–29953.
229. Hartmann T, Prinetti A. Going the wrong road: Fyn and targeting of amyloid precursor protein to lipid rafts. *J Neurochem.* 2011;118:677–679.
230. Manna M, Mukhopadhyay C. Binding, conformational transition and dimerization of amyloid-beta peptide on GM1-containing ternary membrane: insights from molecular dynamics simulation. *PLoS One.* 2013;8:e71308.

231. Beel AJ, Sakakura M, Barrett PJ, Sanders CR. Direct binding of cholesterol to the amyloid precursor protein: an important interaction in lipid-Alzheimer's disease relationships? *Biochim Biophys Acta.* 2010;1801:975–982.
232. Fantini J, Yahi N, Garmy N. Cholesterol accelerates the binding of Alzheimer's beta-amyloid peptide to ganglioside GM1 through a universal hydrogen-bond-dependent sterol tuning of glycolipid conformation. *Front Physiol.* 2013;4:120.
233. Di Scala C, Yahi N, Lelievre C, Garmy N, Chahinian H, Fantini J. Biochemical identification of a linear cholesterol-binding domain within Alzheimer's beta amyloid peptide. *ACS Chem Neurosci.* 2013;4:509–517.
234. Yanagisawa K, Odaka A, Suzuki N, Ihara Y. GM1 ganglioside-bound amyloid beta-protein (A beta): a possible form of preamyloid in Alzheimer's disease. *Nat Med.* 1995;1:1062–1066.
235. Kakio A, Nishimoto SI, Yanagisawa K, Kozutsumi Y, Matsuzaki K. Cholesterol-dependent formation of GM1 ganglioside-bound amyloid beta-protein, an endogenous seed for Alzheimer amyloid. *J Biol Chem.* 2001;276:24985–24990.
236. Kim SI, Yi JS, Ko YG. Amyloid beta oligomerization is induced by brain lipid rafts. *J Cell Biochem.* 2006;99:878–889.
237. Yanagisawa K, Fantini J, Chakrabartty A, Eckert A. Abeta behavior on neuronal membranes: aggregation and toxicities. *Int J Alzheimers Dis.* 2011;2011:286536.
238. Ikeda K, Yamaguchi T, Fukunaga S, Hoshino M, Matsuzaki K. Mechanism of amyloid beta-protein aggregation mediated by GM1 ganglioside clusters. *Biochemistry.* 2011;50:6433–6440.
239. Rushworth JV, Griffiths HH, Watt NT, Hooper NM. Prion protein-mediated toxicity of amyloid-beta oligomers requires lipid rafts and the transmembrane LRP1. *J Biol Chem.* 2013;288:8935–8951.
240. Tagami S, Inokuchi Ji J, Kabayama K, Yoshimura H, Kitamura F, Uemura S, Ogawa C, Ishii A, Saito M, Ohtsuka Y, Sakaue S, Igarashi Y. Ganglioside GM3 participates in the pathological conditions of insulin resistance. *J Biol Chem.* 2002;277:3085–3092.
241. Kabayama K, Sato T, Kitamura F, Uemura S, Kang BW, Igarashi Y, Inokuchi J. TNFalpha-induced insulin resistance in adipocytes as a membrane microdomain disorder: involvement of ganglioside GM3. *Glycobiology.* 2005;15:21–29.
242. Gustavsson J, Parpal S, Karlsson M, Ramsing C, Thorn H, Borg M, Lindroth M, Peterson KH, Magnusson KE, Stralfors P. Localization of the insulin receptor in caveolae of adipocyte plasma membrane. *Faseb J.* 1999;13:1961–1971.
243. Couet J, Li S, Okamoto T, Ikezu T, Lisanti MP. Identification of peptide and protein ligands for the caveolin-scaffolding domain. Implications for the interaction of caveolin with caveolae-associated proteins. *J Biol Chem.* 1997;272:6525–6533.
244. Kabayama K, Sato T, Saito K, Loberto N, Prinetti A, Sonnino S, Kinjo M, Igarashi Y, Inokuchi J. Dissociation of the insulin receptor and caveolin-1 complex by ganglioside GM3 in the state of insulin resistance. *Proc Natl Acad Sci USA.* 2007;104:13678–13683.
245. Veillon L, Go S, Matsuyama W, Suzuki A, Nagasaki M, Yatomi Y, Inokuchi J. Identification of ganglioside GM3 molecular species in human serum associated with risk factors of metabolic syndrome. *PLoS One.* 2015;10:e0129645.
246. Inokuchi J. GM3 and diabetes. *Glycoconj J.* 2014;31:193–197.

CHAPTER FOUR

# Biological and Pathological Roles of Ganglioside Sialidases

**Taeko Miyagi*[,1], Kohta Takahashi**, Koji Yamamoto*[,†], Kazuhiro Shiozaki[‡], Kazunori Yamaguchi***

[*] Miyagi Cancer Center Research Institute, Natori, Japan
[**] Tohoku Medical and Pharmaceutical University, Sendai, Japan
[†] Saitama Medical University, Moroyama, Saitama, Japan
[‡] Kagoshima University, Kagoshima, Japan
[1] Corresponding author. E-mail address: miyagi-ta@nifty.com

## Contents



## Abstract

Sialidases are glycosidases responsible for the removal of α-glycosidically linked sialic acid residues from carbohydrate portions of glycoproteins and glycolipids, this process being the initial step in the degradation of such glycoconjugates. Sialic acids are considered to play important roles in various biological processes largely in two ways, one related to their hydrophilic and acidic properties exerting physicochemical effects on the glycoconjugates to which they are attached, and the other as recognition sites or in an opposing fashion as masking sites. The removal of sialic acids catalyzed by a sialidase, therefore greatly influences many biological processes through changing the

*Progress in Molecular Biology and Translational Science*, Volume 156
ISSN 1877-1173
http://dx.doi.org/10.1016/bs.pmbts.2017.12.005

conformation of glycoproteins and through recognition and masking of biological sites of functional molecules. Sialidases are found widely distributed in metazoan animals, from echinoderms to mammals, and are also present in viruses and other microorganisms, including fungi, protozoa, and bacteria even mostly lacking sialic acids. In mammals, there are four forms of sialidase (Neu1, Neu2, Neu3, and Neu4), differing in their major subcellular localization and enzymatic properties. They have been implicated in regulation of various cellular activities, such as cell differentiation, cell growth, and cell adhesion and motility, depending on their particular properties. In contrast, in microorganisms the enzymes appear to play roles limited to nutrition and pathogenesis. In this chapter, the focus is on mammalian sialidases preferentially hydrolyzing gangliosides, mostly Neu3 and Neu4, with an attempt to provide a brief overview of their physiological and pathological roles.

# 1. INTRODUCTION

## 1.1 General Properties of Mammalian Sialidases

Sialidases of mammalian origin have been found to differ from those of microbes in their overall primary sequences and enzymatic properties, although they do contain a RIP (−Phe (Tyr)-Arg-Ileu-Pro-) sequence and Asp boxes (−Ser-X-Asp-X-Gly-X-Thr-Trp-) in the primary sequences, conserved in viral and bacterial sialidases. Distinctive behavioral differences in response to inhibitors between mammalian and microbial sialidases are suggestive of molecular variation. For example, antiinfluenza drugs, Tamiflu (oseltamivir) and Relenza (zanamivir), are known to be effective in vitro at nanomolar concentrations for the viral sialidase, but Tamiflu hardly affects any of the human sialidases, even at 1 mM, and Relenza inhibits only in the micromolar range.[1,2] This indicates major differences between human and viral sialidases. On the other hand, a sialic acid analogue 2-deoxy-2,3-dehydro-N acetylneuraminic acid, Neu5Ac2en (DANA), is known to be a common inhibitor of almost all members of sialidases, although viral and bacterial forms are 10 times more sensitive than their mammalian counterparts.

As sialidase activity in higher organisms was first detected in preparations of Cohn Fraction VI,[3] numerous papers have documented presence in a wide variety of mammalian cells and tissues. Molecular cloning has validated four forms of mammalian sialidase and facilitated elucidation of their functional roles.[4] General properties of the four sialidases are compared in Table 1. They differ in their major subcellular localization and enzymatic properties, as well

**Table 1** General Properties of Four Mammalian Sialidases.

| | Neu1 | Neu2 | Neu3 | Neu4 |
|---|---|---|---|---|
| Major subcellular localization | Lysosomes | Cytosol | Plasma membrane | Lysosomes,[a] mitochondria,[a] and ER |
| Good substrates in activity assays | Oligosaccharides<br>Glycopeptides | Oligosaccharides<br>Glycoproteins<br>Gangliosides | Gangliosides | Oligosaccharides<br>Glycoproteins<br>Gangliosides |
| Endogenous substrates identified in cellular level | LAMP-1<br>IR, IGF-1R<br>PDGF-R<br>Integrinβ4 | Sialyl-Le$^x$<br>GM3 | GM3<br>GD3, GD1a | Sialyl-Le$^x$<br>Sialyl-Le$^a$<br>PolySia-NCAM<br>GD1a |
| Optimal pH | 4.4–4.6 | 6.0–6.5 | 4.5–4.8<br>6.0–6.5 | 4.5–4.7 |
| *Total amino acids* | | | | |
| (Human) | 415 | 380 | 428 | 496 (484)[b] |
| (Mouse) | 409 | 379 | 418 | 501 (478)[b] |
| *Chromosomal location* | | | | |
| (Human) | 6p 21.3 | 2q 37 | 11q13.5 | 2q37.3 |
| (Mouse) | 17 | 1 | 7 | 10 |
| Proposed functions | Lysosomal catabolism<br>Exocytosis<br>Phagocytosis<br>Immune function<br>Elastogenesis<br>Muscle differentiation | Myoblast differentiation<br>Neural differentiation | Neuronal differentiation<br>Apoptosis<br>Adhesion | Neuronal differentiation<br>Apoptosis<br>Adhesion |
| Neoplastic alterations (frequent changes in expression) | Downregulation | Downregulation | Upregulation | Downregulation |

[a]For the subcellular localization of human NEU4, two different descriptions have been reported on the basis of gene transfection studies.

[b]Human NEU4 consists of two forms, long (L) and short (S), which differ in the presence of 12 N-terminal amino acid residues, possibly acting in mitochondrial targeting. Mouse NEU4 also presents as two isoforms, designated Neu4a and Neu4b. Neu4a has an additional 23 amino acid stretch at the N-terminus.

as being encoded by different genes. Regarding subcellular localization, Neu1, Neu2, and Neu3 are now known to be found localized mainly in the lysosomes, cytosol, and plasma membranes, respectively, and Neu4 is found in lysosomes or in mitochondria/endoplasmic reticulum. However, recent observations have indicated that the subcellular localization can vary with particular cellular stimuli. In fact, Neu1 and Neu4 as well as Neu3 are also observed at cell surfaces depending on localization of their accessible natural substrates. There is further evidences of Neu1 presence in the outer membrane and Neu3 in the inner membrane of the nuclear envelope,[5] and Neu2 in nucleoplasm of muscle fibers other than in cytosol as assessed by immuno-histochemical analysis on electron microscopy probably due to the presence of a nuclear localization signal near the N-terminus.[6] Three dimensional structure has been determined only for human Neu2 (NEU2) by X-ray crystallography in comparison with that of *S. typhimurium* sialidase,[7] has allowed prediction of the structures of other human sialidases NEU1, NEU3, and NEU4 using a homology modeling approach. Among human sialidases, the overall amino acid identity of NEU1 to the other forms is relatively low (19%–24%), whereas NEU2, NEU3, and NEU4 show 34%–40% homology with each other. Regarding comparative expression levels, NEU1 generally shows the strongest expression, 10–20 times greater than those of NEU3 and NEU4, while NEU2 expression is extremely low, at only four- to ten-thousandth of the NEU1 value at most in a range of tissues,[1,8] although profiles differ among the human, rat, and mouse. During development, the four forms also display different patterns as compared in the developing mouse brain (Fig. 1).

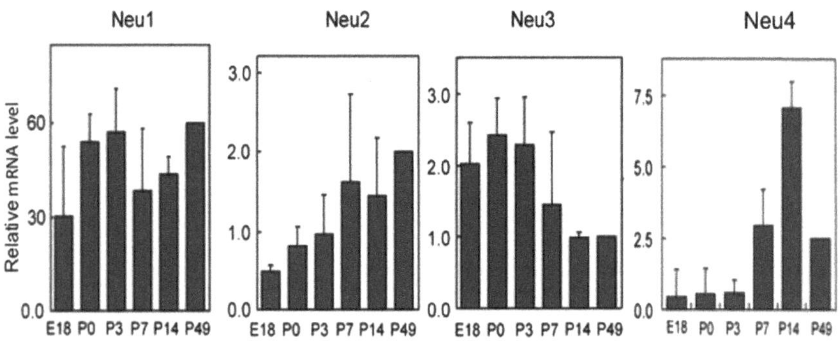


**Fig. 1** Relative Neu1, Neu2, Neu3, and Neu4 expression levels in the developing mouse brain as a function of embryonic (E) and postnatal (P) age.[9]

## 1.2 Enzymatic Properties of Mammalian Sialidases

### 1.2.1 Sialidase Neu1

Taking advantage of the SM/J inbred strain of mice carrying a defective sialidase allele, the mouse *Neu1* gene[10–12] was mapped by linkage analysis near the H-2D end of the major histocompatibility complex (MHC) on chromosome 17, a region which is syntenic to human MHC on chromosome 6. The human lysosomal sialidase, NEU1,[13–15] has been extensively investigated as a target in sialidosis, and has been found to be associated with a protective protein (carboxypeptidase A) and β-galactosidase as a complex in lysosomes, dissociation of the complex leading to sialidase inactivation.[16,17] NEU1 features a lysosomal C-terminal targeting motif and there is evidence that a protective protein is responsible for NEU1 transport to lysosomes, but recent research has revealed an intracellular distribution at plasma membranes as well as lysosomes under certain conditions. In sialidase assays in vitro, Neu1 possesses narrow substrate specificity, with oligosaccharides and glycopeptides and a synthetic substrate, 4-methylumbelliferyl-neuraminic acid (4MU-NeuAc) serving as good substrates, but hardly acts on gangliosides with an optimal pH of 4.5–4.7. Oligosaccharide substrates possessing α2–3 sialyl linkage can be hydrolyzed faster than those with α2–6 and α2–8 linkages. NEU1 is linked to two neurodegenerative lysosomal storage disorders, sialidosis and galactosialidosis,[18] the former caused by defects in genomic DNA, including frameshift insertions and missense mutations, and the latter featuring a combined deficiency of the sialidase and β-galactosidase due to the absence of a functional protective protein. In addition to glycoconjugate catabolism in lysosomes, Neu1 has been implicated in cellular signaling for immune responses and elastic fiber assembly during transportation to plasma membranes, and in negative regulation of lysosomal exocytosis, a cellular process for the recruitment of lysosomes to the plasma membrane, resulting in an increase in extracellular proteolytic activity.[19] Neu1 also contributes to the regulation of phagocytosis in macrophages and dendritic cells through the desialylation of surface receptors.[20] In cancer cells, human NEU1 has been described to play crucial roles in determination of metastatic potential.[21]

### 1.2.2 Sialidase Neu2

Neu2 was the first mammalian sialidase for which cDNA cloning was achieved[22] and the homologues subsequently cloned from cDNA libraries of CHO,[23] mouse brain,[24,25] rat thymus,[26] and from a genomic library of human skeletal muscle,[27] showed high amino acid identity (70%–98%) to the rat gene. The structure of the human ortholog NEU2 determined by X-ray crystallography,[7] provided evidence for a canonical six-blade β-propeller with the active site in a shallow crevice. Residues exist recognizing the N-acetyl and glycerol moieties of DANA, as observed for viral and bacterial sialidases. Unlike Neu1 sialidase, Neu2 possesses broad substrate specificity in vitro at nearly neutral pH, hydrolyzing glycoprotein and oligosaccharides, and also gangliosides in the presence of sodium cholate. It participates in muscle cell and neuronal differentiation in murine cells, and the rat *Neu2* gene contains two E-box pairs in the 5′-flanking enhancer/promoter region,[28] known to be consensus binding sites for muscle-specific transcription factors, and exhibits transcriptional activity in murine myogenic cells. In PC12 cells the sialidase has been suggested to participate in neuronal differentiation on the basis of nerve-growth factor-induced transcriptional activation.[29] However, it is uncertain what glycoconjugate molecules are involved in these phenomena as the results of enzymatic reaction by Neu2. In man, NEU2 expression was found to be extremely low or undetectable in many tissues and cells, even in cancers, with notable exceptions like the placenta and testes.[30]

### 1.2.3 Sialidase Neu3

Neu3 was first cloned from a bovine brain library[31] based on peptide sequence information from purified enzyme protein, and later from a human cDNA and genomic library.[32,33] The primary sequences covering the entire coding region of the corresponding human, mouse, and rat genes display 83%, 79%, and 78% overall identity, respectively, with the bovine gene. The major subcellular localization of the bovine sialidase proved to be plasma membranes on Percoll density gradient centrifugation of cell homogenates and by immunofluorescence staining.[31] With administration of the radiolabeled ganglioside GD1a to murine Neu3-transfected cells, Neu3 was shown to hydrolyze ganglioside substrates in intact living cells at a neutral pH mainly through cell-to-cell interactions. Unlike the bovine and mouse Neu3, the human ortholog NEU3 is not always detected on the cell surface but may exist on other cellular membrane components and can be moved to and concentrated at leading edges in response to growth stimuli.[34] An analysis of membrane topology has suggested that the sialidase might be localized partially on the cell surface as a peripheral membrane protein and also in endosomal structures.[35] Catalytic hydrolysis is essentially specific for gangliosides other than GM1 and GM2 in the presence of Triton X-100. The bovine and human enzymes specifically hydrolyze gangliosides, and the murine enzyme acts on oligosaccharides, 4MU-NeuAc, and glycoproteins to a certain extent. Desialylation of GM1 and GM2 by murine enzymes is evident in the presence of GM2 activator protein,[36] supporting the existence of an asialo-derivative GA2 pathway for catabolism of GM2 in the mouse. The human enzyme acts on GM3 but hardly impacts on sialyllactose, suggesting that a ceramide moiety might be critical for catalysis. However, it was also proposed that only a hydrophobic aglycon is required for the strict ganglioside preference,[37] not ceramide itself. With regard to the glycosidic linkage specificity, Neu3 hydrolyzes gangliosides with α 2–3 (GM3) and α 2–8 (GD3) linkages almost equally and faster than those with an α 2–6 linkage (synthetic GM3). GM3 containing N-glycolylneuraminic acid is hydrolyzed by Neu3 at a lower rate. Unlike its bovine and murine counterparts, the human enzyme shows two peaks of activity at pH 4.5–4.8 and at pH 6.0–6.5, and Northern blot analysis revealed high expression in skeletal muscle and testis, but only low levels in kidney, placenta, lung, and digestive organs.[32]

### *1.2.4 Sialidase Neu4*

The fourth sialidase, Neu4, was identified based on cDNA sequences in public databases.[8,38–40] With regard to the subcellular localization of the human form, NEU4, two differing descriptions have been reported on the basis of gene transfection studies: one featuring targeting to the lysosomal lumen,[40] and the other to mitochondria and endoplasmic reticulum.[8,41] NEU4 appears to consist of isoforms which differ in their possession of 12 N-terminal amino acid residues for mitochondrial targeting. The isoforms are also differentially expressed in a tissue-specific manner, brain, muscle and kidney containing both, and the liver and colon possessing predominantly the short form as assessed by RT-PCR.[8] There are similarly two isoforms of mouse Neu4 generated by alternative splicing, designated Neu4a and Neu4b.[9] Mouse Neu4a has an additional 23 amino acid stretch at the N-terminus and exhibits lower enzymatic activity than Neu4b. The mouse *Neu4* gene is expressed predominantly in brain[40] and is only found at very low levels in other tissues. Neu4 possesses broad substrate specificity, hydrolyzing glycoprotein and oligosaccharides as well as gangliosides at acidic pH, and appears to be only a sialidase hydrolyzing mucins. Unlike other sialidases, human NEU4 hydrolyzes sialyl-Lewis antigens expressed on *O*-glycans, such as sialyl-Le$^x$ and sialyl-Le$^a$,[42] and polysialic acid (polySia) on the neural cell adhesion molecule (NCAM).[43] Herein, we focus on physiological and pathological cellular functions of mammalian ganglioside sialidases, extending the earlier literature with recent studies.[4,44–46]

# 2. BIOLOGICAL ROLES OF GANGLIOSIDE SIALIDASES

## 2.1 Activation of Sialidase Neu3

Although Neu3 has been considered as a crucial signaling regulator due to its modulation of gangliosides and colocalization at cell surfaces, the mechanisms underlying its activation are not fully understood. Chemical detergents, such as Triton X-100, are necessary to detect its sialidase activity in vitro described earlier,[31,32] because a natural activator(s) for Neu3 was not identified. Human Neu3 (NEU3) possesses a long hydrophobic stretch in the molecule, but is not always detected on cell surfaces, often changing its subcellular localization to plasma membranes through the action of specific stimuli.[34] Among various phospholipids thought to be potential activators, because of their detergent properties and recent increasing recognition as important signaling lipids, phosphatidic acid was found to be a possible activator for the sialidase at neutral pH, and interestingly, also a regulator of its translocation to and functions at plasma membranes.[47] Neu3 is activated under hypoxic conditions in cultured mouse skeletal muscle cells, resulting in enhanced cell survival and proliferation through activation of the epidermal growth factor receptor (EGFR) signaling and the hypoxia-inducible factor 1 alpha.[48]

A study[49] on transcriptional activation of the *NEU3* gene provided evidence of diverse regulation by Sp1/Sp3 transcription factors binding to alternative promoters. Multiple transcriptional start sites (TSSs) have been identified in two clusters, one preferentially utilized in various tissues, and the other preferentially utilized only in brain. Recent annotation of TSSs in the human genome indicated the existence of alternative promoters in many tissue specific genes and examples encoding molecules involved in signal transduction.[50] Such transcriptional control might also account for the upregulation of NEU3 in various human cancers as described below, because Sp1 and Sp3 were initially considered as constitutive activators of housekeeping genes but have recently been documented to play critical roles in regulating the transcription of genes involved in cell growth control and tumorigenesis.[51]

## 2.2 Biological Roles of Sialidase Neu3

Because of its unique substrate specificity and association with the plasma membrane, Neu3 is presumed to be a potentially important molecule for various cell surface events. Murine Neu3 can regulate cell apoptosis in human fibroblasts by producing ceramides from GM3 in plasma membranes through further degradation of the sugar unit.[52] Previous biochemical studies suggested a cell surface presence and essential roles of ganglioside sialidases in the control of proliferation and differentiation in neuroblastoma cells,[53–55] and later using a cDNA clone, Neu3 was indeed confirmed to positively participate in neurite formation in mice,[25] in differentiation of human neuroblastoma cells,[56] and in neuronal differentiation and regeneration of rat hippocampal neurons.[57,58] In a recent report[59] on axon regeneration of rat neurons it was concluded that Neu3 activation, initiated by extracellular calcium influx and upregulation of P38MAPK, and the consequent Neu3-mediated conversion of GD1a to GM1 are necessary for regeneration of PNS (sensory), but not in CNS (retinal) axons. Consistent with these results, Neu3 activity remained high with differentiation and aging in culture of rat cerebellar granule cells, accompanied by increased β-galactosidase and β-glucosidase activity and paralleled ceramide levels in plasma membranes.[60] An involvement of Neu3 in skeletal muscle differentiation and apoptosis has been indicated by gene knock down in murine C2C12 cells causing inhibition of myotube formation and sensitization to apoptotic stimuli, with accumulation of GM3 subsequent to inhibition of EGFR actions.[61] *NEU3* gene expression leads to inhibition of TNF-α-induced matrix metalloproteinase-9 expression in vascular smooth muscle cells by decreasing the promoter activity in response to TNF-α, suggesting contributions to plaque instability in atherosclerosis as a physiological modulator of smooth muscle cell responses.[62] Neu3 is located in rafts of neuroblastoma cells[63] and in caveolae of HeLa cells,[64] closely associated with caveolin-1, probably through the caveolin binding region. In caveolae, Neu3 has been proposed to control platelet-derived growth factor (PDGF)-induced Src mitogenic signaling and DNA synthesis by modification of GM1 levels at cell surfaces. A member of the transmembrane adaptor protein family, PAG, induces Neu3 sialidase activity and GM1 accumulation, thereby excluding PDGF receptors from caveolae and inhibiting Src mitogenic signaling, although it remains unclear how PAG affects NEU3 activity.[65] Depletion of ganglioside GM3 by Neu3 causes increased EGFR phosphorylation and inhibition of the EGF-induced tyrosine phosphorylation of caveolin-1, leading to activation of EGFR signaling through retention of caveolin-1 in caveolae.[66] NEU3 has been found to contribute to development of adaptive immune responses, being upregulated together with NEU1 during differentiation of monocytes into dendritic cells. LPS-induced production of cytokines, including IL-6,

IL-12p40, and TNF-α in mature dendritic cells was reduced by zanamivir, an efficient inhibitor for NEU3, and also by DANA, an inhibitor of NEU1 and NEU3, in cells from *Neu1* (−/−) and wild-type mice, suggesting an involvement of NEU3 in cytokine production by desialylation of specific cell surface glycoconjugates.[67] NEU3 is now considered to play important roles in regulation of transmembrane signaling by interacting with signaling molecules, including caveolin-1, Rac-1, integrin β4, Grb-2, and EGFR as well as by modulation of gangliosides as an enzyme,[68] as described below.

NEU3 has been observed in human lung epithelial cells. Superficial structures including ciliated brush borders as well as nuclei in lung tissue were found to express catalytically active NEU3 sialidase as assessed by immunostaining, mRNA level, and activity assays. NEU3 is thus strategically positioned to influence airway pathogen interactions.[69] In human red cells, NEU1 and NEU3 detection at plasma membranes, as peripheral proteins, is associated with external leaflets. The sialidases appear to be progressively lost however, during erythrocyte life, which could be a tool to preserve the cellular sialic acid content to avoid early ageing and processes of cell aggregation in the capillaries.[70]

## 2.3 Biological Roles of Neu4

Human NEU4 may be involved in neuronal cell apoptosis, based on the observation that its long form probably regulates levels of GD3, known to be an apoptosis-related ganglioside, in mitochondria of neuroblastoma cells.[71] In contrast to Neu3, murine Neu4 appears to negatively regulate neurite formation by Neuro2a cells and hippocampal neurons.[9,43] As illustrated in Fig. 2, Neu3 and Neu4 thus may function in neurite formation in opposing directions. The two enzymes can both act on gangliosides, but the endogenous substrate for Neu4 in this case seems to be an approximately 95 kDa glycoprotein. However, definite target molecules have yet to be identified. In one report it was suggested that Neu4 is activated by thymoquinone from the *Nigella sativa* plant in various cells including macrophages, and that the induced Neu4 activity is related to activation of NF-kB and proinflammatory cytokine production.[72] Expression of human NEU4 in cells of sialidosis patients results in clearance of storage materials from lysosomes and thus NEU4 may be useful for novel therapeutic purposes.[40]

Despite crucial roles of polySia in neural function, whether NEU4 involved in its degradation and physiological turnover has remained uncertain. PolySia has been implicated in control of synaptic plasticity, neuronal

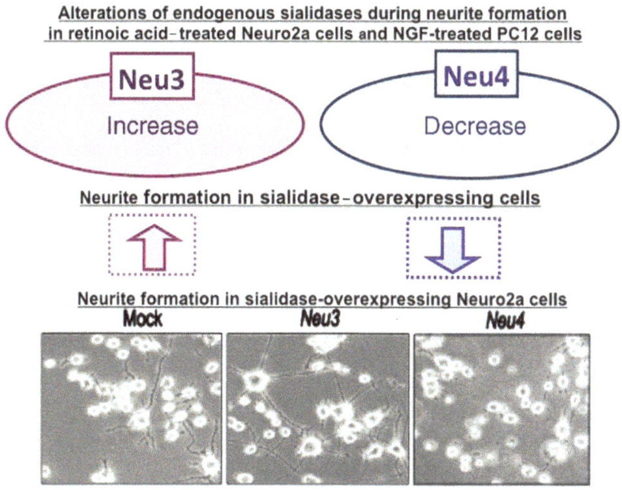


**Fig. 2** Opposite roles of two murine sialidases, Neu3 and Neu4, in neurite formation during retinoic acid-induced neuronal differentiation in Neuro2a cells. Neu3 is upregulated during neuronal differentiation and accelerates neurite formation,[25] whereas Neu4 attenuates retinoic acid-induced differentiation.[9] Photographs of Neuro2a cells treated with 20 μM retinoic acid for 48 hours at 100× magnification.

differentiation, and cell migration.[73,74] Levels are high during embryonic development, whereas its expression in the adult is restricted to brain regions featuring persistent neural plasticity or neural development, such as the hippocampus.[75] Regarding the biosynthesis of polySia, two polysialyltransferases, ST8SiaII and ST8SiaIV, have been identified,[76] but physiological turnover remains to be clarified. Mouse Neu4a possessing an additional 23 amino acid stretch at the N-terminus demonstrated lower enzymatic activity than Neu4b but interestingly more efficient hydrolysis of polySia on the neural cell adhesion molecule (polySia-NCAM). Although a bacteriophage-derived endosialidase essential for bacterial infection is well known, Neu4 was found to be the major degradation enzyme for polySia in vertebrates, negatively regulating neurite outgrowth of hippocampal neurons[43] as illustrated in Fig. 3.

## 2.4 Neu3 and Neu4 Knockout Mice

Although abnormal accumulation of gangliosides was expected with Neu3 deficiency, given its preferential hydrolysis of these as substrates, Neu3

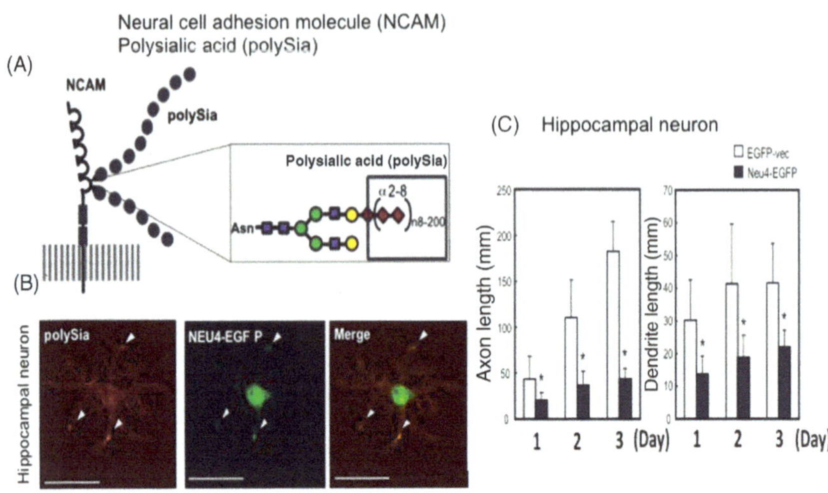


**Fig. 3** Suppression of neurite outgrowth from hippocampal primary neurons by Neu4.[43] Structure of polysialic acids (polySia) on the neural cell adhesion molecule (A) and partial colocalization of Neu4 in the growth cones of neurons expressing polySia (B). Suppression of neurite outgrowth by transfection of Neu4 in mouse hippocampal neurons with decreased polySia expression. Longest neurites were measured as axon, and other neurites as dendrite.

deficient mice showed no conspicuous accumulation of gangliosides in the brain or colon mucosa, or overt abnormalities in their growth, development, behavior, or fertility.[77] In the case of Neu4 deficiency, the mice lacks gross morphological abnormalities, and the ganglioside pattern demonstrated increased GD1a and decreased GM1 in the brain.[78] In Neu3 and Neu4 double-knockout mice, $G_{M3}$ gangliosides could be found stored in microglia, vascular pericytes, and neurons, causing micro- and astrogliosis, neuroinflammation, accumulation of lipofuscin bodies, and memory loss, with cortical and hippocampal neurons exhibiting lower rates of neuritogenesis in vitro. The mice also featured reduced levels of $G_{M1}$ ganglioside and myelin in neuronal axons.[79]

β-Hexosaminidase A catabolizes the GM2 ganglioside by converting it to GM3, and genetic deficiency of this enzyme causes GM2 accumulation in Tay-Sachs disease. However, *Hexa*−/− mice remain asymptomatic up to 1 year of age because GM2 gangliosides are desialylated into the GA2 glycolipid, which is further processed by β-hexosaminidase B to lactosylceramide

(LacCer), thereby bypassing the β-hexosaminidase A defect.[80] Because this bypass is not effective in humans, infantile Tay-Sachs disease is fatal in the first years of life. To identify what sialidase is responsible for this bypassing function, recent studies on double knockout mice of β-hexosaminidase A and Neu4 or Neu3 double knockout mice were performed. The Neu4 and β-hexosaminidase A double deficiency resulted in epileptic seizures and rapid neuronal loss accompanied by GM2 accumulation. Unlike the case with human Tay-Sachs disease, hexosaminidase A deficient mice remain asymptomatic for over 1 year, because of the expected GM2 catabolism. In this context, the results with the double deficient mice indicate that Neu4 is a sialidase involved in the mouse model of Tay-Sachs disease, possibly acting as a ganglioside-metabolizing enzyme in vivo.[81] However, while disease severity in the double mutant was increased, it was not profound, suggesting that Neu4 was not the only sialidase contributing to the metabolic bypass in *Hexa*(−/−) mice, even though it is dominantly expressed in the mouse brain, but rather that Neu3 is possibly also involved. Recent observations with Neu3 and β-hexosaminidase A double deficiency included drastically increased storage of $G_{M2}$ in brain tissues of this asymptomatic mouse model of Tay-Sachs disease, indicating that Neu3 is in fact responsible for the metabolic bypass of β-hexosaminidase A deficiency.[79,82] The results have also verified the difference in the substrate specificity in vitro assay between murine and human Neu3 sialidases, the former possessing broader specificity. Deficient mice have indeed demonstrated that Neu3 and Neu4 have important roles in regulation of neuronal function by catabolizing brain gangliosides.

## 3. PATHOLOGICAL ROLES OF GANGLIOSIDE SIALIDASES

### 3.1 Possible Involvement in Neurological Diseases

Preclinical and clinical studies have shown that systemic administration of GM1 ganglioside has neuroprotective and neurorestorative properties in Parkinson's disease (PD) mouse models[83] and in PD patients.[84] In an attempt to increase the therapeutic efficiency, instead of bovine brain GM1 used, intraventricular administration of *Vibrio cholera* sialidase was tried to enhance GM1 levels in the brain via enzymatic conversion of polysialogangliosides.[85] Previous experimental studies have also demonstrated that sialidases of bacterial origin increase GM1 levels and exert a neuroprotective effects.

Dhanushkodi and McDonald[86] showed that brain ganglioside profiles can be altered in vivo by intraventricular infusion of *V. cholera* sialidase and that this protects against excitotoxic neurodegeneration. Yang *et al.*[87] also reported that infusion of *Clostridium perfringens* sialidase enhanced spinal axon outgrowth into implanted peripheral nerve grafts in a rat model of brachial plexus avulsion. The experimental studies strongly suggest that Neu3 and Neu4 would be good therapeutic targets for these neurodegenerative diseases, if any activators for them could be discovered to enhance the production of neuroprotective GM1.

## 3.2 Possible Involvement in Inflammation and in Hepatic Steatosis

A recent report described a possible role of bacterial sialidases in development of intestinal inflammation.[88] Using mice with dextran sulfate-induced colitis, outburst of *Escherichia coli* likely directly depends on sialic acid release from host glycans through sialidase activity originating from *Bacteroides vulgatus*, and this exacerbates the inflammatory response by stimulating the production of proinflammatory cytokines by intestinal dendritic cells. During intestinal inflammation in mice, oral administration of a sialidase inhibitor, DANA, decreased outgrowth of bacteria and severity of colitis. Regulation of sialic acid catabolism thus may be helpful for the treatment of intestinal inflammation. The experimental observations give an indication of involvement of mammalian ganglioside sialidases in inflammatory responses. In fact, another recent report[89] demonstrated that in human intestine inflammatory bowel diseases, the relative content of β-hexosaminidase A was increased 1.7-fold ($P < 0.05$) and NEU3 was increased 8.3-fold ($P < 0.01$) compared to normal intestine assessed by Western blotting. As a consequence, intestinal mucosa featured increased GM3 content, decreased GD1a, and a reduction in polyunsaturated fatty acid constituents in GD3, GD1a and phosphatidylcholine.

Quite recently, a study[90] on the most common chronic liver disease, nonalcoholic fatty liver disease, demonstrated activation of intestinal hypoxia-inducible factor 2α (HIF-2α) to contribute to the hepatic steatosis by regulating ceramide metabolism mainly through increased NEU3 expression. Human intestine biopsies from individuals with obesity showed increased HIF-2α signaling. This correlation was verified in mice with an intestine-specific disruption of *Hif2α*. PT2385, a HIF-2α specific inhibitor, in fact, had preventive and therapeutic effects on the metabolic disorders. This could be a new important role of NEU3 for regulation of intestine and serum ceramide levels by supplying LacCer, as a precursor of ceramide, from GM3.

## 3.3 Involvement of Neu3 in Impaired Glucose and Insulin Tolerance

Interestingly, NEU3 transgenic mice develop impaired insulin signaling and insulin-resistant diabetes mellitus by 18–22 weeks,[91] associated with hyperinsulinemia, islet hyperplasia and increase in the β-cell mass. As compared to the wild type, insulin-stimulated phosphorylation of the insulin receptor and insulin receptor substrate I are significantly reduced, and activities of phosphatidylinositol 3-kinase and glycogen synthase are also decreased. In muscle extracts, an association of tyrosine-phosphorylated NEU3 with Grb2 was observed in response to insulin, together with accumulation of GM1 and GM2. These findings suggest an involvement of NEU3 in regulation of signal transduction through ganglioside modulation and interaction with signaling molecules. The apparent involvement of NEU3 in both cancer progression (as described below) and development of diabetes supports the conclusion that these diseases might be closely related to each other in pathogenesis, in line with recent epidemiological reports of higher cancer risk in diabetic patients than in controls.

When *NEU3* was transiently expressed, however, using adenoviral vectors in the livers of C57BL/6 mice on standard and high-fat diets and insulin-resistant KKAy mice on a standard diet, paradoxically, hepatic NEU3 overexpression improved glucose tolerance and insulin sensitivity. Increased levels of GM1 and markedly reduced GM3 were observed in the livers and basal and insulin-stimulated tyrosine phosphorylation of insulin receptor substrate 1 was significantly increased, but that of the insulin receptor and insulin receptor substrate 2 was unchanged in liver and adipose tissues.[92] The detailed mechanisms underlying these opposing effects of NEU3 on glucose and insulin tolerance is uncertain, but there was a difference between the two experiments in the period of overexpression as well as in the target organ examined. Transient expression of NEU3 may attenuate glucose tolerance by reduction of GM3, consistent with experiments on GM3 synthase knockout mice, demonstrating that GM3 accumulation influenced glucose tolerance,[93] but long-term exposure for 18–20 weeks may induce islet hyperplasia and increase in the β-cell mass, resulting in hyperinsulinemia and insulin resistance. The findings however, do provide evidence that NEU3 is an important regulator of insulin sensitivity and glucose tolerance.

## 3.4 Involvement of Neu3 in Cancer Initiation, Promotion and Progression

### *3.4.1 Alteration of Neu3 Expression in Cancer*

After validation of the four types of mammalian sialidase through gene cloning, a great deal of attention has been paid to sialidase expression levels in various human cancers determined by activity assays and quantitative RT-PCR. NEU3 has exhibited remarkable upregulation in colon, prostate, ovary, renal, and head and neck cancers,[45] although downregulation has been reported for acute lymphoblastic leukemia[94] and glioblastoma.[95] The mRNA level of NEU3 was found to be increased in all surgical tissue specimens ($n = 50$) of human colon cancer by 3- to 100-fold as compared with that in the adjacent noncancerous mucosa.[96] In-situ hybridization analysis showed NEU3 expression in the epithelial elements of adenocarcinomas. On the other hand, NEU1 and NEU4 showed a tendency for reduced expression in colon cancer, and NEU2 proved to be hardly detectable in both human normal and cancer samples. Significant elevation of sialidase activity against gangliosides has also been observed in tumor tissues ($8.9 \pm 0.2$ to $5.7 \pm 0.1$ units/mg protein, relative to mucosa) and although nontumor mucosa contained higher ganglioside activity than expected, this was probably due to relatively high NEU4 expression. These results imply that most of the ganglioside sialidase activity observed in cancer tissues might actually be accounted for by NEU3. As described earlier,[49] NEU3 gene expression is under diverse regulation by Sp1/Sp3 transcription factors, so that altered to transcriptional regulation of these might be one of the mechanisms behind NEU3 upregulation in neoplasia, In fact, NEU3 expression exhibits good correlations with those of Sp1 or Sp3 in many cancers.

With regard to NEU3 downregulation in human glioblastoma,[95] it is uncertain at present what factors are responsible for the contrast from epithelial cancers. NEU3 effects different from those in other epithelial cancers. However, it may be attributed to the origin of most gliomas. According to the current consensus,[97] neural stem cells located in the subventricular zone of the brain are thought to be the source of glioma. They accumulate enough genomic mutations leading to diverse forms of glioma, and undergo epithelial to mesenchymal transition, which is characterized by loss of epithelial morphology. This feature is likely to be one of the factors for the differential effects of NEU3. NEU3 downregulation in acute lymphoblastic leukemia may also be similarly due to characteristic features of this tumor type.

### 3.4.2 Roles of NEU3 in Cancer Initiation and Promotion

Observations in genetically engineered mouse models have suggested pathological roles of Neu3 in both cancer initiation and promotion. When human NEU3 transgenic mice were treated with a carcinogen, azoxymethane, induction of precancerous colonic aberrant crypt foci (ACF) was significantly increased.[98] Enhanced phosphorylation of EGFR, Akt, and ERK and upregulation of Bcl-xL protein were observed in the transgenic colon mucosa, but no changes were found in cell proliferation, suggesting that the increased ACF formation might have been related to suppression of apoptosis. When *Neu3*-deficient mice were exposed to dimethylhydrazine, although there were no differences from wild-type mice in the incidence or growth of tumors, the *Neu3*-deficiency was associated with lowered susceptibility to colitis-associated colon carcinogenesis induced by azoxymethane and dextran sodium sulfate, indicating an involvement of Neu3 in inflammation-dependent tumor development.[77]

In human colon cancer cells, NEU3 was found to accelerate initiation of anchorage-independent growth,[99] whereas *Neu1* gene transfection into murine melanoma B16 cells resulted in diminished overall and anchorage-independent growth, and increased sensitivity to apoptosis induced by suspension culture or serum depletion, compared with parental and control cells.[100] Downregulation of NEU3 in colon cancer cells was found to cause decreased colony formation, chemosensitivity, anchorage-independent growth and in vivo tumor growth in nude mice. A close link appears to exist between NEU3 expression and Wnt/β-catenin signaling in colon cancer cells, whereby NEU3 enhanced phosphorylation of the Wnt receptor LRP6 and consequently β-catenin activation by accelerating complex formation with LRP6 and recruitment of GSK3β and Axin. NEU3 activity-null mutants failed to demonstrate such activation, indicating a requirement of ganglioside modulation by the sialidase. The data strongly suggest NEU3 participation in tumor initiation and promotion, as illustrated in Fig. 4, as constitutive activation of Wnt/β-catenin signaling is implicated in the maintenance of cancer stem cells and initiation of the process of colon carcinogenesis.[101]

To further address whether activation of EGF signaling by the sialidase is associated with oncogenic transformation, the effects of overexpression of NEU3 and EGFR were analyzed in NIH-3T3 cells.[102] When NEU3 was stably transfected with or without EGFR, it was associated with significant increase in clonogenic growth, clonogenicity in soft agar and in vivo tumor

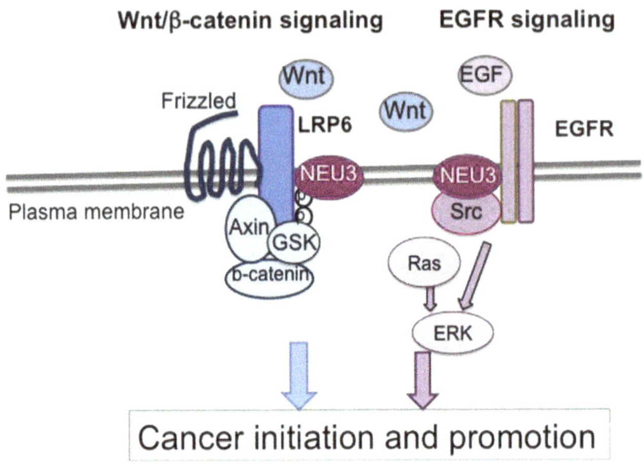


**Fig. 4** NEU3 participates in cancer initiation, promotion, and progression by regulating Wnt/β-catenin and epidermal growth factor receptor signaling pathways often aberrantly activated in human cancers.[99,102,103]

growth in nude mice either with or without the receptor overexpression in the presence of EGF, compared with levels in vector controls. Despite the fact that the endogenous level of EGFR is known to be extremely low in these cells, NEU3 significantly enhanced the phosphorylation of Akt and ERK, as well as that of the receptor. The NEU3-mediated activation was largely abrogated by the EGFR inhibitor AG1478 or PD153035, but significant clonogenic growth still remained. NEU3 was then found to activate Src kinase, and the clonogenicity was completely suppressed by an Src inhibitor, PP2. The activity-null mutants failed to activate Src and EGFR, indicating that ganglioside modulation by NEU3 may be necessary for the activation. NEU3 and Src coimmunoprecipitated with EGFR in NEU3- and EGFR-transfected cells. These findings together point to NEU3 as an essential participant in tumorigenesis through activation of the EGFR/Src and Wnt/β-catenin signaling pathways.

### *3.4.3 Roles of NEU3 in Acceleration of Cancer Progression*

Investigation of NEU3 upregulation in cancers has also suggested involvement in cancer progression. In colon cancer cells, increase caused suppression of cell differentiation and apoptosis, accompanied by increased Bcl-2 and decreased caspase 3 expression. Endogenous sialidase levels are known to be downregulated in the process of differentiation and apoptosis induced by sodium butyrate. Compared to nontumor mucosa, colon cancer tissues exhibited a marked accumulation of LacCer, a possible NEU3 product, and addition of LacCer to the culture confirmed reduction of apoptotic cells during sodium butyrate treatment, indicating that high expression of NEU3 can protect against programmed cell death, probably through modulation of gangliosides.[96]

NEU3-transfected colon cancer cells exhibited increased adhesion to laminins and consequent cell proliferation, but decreased adhesion to fibronectin, collagen I and IV, as compared to control cells.[104] On laminin-5, NEU3 clearly stimulated phosphorylation of FAK and ERK, and markedly enhanced tyrosine phosphorylation of integrin β4, with recruitment of Shc and Grb-2. These results indicate that NEU3 differentially regulates cell proliferation through integrin mediated signaling depending on the extracellular matrix. This selective NEU3 effect may be favorable for cancer cell growth, because laminin-5 has been reported to increase and in contrast fibronectin to reduce progression of carcinomas. It is interesting to note here that the two sialidases NEU1 and NEU3 appear to have unique opposing functions under various conditions, through preferentially desialylating glycoproteins and gangliosides, respectively, even if targeting the same molecule, such as integrin β4 in the same cells. For example, in human colon cancer cells, NEU1 causes attenuation of in vivo liver metastasis through the removal of sialic acids from the integrin *O*-glycans and reduction of phosphorylation of integrin β4.[105]

NEU3 expression was found to be increased in renal cell carcinomas (RCCs) compared to adjacent nontumor tissues, significantly correlating with elevation of interleukin (IL)-6, a pleiotropic cytokine which has been implicated in immune responses and pathogenesis of several cancers.[106] Upregulation of NEU3 in the tumor tissues was strongly linked to the IL-6 expression level and NEU3 in renal cancer ACHN cells was activated by IL-6 in a positive feedback manner, mainly through the PI3K/Akt pathway, resulting in suppression of apoptosis and promotion of migration. Both NEU3 transfection and IL-6 treatment resulted in suppression of apoptosis and promotion of cell motility, and their combination resulted in synergistic effects. NEU3 hardly affected MAPK or IL-6-induced STAT3 activation but promoted the PI3K/Akt cascade in both IL-6 dependent and independent ways. Furthermore, IL-6 promoted Rho activation and this was potentiated by NEU3, leading to increased cell motility, whereas NEU3 silencing resulted in decreased Akt phosphorylation and inhibition of Rho activation. Glycolipid analysis showed a decrease in the ganglioside GM3 and an increase in LacCer after NEU3 transfection, and addition of these lipids to the culture apparently affected cell apoptosis and motility, consistent with the observations in colon cancer cells. The results indicate that NEU3 activated by IL-6 contributes to IL-6-mediated signaling largely via the PI3K/Akt cascade in a positive feedback manner, reinforcing a malignant phenotype in RCCs.

Upregulation of NEU3 was also detected in prostate cancer, significantly correlated with malignancy as assessed in terms of the Gleason score.[107] In androgen-sensitive LNCaP cells, enforced overexpression of NEU3 significantly induced progression-related transcription factor EGR-1, androgen receptors and prostate specific antigen, PSA, both with and without androgen, the cells becoming hormone sensitive. This NEU3-mediated induction was abrogated by inhibitors of PI3K and MAPK, in line with increased phosphorylation of Akt and ERK1/2 in NEU3-overexpressing cells. NEU3 silencing moreover caused a reduction in the cell growth of androgen-independent PC-3 cells in culture and of transplanted tumors in nude mice, these data strongly suggesting that NEU3 regulates progression of prostate cancer through androgen receptor signaling. Further evidence of involvement in malignancy was provided by decreased cell motility, invasion and in vivo bone metastasis with decreased expression of the matrix metalloproteinases MMP-2 and MMP-9, in PC-3 cells on NEU3 knockdown.[108] The data suggest that NEU3 regulates tumor progression of prostate cancer through androgen receptor signaling, and is likely to be involved in bone metastasis.

In the serum of patients with prostate cancer, a significant increase of sialidase was found compared with the low, if any, activity, in healthy subjects.[109] As this was apparent with gangliosides as substrates, but only to a very limited extent with 4-MU-NeuAc, a good synthetic substrate for sialidases other than human NEU3. The serum sialidase was also almost entirely immunoprecipitated with anti-NEU3 antibody, but not with antibodies for other sialidases. Results of a sandwich ELISA method using two anti-NEU3 antibodies also provided strong evidence that the serum sialidase is, in fact, NEU3, and this subtype may, therefore have potential utility as a novel diagnostic marker.

Regional lymph node metastasis with a head and neck squamous cell carcinoma (HNSCC) is a crucial event for progression, associated with a high rate of mortality. NEU3 increase at mRNA and activity levels significantly correlated with histological differentiation grade, lymphatic invasion, and lymph node metastasis.[110] With HSC-2 and SAS squamous carcinoma cells, NEU3 was found to promote cell motility and invasion, accompanied by increased expression of MMP-9, through enhancing phosphorylation of EGFR.

Glioblastoma multiforme is one of the most malignant tumors of the human central nervous system characterized by a high degree of invasiveness. When NEU3 expression was investigated because Gangliosides are major components of brain tissues, but unlike in epithelial cancers, NEU3 expression was found to be downregulated in human glioblastoma specimens. In the human glioblastoma cell lines, U251, A172, and T98G, NEU3 overexpression reduced invasion and migration by promoting the assembly of focal adhesions through reduced calpain-dependent proteolysis, whereas NEU3 silencing resulted in accelerating cell invasion via disassembly of focal adhesions. In NEU3-silenced cells, elevation of calpain activity and GM3 accumulation were observed, as a consequence of reduced sialidase hydrolysis, localization of calpain and GM3 at cell lamellipodia being demonstrated by immunofluorescence microscopy. Indeed, downregulation of NEU3 was found to exert a major influence on cell invasion through regulation of calpain activity and focal adhesion disassembly and consequently invasive potential of glioblastoma cells.[95] The opposite effects of NEU3 downregulation to those of other epithelial cancers may arise from the characteristics of the glioma cells with loss of epithelial morphology.

To define further the molecular mechanisms of NEU3 influence and its possible targets, the encoding gene has been silenced by siRNA or overexpressed in various human cancer cells.[103] Silencing caused apoptosis without specific stimuli, accompanied by decreased Bcl-xL and increased mda7 and GM3 synthase mRNA levels in HeLa cells, whereas overexpression resulted in the opposite effects. Human colon and breast carcinoma cell lines, HT-29 and MCF-7, appeared to be similarly affected by treatment with the NEU3 siRNA, but interestingly noncancerous human fibroblasts and keratinocytes showed no significant changes. NEU3 siRNA was found to inhibit and NEU3 overexpression to stimulate Ras activation with consequent influence on ERK and Akt. Ras activation by NEU3 was largely abrogated by PP2 (a Src inhibitor) or AG1478 (an EGFR inhibitor). In fact, siRNA introduction reduced while overexpression promoted phosphorylation of EGFR in response to EGF. NEU3 could be shown to coimmunoprecipitate with EGFR, and EGF-stimulation yielded a higher amount of immunoprecipitable NEU3. In addition to enhancing tumorigenesis at the stage of the tumor initiation as described earlier, the results indicate that NEU3 suppresses apoptosis of cancer cells by promoting EGFR phosphorylation, probably through its association with EGFR and consequent activation of Ras cascades, especially via the Ras/ERK pathway.

## 3.5 Involvement of Neu4 in Cancer Progression

In contrast to NEU3, NEU1 and NEU4 downregulation in cancer is likely to be related to increased progression and metastasis, for example, to venous invasion involved in hematogenous metastasis. When NEU4 mRNA levels were compared between human colon cancer and adjacent noncancer tissues, marked decreases in expression were noted in the tumors,[111] in clear contrast to the NEU3 case. Levels did not significantly correlate with the histological differentiation or the pathological stage, but the tumor to non-tumor expression ratio significantly correlated with the degree of venous invasion ($P = 0.025$). In cultured human colon cancer cells, the enzyme was upregulated in the early stage of apoptosis induced by either the death ligand TRAIL, or serum-depletion. Transfection of NEU4 into DLD-1 and HT-15 colon adenocarcinoma cells resulted in enhanced apoptosis and decreased invasiveness and cellular motility. siRNA-mediated NEU4 targeting, on the other hand, caused a significant inhibition of apoptosis and promotion of invasiveness and motility. Lectin blot analyses revealed that desialylated forms of approximately 100 kDa glycoproteins were prominently increased with peanut agglutinin (PNA) in the NEU4 transfectants, whereas only slight changes in glycolipids were detected on thin layer chromatography. These results suggest that NEU4 plays important roles in the maintenance of normal mucosa, mostly through desialylation of glycoproteins and that downregulation may contribute to invasive properties and protect against programmed cell death in colon cancers.

To further elucidate the significance of human NEU4 downregulation and its close relation with venous invasion in colon cancer, sialyl-Lewis antigens, sialyl-Le$^a$ and sialyl-Le$^x$, as endogenous substrates for the sialidase, were investigated,[42] because they have been utilized as tumor markers, and their increase in cancer is associated with tumor progression. NEU4 was found to hydrolyze the antigens in vitro and decrease cell surface levels much more effectively than other sialidases. Western blotting, thin layer chromatography and metabolic inhibition studies of desialylation products revealed NEU4 to preferentially catalyze sialyl-Lewis antigens expressed on O-glycans. Cell adhesion to and motility and growth on E-selectin were significantly reduced by NEU4. In addition, E-selectin stimulation of colon cancer cells enhanced cell motility through activation of the p38/Hsp27/actin reorganization pathway, whereas NEU4 attenuated such signaling. It is interesting to note here that the sialidase did not change the level of a normal glycan, disialyl-Le$^{a}$, generally expressed in nonmalignant epithelial cells. Although it has been proposed that glycosyltransferases are responsible for synthesis of these antigens, expression levels of the encoding genes have not always been found to correlate with sialyl-Lewis antigen contents, with even contradictory expression noted in various cells.

Considering regulation of polySia on NCAM by Neu4,[43] it is feasible that downregulation of the enzyme in colon cancer might be involved in polySia presence in cancers, probably enhancing tumor development, as it has been reported to be highly expressed in malignant tumors, including gliomas, lung, and colon cancers, as well as nervous tissue. A correlation of polySia existence with tumor development, invasion and poor prognosis has also been suggested.[112,113] These data overall suggest that Neu3 and Neu4 play mostly opposing roles in cancer progression as exemplified in colon cancer (Fig. 5).

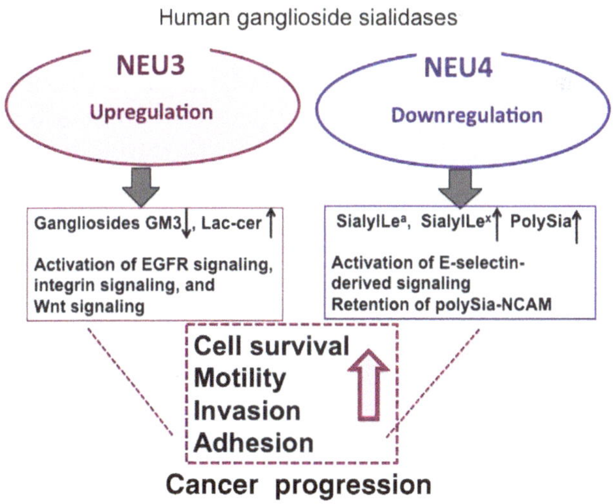


**Fig. 5** Involvement of human ganglioside sialidases in tumor progression of colon cancer[45] by disturbing cellular signaling. It should be noted here that interestingly, opposite to the case of NEU3, NEU1 desialylates integrin β4, and decreases its phosphorylation, leading to suppression of metastasis, as described in the text.

## 4. CONCLUDING REMARKS

Recent progress in mammalian sialidase research has provided a growing body of evidence for great importance in cellular functions, with clarification of the molecular bases and significance of physiological and pathological desialylation.

Remodeling of sialic acid residues in glycoconjugates by modulating expression of these sialidase genes should provide important clues to how control of the degree of sialylation impacts on functional molecules in cells. However, this area remains largely unexplored, partly due to relatively low expression levels and instability of endogenous sialidases relative to other proteins. Although transfection studies with designated r cDNAs have provided evidence of subcellular sites of action, overexpression of recombinant proteins might result in artificial patterns. At present, the detailed physiological topology of human NEU3 at plasma membranes is still uncertain, although intracellular movement is to be expected with growth stimuli. Regarding functional aspects, each sialidase seems to play unique roles depending on its particular enzyme properties, including substrate specificity. Human NEU3 essentially specifically hydrolyzes gangliosides, whereas mouse Neu3 possesses a much broader substrate specificity even acting on GM2, and this difference has recently clarified the pathogenesis of an asymptomatic mouse model of Tay-Sachs disease. It has now been validated in double knockout mice that two ganglioside sialidases, Neu3 and Neu4, have important roles in regulation of neuronal function by catabolizing brain gangliosides. The characteristic substrate specificity of Neu4, in contrast to other sialidases, acting on mucin and polySia will hopefully provide clues to further discover its functions in the future.

Altered expression of sialidases indeed exerts marked influence on malignancy, through desialylation of endogenous glycoconjugates, probably not simply dependent on the sialic acid content. In general, the two human sialidases NEU1 and NEU4 appear to be downregulated in several cancers and this facilitates metastasis by changing mainly adhesion mechanisms, whereas NEU3 upregulation observed in various carcinomas probably augments all the processes contributing to tumorigenesis, including cancer initiation, promotion, and progression (Fig. 5). Further investigation of mammalian sialidases should clarify the molecular basis of various pathological phenomena due to aberrant sialylation. Alteration of endogenous sialidase expression by treatment with specific siRNAs, antibodies, activators, or inhibitors has promise for clinical applications, especially against cancers and neurological diseases.

## REFERENCES


1. Hata K, Koseki K, Yamaguchi K, et al. Limited inhibitory effects of oseltamivir and zanamivir on human sialidases. *Antimicrob Agents Chemother.* 2008;52:3484–3491.
2. Chavas LM, Kato R, Suzuki N, et al. Complexity in influenza virus targeted drug design: interaction with human sialidases. *J Med Chem.* 2010;53:2998–3002.
3. Warren L, Spearing CW. Mammalian sialidase (neuraminidase). *Biochem Biophys Res Commun.* 1960;3:489–492.
4. Miyagi T, Yamaguchi K. Mammalian sialidases: physiological and pathological roles in cellular functions. *Glycobiology.* 2012;22:880–896.
5. Wang J, Wu G, Miyagi T, Lu ZH, Ledeen RW. Sialidase occurs in both membranes of the nuclear envelope and hydrolyzes endogenous GD1a. *J Neurochem.* 2009;111: 547–554.
6. Akita H, Miyagi T, Hata K, Kagayama M. Immunohistochemical evidence for the existence of rat cytosolic sialidase in rat skeletal muscles. *Histochem Cell Biol.* 1997;107: 495–503.
7. Chavas LM, Tringali C, Fusi P, et al. Crystal structure of the human cytosolic sialidase Neu2. Evidence for the dynamic nature of substrate recognition. *J Biol Chem.* 2005;280: 469–475.
8. Yamaguchi K, Hata K, Koseki K, et al. Evidence for mitochondrial localization of a novel human sialidase (NEU4). *Biochem J.* 2005;390:85–93.
9. Shiozaki K, Koseki K, Yamaguchi K, Shiozaki M, Narimatsu H, Miyagi T. Developmental change of sialidase neu4 expression in murine brain and its involvement in the regulation of neuronal cell differentiation. *J Biol Chem.* 2009;284: 21157–21164.
10. Carrillo MB, Milner CM, Ball ST, Snoek M, Campbell RD. Cloning and characterization of a sialidase from the murine histocompatibility-2 complex: low levels of mRNA and a single amino acid mutation are responsible for reduced sialidase activity in mice carrying the Neu1a allele. *Glycobiology.* 1997;7:975–986.
11. Igdoura SA, Gafuik C, Mertineit C. Cloning of the cDNA and gene encoding mouse lysosomal sialidase and correction of sialidase deficiency in human sialidosis and mouse SM/J fibroblasts. *Hum Mol Genet.* 1998;7:115–121.
12. Rottier RJ, Bonten E, d'Azzo A. A point mutation in the neu-1 locus causes the neuraminidase defect in the SM/J mouse. *Hum Mol Genet.* 1998;7:313–321.
13. Bonten E, van der Spoel A, Fornerod M, Grosveld G, d'Azzo A. Characterization of human lysosomal neuraminidase defines the molecular basis of the metabolic storage disorder sialidosis. *Genes Dev.* 1996;10:3156–3169.
14. Pshezhetsky AV, Richard C, Michaud L, et al. Cloning, expression and chromosomal mapping of human lysosomal sialidase and characterization of mutations in sialidosis. *Nat Genet.* 1997;15:316–320.
15. Milner CM, Smith SV, Carrillo MB, Taylor GL, Hollinshead M, Campbell RD. Identification of a sialidase encoded in the human major histocompatibility complex. *J Biol Chem.* 1997;272:4549–4558.
16. D'Azzo A, Hoogeveen A, Reuser AJ, Robinson D, Galjaard H. Molecular defect in combined beta-galactosidase and neuraminidase deficiency in man. *Proc Natl Acad Sci U S A.* 1982;79:4535–4539.
17. Galjart NJ, Gillemans N, Harris A, et al. Expression of cDNA encoding the human "protective protein" associated with lysosomal beta-galactosidase and neuraminidase: homology to yeast proteases. *Cell.* 1988;54:755–764.

18. D'Azzo A, Bonten E. Molecular mechanisms of pathogenesis in a glycosphingolipid and a glycoprotein storage disease. *Biochem Soc Trans.* 2010;38:1453–1457.
19. Yogalingam G, Bonten EJ, van de Vlekkert D, et al. Neuraminidase 1 is a negative regulator of lysosomal exocytosis. *Dev Cell.* 2008;15:74–86.
20. Pshezhetsky AV, Hinek A. Where catabolism meets signalling: neuraminidase 1 as a modulator of cell receptors. *Glycoconj J.* 2011;28:441–452.
21. Uemura T, Shiozaki K, Yamaguchi K, et al. Contribution of sialidase NEU1 to suppression of metastasis of human colon cancer cells through desialylation of integrin beta4. *Oncogene.* 2009;28:1218–1229.
22. Miyagi T, Konno K, Emori Y, et al. Molecular cloning and expression of cDNA encoding rat skeletal muscle cytosolic sialidase. *J Biol Chem.* 1993;268:26435–26440.
23. Ferrari J, Harris R, Warner TG. Cloning and expression of a soluble sialidase from Chinese hamster ovary cells: sequence alignment similarities to bacterial sialidases. *Glycobiology.* 1994;4:367–373.
24. Fronda CL, Zeng G, Gao L, Yu RK. Molecular cloning and expression of mouse brain sialidase. *Biochem Biophys Res Commun.* 1999;258:727–731.
25. Hasegawa T, Yamaguchi K, Wada T, Takeda A, Itoyama Y, Miyagi T. Molecular cloning of mouse ganglioside sialidase and its increased expression in Neuro2a cell differentiation. *J Biol Chem.* 2000;275:8007–8015.
26. Kotani K, Kuroiwa A, Saito T, Matsuda Y, Koda T, Kijimoto-Ochiai S. Cloning, chromosomal mapping, and characteristic 5′-UTR sequence of murine cytosolic sialidase. *Biochem Biophys Res Commun.* 2001;286:250–258.
27. Monti E, Preti A, Rossi E, Ballabio A, Borsani G. Cloning and characterization of NEU2, a human gene homologous to rodent soluble sialidases. *Genomics.* 1999;57: 137–143.
28. Sato K, Miyagi T. Genomic organization and the 5′-upstream sequence of the rat cytosolic sialidase gene. *Glycobiology.* 1995;5:511–516.
29. Fanzani A, Colombo F, Giuliani R, Preti A, Marchesini S. Cytosolic sialidase Neu2 upregulation during PC12 cells differentiation. *FEBS Lett.* 2004;566:178–182.
30. Koseki K, Wada T, Hosono M, et al. Human cytosolic sialidase NEU2 – low general tissue expression but involvement in PC-3 prostate cancer cell survival. *Biochem Biophys Res Commun.* 2012;428:142–149.
31. Miyagi T, Wada T, Iwamatsu A, et al. Molecular cloning and characterisation of a plasma membrane-associated sialidase specific for gangliosides. *J Biol Chem.* 1999;274:5004–5011.
32. Wada T, Yoshikawa Y, Tokuyama S, Kuwabara M, Akita H, Miyagi T. Cloning, expression, and chromosomal mapping of a human ganglioside sialidase. *Biochem Biophys Res Commun.* 1999;261:21–27.
33. Monti E, Bassi MT, Papini N, et al. Identification and expression of NEU3, a novel human sialidase associated to the plasma membrane. *Biochem J.* 2000;349:343–351.
34. Yamaguchi K, Hata K, Wada T, Moriya S, Miyagi T. Epidermal growth factor-induced mobilization of a ganglioside-specific sialidase (NEU3) to membrane ruffles. *Biochem Biophys Res Commun.* 2006;346:484–490.
35. Zanchetti G, Colombi P, Manzoni M, et al. Sialidase NEU3 is a peripheral membrane protein localized on the cell surface and in endosomal structures. *Biochem J.* 2007;408: 211–219.
36. Li SC, Li YT, Moriya S, Miyagi T. Degradation of G(M1) and G(M2) by mammalian sialidases. *Biochem J.* 2001;360:233–237.
37. Sandbhor MS, Soya N, Albohy A, et al. Substrate recognition of the membrane-associated sialidase NEU3 requires a hydrophobic aglycone. *Biochemistry.* 2011;50: 6753–6762.
38. Comelli EM, Amado M, Lustig SR, Paulson JC. Identification and expression of Neu4, a novel murine sialidase. *Gene.* 2003;321:155–161.

39. Monti E, Bassi MT, Bresciani R, et al. Molecular cloning and characterization of NEU4, the fourth member of the human sialidase gene family. *Genomics*. 2004;83: 445–453.
40. Seyrantepe V, Landry K, Trudel S, Hassan JA, Morales CR, Pshezhetsky AV. Neu4, a novel human lysosomal lumen sialidase, confers normal phenotype to sialidosis and galactosialidosis cells. *J Biol Chem*. 2004;279:37021–37029.
41. Bigi A, Morosi L, Pozzi C, et al. Human sialidase NEU4 long and short are extrinsic proteins bound to outer mitochondrial membrane and the endoplasmic reticulum, respectively. *Glycobiology*. 2010;20:148–157.
42. Shiozaki K, Yamaguchi K, Takahashi K, Moriya S, Miyagi T. Regulation of sialyl Lewis antigen expression in colon cancer cells by sialidase NEU4. *J Biol Chem*. 2011;286: 21052–21061.
43. Takahashi K, Mitoma J, Shiozaki K, et al. Sialidase NEU4 hydrolyzes polysialic acids of neural cell adhesion molecules and negatively regulates neurite formation by hippocampal neurons. *J Biol Chem*. 2012;287:14816–14826.
44. Monti E, Bonten E, D'Azzo A, et al. Sialidases in vertebrates: a family of enzymes tailored for several cell functions. *Adv Carbohydr Chem Biochem*. 2010;64:403–479.
45. Miyagi T, Takahashi K, Hata K, Shiozaki K, Yamaguchi K. Sialidase significance for cancer progression. *Glycoconjugate J*. 2012;29:567–577.
46. Monti E, Miyagi T, Structure and function of mammalian sialidases. Gerardy-Schahn R, Delannoy P, von Itzstein M, eds. *Top Curr Chem*. 366. Springer; 2015: 183–208.
47. Shiozaki K, Takahashi K, Hosono M, et al. Phosphatidic acid-mediated activation and translocation to the cell surface of sialidase NEU3, promoting signaling for cell migration. *FASEB J*. 2015;29:2099–2111.
48. Scaringi R, Piccoli M, Papini N, et al. NEU3 sialidase is activated under hypoxia and protects skeletal muscle cells from apoptosiss through the activation of the epidermal growth factor receptor signaling pathway and hypoxia-inducible factor (HIF)-1α. *J Biol Chem*. 2013;88:3153–3162.
49. Yamaguchi K, Koseki K, Shiozaki M, Shimada Y, Wada T, Miyagi T. Regulation of plasma-membrane-associated sialidase NEU3 gene by Sp1/Sp3 transcription factors. *Biochem J*. 2010;430:107–117.
50. Kimura K, Wakamatsu A, Suzuki Y, et al. Diversification of transcriptional modulation: large-scale identification and characterization of putative alternative promoters of human genes. *Genome Res*. 2006;16:55–65.
51. Wierstra I. Sp1:emerging roles—beyond constitutive activation of TATA-less housekeeping genes. *Biochem Biophys Res Commun*. 2008;372:1–13.
52. Valaperta R, Chigorno V, Basso L, et al. Plasma membrane production of ceramide from ganglioside GM3 in human fibroblasts. *FASEB J*. 2006;20:1227–1229.
53. Kopitz J, von Reitzenstein C, Muhl C, Cantz M. Role of plasma membrane ganglioside sialidase of human neuroblastoma cells in growth control and differentiation. *Biochem Biophys Res Commun*. 1994;199:1188–1193.
54. Kopitz J, von Reitzenstein C, Burchert M, Cantz M, Gabius HJ. Galectin-1 is a major receptor for ganglioside GM1, a product of the growth-controlling activity of a cell surface ganglioside sialidase, on human neuroblastoma cells in culture. *J Biol Chem*. 1998;273:11205–11211.
55. Kopitz J, Oehler C, Cantz M. Desialylation of extracellular GD1a-neoganglioprotein suggests cell surface orientation of the plasma membrane-bound ganglioside sialidase activity in human neuroblastoma cells. *FEBS Lett*. 2001;491:233–236.
56. Proshin S, Yamaguchi K, Wada T, Miyagi T. Modulation of neuritogenesis by ganglioside-specific sialidase (Neu 3) in human neuroblastoma NB-1 cells. *Neurochem Res*. 2002;27:841–846.

57. Rodriguez JA, Piddini E, Hasegawa T, Miyagi T, Dotti CG. Plasma membrane ganglioside sialidase regulates axonal growth and regeneration in hippocampal neurons in culture. *J Neurosci.* 2001;21:8387–8395.
58. Da Silva JS, Hasegawa T, Miyagi T, Dotti CG, Abad-Rodriguez J. Asymmetric membrane ganglioside sialidase activity specifies axonal fate. *Nat Neurosci.* 2005;8:606–615.
59. Kappagantula S, Andrews MR, Cheah M, Abad-Rodriguez J, Dotti CG, Fawcett JW. Neu3 sialidase-mediated ganglioside conversion is necessary for axon regeneration and is blocked in CNS axons. *J Neurosci.* 2014;34(7):2477–2492.
60. Aureli M, Loberto N, Lanteri P, Chigorno V, Prinetti A, Sonnino S. Cell surface sphingolipid glycohydrolases in neuronal differentiation and aging in culture. *J Neurochem.* 2010;116:891–899.
61. Anastasia L, Papini N, Colazzo F, et al. NEU3 sialidase strictly modulates GM3 levels in skeletal myoblasts C2C12 thus favoring their differentiation and protecting them from apoptosis. *J Biol Chem.* 2008;283:36265–36271.
62. Moon SK, Cho SH, Kim KW, et al. Overexpression of membrane sialic acid-specific sialidase Neu3 inhibits matrix metalloproteinase-9 expression in vascular smooth muscle cells. *Biochem Biophys Res Commun.* 2007;356:542–547.
63. Kalka D, von Reitzenstein C, Kopitz J, Cantz M. The plasma membrane ganglioside sialidase cofractionates with markers of lipid rafts. *Biochem Biophys Res Commun.* 2001;283:989–993.
64. Wang Y, Yamaguchi K, Wada T, et al. A close association of the ganglioside-specific sialidase Neu3 with caveolin in membrane microdomains. *J Biol Chem.* 2002;277: 26252–26259.
65. Veracini L, Simon V, Richard V, et al. The Csk-binding protein PAG regulates PDGF-induced Src mitogenic signaling via GM1. *J Cell Biol.* 2008;182:603–614.
66. Wang XQ, Sun P, Paller AS. Ganglioside induces caveolin-1 redistribution and interaction with the epidermal growth factor receptor. *J Biol Chem.* 2002;277: 47028–47034.
67. Stamatos NM, Carubelli I, van de Vlekkert D, et al. LPS-induced cytokine production in human dendritic cells is regulated by sialidase activity. *J Leukoc Biol.* 2010;88: 1227–1239.
68. Miyagi T, Wada T, Yamaguchi K, Hata K, Shiozaki K. Plasma membrane-associated sialidase as a crucial regulator of transmembrane signalling. *J Biochem.* 2008;144: 279–285.
69. Lillehoj EP, Hyun SW, Feng C, et al. Human airway epithelia express catalytically active NEU3 sialidase. *Am J Physiol Lung Cell Mol Physiol.* 2014;306:L876–86.71.
70. D'Avila F1. Tringali C, Papini N, et al. Identification of lysosomal sialidase NEU1 and plasma membrane sialidase NEU3 in human erythrocytes. *J Cell Biochem.* 2013;114: 204–211.
71. Hasegawa T, Sugeno N, Takeda A, et al. Role of Neu4L sialidase and its substrate ganglioside GD3 in neuronal apoptosis induced by catechol metabolites. *FEBS Lett.* 2007;581:406–412.
72. Finlay TM, Abdulkhalek S, Gilmour A, et al. Thymoquinone-induced Neu4 sialidase activates NFkappaB in macrophage cells and pro-inflammatory cytokines in vivo. *Glycoconj J.* 2010;27:583–600.
73. Kleene R, Schachner M. Glycans and neural cell interactions. *Nat Rev Neurosci.* 2004;5: 195–208.
74. Rutishauser U. Polysialic acid in the plasticity of the developing and adult vertebrate nervous system. *Nat Rev Neurosci.* 2008;9:26–35.
75. Seki T, Arai Y. Distribution and possible roles of the highly polysialylated neural cell adhesion molecule (NCAM-H) in the developing and adult central nervous system. *Neurosci Res.* 1993;17:265–290.

76. Angata K, Fukuda M. Polysialyltransferases: major players in polysialic acid synthesis on the neural cell adhesion molecule. *Biochimie.* 2003;85:195–206.
77. Yamaguchi K, Shiozaki K, Moriya S, et al. Reduced susceptibility to colitis-associated colon carcinogenesis in mice lacking plasma membrane-associated sialidase. *PLoS ONE.* 2012;7:e41132.
78. Seyrantepe V, Canuel M, Carpentier S, et al. Mice deficient in Neu4 sialidase exhibit abnormal ganglioside catabolism and lysosomal storage. *Hum Mol Genet.* 2008;17: 1556–1568.
79. Pan X, De Aragão CBP, Velasco-Martin JP, et al. Neuraminidases 3 and 4 regulate neuronal function by catabolizing brain gangliosides. *FASEB J.* 2017;31:3467–3483.
80. Gravel R, Kaback MM, Proia RL, Sandhoff K, Suzuki K. The GM2 gangliosidoses. In: Scriver CR, Beaudet AL, Sly WS, Valle D, eds. *The Metabolic and Molecular Bases of Inherited Diseases. New York: McGraw-Hill; 2001:3827–3876.*
81. Seyrantepe V, Lema P, Caqueret A, et al. Mice doubly-deficient in lysosomal hexosaminidase A and neuraminidase 4 show epileptic crises and rapid neuronal loss. *PLoS Genet.* 2010;6:e1001118.
82. Seyrantepe V, Demir SA, Timur ZK, et al. Murine sialidase Neu3 facilitates GM2 degradation and bypass in mouse model of Tay-Sachs disease. *Exp Neurol.* 2018;299: 26–41.
83. Schneider JS, Yuwiler A. GM1 ganglioside treatment promotes recovery of striatal dopamine concentrations in the mouse model of MPTP-induced parkinsonism. *Exp Neurol.* 1989;105:177–183.
84. Schneider JS, Gollomp SM, Sendek S, Colcher A, Cambi F, Du W. A randomized, controlled, delayed start trial of GM1 ganglioside in treated Parkinson's disease patients. *J Neurol Sci.* 2013;324:140–148.
85. Schneider JS, Seyfried TN, Choi HS. Intraventricular sialidase administration enhances GM1 ganglioside expression and is partially neuroprotective in a mouse model of Parkinson's disease. *PLOS ONE.* 2015;10:e0143351.
86. Dhanushkodi A, McDonald MP. Intracranial *V. cholerae* sialidase protects against excitotoxic neurodegeneration. *PLoS ONE.* 2011;6:e29285.
87. Yang LJ, Lorenzini I, Vajn K, Mountney A, Schramm LP, Schnaar RL. Sialidase enhances spinal axon outgrowth in vivo. *Proc Natl Acad Sci U S A.* 2006;103:11057–11062.
88. Huang YL, Chassard C, Hausmann M, von Itzstein M, Hennet T. Sialic acid catabolism drives intestinal inflammation and microbial dysbiosis in mice. *Nat Commun.* 2015;6:8141.
89. Miklavcic JJ, Hart TD, Lees GM, et al. Increased catabolism and decreased unsaturation of ganglioside in patients with inflammatory bowel disease. *World J Gastroenterol.* 2015;21:10080–10090.
90. Xie C, Yagai T, Luo Y, et al. Activation of intestinal hypoxia-inducible factor 2α during obesity contributes to hepatic steatosis. *Nat Med.* 2017;23(11):1298–1308.
91. Sasaki A, Hata K, Suzuki S, et al. Overexpression of plasma membrane-associated sialidase attenuates insulin signaling in transgenic mice. *J Biol Chem.* 2003;278: 27896–27902.
92. Yoshizumi S, Suzuki S, Hirai M, et al. Increased hepatic expression of ganglioside-specific sialidase, NEU3, improves insulin sensitivity and glucose tolerance in mice. *Metabolism.* 2007;56:420–429.
93. Kabayama K, Sato T, Saito K, et al. Dissociation of the insulin receptor and caveolin-1 complex by ganglioside GM3 in the state of insulin resistance. *Proc Natl Acad Sci U S A.* 2007;104:13678–13683.
94. Mandal C, Tringali C, Mondal S, Anastasia L, Chandra S, Venerando B. Down regulation of membrane-bound Neu3 constitutes a new potential marker for childhood acute lymphoblastic leukemia and induces apoptosis suppression of neoplastic cells. *Int J Cancer.* 2010;126:337–349.

95. Takahashi K, Proshin S, Yamaguchi K, et al. Sialidase NEU3 defines invasive potential of human glioblastoma cells by regulating calpain-mediated proteolysis of focal adhesion proteins. *Biochim Biophys Acta.* 2017;1861(11 Pt A):2778–2788.
96. Kakugawa Y, Wada T, Yamaguchi K, et al. Up-regulation of plasma membrane-associated ganglioside sialidase (Neu3) in human colon cancer and its involvement in apoptosis suppression. *Proc Natl Acad Sci U S A.* 2002;99:10718–10723.
97. De Weille J. On the genesis of neuroblastoma and glioma. *Int J Brain Sci.* 2014;2014:14 p. Article ID 217503.
98. Shiozaki K, Yamaguchi K, Sato I, Miyagi T. Plasma membrane-associated sialidase (NEU3) promotes formation of colonic aberrant crypt foci in azoxymethane-treated transgenic mice. *Cancer Sci.* 2009;100:588–594.
99. Takahashi K, Hosono M, Sato I, et al. Sialidase NEU3 contributes neoplastic potential on colon cancer cells as a key modulator of gangliosides by regulating Wnt signaling. *Int J Cancer.* 2015;137:1560–1573.
100. Kato T, Wang Y, Yamaguchi K, et al. Overexpression of lysosomal-type sialidase leads to suppression of metastasis associated with reversion of malignant phenotype in murine B16 melanoma cells. *Int J Cancer.* 2001;92:797–804.
101. Clevers H. Wnt/beta-catenin signaling in development and disease. *Cell.* 2006;127: 469–480.
102. Yamamoto K, Takahashi K, Shiozaki K, et al. Potentiation of epidermal growth factor-mediated oncogenic transformation by sialidase NEU3 leading to Src activation. *PLOS ONE.* 2015;10:e0120578.
103. Wada T, Hata K, Yamaguchi K, et al. A crucial role of plasma membrane-associated sialidase in the survival of human cancer cells. *Oncogene.* 2007;26:2483–2490.
104. Kato K, Shiga K, Yamaguchi K, et al. Plasma-membrane-associated sialidase (NEU3) differentially regulates integrin-mediated cell proliferation through laminin- and fibronectin-derived signalling. *Biochem J.* 2006;394:647–656.
105. Uemura T, Shiozaki K, Yamaguchi K, et al. Contribution of sialidase NEU1 to suppression of metastasis of human colon cancer cells through desialylation of integrin beta4. *Oncogene.* 2009;28:1218–1229.
106. Ueno S, Saito S, Wada T, et al. Plasma membrane-associated sialidase is up-regulated in renal cell carcinoma and promotes interleukin-6-induced apoptosis suppression and cell motility. *J Biol Chem.* 2006;281:7756–7764.
107. Kawamura S, Sato I, Wada T, et al. Plasma membrane-associated sialidase (NEU3) regulates progression of prostate cancer to androgen-independent growth through modulation of androgen receptor signaling. *Cell Death Differ.* 2012;19:170–179.
108. Li X, Zhang L, Shao Y, et al. Effects of a human plasma membrane-associated sialidase siRNA on prostate cancer invasion. *Biochem Biophys Res Commun.* 2011;416:270–276.
109. Hata K, Tochigi T, Sato I, et al. Increased sialidase activity in serum of cancer patients: identification of sialidase and inhibitor activities in human serum. *Cancer Sci.* 2015;106:383–389.
110. Shiga K, Takahashi K, Sato I, et al. Up-regulation of sialidase NEU3 in head and neck squamous cell carcinoma associated with lymph node metastasis. *Cancer Sci.* 2015;106:1544–1553.
111. Yamanami H, Shiozaki K, Wada T, et al. Down-regulation of sialidase NEU4 may contribute to invasive properties of human colon cancers. *Cancer Sci.* 2007;98:299–307.
112. Tanaka F, Otake Y, Nakagawa T, et al. Prognostic significance of polysialic acid expression in resected non-small cell lung cancer. *Cancer Res.* 2001;161:1666–16670.
113. Suzuki M, Nakayama J, Suzuki A, et al. Polysialic acid facilitates tumor invasion by glioma cells. *Glycobiology.* 2005;15:887–894.

CHAPTER FIVE

# Biology of GM3 Ganglioside

**Jin-ichi Inokuchi*[,1], Kei-ichiro Inamori*, Kazuya Kabayama[†], Masakazu Nagafuku*, Satoshi Uemura[‡], Shinji Go*, Akemi Suzuki*, Isao Ohno[§], Hirotaka Kanoh*, Fumi Shishido***

[*] Institute of Molecular Biomembrane and Glycobiology, Tohoku Medical and Pharmaceutical University, Sendai, Miyagi, Japan
[†] Osaka University, Toyonaka, Osaka, Japan
[‡] Tohoku Medical and Pharmaceutical University, Sendai, Miyagi, Japan
[§] Center for Medical Education, Tohoku Medical and Pharmaceutical University, Sendai, Miyagi, Japan
[1] Corresponding author. E-mail address: jin@tohoku-mpu.ac.jp

## Contents



*Progress in Molecular Biology and Translational Science,* Volume 156
ISSN 1877-1173
http://dx.doi.org/10.1016/bs.pmbts.2017.10.004

## Abstract

Since the successful molecular cloning in 1998 of GM3 synthase (GM3S, ST3GAL5), the enzyme responsible for initiating biosynthesis of all complex gangliosides, the efforts of our research group have been focused on clarifying the physiological and pathological implications of gangliosides, particularly GM3. We have identified isoforms of GM3S proteins having distinctive lengths of N-terminal cytoplasmic tails, and found that these cytoplasmic tails define subcellular localization, stability, and in vivo activity of GM3S isoforms. Our studies of the molecular pathogenesis of type 2 diabetes, focused on interaction between insulin receptor and GM3 in membrane microdomains, led to a novel concept: type 2 diabetes and certain other lifestyle-related diseases are membrane microdomain disorders resulting from aberrant expression of gangliosides. This concept has enhanced our understanding of the pathophysiological roles of GM3 and related gangliosides in various diseases involving chronic inflammation, such as insulin resistance, leptin resistance, and T-cell function and immune disorders (e.g., allergic asthma). We also demonstrated an essential role of GM3 in murine and human auditory systems; a common pathological feature of GM3S deficiency is deafness. This is the first direct link reported between gangliosides and auditory functions.

## 1. INTRODUCTION

The presence of sialic acid at the surface of animal cell membranes was first demonstrated in 1951 by Tamio Yamakawa.[1] He isolated a sialic acid-containing glycolipid, which he called "hematoside," from 100 g of dried ghosts obtained from 10 L of packed horse erythrocytes.[1,2] Hematoside (now termed "GM3 ganglioside") is the first product in the biosynthetic pathway of ganglio-series gangliosides, and the common precursor of complex gangliosides, such as a- and b-series gangliosides (Fig. 1). Gangliosides are expressed in a cell type-specific manner on the outer leaflet of cell membranes, and interact with various molecules on plasma membranes and certain intracellular membranes based on potentials of noncovalent bondings, for example,

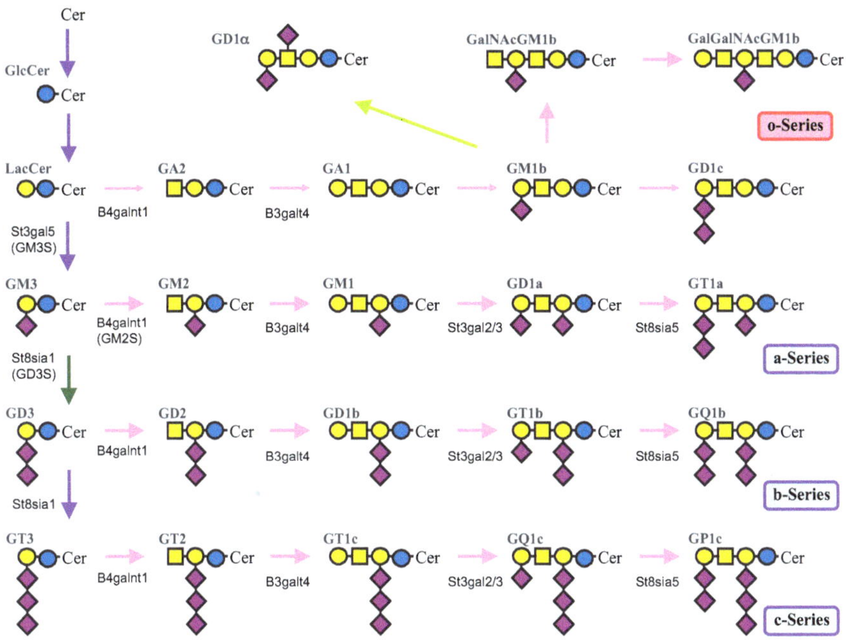


**Fig. 1** Biosynthetic pathway of gangliosides.

hydrogen bonding and electrostatic and hydrophobic interactions. Members of the various ganglioside families participate in numerous cellular activities by forming dynamic functional complexes termed "membrane microdomains" or "lipid rafts."[3,4] Expression levels of cellular gangliosides are affected by various extracellular stimuli, including inflammatory cytokines. The presence of gangliosides in membrane microdomains thus reflects characteristics of individual cells under various physiological and pathological environments. Since molecular cloning of GM3S was achieved in 1988,[5] the activity of our research group has been focused on elucidating the biological functions of GM3 and related gangliosides. Here, we review our recent discoveries and advances regarding the molecular biology of GM3S, and the pathophysiological significance of GM3 and related gangliosides.[6–8]

## 2. MOLECULAR BIOLOGY OF GM3S/ST3GAL5

GM3S is a type II membrane protein having a catalytic domain in the luminal side (Fig. 2). Human and mouse GM3Ss initially undergo modification with two or three *N*-glycans, respectively, in the endoplasmic

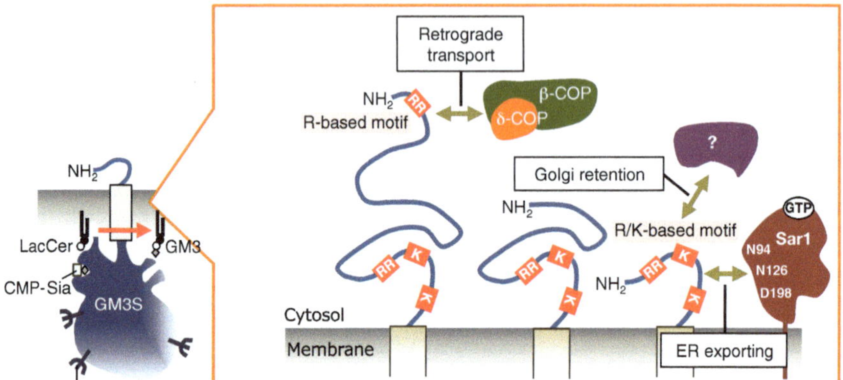


| Isoform names | M1 | M2 | M3 |
|---|---|---|---|
| Localization | ER | Golgi | Golgi |
| Stability | Higher | Lower | Higher |
| Structure of *N*-glycans | High-mannose type | High-mannose type | Complex type |

**Fig. 2** Regulation of intracellular dynamics of GM3S isoforms by R-based and R/K-based motifs located within cytoplasmic tails. All isoforms have an R/K-based motif, which functions as an endoplasmic reticulum (ER) export signal by interacting with Sar1 protein. The R/K-based motif also functions as a Golgi retention signal in M3 isoform only, leading to stabilization of this isoform in Golgi. R/K-based motifs of M1 and M2 isoforms are presumably unable to function as Golgi retention signals because of their long cytoplasmic tails. Only M3 carries highly matured *N*-glycans. M1, which has R-based motif in its cytoplasmic tail, is targeted for ER by interaction with β-COP-δ-COP complex. The significance of ER targeting of M1 is unclear.

reticulum (ER).[9] *N*-glycans are further modified in the Golgi apparatus, and glycan structures are converted from high-mannose type (immature form) to complex type (mature form). The presence of high-mannose type sugar chains is essential for enzyme activity.[9] Lactosylceramide (LacCer), the acceptor substrate for GM3S, is synthesized in the luminal side of Golgi.[10] CMP-sialic acid, a donor substrate for GM3 synthesis, is synthesized in cytosol and transported to the luminal side by CMP-sialic acid transporter localized in Golgi.[11] GM3 synthesis occurs subsequently in the luminal side of *trans*-Golgi.[12] This section describes regulation of intracellular dynamics of three GM3S isoforms (M1, M2, and M3) we discovered.

## 2.1 Tissue Distribution of GM3S mRNATranscriptional Variants

Human GM3S has five mRNA variants,[5,13–15] which are categorized into a-, b-, and c-type based on differences in transcription start sites (TSSs).[16] a-Type and b-type are further subdivided into a1- (type 1), a2- (type 2), b1- (type 3), and b2-type (type 4) based on differences in splicing. a2-Type is expressed in a variety of tissues.[16] c-Type (type 5) is expressed specifically and exclusively in pancreas.[15]

Mouse GM3S has two transcriptional variants: a- and b-type.[13,17] The TSSs of these variants are very close, and their transcriptional regulation is therefore considered to be similar. Tissue distribution of the two variants is nearly identical.[16] We recently discovered a liver-specific transcriptional variant, c-type.[18] The c-type TSS is located downstream and far away from the TSSs of a- and b-type. We are currently investigating the liver-specific expression mechanism of c-type.

## 2.2 Translation of GM3S Isoforms by Leaky Scanning System

In general, the first AUG codon is recognized by ribosomes as an initiation codon (first-AUG rule).[19] In mammals, the optimal context for recognition of the AUG start codon is GCCRCCAUGG (underline: initiation codon; R: A or G). When the first AUG residue is presented in a weak context, in particular, lacking both R in position -3 and G in position +4, the first AUG codon is sometimes skipped and a downstream AUG is recognized as an initiation codon. This process is termed a leaky scanning system.[19] By using this system, GM3S isoforms (M1, M2, M2′, and M3 in human; M1, M2, and M3 in mice), each having a distinctive length of N-terminal cytoplasmic tail, are translated from each mRNA variant. The lengths of isoforms M1, M2, M2′, and M3 are respectively 69, 42, 47, and 14 aa.

Three isoforms (M1, M2, M3) or two isoforms (M2, M3) were detected, respectively, in cells expressing mouse a- or b-type variant. In contrast, one isoform (M3) or two isoforms (M1 and M3) were detected respectively in cells expressing human a1- or a2-type variant.[16] M2′ and M3 isoforms were detected in cells expressing human c-type variant.[20] Thus, the leaky scanning system seems to elicit coexistence of various GM3S isoforms in cells.

## 2.3 Regulation of ER Export of GM3S by R/K-Based Motif

Transport of GM3S from ER to Golgi is the first regulatory step for GM3 synthesis. Glycosyltransferases are loaded efficiently into COPII vesicles through interaction between [R/K](X)[R/K] motifs in the cytoplasmic tail and small GTPase Sar1, a COPII coated-protein.[21,22] All GM3S isoforms carry RR sequence just above the transmembrane domain. However, two lysine residues located downstream of the RR sequence are also involved in ER export signal [$^{2}R^{3}R(X)_5{}^{9}K(X)_3{}^{13}K$] in M3-GM3S.[23] We therefore propose this "R/K-based motif" as a new ER export signal (Fig. 2).

Formation of COPII-coated vesicles begins with conversion from Sar1-GDP to Sar1-GTP by the guanine nucleotide exchange factor, Sec12.[24] Sar1-GTP is anchored to ER membrane and recruits Sec23–Sec24 complex by interacting with Sec23.[25] The Sar1-Sec23-Sec24 complex recruits many cargo proteins containing ER export signals (DXE and FF sequences) through interaction with Sec23 or Sec24 to generate a prebudding complex.[26–30] Sec23 also exhibits GTPase-activating protein activity for Sar1-GTP. Once Sar1 forms a prebudding complex, Sar1-GTP is hydrolyzed by Sec23 and is dissociated from prebudding complex.[31] Prebudding complex without Sar1 is unstable, but Sar1 is immediately reactivated by Sec12 and incorporated into the complex. When prebudding complex without Sar1 contains unfolded proteins, it breaks before reassembly of Sar1-GTP because of weak interaction between the proteins and Sec23/Sec24. The GTP/GDP cycle of Sar1 thus functions as a selective sorting system for cargo proteins.[31] If R/K-based motif interacts only with Sar1, the cargo proteins dissociate from prebudding complex when Sar1-GTP is converted to Sar1-GDP. The interaction by itself is therefore insufficient for glycosyltransferases to be efficiently concentrated in COPII-coated vesicles. Our current hypothesis is that R/K-based motif functions as a guide to prebudding complex through interaction with Sar1.[23] Glycosyltransferases carrying R/K-based motif may be transferred to other cargo proteins (receptors) in prebudding complex.

## 2.4 Regulation of Retrograde Transport of M1-GM3S Isoform by R-Based Motif

Among three isoforms, only M1-GM3S has a retrograde transport signal (R-based motif) in its N-terminal cytoplasmic tail (Fig. 2).[16] Accordingly, most M1-GM3S is transported from Golgi to ER by R-based motif, and localized in ER. As R-based motifs interact with an acidic pocket of β-COP-δ-COP complex in COPI subunits, proteins carrying R-based motif are efficiently concentrated in COPI-coated vesicles involved in retrograde transport.[32] R-based motifs must be positioned at a certain distance from the lipid bilayer to function,[33] indicating that R/K-based motif (ER export signal) just above transmembrane domain cannot function as a retrograde transport signal. In vivo enzymatic activity of M1-GM3S is lower than those of M2-GM3S and M3-GM3S. However, GM3 synthesis is not completely eliminated when GM3S-KO cells express only M1-GM3S, suggesting that some M1-GM3S is sorted to *trans*-Golgi by escaping retrograde transport.

What is the biological significance of ER targeting of M1-GM3S? One possibility is that M1-GM3S has functions other than GM3 synthesis in ER. For instance, protein O-fucosyltransferase (OFUT1) is known as a glycosyltransferase localized in ER, similar to M1-GM3S.[34] OFUT1 has an ER targeting signal (KDEL-like motif) in its C-terminus and is localized in ER. OFUT1 glycosylates the epidermal growth factor (EGF)-like domain of Notch in Golgi, but displays Notch chaperone activity in ER.[34] Glycosylation activity of OFUT1 is not necessary for chaperone activity. The strong protein-protein interaction between OFUT1 and EGF-like domain of Notch presumably allows the unusual function. In contrast, GM3S recognizes the glycosphingolipids (GSLs), LacCer and galactosylceramide (GalCer) as substrates. GM3S does not display sialylation activity toward lactose and *N*-glycans of glycoproteins,[17] so the active site of GM3S presumably recognizes both carbohydrate (hydrophilic) and ceramide (hydrophobic) portions. The partial hydrophobic pocket may interact with unfolded proteins, but the folding site is limited within the lipid bilayer. It therefore seems unlikely that M1-GM3S is involved in protein folding, in contrast to OFUT1.

Another possibility is that M1-GM3S regulates expression levels of M2-GM3S and M3-GM3S in Golgi. As indicated by expression patterns of mRNA variants in tissues, three GM3S isoforms usually coexist within a single cell. All GM3S isoforms are capable of interacting with each other (unpublished data); therefore, M2-GM3S and M3-GM3S may partially localize in ER by coexpression with M1-GM3S. Total amounts of GM3S isoforms in Golgi are consequently maintained at low level. We were able to establish stable clones expressing M2-GM3S or M3-GM3S only at low level, but easily established stable clones expressing M1-GM3S at high level. These findings suggest that excessive amounts of GM3S in Golgi are cytotoxic. M1-GM3S may promote cell survival by suppressing accumulation of M2-GM3S and M3-GM3S in Golgi.

## 2.5 Golgi Retention and Maturation of *N*-Glycans in GM3S

Golgi cisternae (*cis*, *medial*, and *trans*) are characterized by Golgi-resident enzymes. In the cisternal maturation model, the Golgi apparatus is regarded as a dynamic structure.[35,36] According to this model, a new *cis* cisterna is continually formed by vesicles supplied from ER, and progressively matures to a *medial* cisterna and then a *trans* cisterna. Retrograde transport of Golgi-resident enzymes by COPI-coated vesicles explains their distinctive distribution and Golgi retention. The (F/L)(L/I)XX(R/K) motif of Kre6, which is involved synthesis of β1-6 glycans synthesis, functions as a Golgi retention signal in *Saccharomyces cerevisiae*.[37] This motif interacts with Vps74 in COPI-coated protein complex. Golph3 and Golph3L, the mammalian homologs of Vps47, cannot function complementarily in *vps47Δ* strain and (F/L)(L/I) XX(R/K) motif is not conserved in mammalian glycosyltransferases. Golgi retention of enzymes in mammals therefore presumably occurs by other mechanisms.

The half-life of M3-GM3S is longer than that of M2-GM3S, although both are localized in Golgi.[16] We recently found that partial mutations of R/K-based motif (R2A, R3A, R2A/R3A, K9A/K13A) in M3-GM3S lead to altered localization from Golgi to endosomes and to rapid degradation,[23] suggesting that R/K-based motif also functions as a Golgi retention signal (Fig. 2). The motif is presumably involved in retrograde transport from Golgi. Masking of R/K-based motif by the long cytoplasmic tail of M2-GM3S may contribute to its low stability.

The differences of *N*-glycan structures among GM3S isoforms are noteworthy. The *N*-glycan structure of M1-GM3S is high-mannose type (immature form), as evidenced by the immediate relocation of M1-GM3S from Golgi to ER by R-based motif. The *N*-glycan structure of M2-GM3S localized in Golgi is also high-mannose type, whereas that of M3-GM3S is complex type (mature form).[16] The proportion of mature *N*-glycans in M3-GM3S is reduced by mutations (R2A, R3A, R2A/R3A, K9A/K13A) of Golgi retention signal (R/K-based motif).[23] Our findings, taken together, suggest that Golgi-retention time of glycoprotein determines maturation level of *N*-glycans. For full maturation of *N*-glycans, glycoproteins must undergo modification by several glycosyltransferases in dynamic Golgi cisternae. A single passage through Golgi is not sufficient to convert all glycosyltransferases to mature forms.

## 2.6 Perspectives

Our studies of GM3S isoforms help clarify the regulatory mechanism of ganglioside synthesis in dynamic Golgi cisternae. Issues to be addressed in the future studies include:

1. Determination of three-dimensional structure of GM3S by X-ray crystallography.
2. Identification of molecules involved in ER export and Golgi retention of GM3S.

3. Examination of the heterooligomer formation by GM3S isoforms, and cytotoxic effects of GM3S by the accumulation in Golgi.
4. Elucidation of *N*-glycan maturation mechanism through analysis of *N*-glycan structures of GM3S isoforms and their mutants, which cannot be tethered to Golgi.

Elucidation of the posttranslational regulation of GM3S will contribute to more effective therapeutic strategies for pathological conditions resulting from abnormal GM3 synthesis.

## 3. INSULIN RESISTANCE IN ADIPOCYTES AND MACROPHAGES

Visceral adipose tissues are comprised of not only adipocytes; but also immune cells, including resident macrophages, T lymphocytes, and other types of cells.[38–41] It remains unclear how interactions between adipocytes and resident macrophages are affected by expression of specific GSLs. We have shown that expression of GSLs and their corresponding synthase genes in adipocytes is maintained by soluble factors secreted from resident macrophages under not only inflammatory states, but also steady-state physiological conditions. Obese *GM3S*-deficient mice fed a high-fat diet were resistant to development of proinflammatory states in adipose tissues. This section focuses on association of GM3 with metabolic syndrome, and the physiological and pathological implications of GM3 present in adipose tissues.

### 3.1 GM3 is an Inducer of Insulin Resistance

Insulin elicits a wide variety of biological activities, which can be categorized into metabolic and mitogenic actions. Binding of insulin-to-insulin receptor (IR) activates IR internal-tyrosine (Tyr) kinase activity. Tyr-phosphorylated and activated IR is able to recruit and phosphorylate adaptor proteins, such as IR substrate (IRS). Phosphorylated IRS activates PI3-kinase (PI3K). Activated PI3K is translocated to lipid rafts and converts $PIP_2$ to $PIP_3$, and $PIP_3$ then recruits PDK1 to phosphorylate Akt. Full activation of Akt may be required for signaling by phosphorylation of the other site by mTORC2 (mTOR complex 2).[42] This IR-IRS-PI3K-Akt signaling cascade is a representative metabolic pathway triggered by insulin, and results in translocation of glucose transporter 4 (GLUT-4) to plasma membrane to facilitate glucose uptake.

When mouse adipocytes were cultured in low concentrations of TNFα that did not cause generalized suppression of expression of adipocyte genes (e.g., IRS-1, GLUT-4), TNFα interfered with insulin action.[43] Prolonged treatment (≥72 h) was required for this effect, in contrast to many acute effects of this cytokine. The slowness of the effect suggests that insulin resistance in adipocytes treated with 0.1 nM TNFα was associated with progressive increases in cellular GM3 content, GM3S activity, and GM3S mRNA content, and that such treatment upregulates GM3 synthesis at the transcriptional level in cultured adipocytes.[44,45] In contrast, ceramide levels under TNFα treatment increased transiently up to 6 h, and returned to normal by 24 h. These observations reflect the distinct and independent roles of GM3 and ceramides in development of insulin resistance in adipocytes.[44,45] To determine whether elevated GM3 in 3T3-L1 adipocytes treated with TNFα is involved in insulin resistance, we used D-*threo*-1-phenyl-2-decanoylamino-3-morpholino-1propanol (D-PDMP), an inhibitor of glucosylceramide synthase,[46] to deplete cellular GSLs derived from GlcCer. D-PDMP treatment counteracted TNFα-induced increase of GM3 content in adipocytes, and reverted to normal the TNFα-induced defect in Tyr phosphorylation of IRS-1 in response to insulin stimulation.[7] These findings are consistent with the observation that insulin signaling was enhanced in GM3S-knockout mice.[47] It has been reported that TNFα treatment of adipocytes induces an increase in serine phosphorylation of IRS-1.[48] Such phosphorylation is important because immunoprecipitated IRS-1 that has been serine phosphorylated in response to TNFα is a direct inhibitor of IR Tyr kinase activity. We found that TNFα-induced serine phosphorylation of IRS-1 in adipocytes was completely suppressed by D-PDMP inhibition of GM3 biosynthesis, suggesting that TNFα-induced increase of GM3 synthesis causes upregulation of serine phosphorylation of IRS-1.[45] An improved PDMP analog and a different type of GlcCer synthase inhibitor,[49,50] had therapeutic value when orally administered in diabetic rodent models.

TNFα levels were fairly high in two obese diabetic rodent models (Zucker *fa/fa* rats and *ob/ob* mice),[48] but much lower in adipose tissues from lean control animals. We, therefore, examined *GM3S* mRNA expression in epididymal fat of the diabetic models. Northern blot analysis showed that *GM3S* mRNA content in adipose tissues of the two models was significantly higher than in lean controls.[45] TLC comparison of GM3 band mobility revealed the appearance of GM3 species with lower mobility (i.e., more hydrophilic) in the models.

We also examined GM3 expression in diet-induced obese (DIO) mice.[51] Following 10 weeks of feeding, mean body weight of standard diet (SD) and high-fat diet (HFD) groups were $31.0 \pm 0.6$ g and $46.0 \pm 0.8$ g and fasted blood glucose levels were 137 and 203 mg/dL, respectively.[51] GM3 level in epididymal fat was threefold higher in HFD than in SD group. mRNA level of *GM3S* gene was also threefold higher in HFD group.[51]

These findings strongly indicate that increased expression of GM3 in abdominal adipose tissues tends to induce malfunctions of adipose tissue, for example, chronic low-grade inflammatory states in obesity. In recent studies of omental adipose tissues in obese and insulin-resistant women, adipocyte hypertrophy, and macrophage infiltration were associated with increases of GM3 content and *GM3S* gene expression.[52] Therefore, drugs that target GM3 biosynthesis may be useful for treating insulin resistance in obesity.

## 3.2 Insulin Resistance as a Membrane Microdomain Disorder

In studies of TNFα-induced insulin resistance in adipocytes, we showed that transformation to a resistant state might result from increased GM3 biosynthesis following upregulation of GM3S gene expression. Increased GM3 level during chronic exposure to TNFα may therefore suppress insulin signaling.[45] GSLs, including GM3, are important components of lipid rafts. We considered the possibility that increased GM3 levels in lipid rafts confer insulin resistance in TNFα-treated adipocytes. We examined GM3-protein interactions within plasma membranes of living cells by performing cross-linking assay with a photoactivatable, radioactive GM3 derivative. Adipocytes were preincubated with [$^3$H]GM3(N3) and then irradiated to induce crosslinking of GM3. Target proteins were separated by SDS-PAGE and visualized by autoradiography. A specific radioactive band corresponding to the 90-kDa IRβ-subunit was immunoprecipitated with anti-IRβ antibodies, indicating direct association of GM3 with IR. We concluded that IR forms complexes with caveolin-1 and GM3 independently in 3T3-L1 adipocytes.[53]

Lipids are asymmetrically distributed in the outer and inner leaflets of plasma membranes. In typical mammalian cells, most acidic phospholipids are located in the inner leaflet, and acidic GSLs, such as sulfatides and gangliosides are in the outer leaflet. Binding of proteins to lipid membranes is often mediated by electrostatic interactions between protein basic domains and acidic lipids. Gangliosides, which bear sialic acid residues, are ubiquitously present in the outer leaflet of vertebrate plasma membranes. GM3 is the most abundant ganglioside, and the primary ganglioside found in adipocytes.[54] Glycan chains of GSLs, including gangliosides are oriented at a defined angle to the axis of ceramide.[55] GM3s spontaneously form clusters via hydrophobic interaction between their own saturated acyl chains, regardless of any repulsion between negatively charged sialic acids in the glycan chains.[56] GM3 clusters, together with other cell surface gangliosides, thus form GSL-enriched microdomains (GEM) and produce a negatively charged environment just above the plasma membrane. On the other hand, IR has a sequence in its transmembrane domain, that is, homologous (conserved) in mammals and presents the basic amino acid lysine (IR944) just above the transmembrane domain. During lateral diffusion, electrostatic interaction may occur between the lysine residue at IR944 and the negative charge of GM3 cluster, because of their close topological proximity on the plasma membrane. In a study using FRAP technique in living cells, we found that the mechanism for dissociation of IR-caveolin-1 complex is based on interaction of a lysine residue localized just above the transmembrane domain of IRβ-subunit with GM3 clusters at the cell surface.[53] Based on these observations, we proposed a mechanism that determines insulin resistance state through localization shift of IR from caveolae to GEM in adipocytes.

Insulin signaling in skeletal muscle was shown to be higher in *GM3S* knockout (*GM3S*-null) mice than in wild-type (WT) C57BL/6 mice.[47] On the other hand, inhibition of insulin signaling by exogenous addition of saturated fatty acids into C2C12 myotubes was not eliminated by treatment with a GlcCer synthase inhibitor (D-PDMP analog).[57] The involvement of GM3 in pathophysiology of insulin resistance in mammalian skeletal muscle requires further study.

We explored the role of acyl chain length of ceramides in insulin signaling using ceramide synthase 2 (CerS2)-null mice that lack the ability to synthesize very long acyl chain (C22-C24) ceramides.[58] These mice did not display IR or Akt phosphorylation in liver in response to insulin. The absence of IR phosphorylation was correlated with its inability to translocate into detergent-resistant membranes (DRMs). The properties of DRMs in the CerS2-null mice differed significantly from those in WT mice, indicating that altered sphingolipid acyl chain length directly affects IR translocation to lipid rafts and subsequent signaling.

### 3.3 GM3 Functions as a Physiological Regulator for Insulin Signaling and Adipogenesis

Visceral adipose tissues, particularly mesenteric adipose tissue, play key roles in the pathogenesis of metabolic syndrome.[38–40,59–63] To investigate the fundamental characteristics of mesenteric adipocytes, we established a physiologically relevant differentiation system in which rat mesenteric-stromal vascular cells (mSVC) were induced to differentiate into mesenteric-visceral adipocytes (mVAC).[64,65] We optimized the concentrations of insulin and insulin-like growth factor (IGF-1) at levels comparable to those in vivo: 0.85 and 200 ng/mL. IGF-1 and insulin were shown to function synergistically; IGF-1 by itself enhanced CCAAT/enhancer binding protein alpha (C/EBPα) and adipocyte lipid binding protein (aP2) mRNA expression, but in the absence of insulin at physiological concentration IGF-1 did not induce lipid droplet accumulation associated with adipocyte maturation.

We used the same differentiation system to elucidate the role of resident macrophages in mesenteric adipose tissue in physiological adipogenesis. Adipogenesis of mSVCs in the system was increased by removal of resident macrophages, along with enhanced insulin signaling and concomitant decrease of GSLs, including GlcCer, LacCer, and GM3. Phosphorylation levels of IR and IRS-1 after insulin stimulation were enhanced by depletion of macrophages, and protein level of IR per se was increased in mSVCs.[51]

Levels of GSLs (particularly GM3) in preadipocytes and mature adipocytes are thus precisely maintained by soluble factors secreted from resident macrophages to control physiological adipogenesis (Fig. 3). Adipogenesis of MEFs prepared from *GM3S*-null mice was accelerated, in combination with enhanced insulin signaling.[51] GM3 is directly involved in insulin signaling and adipogenesis.

## 3.4 Serum GM3 as a New Biomarker of Metabolic Syndrome

GM3 is the major ganglioside in serum and is associated with serum lipoproteins.[66] We examined the relationships of serum GM3 levels with: (i) adiposity indices, and (ii) metabolic risk variables.[66] Serum GM3 levels were significantly elevated in type 2 diabetes patients with severe obesity (visceral fat area > 200 $cm^2$, BMI > 30). GM3 level was positively correlated with LDL-c (0.403, P = 0.012) in type 2 diabetes mellitus, but not related to blood pressure. High levels of small dense LDL (>10 mg/dL) were associated with increased GM3. Serum GM3 levels were affected by abnormalities of glucose and lipid metabolism, and by visceral obesity. Small dense LDL levels are associated with atherosclerosis development,[67,68] and GM3 has been detected in atherosclerotic lesions.[67,68] Thus, GM3 is a potentially useful marker for management of metabolic syndrome, including insulin resistance, and for early diagnosis of atherosclerosis.

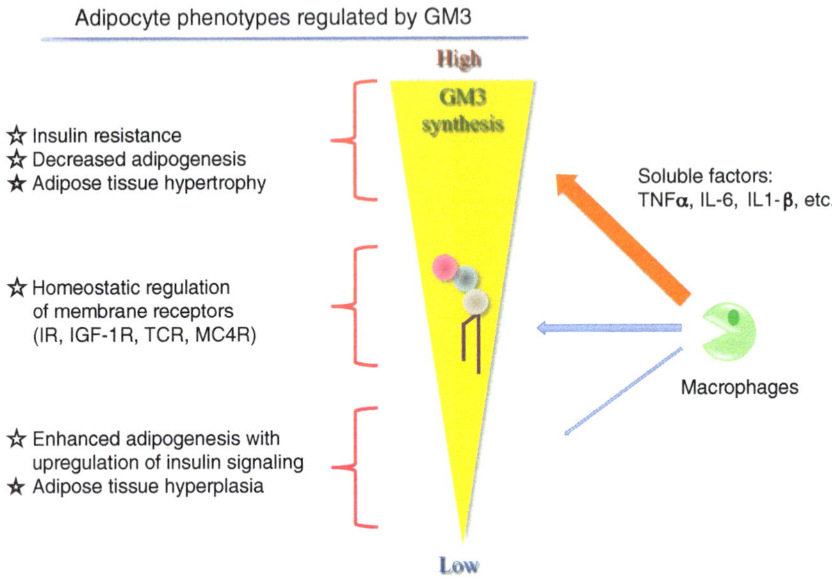


**Fig. 3** GM3 as homeostatic and pathogenic mediator in adipogenesis and insulin signaling. Interactions between macrophages and adipose progenitor cells may influence the number of preadipocytes and/or their differentiation capacity, and induce adipose tissue dysfunction by inhibiting overall adipogenic capacity. Control of GM3 levels is a potential therapeutic intervention strategy for restoring healthy adipose tissue function in obese individuals, distinct from weight-reduction strategies.

The structural diversity of ceramide species arises from several types of modification in *N*-acyl chains [chain length (C16–24); alpha hydroxylation; desaturation] and in sphingoid bases (d18:1, d18:0; hydroxylation at C4), resulting in a large number of possible molecular species. We performed LC-MS/MS analysis to identify serum GM3 molecular species in 125 Japanese residents.[69] Individuals were categorized based on presence versus absence of metabolic disease risk factors, including visceral fat accumulation, hyperglycemia, and dyslipidemia. A total of 23 GM3 molecular species were identified, of which 8 were significantly elevated in individuals with visceral fat accumulation and metabolic disease (defined as the presence of hyperglycemia and dyslipidemia). All of the GM3 molecular species were composed of sphingenine [d18:1(Δ4)], and six of the eight elevated species contained α-hydroxylated acyl chains. Hydroxylated GM3 species were, in order of decreasing abundance, d18:1-h24:0 = d18:1-h24:1 > d18:1-h22:0 >> d18:1-h20:0 > d18:1-h21:0 > d18:1-h18:1. We conducted univariate and multiple linear regression analyses using numerous clinical health variables associated with obesity, type 2 diabetes, metabolic disease, atherosclerosis, and hypertension. GM3 (d18:1-h24:1) was selected as the best candidate for metabolic screening. This molecular species showed significant correlation with intima media thickness (a criterion for diagnosis of atherosclerotic disease in humans) and other metabolic disease risk factors, including autotaxin, LDL-c, and homeostatic model assessment insulin resistance (HOMA-IR).[69]

## 3.5 Pathogenic Control of Adipocytes by Increased Expression of GM3

Adipose tissue macrophages are classified into two major subtypes: M1 and M2. A concept of M1/M2 polarization has been developed based on expression patterns of cytokines, surface markers, and metabolic enzymes.[70] M1 macrophages are potent effector cells that produce inflammatory cytokines, such as TNF-α, IL1-β, and IL-6. M2 macrophages are present as resident cells in nearly all organs under physiological conditions, and help maintain tissue homeostasis.[71,72] They exert antiinflammatory effects by producing IL-10 and arginase I enzyme (Arg1). IL-10 potentiates insulin signaling in adipocytes,[72,73] and Arg1 reduces nitric oxide synthesis and inflammation by metabolizing arginine to ornithine.[71] Consumption of HFD shifts cytokine expression of murine adipose tissue macrophages from M2- to M1-like patterns by decreasing expression of IL-10 and Arg1 and increasing expression of TNFα and iNOS.[73]

We analyzed gene expression profiles in epididymal adipose tissue of *GM3S*-null mice fed with HFD, but not with SD.[51] In comparison with expression levels in *GM3S*$^{+/-}$ mice, proinflammatory cytokine TNFα was significantly decreased, whereas antiinflammatory cytokine IL-10 was increased. Levels of adiponectin were increased, while those of atherogenic genes, such as *PAI-1* and *iNOS* were reduced. M2 signature genes, such as *MGL1* and *Arg1* showed increased expression in the *GM3S*-null mice. Glucose tolerance test (GTT) and insulin tolerance test (ITT) both revealed significant improvement of insulin resistance under HFD condition. Thus, GM3 plays key roles in development of obesity-induced chronic low-grade inflammatory states and of insulin resistance (Fig. 4).[51]

## 3.6 Perspectives

Adipocytes are more than inert energy depots; adipose tissue is a biologically active organ responsible for important physiological processes, including energy homeostasis and whole-body insulin sensitivity. Dynamic remodeling

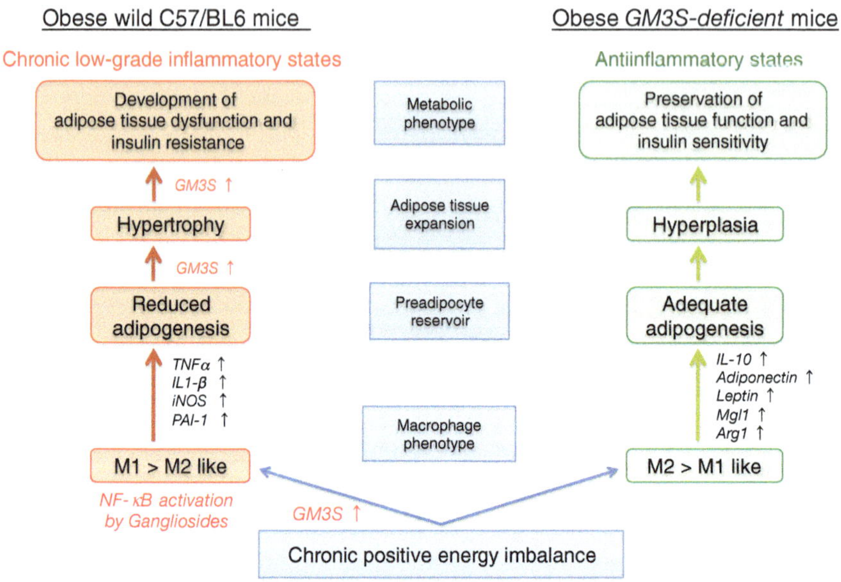


**Fig. 4** Regulation of adipose tissue remodeling by GM3 during chronic positive energy imbalance. Chronic positive energy imbalance [high fat diet (HFD)] leads to obesity. Macrophage phenotype may affect the mechanism whereby adipose tissue expands. In the obese state, an adipose tissue mass increase through hyperplasia and hypertrophy, and the latter is associated with activation of stress signaling. When proinflammatory M1 macrophages are predominant, the preadipocyte reservoir may be inadequate because of reduced preadipocyte survival, proliferation, and/or adipogenic capacity. Energy storage occurs via exaggerated adipose hypertrophy, resulting in dysfunctional adipose tissue and contributing to an inflamed insulin-resistant state. Chronic increase of GM3 through upregulation of *St3gal5* gene by proinflammatory cytokines, such as TNFα and IL-β may promote development of insulin resistance.[45,53] In obese *St3gal5*-deficient mice, insulin action was enhanced without any significant impact on diet-induced obesity.[51] These mice showed anti-inflammatory M2-like phenotypes in visceral adipose tissue (epididymal fat),[51] and significant increases of adiponectin and interleukin-10 (IL-10) in comparison with obese WT mice. IL-10 and adiponectin play essential roles in maintaining insulin sensitivity of adipocytes.[72,73]

of adipose tissue architecture occurs during its expansion. Under positive caloric balance, development of metabolic disease is more closely related to how fat is stored (through adipocyte hypertrophy vs. hyperplasia) than simply to the amount of stored fat. Inhibitors of adipogenesis were frequently proposed in the past as potential antiobesity drugs. However, results from a variety of recent mouse and human experiments indicate that adipogenesis inhibitors are poor candidates for amelioration of metabolic disease states, because limitation of fat cell expansion is associated with insulin resistance. It was proposed over a decade ago that type 2 diabetes might result from failure of adipocyte differentiation.[74] This hypothesis is now generally accepted, and has been supported by several independent lines of investigation in adipocyte biology.

Interactions between macrophages and adipose progenitor cells must be considered, because macrophages affect numbers of preadipocytes and/or their differentiation capacity, and induce adipose tissue dysfunction by inhibiting overall adipogenic capacity. The existence of resident and infiltrating macrophages is well documented, and studies during the past decade suggest that such macrophages are modulated in obesity and type 2 diabetes. Proinflammatory TNFα[75] and IL1-β[76] are potent inhibitors of adipocyte differentiation, and both of these cytokines induce insulin resistance in adipocytes.[77,78] TNFα expression is induced in adipose tissue of obese diabetic rodents[48] and humans.[79] TNFα detected in adipose tissue has been shown to be secreted from macrophages.[80] These studies, taken together, indicate that macrophages in adipose tissue produce TNFα and IL1-β, which can inhibit differentiation of preadipocytes and induce insulin resistance in mature adipocytes. Less attention has been paid to the ability of these cytokines to induce insulin resistance by inhibiting adipogenesis. Recent studies using new model systems demonstrate that limitation of adipose tissue expansion is associated with insulin resistance. Mice that are very obese but have unlimited adipose tissue expansion are metabolically healthy and insulin-sensitive.[81] Overall, these studies support the concept that adequate numbers of differentiation-competent preadipocytes allow for hyperplastic growth by preserving metabolic function in obesity.

As described earlier, our studies using *GM3S* gene and *GM3S*-knockout mice have shown that GM3 is crucially involved in both homeostatic adipogenesis (by controlling insulin signaling) and development of obesity-induced chronic low-grade inflammatory states and insulin resistance in adipose tissue. GM3 expression in adipocytes is controlled by soluble factors secreted from resident macrophages, to execute physiological adipogenesis. GM3 expression in adipose tissue is further increased under obesity-induced proinflammatory conditions, and GM3S-deficient mice fed a high-fat diet are less likely to develop insulin resistance or chronic low-grade inflammatory states (Fig. 4).[51] GM3 thus functions as a novel homeostatic and pathogenic mediator in adipose tissue (Fig. 3).

Our findings provide a rationale for design of novel therapies against metabolic syndrome (including type 2 diabetes) through inhibition of GM3 biosynthesis to maintain homeostatic insulin signaling. Extensive reduction of all complex gangliosides by inhibition of GM3 biosynthesis produces widespread physical and chemical modifications of all cellular plasma membranes, particularly of lipid microdomains, and is therefore drastic and of little therapeutic value. Such extensive ganglioside depletion is not necessary for treatment of metabolic disorders. We have demonstrated the effectiveness of D-PDMP on impaired insulin signaling in TNFα-treated 3T3-L1 adipocytes, and that normalization of elevated GM3 levels was sufficient to ameliorate the state of insulin resistance.[45]

In humans, GM3 is expressed predominantly in insulin-responsive organs (skeletal muscle, liver, adipose tissue) and lymphocytes. The presence of GM3-enriched membrane microdomains (lipid rafts) reflects the characteristics of individual cells. For accumulation of gangliosides in lipid rafts, critical roles are played by hydrogen donors and acceptors; and the abundance of (relative to phospholipids) long, saturated acyl chains in the ceramide backbone of gangliosides. Structural diversity of sphingoid bases and *N*-acyl chains in the ceramide moiety is essential in determining the behavior of gangliosides in cell membranes, and the localization of lipid rafts. A "sphingolipidomics" approach must be used to precisely characterize the ceramide structures actually present in cell membranes.

## 4. LEPTIN SIGNALING AND RESISTANCE IN HYPOTHALAMUS

Leptin is a 16-kDa protein required for maintenance of energy homeostasis and body weight.[82] White adipose tissue produces and secretes leptin into the bloodstream, and circulating levels are correlated with body fat content.[83] Leptin signals adiposity status to the brain, particularly to the mediobasal hypothalamus (MBH), which plays a key role in control of feeding, body weight, and energy expenditure. Within the MBH, the arcuate nucleus (ARC) is an important site of leptin action. The ARC contains two interconnected groups of neurons. One group is orexigenic and releases agouti-related peptide (AgRP) and neuropeptide Y (NPY). The other group is anorexigenic and releases α-melanocyte-stimulating hormone (α-MSH), generated from proopiomelanocortin (POMC) precursors, which binds to melanocortin receptor 4 (MC4R) in the paraventricular nucleus (PVH) and other hypothalamic regions, and exerts anorectic effects. NPY binds to its cognate receptor, thereby stimulating feeding and reducing basal energy expenditure. AgRP acts as an inverse agonist of MC4R and inhibits α-MSH activity.[84] AgRP/NPY neurons and POMC neurons thereby regulate melanocortin signaling in a coordinated manner to maintain energy balance.

Elevated levels of leptin reduce food intake and body weight through the long form of the leptin receptor, LepRb,[85] which is highly expressed in certain hypothalamic nuclei, including the ARC and ventromedial hypothalamus (VMH). LepRb is expressed in both AgRP/NPY neurons and POMC neurons. Leptin regulates energy balance by inhibiting neuronal activity and expression of NPY and AgRP in AgRP/NPY neurons, and promoting synthesis of α-MSH in POMC neurons.[86] Leptin increase can generate a strong, beneficial signal to prevent or reduce obesity. However, leptin activity may be disrupted by development of leptin resistance, a phenomenon defined by reduced ability to suppress food intake and body weight gain.[87]

Gangliosides play important roles in neural development and functions.[88,89] Recent studies using ganglioside-deficient mouse models have helped elucidate the functions of specific gangliosides in hypothalamic leptin signaling.

## 4.1 Leptin Receptor Signaling

Leptin action is transmitted by long form receptor LepRb, which is highly expressed in brain regions involved in control of feeding and energy expenditure.[90] LepRb is a transmembrane receptor and a member of the class I cytokine receptor family.[85] Six LepR isoforms (LepRa, b, c, d, e, f) have been identified in mice; they arise from the single *Lepr* gene by alternative mRNA splicing,[91,92] and have been categorized into three classes: secreted (LepRe), short (LepRa, LepRc, LepRd, LepRf), and long (LepRb) forms. The six isoforms share an N-terminal extracellular domain with the ability to bind leptin, but differ in their C-terminal cytoplasmic domains. Only long-form LepRb is capable of mediating leptin signaling. LepRb-deficient *db/db* mice have an obese phenotype similar to that of leptin-deficient *ob/ob* mice.[93] LepRb has no inherent enzymatic activity, but produces a signal through binding to a cytoplasmic Tyr kinase, Janus kinase 2 (JAK2).[94] Leptin binding to the extracellular domain of LepRb results in conformational change of the receptor and JAK2 activation, and JAK2 subsequently phosphorylates other Tyr residues within LepRb and JAK2 itself.[95] Three tyrosine residues in LepRb ($Tyr^{985}$, $Tyr^{1077}$, $Tyr^{1138}$) undergo phosphorylation, and recruit Src homology 2 (SH2) domain-containing proteins to the LepRb-JAK2 complex, thereby enabling JAK2 to phosphorylate these signaling proteins.[96]

In addition to JAK2, certain members of the Src Tyr kinase family are involved in control of leptin signaling. Phospho-Tyr[1138] binds to the SH2 domain of signal transducer and activator of transcription 3 (STAT3). STAT3 is subsequently phosphorylated by JAK2, resulting in nuclear translocation and transcriptional initiation of target genes including POMC and suppressor of cytokine signaling 3 (SOCS3), a negative-feedback regulator of leptin signaling.[97] Numerous studies have demonstrated the essential role of the JAK2/STAT3 pathway in antiobesity action of leptin in the brain. STAT5 is similarly recruited to phospho-Tyr[1077] in response to leptin, and phosphorylated and activated by JAK2. Phospho-Tyr[985] serves as a binding site for the SH2 domain of protein Tyr phosphatase 2 (SHP2), which mediates leptin-dependent activation of the extracellular signal-regulated kinase (ERK) pathway. The SHP2/ERK pathway is also involved in the antiobesity effect of leptin.[98,99]

## 4.2 Leptin Resistance

Leptin resistance is a key risk factor for obesity. Diet-induced obese (DIO) mice and obese human patients typically show high circulating leptin levels and reduced responsiveness to leptin's antiobesity effects.[87] Even after exogenous leptin administration, obese subjects do not show reductions of food intake or body weight.[100] Various molecular mechanisms have been proposed to explain leptin resistance; these include impairment of leptin transport across the blood–brain barrier (BBB), suppressed leptin signaling, and impaired leptin-targeted neural circuits.[96,101]

### *4.2.1 The BBB and Leptin Resistance*

Circulating leptin, as an energy regulator, must across the BBB to reach the neurons responsible for suppressing food intake and increasing energy expenditure. Impaired leptin transport across the BBB has been observed in obese mice,[102,103] suggesting a possible cause of leptin resistance in obese humans and rodents.[104] The median eminence (ME), located on the floor of the third ventricle, forms a BBB structure that regulates passage of metabolic hormones and nutrients from blood to cerebrospinal fluid and energy-sensing MBH neurons.[105] Tanycytes, which are specialized radial glial cells that line the walls of the third ventricle, have been identified as gatekeepers controlling entry of leptin into the hypothalamus.[106] Tanycytes express all six LepR isoforms, activate STAT3 and ERK in response to leptin, and thereby take up leptin through clathrin-coated vesicles. Internalized leptin is then released into the MBH under the control of LepRb-ERK signaling. This pathway is disrupted in DIO mice.

### 4.2.2 Suppression of LepRb Signaling

In obese rodents and humans, chronically elevated levels of circulating leptin activate an inhibitory negative feedback mechanism, resulting in impairment of LepRb signaling. SOCS3 is a member of a large family of cytokine-inducible inhibitors of signaling. Its gene expression is induced by leptin, and also by cytokines, such as interleukin-6.[97,107] SOCS3 binds to JAK2 and inhibits its kinase activity,[108] and also inhibits LepRb signaling through binding to phospho-$Tyr^{985}$ of LepRb.[109] Peripheral leptin administration rapidly induces SOCS3 mRNA in the hypothalamus of leptin-deficient *ob/ob* mice, but not of LepRb-deficient *db/db* mice.[110] In leptin-resistant obese mouse model *Ay/a*, SOCS3 expression in the ARC and dorsomedial hypothalamic nuclei is elevated, suggesting that SOCS3 is a mediator of leptin resistance in obesity. Mice with heterozygous *Socs3* deficiency or neuron-specific *Socs3* knockout showed increased leptin sensitivity and resistance to DIO,[111,112] supporting the concept that SOCS3 is a physiological negative regulator of LepRb signaling.

LepRb signaling is also suppressed by many other molecules, including protein Tyr phosphatases (PTPs). PTP1B, a nonreceptor Tyr phosphatase known to inhibit insulin signaling, suppresses LepRb signaling in vivo by dephosphorylating and thereby inhibiting JAK2.[113] PTP1B-null mice are hypersensitive to leptin and resistant to DIO.[113,114] Mice with neuronal deletion or POMC-neuron-specific deletion of PTP1B show increased leptin sensitivity and energy expenditure, and are less susceptible to DIO.[115,116] These findings suggest that increased PTP1B expression mediates leptin resistance in obesity.

### 4.2.3 Impairment of Downstream Signaling of LepRb

Melanocortin receptor 4 (MC4R) plays a central role as a leptin-targeted neural circuit in energy homeostasis. MC4R gene mutations with associated loss of function are the most common monogenic form of obesity in humans.[117] *Mc4r*-deficient mice develop obesity associated with hyperphagia,[118] and do not respond to the anorectic action of MTII, a synthetic analogue of α-MSH.[119] In *Ay/a* mice, melanocortin signaling is systemically disrupted by ectopic expression of agouti signaling protein (ASP), a homologue of AgRP.[120] Constitutive expression of ASP in the *Ay/a* hypothalamus blocks the effect of α-MSH and inhibits activation of MC4R, resulting in disruption of hypothalamic control of feeding and energy expenditure, and in development of leptin resistance.[100] Downstream of MC4R, expression of brain-derived neurotrophic factor (BDNF) in the VMH is regulated by the leptin-MC4R axis, and inhibition of the BDNF/TrkB pathway results in leptin resistance.[121,122]

## 4.3 Leptin Signaling in Ganglioside-Deficient Mice

Adult mammalian brain gangliosides consist mainly of a- and b-series gangliosides, such as GM1, GD1a, GD1b, and GT1b (Fig. 1).[89] GM3S-deficient mice that lack these ganglioside species have enhanced insulin signaling and are less susceptible to DIO.[47,51] On the other hand, GD1a has an enhancing effect on EGFR activation.[123] In a 2013 study of mice with deletion of inducible neuron-specific *Ugcg* (the gene encoding GlcCer synthase), Nordstrom *et al.* reported possible functions of neuronal ganglioside species in LepR.[124] These mice lacked gangliosides, as well as GlcCer and LacCer in distinct populations of forebrain neurons following administration of tamoxifen. Loss of GlcCer synthase in hypothalamic neurons inhibited LepR signaling, and the mice developed progressive obesity. The authors demonstrated that GM1 and GD1a in hypothalamic neuronal cell line N-41 interact closely with LepR, suggesting that these a-series gangliosides enhance the action of leptin in hypothalamic neurons. Ji *et al.*, in 2015 and 2016 studies, showed that GD3 synthase (GD3S)-null mice, which lack all b-series gangliosides, show reduced leptin secretion from adipocytes and increased leptin sensitivity.[125,126] Serum leptin levels in the *GD3S*-null mice were strongly reduced and leptin was accumulated in adipose tissues, indicating impaired leptin secretion. Addition of exogenous b-series gangliosides into medium of primary adipocyte culture restored leptin secretion.[125] These findings suggest that b-series gangliosides are involved in leptin secretion from adipocytes, although the mechanism is unknown. *GD3S*-null mice did not develop obesity, even though levels of circulating leptin were greatly reduced. Leptin signaling was altered in the hypothalamus of these mice. LepRb signaling in the hypothalamus, as measured by phosphorylated STAT3 levels, was enhanced in the mice, and GD3S overexpression in N-41 cells reduced responsiveness to leptin.[126] Increased levels of a-series gangliosides in *GD3S*-null mice enhance leptin sensitivity, which may explain the apparently normal energy expenditure observed.

An obese mouse model, KK-Ay, was established by introducing *Ay* mutation into KK mice, resulting in more severe and earlier onset of obesity and diabetic pathology.[127] Based on our finding that *GM3S*-null mutation in C57BL/6 mice resulted in no overt difference in obesity phenotype relative to WT,[47,51] we introduced *GM3S*-null mutation in KK-Ay mice to evaluate the pathophysiological roles of GM3 and related gangliosides in obesity.

Whereas KK-Ay mice were hyperphagic and developed morbid obesity, the KK-Ay/*GM3S*-null mice had much lower body weight and food intake, comparable to values in control KK mice (unpublished data). The KK-Ay/*GM3S*-null mice were less susceptible to central leptin resistance, suggesting that GM3 and/or related ganglioside species negatively regulate hypothalamic leptin signaling.

## 4.4 Perspectives

Numerous recent studies indicate that GM3 and related gangliosides play essential roles in hypothalamic control of energy homeostasis. GM1 and GD1a have been implicated as positive regulators of leptin signaling. KK-Ay/*GM3S*-null mice, which do not express a- or b-series gangliosides, are less susceptible to leptin resistance. We observed enhancement of leptin-dependent ERK phosphorylation in GM3S-deficient hypothalamic neuronal cells (unpublished data), suggesting differential roles of specific gangliosides in LepR signaling (Fig. 5). Altered expression of gangliosides may also affect other signaling pathways (e.g., insulin signaling) involved in hypothalamic control of feeding and energy balance.

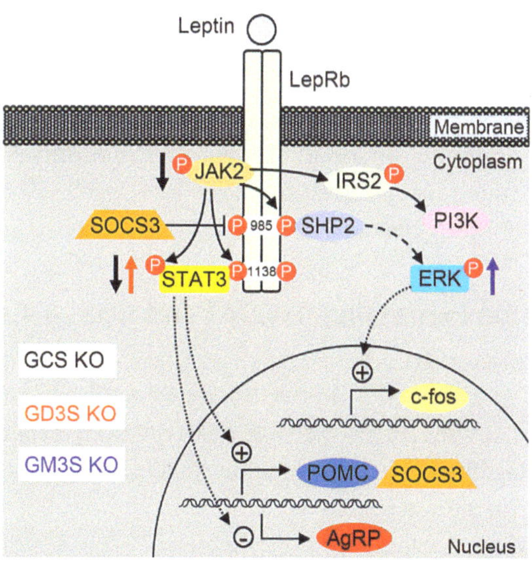


**Fig. 5** Leptin receptor-signaling pathway, and alterations in signals observed in ganglioside-deficient models. Increased or decreased activation of each pathway component is indicated by an *arrow* with respective color.

There is increasing evidence for association of hypothalamic inflammation with development of leptin resistance and obesity.[128] Activation of inflammatory pathways in the hypothalamus is increased by DIO, even without overt weight gain. GM3 levels are elevated in visceral fat of obese model animals and in sera of obese patients.[45,64,69] It is therefore conceivable that inflammation resulting from DIO alters ganglioside expression in the hypothalamus. Such alteration could affect interactions between neurons and nonneuronal cell populations and play a pathophysiological role in regulation of energy homeostasis.

## 5. T-CELL IMMUNITY AND IMMUNE DISORDERS

T cells are lymphocytes that have differentiated under the influence of the thymus and are responsible for cell-mediated immunity. T-cell activation, mediated by T cell antigen receptors (TCRs) in combination with costimulatory receptors, requires recruitment of both extracellular and intracellular molecules into specific cell membrane regions termed "lipid rafts."[129] T cells are functionally subdivided into T helper (Th) cells and T cytotoxic (Tc) cells, which typically express specific cell surface molecules CD4 and CD8, respectively. $CD4^+$ Th cells and $CD8^+$ Tc cells have distinctive immunological functions, but are similar in terms of mechanisms of TCR-mediated signaling. Th and Tc cells have differing ganglioside levels and ganglioside species composition, providing distinctive intracellular signaling events for their respective immune functions.[130] This section summarizes ganglioside species-specific functions in activation of the two T cell subsets, and roles of gangliosides in immune diseases.

### 5.1 Ganglioside Expression in $CD4^+$ and $CD8^+$ T-Cell Subsets

Gangliosides are classified into several series based on the absence (o-series) or presence of one (a-series) or two (b-series) sialic acid residues linked to the galactose residue in the second position from ceramide backbone (Fig. 1).[131,132] GM3, the simplest member of the a-series, is synthesized by GM3S. GM3 can be altered by GM2/GD2 synthase (GM2/GD2S, also termed B4galnt1) to form GM2 (a downstream a-series ganglioside), or by GD3S to form GD3 (the simplest b-series ganglioside). GM2/GD2S also elongates LacCer to form GA2, the simplest precursor of the o-series gangliosides. At each branching point in the various ganglioside biosynthetic pathways, competition among responsible enzymes determines the relative expression levels of o-, a-, and b-series gangliosides.

T cells interact with and help activate antigen-presenting cells (APCs) through a unique immunological synapse, and lipid rafts are involved in this interaction.[133–137] GM1 (a downstream a-series ganglioside) accumulates at the site of contact between TCRs and beads coated with TCR-stimulating antibodies in $CD4^+$ T cells, but not in $CD8^+$ T cytotoxic cells.[138] GM1 and GM3 define different types of raft membrane domains that segregate respectively to the leading pole or the trailing uropod of polarized human T cells.[139] Expression of GM1 varies depending on cell type and developmental stage, and staining with cholera toxin B subunit (CTx-B) is commonly used for visualization of rafts. CTx-B also reacts with an o-series ganglioside termed extended-GM1b (Fig. 1). Elucidating the ganglioside compositions of $CD4^+$ and $CD8^+$ T cells is essential for understanding the role of rafts in differentiation, maturation, and activation processes of these two T-cell subsets. Differences have been reported in ganglioside composition of primary T-cell subsets freshly isolated from lymphoid organs versus cultured T cells, including cloned cell lines and blasted cells stimulated by concanavalin A.

Biochemical analyses (TLC, HPLC) of whole T cell populations have traditionally been used to evaluate ganglioside expression patterns in primary T cells. However, whole T cell populations (i.e., all cells that express TCRs) are highly mixed populations, and such results are therefore of limited value for studies of specific T cell subsets. Flow cytometric techniques allow multiparameter analysis of target cells at the single-cell level, and therefore provide a valuable tool for determining levels of specific gangliosides in each T cell subset. Flow cytometric analyses using mAbs specific for various gangliosides revealed differential ganglioside expression by mature $CD4^+$ and $CD8^+$ T cells.[140,141] We used LC-MS/MS for structural analysis of gangliosides in immature thymocytes, and in $CD4^+$ and $CD8^+$ T cells isolated from mouse lymphoid organs.[130] The two T cell subsets both express six ganglioside species (GM1a, GM1b, GD1b, GD1c, GalNAcGM1b, extended-GM1b), but differ markedly in their expression levels of these six species. Expression of o-series gangliosides (GalNAcGM1b, extended-GM1b) increases greatly following differentiation of thymocytes to $CD4^+$ and $CD8^+$ T cells. Most of the gangliosides expressed in $CD8^+$ T cells are o-series. GM1b is expressed strongly in both subsets, whereas GM1a is expressed strongly in thymocytes and $CD4^+$ T cells; but at only trace levels in $CD8^+$ T cells.

*GM2/GD2S* gene expression is markedly higher in $CD4^+$ and $CD8^+$ T cells than in thymocytes. *GM3S* gene expression is higher in $CD4^+$ T cells and lower in $CD8^+$ T cells relative to thymocytes.[130] These gene expression patterns account in part for the differential ganglioside expression profiles during the differentiation processes from immature thymocytes to mature $CD4^+$ and $CD8^+$ T cells. In regard to GD1c, $CD4^+$ T cells can be separated into GD1c-positive IL-2-producing Th1-like cells and GD1c-negative IL-4-producing Th2-like cells.[141] Thus, it is possible to classify T cell subsets into distinctive functional subpopulations based on differences in ganglioside expression profiles.

## 5.2 Ganglioside Expression in TCR-Mediated T Cell Activation

T cell activation is initiated and maintained by engagement of TCRs with major histocompatibility complex (MHC)-peptide complex on APCs, and by engagement of costimulatory molecules. CD4 and CD8 are coreceptors that bind to nonpolymorphic regions of MHC and facilitate signaling by TCRs during T cell activation. T cell activation requires recruitment of both extracellular and intracellular molecules into specific cell membrane regions (lipid rafts).

Cholesterol, a major raft lipid component, plays a key role in maintaining raft structure. It fills the space between hydrocarbon chains of sphingolipids, and functions as a "glue" holding the raft assembly together. Treatment with drugs (e.g., methyl-β-cyclodextrin) that deplete cholesterol from cell membranes typically results in disruption of raft-mediated cellular functions.[142] In spite of some concerns regarding side effects of the drug on signaling events,[143] experimental approaches involving methyl-β-cyclodextrin have been widely used to document the important functions of cholesterol in rafts. Many studies have been based on manipulation of cholesterol level in lipid rafts, but a much smaller number have addressed the role of gangliosides in raft functions.

We highlighted one important aspect of ganglioside expression in T cells in our 2012 study of two strains of mice with distinctive genetic characteristics.[130] One strain, carrying a defective GM3S enzyme, expressed o-series, but not a- or b-series gangliosides. The other strain had an altered form of GM2/GD2S enzyme, and expressed GM3 and GD3; but no other gangliosides (Fig. 1). In WT mice, expression of a-series GM1a is higher in $CD4^+$ than in $CD8^+$ T cells. In *GM3S*-null mice, $CD4^+$ T cells show severe impairment of TCR-mediated proliferation and cytokine production. These defects are reversed by preincubation of cells with a-series

(GM3, GM1a), but not with b-series gangliosides. Most gangliosides expressed in WT $CD8^+$ T cells are o-series (GA1, GM1b, GalNAcGM1b, extended-GM1b). T cells from *GM2/GD2S*-null mice also show severe impairment of TCR-mediated proliferation and cytokine production. These defects are reversed by preincubation of cells with o-series GA1 and GM1b, but not with a- or b-series gangliosides. These findings indicate that the two T cell subsets have specific raft microenvironments involving different gangliosides, and that these raft microenvironments are the basis of distinctive functions following TCR stimulation (Fig. 6). The two T cell subsets, although they have common mechanisms of TCR-mediated signaling, have differing cellular and molecular modifications. CD4 and CD8 are

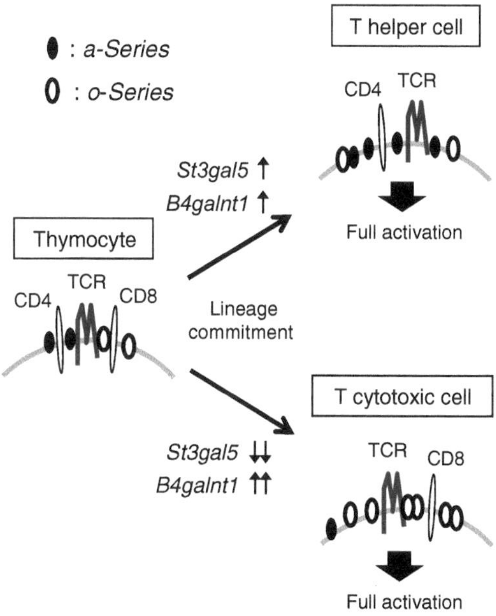


**Fig. 6** Distinct expression signatures of ganglioside species during T-cell development. Lineage commitment from thymocytes (CD4/CD8 double positive) to mature single positive T cell subsets is accompanied by selective ganglioside expression. $CD4^+$ T helper (Th) cells express predominantly a-series gangliosides because of upregulation of *St3gal5* gene, while $CD8^+$ T cytotoxic (Tc) cells carry o-series gangliosides because of downregulation of *St3gal5* and upregulation of *B4galnt1* genes. Evidently, each T cell subset has unique lipid rafts composed of differing ganglioside species, and these rafts are the basis for distinctive functions in intracellular events following TCR-mediated stimulation. The ganglioside selection process is thus essential for formation of distinct, functional lipid rafts in mature T cells.

localized to lipid rafts by palmitoylation (a process whereby lipids are attached to other molecules), but raft localization is not determined solely by this process.[144,145] To ensure movement of CD4 and CD8 to specific, correct locations on the membrane, it may be essential that these molecules interact with rafts carrying specific ganglioside species. Specific roles of the various gangliosides in regulation of membrane microenvironments remain to be elucidated.

## 5.3 T Cell Gangliosides in Autoimmune and Allergic Diseases

Almost all T cells that recognize self or harmless antigens are killed or inactivated in normal individuals. Failure of normal self-tolerance mechanisms leads to development of autoimmune and allergic diseases. The four T helper cell subpopulations (Th1, Th2, Th17, Treg) produce different cytokines and have distinctive effector functions, presumably based on differing organization of TCR signaling complexes in lipid rafts.[146] Lipid composition of lipid rafts and TCR signaling ability differ in Th1 versus Th2 cells.[147–150] Studies of clinical cases and mouse models indicate that some of these disease processes are related to ganglioside composition of T cells; we will review three of these diseases in this section.

Allergic asthma, a type 1 hypersensitivity reaction, is a manifestation of immediate and late-phase reactions in the lung. Innate immune cells, such as mast cells, basophils, and eosinophils are the effector cells of airway inflammation.[151,152] Adaptive immunity contributes to initiation of allergic reactions. $CD4^+$ T cells play a crucial role in production of Th2 cytokine (IL-4 and IL-13), which then help B cells produce IgE antibodies specific for the "harmless" antigens.[153,154] We examined mechanisms of allergic airway inflammation and airway hyperresponsiveness triggered by inhalation of ovalbumin (OVA) antigen in OVA-sensitized mice. *GM3S*-null mice showed striking reduction of allergic airway responses normally induced by OVA inhalation, that is, extensive mucus hypersecretion (characteristic of asthma), airway infiltration of inflammatory cells, OVA-specific IgE production, and increased Th2 cytokine levels in serum.[130] $CD4^+$ T cells from these mutant mice lack GM1a, and self-reactive $CD4^+$ T cells display enhanced GM1a expression resulting in persistence of abnormal cell activation.[155,156] Karman *et al.* demonstrated suppression of airway inflammation by administration of antisense oligonucleotides against *GM3S* gene.[157] These findings indicate that a-series ganglioside GM1a is essential for T helper cell function. The novel $CD4^+$ T cell subsets Th17 and Treg were described recently. In allergic airway inflammation, the balance between effector Th2 cells and suppressive Treg cells is skewed toward Th2 predominance.[158] It has been suggested that Th17 cells contribute to neutrophilic, steroid-resistant severe asthma and enhance Th2-mediated airway inflammation, although the precise role of Th17 cells in asthma remains unclear.[159] *GM3S*-null mice displayed a reduction of Th17 cell number induced by in vitro culture.[160]

Systemic lupus erythematosus (SLE), one of the most common autoimmune diseases, presents with multisystem clinical manifestations, including rashes, arthritis, glomerulonephritis, hemolytic anemia, thrombocytopenia, and central nervous system involvement.[161,162] A variety of autoantibodies are typically found in sera of SLE patients, and autoreactive T and B cells are involved in pathogenesis of this disease. T cells from SLE patients have intrinsic alterations of lipid components of lipid rafts.[155,163,164] GM1a expression levels in $CD4^+$ T cells (but not $CD8^+$ T cells) from SLE patients, in comparison with healthy controls, are significantly higher. LacCer, GA2, Gb3, GM1a, and GD1a are also upregulated in SLE $CD4^+$ T cells.[165] Increased GM1a expression is more marked in CD45RO-positive $CD4^+$ T cells (memory T cells) from active SLE patients than from inactive SLE patients.[155] Activated human T cells show upregulation of GM1a levels.[155,164,166,167] GM1a expression is also enhanced in self-reactive $CD4^+$ T cells, resulting in persistence of abnormal cell activation.[155,156] These findings, taken together, clearly indicate that gangliosides are key factors in SLE pathogenesis and pathology. In a 2014 study by McDonald *et al.*, SLE $CD4^+$ T cells were stimulated for 72 h with a combination of anti-CD3 mAb, anti-CD28 mAb, and GlcCer synthase inhibitor NB-DNJ. In the treated cells, expression levels of GM1a and LacCer were similar to those observed in healthy controls, and defects in TCR signaling were partially reversed.[165] Expression levels of cellular GSLs are determined by the coordinated effects of de novo synthesis, turnover, and recycling.[168] The increased GSL expression in SLE $CD4^+$ T cells was associated with accelerated internalization of GSLs from plasma membrane into intracellular compartments and rapid recycling of GSLs back to plasma membrane, resulting in a net increase in plasma membrane expression. SLE T cells display both increased GSL biosynthesis and accelerated trafficking to and from plasma membrane, which presumably lead to aberrant accumulation of gangliosides in lipid rafts.[169]

Rheumatoid arthritis (RA), another common autoimmune disease, is characterized by synovial inflammation associated with destruction of joint cartilage and bone.[170] Both cell-mediated and humoral immune responses evidently contribute to RA development. Fluid around inflamed synovia and joints contains numerous types of inflammatory cells (Th1 cells, Th17 cells, activated B cells, macrophages) and cytokines (IL-1, IL-8, TNFα, IL-6, IL-17, IFNγ).[171,172] GM3 and *GM3S* gene expression levels were higher in synovia of RA patients than in those of patients with osteoarthritis, which is not an autoimmune disease.[173] Progression of collagen-induced inflammatory arthritis, a mouse model of RA, was accelerated in *GM3S*-null mice, and induction of IL-17-producing cells in regional lymph nodes was enhanced by collagen immunization.[173] In another report, on the other hand, *GM3S*-null mice exhibit a decreased number of Th17 cells induced by in vitro culture.[160] Future studies will clarify whether a-series gangliosides expressed in Th17 cells are involved in RA development. $CD4^+$ T cells can be separated into GD1c-positive IL-2 producing Th1-like cells and GD1c-nagative IL-4 producing Th2-like cells.[141] It appears likely that each subpopulation of $CD4^+$ T cells has a unique, characteristic ganglioside expression pattern in its lipid rafts, and these patterns are responsible for specific functions of Th effector cells.

### 5.4 Perspectives

Results of numerous recent studies indicate that pathogenesis of allergic and autoimmune diseases are controlled in part by a variety of lipid rafts with differing ganglioside profiles in individual $CD4^+$ T cell subsets. Differentiation events of T cells in thymus are associated with differential ganglioside expression patterns in individual T cell subsets. The various subpopulations of effector $CD4^+$ T cells in peripheral lymphoid organs appear to have unique ganglioside expression patterns in their rafts, which are responsible for specific Th functions. Thus, the ganglioside selection process is essential for formation of distinct, functional lipid rafts in mature T cells. Although allergic reactions share certain common features, they differ greatly in the types of antigens involved. Reduction of a-series gangliosides by GM3S inhibition, and consequent modulation of ganglioside expression in lipid rafts, is potentially a powerful technique for therapy of immune system disorders.

## 6. GM3S DEFICIENCY AND HEARING LOSS

### 6.1 GM3S Deficiency in Humans

GM3S deficiency is a rare neurological disorder in which mutations of this enzyme result in a complete lack of GM3 and its downstream biosynthetic derivatives. The clinical manifestations of GM3S deficiency include infantile-onset epilepsy, severe intellectual disability, irritability, failure to thrive, blindness, choreoathetosis, and cutaneous dyspigmentation. The disorder was first reported in Old Order Amish[174,175] and more recently in French and Korean families or individuals.[176,177]

Infants with GM3S deficiency display feeding difficulties and vomiting within the first few weeks after birth, which prevent normal growth and weight gain. Seizures begin to occur within the first year and worsen over time. Multiple types of seizures have been reported, and include symptoms, such as muscle rigidity, convulsions, and loss of consciousness. Affected children typically have severe intellectual disability and do not develop skills, such as speaking, unsupported sitting, walking, and reaching for objects. Vision and hearing abilities decline over time, although they are present at birth.[174–182] In some cases, there are alterations of skin coloring on the arms, legs, and face (e.g., dark freckle-like spots or light patches),[178] or respiratory chain dysfunction.

Two homozygous mutations have been reported: c.862C>T (p.Arg288*) and c.994G>A (p.Glu332Lys). The nonsense mutation c.862C>T in Old Order Amish patients results in protein truncation between sialyl motif L and S. The c.994G>A mutation in an African–American family, within sialyl motif S, results in disruption of sialyl motif S function (interaction with LacCer acceptor and a sugar donor). The reported Korean patients had compound heterozygous and missense mutations in sialyl motif L (c.584G>C/p.Cys195Ser, and c.601G>A/p.Gly201Arg). These patients displayed various Rett-like phenotypes, but not epilepsy or blindness. These differences in symptoms may reflect differences in the mutation region of *ST3GAL5*. Humans who lack all a- and b-series gangliosides because of *ST3GAL5* c.694C>T mutations display major disruptions of auditory function, neonatal-onset hearing impairment, weak or absent middle ear muscle reflexes (MEMRs) and distortion product otoacoustic emissions (DPOAEs), abnormal auditory brainstem response (ABR) morphology and thresholds, absent cochlear microphonics (CMs), and delayed cortical auditory evoked potentials (CAEPs). Absent DPOAEs and CMs indicate severe outer hair cell (OHC) impairment. The presence of ABRs in seven of eight patients in our 2015 study suggested at least partial inner hair cell (IHC) function. However, numerous abnormalities revealed by MEMR and ABR testing indicated significant dysfunction of IHCs and/or central auditory pathways.[181]

## 6.2 Significant Increase of GSLs During Postnatal Maturation of Murine Cochlea

We investigated GSL expression in WT murine cochlea during the postnatal period of auditory system maturation. Mice began to recognize sound by postnatal day 12 (P12), designated the "onset of hearing." On P1, GM3 was the dominant cochlear GSL. After P3, there were marked increases of cochlear GM3 GlcCer, complex gangliosides (GM1, GD1a, GD3, GD1b, GT1b), and sulfatides (SM3, SM4). In contrast, cerebral gangliosides achieved stable, mature expression as early as P3.[181]

Significant alterations of ganglioside expression in murine cochlea after P3 reflect their essential role in hearing onset.[181] Following auditory maturation, GM3 and GM1 are distributed discretely among IHCs, OHCs, Deiters cells, and pillar cells of the organ of Corti (Table 1). Both IHCs and OHCs express GM3, which is localized on the apical surface, cuticular plate, and stereocilia. GM1 is the primary IHC ganglioside. It is also the predominant ganglioside on the apical surface of supporting cells, but is largely absent from OHCs.[181]

## 6.3 Glycocalyx Integrity and Membrane Cycling in GM3S-Deficient Mice

Apical membranes of stereocilia are covered with a glycocalyx composed of sialic acid-containing glycoproteins and glycolipids (including gangliosides),[183] creating a dense negative charge field that normally prevents fusion of adjacent stereocilia. In animal models, aminoglycoside administration reduced expression of sialoglycoconjugates in the OHC glycocalyx and enhanced fusion of stereocili.[184,185]

To elucidate the roles of various ganglioside species in hearing, we compared auditory function in *GM3S*-null versus *GM2/GD2S*-null mice, which have distinctive GSL expression patterns within the organ of Corti. *GM2/GD2S*-null cochlea had high GM3 content and was associated with

**Table 1** Comparison of GM3 and GM1 Expression of Specific Regions in the Organ of Corti at P14.

| | OHC | | | IHC | | | DC | | PC |
|---|---|---|---|---|---|---|---|---|---|
| | Stereocilia | Cuticular Plate | Cell Body | Stereocilia | Cuticular Plate | Cell Body | Cell Surface | Cell Body | Cell Body |
| GM3 | ++ | + | – | ++ | + | – | – | + | – |
| GM1 | – | – | – | ++ | + | + | ++ | + | + |

DC, Deiters cell; IHC, inner hair cell; OHC, outer hair cell; PC, pillar cell.

intact hair bundles and normal auditory function; whereas *GM3S*-null cochlea had no GM3 and was associated with deafness. Confocal laser microscopy with phalloidin staining revealed normal IHC and OHC morphology in 4-week-old *GM2/GD2S*-null mice. In contrast, OHCs in *GM3S*-null mice had blebs and intracellular vesicles composed of membranous structures, indicative of lack of balance between endocytosis and exocytosis.[181] GM3 is expressed in IHCs and OHCs of WT mice, whereas only IHCs express GM1. This may account for the observation that OHCs in *GM3S*-null mice degenerate prior to IHCs.

## 6.4 GM3-Enriched Membrane Organization, PTPRQ-Myosin VI Complex Localization and Hair Cell Morphology

Cochlear hair cells are specialized for auditory and vestibular transduction. Projecting from their apical surface are filopodial processes (stereocilia) that contain hundreds of cross-linked actin filaments. Proteins, such as unconventional myosins, Usher proteins, and deafness-related gene products [protein Tyr phosphatase receptor Q (PTPRQ), cadherin23, protocadherin15, usherin, VLGR1] are expressed predominantly or exclusively in stereocilia, forming a structured interactive network for mechanoelectrical transduction.[186]

In the normal organ of Corti, functional PTPRQ-myosin VI complexes are essential to maintain integrity of this network.[187] PTPRQ, a shaft connector at the tapered base of stereocilia, consists of an extracellular domain with 18 fibronectin III (FNIII) repeats, a membrane spanning domain, and a cytoplasmic domain having phosphatidylinositol and Tyr phosphatase activities.[187,188] Myosin VI is an actin-based motor protein whose complex with PTPRQ mediates interactions between plasma membrane and cytoskeleton.[189] PTPRQ-deficient mice are deaf and have fused stereocilia that do not taper at the base.[190] Myosin VI-deficient mice are also deaf and have comparable structural changes of hair cells, accompanied by maldistribution of PTPRQ along the stereocilia.[190–192]

In bullfrogs (*Rana catesbeiana*), PTPRQ is colocalized with ganglioside-rich membrane domains in basal stereocilia.[193] We examined the relationship between gangliosides and basal PTPRQ-myosin VI complex by immunostaining (Fig. 7). PTPRQ is localized exclusively in bases of stereocilia in WT murine IHCs. In contrast, in *GM3S*-null mice PTPRQ is maldistributed along shafts of fused stereocilia (Fig. 7B), and myosin VI is present from base to midshaft but absent from distal regions (Fig. 7A). These structural disruptions presumably result in loss of normal ciliary motor action. In

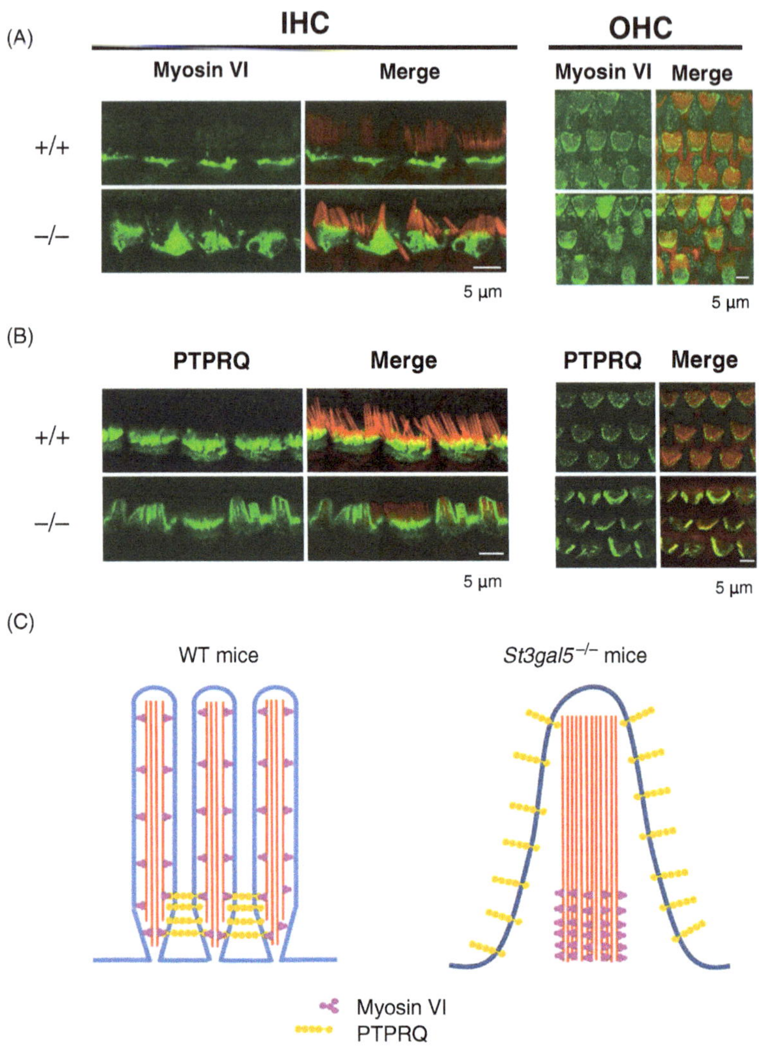


**Fig. 7** Maldistribution of protein Tyr phosphatase receptor Q (PTPRQ) and myosin VI in stereocilia of GM3S-null mice. (A) Confocal images show stereocilia of inner hair cells (IHCs) and outer hair cells (OHCs) from WT and *GM3S*-null mice stained for myosin VI, (B) PTPRQ (green), and F-actin (phalloidin; magenta). (C) Schematic depiction of maldistributed PTPRQ and myosin VI in *GM3S*-null IHC stereocilia.[181]

OHCs of *GM3S*-null mice, myosin VI is expressed mostly at the cuticular plate surface, close to vestigial kinocilia.

These findings, taken together, indicate that GM3-enriched membrane microdomains are essential for formation and proper localization of

PTPRQ-myosin VI complexes in hair cells. Aberrant expression of PTPRQ-myosin VI complexes resulting from absence of GM3 alters the structure of stereocilia and impairs their ability to transduce auditory signals (Fig. 7C).

### 6.5 Perspectives

GM3 plays a specific and essential role in development, function, and viability of cochlear hair cells. In humans with GM3S deficiency, hearing is already impaired at birth. However, studies of auditory phenotype in *GM3S*-null mice suggest that ganglioside addition may partially restore auditory function during a critical postnatal "window." These findings are relevant to our understanding of hearing loss in humans with GM3S deficiency, but do not explain the broader neurological phenotype associated with this condition. Further studies are needed to more precisely characterize the role of lipid rafts generally, and GM3 specifically, in development and function of the central nervous system.

New concepts and techniques are giving rise to novel strategies to reverse hearing impairment. Oshima *et al.* generated functional hair cell-like cells from embryonic and induced pluripotent stem cells.[194] Such strategies will help promote research on the auditory functions of GSL-enriched microdomains, and lead to new, more effective therapies to overcome hearing loss in humans.

## 7. CONCLUSIONS AND FUTURE DIRECTIONS

Functional roles of GM3 have been investigated indirectly based on addition of exogenous GM3 to culture medium, or on depletion of precursor GSLs by GlcCer synthase inhibitors.[195] Results of these studies suggested that GM3 is involved in transmembrane signaling through regulation of growth factor receptor activities, and in cell adhesion and motility through interaction with integrins and/or integrin-related molecule(s). However, it was not possible to definitively conclude that the effects of exogenously added or pharmacologically depleted gangliosides reflected the physiological functions of endogenous GM3.

Our studies during the past decade using *GM3S*-null mice have helped demonstrate the crucial roles of GM3 in diabetes, insulin resistance, leptin resistance, helper T cell function, and hearing function. Our findings provide a foundation for novel therapies against metabolic syndrome, type 2 diabetes, and immunological disorders (e.g., allergies), based on inhibition of GM3 biosynthesis. Extensive depletion of all gangliosides by GM3 biosynthesis inhibition results in excessive physical and chemical modifications of all plasma membranes, and particularly of lipid microdomains in membranes. However, effective treatment of metabolic disorders will not require such extensive ganglioside depletion. In our study of D-PDMP effect on impaired insulin signaling in TNFα-treated 3T3-L1 adipocytes, normalization of elevated GM3 levels was sufficient to restore proper insulin resistance status.[130] We expect that comprehensive studies of functional suprabiomolecular complexes consisting of gangliosides and functional proteins in microdomains will lead to a novel concept and strategy of "membrane microdomain orthosignaling therapy."

Our studies have shown that GM3 functions as a physiological regulatory factor to maintain homeostasis in adipocytes by modulating insulin signaling,[45,51,53,130] and that serum GM3 concentration is elevated in patients with metabolic syndrome.[69] Further analysis of these sera by LC-MS/MS revealed increased levels of GM3 containing very long acyl chains with or without α-hydroxy modification (C24:0, αOH-C24:0), and reduced levels of GM3 carrying long acyl chains (C16:0, C18:0).[69] Inflammatory cytokines, hypoxic conditions, and ER stress increase the biosynthesis of GM3 in the adipocytes, and inflammatory reactions in GM3S null mice are suppressed when these mice are fed with high fat diets.[51] Our ongoing studies are focused on the molecular mechanisms underlying these responses. We have observed that C24:0-GM3 and αOH-C24:0-GM3 induce secretion of inflammatory cytokines when monocytes or macrophages are stimulated by TLR4 ligand, LPS, or HMGB1; whereas C16:0-GM3, 18:0-GM3, and 24:1-GM3 suppress secretion of inflammatory cytokines. The effects are triggered by induction of TLR4/MD2 supramolecular complex formation and upregulation of signal transduction. Based on these findings, we propose a novel inflammation loop triggered by specific GM3 molecular species (MS submitted).

## ACKNOWLEDGMENTS

The authors' studies described in this review were supported by the Core Research for Evolutional Science and Technology division of the Japan Science and Technology Agency (to J.I.), a Grant-in-Aid for Scientific Research (B) from MEXT (J.I.) (16H04767), Mizutani Foundation for Glycoscience (J.I.), Naito Foundation (J.I.), ONO Medical Research Foundation (J.I.), Uehara Memorial Foundation (J.I.), Yamada Science Foundation (J.I.), and the MEXT-Supported Program for the Strategic Research Foundation at Private Universities (S1201031). The authors are grateful to Dr. S. Anderson for English editing of the manuscript.

## REFERENCES


1. Yamakawa T, Suzuki S. The chemistry of the lipids of posthemolytic residue or stroma of erythrocytes. Concerning the ether-insoluble lipids of lyophilized horse blood stroma. *J Biochem.* 1951;38:199–212.
2. Yamakawa T. Studies on erythrocyte glycolipids. *Proc Jap Acad Ser B.* 2005;81:52–63.
3. Todeschini AR, Hakomori S. Functional role of glycosphingolipids and gangliosides in control of cell adhesion, motility, and growth, through glycosynaptic microdomains. *Biochim Biophys Acta.* 2008;1780:421–433.
4. Sabourdy F, et al. Functions of sphingolipid metabolism in mammals—lessons from genetic defects. *Biochim Biophys Acta.* 2008;1781:145–183.
5. Ishii A, et al. Expression cloning and functional characterization of human cDNA for ganglioside GM3 synthase. *J Biol Chem.* 1998;273:31652–31655.
6. Inokuchi J. Membrane microdomains and insulin resistance. *FEBS Lett.* 2010;584:1864–1871.
7. Inokuchi J. Physiopathological function of hematoside (GM3 ganglioside). *Proc Jpn Acad Ser B.* 2011;87:179–198.
8. Inokuchi J. GM3 and diabetes. *Glycoconj J.* 2014;31:193–197.
9. Uemura S, et al. Substitution of the N-glycan function in glycosyltransferases by specific amino acids: ST3Gal-V as a model enzyme. *Glycobiology.* 2006;16:258–270.
10. Burger KN, van der Bijl P, van Meer G. Topology of sphingolipid galactosyltransferases in ER and Golgi: transbilayer movement of monohexosyl sphingolipids is required for higher glycosphingolipid biosynthesis. *J Cell Biol.* 1996;133:15–28.
11. Eckhardt M, Muhlenhoff M, Bethe A, Gerardy-Schahn R. Expression cloning of the Golgi CMP-sialic acid transporter. *Proc Natl Acad Sci USA.* 1996;93:7572–7576.
12. Stern CA, Braverman TR, Tiemeyer M. Molecular identification, tissue distribution and subcellular localization of mST3GalV/GM3 synthase. *Glycobiology.* 2000;10: 365–374.
13. Kapitonov D, Bieberich E, Yu RK. Combinatorial PCR approach to homology-based cloning: cloning and expression of mouse and human GM3-synthase. *Glycoconj J.* 1999;16:337–350.
14. Kim KW, Kim SW, Min KS, Kim CH, Lee YC. Genomic structure of human GM3 synthase gene (hST3 Gal V) and identification of mRNA isoforms in the 5′-untranslated region. *Gene.* 2001;273:163–171.
15. Berselli P, et al. Human GM3 synthase: a new mRNA variant encodes an NH2-terminal extended form of the protein. *Biochim Biophys Acta.* 2006;1759:348–358.
16. Uemura S, Yoshida S, Shishido F, Inokuchi J. The cytoplasmic tail of GM3 synthase defines its subcellular localization, stability, and in vivo activity. *Mol Biol Cell.* 2009;20:3088–3100.
17. Kono M, et al. Molecular cloning and functional expression of a fifth-type alpha 2,3-sialyltransferase (mST3 Gal V: GM3 synthase). *Biochem Biophys Res Commun.* 1998;253:170–175.
18. Shishido F, Uemura S, Nitta T, Inokuchi J. Identification of a new liver-specific c-type mRNA transcriptional variant for mouse ST3GAL5 (GM3/GM4 synthase). *Glycoconj J.* 2017;34(5):651–659.
19. Kozak M. Pushing the limits of the scanning mechanism for initiation of translation. *Gene.* 2002;299:1–34.
20. Zava S, Milani S, Sottocornola E, Berra B, Colombo I. Two active and differently N-glycosylated isoforms of human ST3Gal-V are produced from the placental mRNA variant by a leaky scanning mechanism. *FEBS Lett.* 2010;584:1476–1480.
21. Giraudo CG, Maccioni HJ. Endoplasmic reticulum export of glycosyltransferases depends on interaction of a cytoplasmic dibasic motif with Sar1. *Mol Biol Cell.* 2003;14:3753–3766.

22. Quintero CA, Giraudo CG, Villarreal M, Montich G, Maccioni HJ. Identification of a site in Sar1 involved in the interaction with the cytoplasmic tail of glycolipid glycosyltransferases. *J Biol Chem.* 2010;285:30340–30346.
23. Uemura S, Shishido F, Kashimura M, Inokuchi J. The regulation of ER export and Golgi retention of ST3Gal5 (GM3/GM4 synthase) and B4GalNAcT1 (GM2/GD2/GA2 synthase) by arginine/lysine-based motif adjacent to the transmembrane domain. *Glycobiology.* 2015;25:1410–1422.
24. Barlowe C, Schekman R. SEC12 encodes a guanine-nucleotide-exchange factor essential for transport vesicle budding from the ER. *Nature.* 1993;365:347–349.
25. Yoshihisa T, Barlowe C, Schekman R. Requirement for a GTPase-activating protein in vesicle budding from the endoplasmic reticulum. *Science.* 1993;259:1466–1468.
26. Aridor M, Weissman J, Bannykh S, Nuoffer C, Balch WE. Cargo selection by the COPII budding machinery during export from the ER. *J Cell Biol.* 1998;141:61–70.
27. Kuehn MJ, Herrmann JM, Schekman R. COPII-cargo interactions direct protein sorting into ER-derived transport vesicles. *Nature.* 1998;391:187–190.
28. Peng R, Grabowski R, De Antoni A, Gallwitz D. Specific interaction of the yeast cis-Golgi syntaxin Sed5p and the coat protein complex II component Sec24p of endoplasmic reticulum-derived transport vesicles. *Proc Natl Acad Sci USA.* 1999;96: 3751–3756.
29. Springer S, Schekman R. Nucleation of COPII vesicular coat complex by endoplasmic reticulum to Golgi vesicle SNAREs. *Science.* 1998;281:698–700.
30. Votsmeier C, Gallwitz D. An acidic sequence of a putative yeast Golgi membrane protein binds COPII and facilitates ER export. *EMBO J.* 2001;20:6742–6750.
31. Sato K, Nakano A. Dissection of COPII subunit-cargo assembly and disassembly kinetics during Sar1p-GTP hydrolysis. *Nat Struct Mol Biol.* 2005;12:167–174.
32. Michelsen K, et al. Novel cargo-binding site in the beta and delta subunits of coatomer. *J Cell Biol.* 2007;179:209–217.
33. Shikano S, Li M. Membrane receptor trafficking: evidence of proximal and distal zones conferred by two independent endoplasmic reticulum localization signals. *Proc Natl Acad Sci USA.* 2003;100:5783–5788.
34. Okajima T, Xu A, Lei L, Irvine KD. Chaperone activity of protein O-fucosyltransferase 1 promotes notch receptor folding. *Science.* 2005;307:1599–1603.
35. Losev E, et al. Golgi maturation visualized in living yeast. *Nature.* 2006;441:1002–1006.
36. Matsuura-Tokita K, Takeuchi M, Ichihara A, Mikuriya K, Nakano A. Live imaging of yeast Golgi cisternal maturation. *Nature.* 2006;441:1007–1010.
37. Tu L, Tai WC, Chen L, Banfield DK. Signal-mediated dynamic retention of glycosyltransferases in the Golgi. *Science.* 2008;321:404–407.
38. Chawla A, Nguyen KD, Goh YP. Macrophage-mediated inflammation in metabolic disease. *Nat Rev Immunol.* 2011;11:738–749.
39. Samaan MC. The macrophage at the intersection of immunity and metabolism in obesity. *Diabetol Metab Syndr.* 2011;3:29.
40. Suganami T, Ogawa Y. Adipose tissue macrophages: their role in adipose tissue remodeling. *J Leukoc Biol.* 2010;88:33–39.
41. Sun S, Ji Y, Kersten S, Qi L. Mechanisms of inflammatory responses in obese adipose tissue. *Annu Rev Nutr.* 2012;32:261–286.
42. Zoncu R, Efeyan A, Sabatini DM. mTOR: from growth signal integration to cancer, diabetes and ageing. *Nat Rev Mol Cell Biol.* 2011;12:21–35.
43. Guo D, Donner DB. Tumor necrosis factor promotes phosphorylation and binding of insulin receptor substrate 1 to phosphatidylinositol 3-kinase in 3T3-L1 adipocytes. *J Biol Chem.* 1996;271:615–618.
44. Kabayama K, et al. TNFalpha-induced insulin resistance in adipocytes as a membrane microdomain disorder: involvement of ganglioside GM3. *Glycobiology.* 2005;15:21–29.

45. Tagami S, et al. Ganglioside GM3 participates in the pathological conditions of insulin resistance. *J Biol Chem*. 2002;277:3085–3092.
46. Inokuchi J, Radin NS. Preparation of the active isomer of 1-phenyl-2-decanoylamino-3-morpholino-1-propanol, inhibitor of murine glucocerebroside synthetase. *J Lipid Res*. 1987;28:565–571.
47. Yamashita T, et al. Enhanced insulin sensitivity in mice lacking ganglioside GM3. *Proc Natl Acad Sci USA*. 2003;100:3445–3449.
48. Hotamisligil GS, Shargill NS, Spiegelman BM. Adipose expression of tumor necrosis factor-alpha: direct role in obesity-linked insulin resistance. *Science*. 1993;259:87–91.
49. Aerts JM, et al. Pharmacological inhibition of glucosylceramide synthase enhances insulin sensitivity. *Diabetes*. 2007;56:1341–1349.
50. Zhao H, et al. Inhibiting glycosphingolipid synthesis improves glycemic control and insulin sensitivity in animal models of type 2 diabetes. *Diabetes*. 2007;56:1210–1218.
51. Nagafuku M, et al. Control of homeostatic and pathogenic balance in adipose tissue by ganglioside GM3. *Glycobiology*. 2015;25:303–318.
52. Wentworth JM, et al. GM3 ganglioside and phosphatidylethanolamine-containing lipids are adipose tissue markers of insulin resistance in obese women. *Int J Obes*. 2016;40:706–713.
53. Kabayama K, et al. Dissociation of the insulin receptor and caveolin-1 complex by ganglioside GM3 in the state of insulin resistance. *Proc Natl Acad Sci USA*. 2007;104:13678–13683.
54. Ohashi M. A comparison of the ganglioside distributions of fat tissues in various animals by two-dimensional thin layer chromatography. *Lipids*. 1979;14:52–57.
55. Hakomori S. The glycosynapse. *Proc Natl Acad Sci USA*. 2002;99:225–232.
56. Sonnino S, Prinetti A, Mauri L, Chigorno V, Tettamanti G. Dynamic and structural properties of sphingolipids as driving forces for the formation of membrane domains. *Chem Rev*. 2006;106:2111–2125.
57. Chavez JA, et al. Ceramides and glucosylceramides are independent antagonists of insulin signaling. *J Biol Chem*. 2014;289:723–734.
58. Park JW, et al. Ablation of very long acyl chain sphingolipids causes hepatic insulin resistance in mice due to altered detergent-resistant membranes. *Hepatology*. 2013;57:525–532.
59. de Ferranti S, Mozaffarian D. The perfect storm: obesity, adipocyte dysfunction, and metabolic consequences. *Clin Chem*. 2008;54:945–955.
60. Matsuzawa Y. Pathophysiology and molecular mechanisms of visceral fat syndrome: the Japanese experience. *Diabetes Metab Rev*. 1997;13:3–13.
61. Saltiel AR. Insulin resistance in the defense against obesity. *Cell Metab*. 2012;15:798–804.
62. Sorisky A, Molgat AS, Gagnon A. Macrophage-induced adipose tissue dysfunction and the preadipocyte: should I stay (and differentiate) or should I go? *Adv Nutr*. 2013;4:67–75.
63. Xu H, et al. Chronic inflammation in fat plays a crucial role in the development of obesity-related insulin resistance. *J Clin Invest*. 2003;112:1821–1830.
64. Sato T, et al. Circulating levels of ganglioside GM3 in metabolic syndrome: A pilot study. *Obes Res Clin Pract*. 2008;2:I–II.
65. Shimizu K, et al. Newly developed primary culture of rat visceral adipocytes and their in vitro characteristics. *Cell Biol Int*. 2006;30:381–388.
66. Senn HJ, Orth M, Fitzke E, Wieland H, Gerok W. Gangliosides in normal human serum. Concentration, pattern and transport by lipoproteins. *Eur J Biochem*. 1989;181:657–662.
67. Bobryshev YV, et al. Phenotype determination of anti-GM3 positive cells in atherosclerotic lesions of the human aorta. Hypothetical role of ganglioside GM3 in foam cell formation. *Biochim Biophys Acta*. 2001;1535:87–99.

68. Bobryshev YV, et al. Incorporation and localisation of ganglioside GM3 in human intimal atherosclerotic lesions. *Biochim Biophys Acta.* 1997;1361:287–294.
69. Veillon L, et al. Identification of ganglioside GM3 molecular species in human serum associated with risk factors of metabolic syndrome. *PLoS One.* 2015;10:e0129645.
70. Konner AC, Bruning JC. Toll-like receptors: linking inflammation to metabolism. *Trends Endocrinol Metab.* 2011;22:16–23.
71. Martinez FO, Helming L, Gordon S. Alternative activation of macrophages: an immunologic functional perspective. *Annu Rev Immunol.* 2009;27:451–483.
72. Odegaard JI, Chawla A. Alternative macrophage activation and metabolism. *Annu Rev Pathol.* 2011;6:275–297.
73. Lumeng CN, Bodzin JL, Saltiel AR. Obesity induces a phenotypic switch in adipose tissue macrophage polarization. *J Clin Invest.* 2007;117:175–184.
74. Danforth Jr E. Failure of adipocyte differentiation causes type II diabetes mellitus? *Nat Genet.* 2000;26:13.
75. Torti FM, Dieckmann B, Beutler B, Cerami A, Ringold GM. A macrophage factor inhibits adipocyte gene expression: an in vitro model of cachexia. *Science.* 1985;229: 867–869.
76. Suzawa M, et al. Cytokines suppress adipogenesis and PPAR-gamma function through the TAK1/TAB1/NIK cascade. *Nat Cell Biol.* 2003;5:224–230.
77. Jager J, Gremeaux T, Cormont M, Le Marchand-Brustel Y, Tanti JF. Interleukin-1beta-induced insulin resistance in adipocytes through down-regulation of insulin receptor substrate-1 expression. *Endocrinology.* 2007;148:241–251.
78. Stephens JM, Butts MD, Pekala PH. Regulation of transcription factor mRNA accumulation during 3T3-L1 preadipocyte differentiation by tumour necrosis factor-alpha. *J Mol Endocrinol.* 1992;9:61–72.
79. Hotamisligil GS, Arner P, Caro JF, Atkinson RL, Spiegelman BM. Increased adipose tissue expression of tumor necrosis factor-alpha in human obesity and insulin resistance. *J Clin Invest.* 1995;95:2409–2415.
80. Weisberg SP, et al. Obesity is associated with macrophage accumulation in adipose tissue. *J Clin Invest.* 2003;112:1796–1808.
81. Kim JY, et al. Obesity-associated improvements in metabolic profile through expansion of adipose tissue. *J Clin Invest.* 2007;117:2621–2637.
82. Zhang Y, et al. Positional cloning of the mouse obese gene and its human homologue. *Nature.* 1994;372:425–432.
83. Friedman JM, Halaas JL. Leptin and the regulation of body weight in mammals. *Nature.* 1998;395:763–770.
84. Ollmann MM, et al. Antagonism of central melanocortin receptors in vitro and in vivo by agouti-related protein. *Science.* 1997;278:135–138.
85. Tartaglia LA, et al. Identification and expression cloning of a leptin receptor, OB-R. *Cell.* 1995;83:1263–1271.
86. Myers Jr MG, Munzberg H, Leinninger GM, Leshan RL. The geometry of leptin action in the brain: more complicated than a simple ARC. *Cell Metab.* 2009;9:117–123.
87. Frederich RC, et al. Leptin levels reflect body lipid content in mice: evidence for diet-induced resistance to leptin action. *Nat Med.* 1995;1:1311–1314.
88. Schengrund CL. Gangliosides: glycosphingolipids essential for normal neural development and function. *Trends Biochem Sci.* 2015;40:397–406.
89. Schnaar RL, Gerardy-Schahn R, Hildebrandt H. Sialic acids in the brain: gangliosides and polysialic acid in nervous system development, stability, disease, and regeneration. *Physiol Rev.* 2014;94:461–518.
90. Scott MM, et al. Leptin targets in the mouse brain. *J Comp Neurol.* 2009;514:518–532.
91. Chua Jr SC et al. Fine structure of the murine leptin receptor gene: splice site suppression is required to form two alternatively spliced transcripts. *Genomics.* 1997;45:264–270.

92. Tartaglia LA. The leptin receptor. *J Biol Chem.* 1997;272:6093–6096.
93. Bates SH, Myers Jr MG. The role of leptin receptor signaling in feeding and neuroendocrine function. *Trends Endocrinol Metab.* 2003;14:447–452.
94. Taniguchi T. Cytokine signaling through nonreceptor protein tyrosine kinases. *Science.* 1995;268:251–255.
95. Couturier C, Jockers R. Activation of the leptin receptor by a ligand-induced conformational change of constitutive receptor dimers. *J Biol Chem.* 2003;278:26604–26611.
96. Zhou Y, Rui L. Leptin signaling and leptin resistance. *Front Med.* 2013;7:207–222.
97. Banks AS, Davis SM, Bates SH, Myers Jr MG. Activation of downstream signals by the long form of the leptin receptor. *J Biol Chem.* 2000;275:14563–14572.
98. Rahmouni K, Sigmund CD, Haynes WG, Mark AL. Hypothalamic ERK mediates the anorectic and thermogenic sympathetic effects of leptin. *Diabetes.* 2009;58:536–542.
99. You J, et al. Signaling through Tyr985 of leptin receptor as an age/diet-dependent switch in the regulation of energy balance. *Mol Cell Biol.* 2010;30:1650–1659.
100. Halaas JL, et al. Physiological response to long-term peripheral and central leptin infusion in lean and obese mice. *Proc Natl Acad Sci USA.* 1997;94:8878–8883.
101. Cui H, Lopez M, Rahmouni K. The cellular and molecular bases of leptin and ghrelin resistance in obesity. *Nat Rev Endocrinol.* 2017;13:338–351.
102. Banks WA, Farrell CL. Impaired transport of leptin across the blood-brain barrier in obesity is acquired and reversible. *Am J Physiol Endocrinol Metab.* 2003;285:E10–E15.
103. Pan W, et al. Astrocyte leptin receptor (ObR) and leptin transport in adult-onset obese mice. *Endocrinology.* 2008;149:2798–2806.
104. Banks WA. Leptin transport across the blood-brain barrier: implications for the cause and treatment of obesity. *Curr Pharm Des.* 2001;7:125–133.
105. Banks WA. The blood-brain barrier as a cause of obesity. *Curr Pharm Des.* 2008;14:1606–1614.
106. Balland E, et al. Hypothalamic tanycytes are an ERK-gated conduit for leptin into the brain. *Cell Metab.* 2014;19:293–301.
107. Starr R, et al. A family of cytokine-inducible inhibitors of signalling. *Nature.* 1997;387:917–921.
108. Bjorbaek C, El-Haschimi K, Frantz JD, Flier JS. The role of SOCS-3 in leptin signaling and leptin resistance. *J Biol Chem.* 1999;274:30059–30065.
109. Bjorbak C, et al. SOCS3 mediates feedback inhibition of the leptin receptor via Tyr985. *J Biol Chem.* 2000;275:40649–40657.
110. Bjorbaek C, Elmquist JK, Frantz JD, Shoelson SE, Flier JS. Identification of SOCS-3 as a potential mediator of central leptin resistance. *Mol Cell.* 1998;1:619–625.
111. Howard JK, et al. Enhanced leptin sensitivity and attenuation of diet-induced obesity in mice with haploinsufficiency of Socs3. *Nat Med.* 2004;10:734–738.
112. Mori H, et al. Socs3 deficiency in the brain elevates leptin sensitivity and confers resistance to diet-induced obesity. *Nat Med.* 2004;10:739–743.
113. Zabolotny JM, et al. PTP1B regulates leptin signal transduction in vivo. *Dev Cell.* 2002;2:489–495.
114. Cheng A, et al. Attenuation of leptin action and regulation of obesity by protein tyrosine phosphatase 1B. *Dev Cell.* 2002;2:497–503.
115. Banno R, et al. PTP1B and SHP2 in POMC neurons reciprocally regulate energy balance in mice. *J Clin Invest.* 2010;120:720–734.
116. Bence KK, et al. Neuronal PTP1B regulates body weight, adiposity and leptin action. *Nat Med.* 2006;12:917–924.
117. Farooqi IS, et al. Clinical spectrum of obesity and mutations in the melanocortin 4 receptor gene. *N Engl J Med.* 2003;348:1085–1095.
118. Huszar D, et al. Targeted disruption of the melanocortin-4 receptor results in obesity in mice. *Cell.* 1997;88:131–141.

119. Marsh DJ, et al. Response of melanocortin-4 receptor-deficient mice to anorectic and orexigenic peptides. *Nat Genet.* 1999;21:119–122.
120. Fan W, Boston BA, Kesterson RA, Hruby VJ, Cone RD. Role of melanocortinergic neurons in feeding and the agouti obesity syndrome. *Nature.* 1997;385:165–168.
121. Liao GY, et al. Dendritically targeted Bdnf mRNA is essential for energy balance and response to leptin. *Nat Med.* 2012;18:564–571.
122. Xu B, et al. Brain-derived neurotrophic factor regulates energy balance downstream of melanocortin-4 receptor. *Nat Neurosci.* 2003;6:736–742.
123. Liu Y, Li R, Ladisch S. Exogenous ganglioside GD1a enhances epidermal growth factor receptor binding and dimerization. *J Biol Chem.* 2004;279:36481–36489.
124. Nordstrom V, et al. Neuronal expression of glucosylceramide synthase in central nervous system regulates body weight and energy homeostasis. *PLoS Biol.* 2013;11: e1001506.
125. Ji S, et al. b-Series gangliosides crucially regulate leptin secretion in adipose tissues. *Biochem Biophys Res Commun.* 2015;459:189–195.
126. Ji S, et al. Increased a-series gangliosides positively regulate leptin/Ob receptor-mediated signals in hypothalamus of GD3 synthase-deficient mice. *Biochem Biophys Res Commun.* 2016;479:453–460.
127. Iwatsuka H, Shino A, Suzuoki Z. General survey of diabetic features of yellow KK mice. *Endocrinol Jpn.* 1970;17:23–35.
128. Jais A, Bruning JC. Hypothalamic inflammation in obesity and metabolic disease. *J Clin Invest.* 2017;127:24–32.
129. Dykstra M, Cherukuri A, Sohn HW, Tzeng SJ, Pierce SK. Location is everything: lipid rafts and immune cell signaling. *Annu Rev Immunol.* 2003;21:457–481.
130. Nagafuku M, et al. CD4 and CD8 T cells require different membrane gangliosides for activation. *Proc Natl Acad Sci USA.* 2012;109:E336–E342.
131. Hakomori S. Structure and function of glycosphingolipids and sphingolipids: recollections and future trends. *Biochim Biophys Acta.* 2008;1780:325–346.
132. Tidhar R, Futerman AH. The complexity of sphingolipid biosynthesis in the endoplasmic reticulum. *Biochim Biophys Acta.* 2013;1833:2511–2518.
133. Davis DM, Dustin ML. What is the importance of the immunological synapse? *Trends Immunol.* 2004;25:323–327.
134. Fooksman DR, et al. Functional anatomy of T cell activation and synapse formation. *Annu Rev Immunol.* 2010;28:79–105.
135. Harder T, Rentero C, Zech T, Gaus K. Plasma membrane segregation during T cell activation: probing the order of domains. *Curr Opin Immunol.* 2007;19: 470–475.
136. Monks CR, Freiberg BA, Kupfer H, Sciaky N, Kupfer A. Three-dimensional segregation of supramolecular activation clusters in T cells. *Nature.* 1998;395:82–86.
137. Saito T, Yokosuka T. Immunological synapse and microclusters: the site for recognition and activation of T cells. *Curr Opin Immunol.* 2006;18:305–313.
138. Kovacs B, et al. Human CD8+ T cells do not require the polarization of lipid rafts for activation and proliferation. *Proc Natl Acad Sci USA.* 2002;99:15006–15011.
139. Gomez-Mouton C, et al. Segregation of leading-edge and uropod components into specific lipid rafts during T cell polarization. *Proc Natl Acad Sci USA.* 2001;98: 9642–9647.
140. Marusic A, Markotic A, Kovacic N, Muthing J. Expression of glycosphingolipids in lymph nodes of mice lacking TNF receptor 1: biochemical and flow cytometry analysis. *Carbohydr Res.* 2004;339:77–86.
141. Nakamura K, Suzuki H, Hirabayashi Y, Suzuki A. IV3 alpha (NeuGc alpha 2-8NeuGc)-Gg4Cer is restricted to CD4+ T cells producing interleukin-2 and a small population of mature thymocytes in mice. *J Biol Chem.* 1995;270:3876–3881.

142. Simons K, Toomre D. Lipid rafts and signal transduction. *Nat Rev Mol Cell Biol.* 2000;1:31–39.
143. Pizzo P, et al. Lipid rafts and T cell receptor signaling: a critical re-evaluation. *Eur J Immunol.* 2002;32:3082–3091.
144. Greenwald RJ, Freeman GJ, Sharpe AH. The B7 family revisited. *Annu Rev Immunol.* 2005;23:515–548.
145. Kroczek RA, Mages HW, Hutloff A. Emerging paradigms of T-cell co-stimulation. *Curr Opin Immunol.* 2004;16:321–327.
146. Leitenberg D, Balamuth F, Bottomly K. Changes in the T cell receptor macromolecular signaling complex and membrane microdomains during T cell development and activation. *Semin Immunol.* 2001;13:129–138.
147. Balamuth F, Leitenberg D, Unternaehrer J, Mellman I, Bottomly K. Distinct patterns of membrane microdomain partitioning in Th1 and th2 cells. *Immunity.* 2001;15: 729–738.
148. Izsepi E, et al. Membrane microdomain organization, calcium signal, and NFAT activation as an important axis in polarized Th cell function. *Cytometry A.* 2013;83:185–196.
149. Sloan-Lancaster J, Steinberg TH, Allen PM. Selective loss of the calcium ion signaling pathway in T cells maturing toward a T helper 2 phenotype. *J Immunol.* 1997;159: 1160–1168.
150. Smith JA, Tang Q, Bluestone JA. Partial TCR signals delivered by FcR-nonbinding anti-CD3 monoclonal antibodies differentially regulate individual Th subsets. *J Immunol.* 1998;160:4841–4849.
151. Holgate ST. Pathophysiology of asthma: what has our current understanding taught us about new therapeutic approaches? *J Allergy Clin Immunol.* 2011;128:495–505.
152. Orihara K, Dil N, Anaparti V, Moqbel R. What's new in asthma pathophysiology and immunopathology? *Expert Rev Respir Med.* 2010;4:605–629.
153. Hamid Q, Tulic M. Immunobiology of asthma. *Annu Rev Physiol.* 2009;71:489–507.
154. Levine SJ, Wenzel SE. Narrative review: the role of Th2 immune pathway modulation in the treatment of severe asthma and its phenotypes. *Ann Intern Med.* 2010;152: 232–237.
155. Dong L, et al. Increased expression of ganglioside GM1 in peripheral CD4+ T cells correlates soluble form of CD30 in systemic lupus erythematosus patients. *J Biomed Biotechnol.* 2010;2010:569053.
156. Kabouridis PS, Jury EC. Lipid rafts and T-lymphocyte function: implications for autoimmunity. *FEBS Lett.* 2008;582:3711–3718.
157. Karman J, et al. Reducing glycosphingolipid biosynthesis in airway cells partially ameliorates disease manifestations in a mouse model of asthma. *Int Immunol.* 2010;22:593–603.
158. Shalaby KH, Martin JG. Overview of asthma; the place of the T cell. *Curr Opin Pharmacol.* 2010;10:218–225.
159. Lloyd CM, Hessel EM. Functions of T cells in asthma: more than just T(H)2 cells. *Nat Rev Immunol.* 2010;10:838–848.
160. Zhu Y, et al. Lowering glycosphingolipid levels in CD4+ T cells attenuates T cell receptor signaling, cytokine production, and differentiation to the Th17 lineage. *J Biol Chem.* 2011;286:14787–14794.
161. Rahman A, Isenberg DA. Systemic lupus erythematosus. *N Engl J Med.* 2008;358: 929–939.
162. Tsokos GC. Systemic lupus erythematosus. *N Engl J Med.* 2011;365:2110–2121.
163. Jury EC, Kabouridis PS, Flores-Borja F, Mageed RA, Isenberg DA. Altered lipid raft-associated signaling and ganglioside expression in T lymphocytes from patients with systemic lupus erythematosus. *J Clin Invest.* 2004;113:1176–1187.

164. Krishnan S, et al. Alterations in lipid raft composition and dynamics contribute to abnormal T cell responses in systemic lupus erythematosus. *J Immunol.* 2004;172: 7821–7831.
165. McDonald G, et al. Normalizing glycosphingolipids restores function in CD4+ T cells from lupus patients. *J Clin Invest.* 2014;124:712–724.
166. Tani-ichi S, et al. Structure and function of lipid rafts in human activated T cells. *Int Immunol.* 2005;17:749–758.
167. Tuosto L, et al. Organization of plasma membrane functional rafts upon T cell activation. *Eur J Immunol.* 2001;31:345–349.
168. Degroote S, Wolthoorn J, van Meer G. The cell biology of glycosphingolipids. *Semin Cell Dev Biol.* 2004;15:375–387.
169. Kidani Y, Bensinger SJ. Lipids rule: resetting lipid metabolism restores T cell function in systemic lupus erythematosus. *J Clin Invest.* 2014;124:482–485.
170. McInnes IB, Schett G. The pathogenesis of rheumatoid arthritis. *N Engl J Med.* 2011;365:2205–2219.
171. Gizinski AM, Fox DA. T cell subsets and their role in the pathogenesis of rheumatic disease. *Curr Opin Rheumatol.* 2014;26:204–210.
172. Kobezda T, Ghassemi-Nejad S, Mikecz K, Glant TT, Szekanecz Z. Of mice and men: how animal models advance our understanding of T-cell function in RA. *Nat Rev Rheumatol.* 2014;10:160–170.
173. Tsukuda Y, et al. Ganglioside GM3 has an essential role in the pathogenesis and progression of rheumatoid arthritis. *PLoS One.* 2012;7:e40136.
174. Farukhi F, Dakkouri C, Wang H, Wiztnitzer M, Traboulsi EI. Etiology of vision loss in ganglioside GM3 synthase deficiency. *Ophthalmic Genet.* 2006;27:89–91.
175. Simpson MA, et al. Infantile-onset symptomatic epilepsy syndrome caused by a homozygous loss-of-function mutation of GM3 synthase. *Nat Genet.* 2004;36:1225–1229.
176. Fragaki K, et al. Refractory epilepsy and mitochondrial dysfunction due to GM3 synthase deficiency. *Eur J Hum Genet.* 2013;21:528–534.
177. Lee JS, et al. GM3 synthase deficiency due to ST3GAL5 variants in two Korean female siblings: Masquerading as Rett syndrome-like phenotype. *Am J Med Genet A.* 2016;170:2200–2205.
178. Boccuto L, et al. A mutation in a ganglioside biosynthetic enzyme, ST3GAL5, results in salt & pepper syndrome, a neurocutaneous disorder with altered glycolipid and glycoprotein glycosylation. *Hum Mol Genet.* 2014;23:418–433.
179. Wang H, Bright A, Xin B, Bockoven JR, Paller AS. Cutaneous dyspigmentation in patients with ganglioside GM3 synthase deficiency. *Am J Med Genet A.* 2013;161A: 875–879.
180. Wang H, et al. Early growth and development impairments in patients with ganglioside GM3 synthase deficiency. *Clin Genet.* 2016;89:625–629.
181. Yoshikawa M, et al. Ganglioside GM3 is essential for the structural integrity and function of cochlear hair cells. *Hum Mol Genet.* 2015;24:2796–2807.
182. Yoshikawa M, et al. Mice lacking ganglioside GM3 synthase exhibit complete hearing loss due to selective degeneration of the organ of Corti. *Proc Natl Acad Sci USA.* 2009;106:9483–9488.
183. Luft JH. Ruthenium red and violet. II. Fine structural localization in animal tissues. *Anat Rec.* 1971;171:369–415.
184. de Groot JC, Hendriksen EG, Smoorenburg GF. Reduced expression of sialoglycoconjugates in the outer hair cell glycocalyx after systemic aminoglycoside administration. *Hear Res.* 2005;205:68–82.
185. Takumida M, Wersall J, Bagger-Sjoback D, Harada Y. Observation of the glycocalyx of the organ of Corti: an investigation by electron microscopy in the normal and gentamicin-treated guinea pig. *J Laryngol Otol.* 1989;103:133–136.

186. Manor U, Kachar B. Dynamic length regulation of sensory stereocilia. *Semin Cell Dev Biol.* 2008;19:502–510.
187. Wright MB, Hugo C, Seifert R, Disteche CM, Bowen-Pope DF. Proliferating and migrating mesangial cells responding to injury express a novel receptor protein-tyrosine phosphatase in experimental mesangial proliferative glomerulonephritis. *J Biol Chem.* 1998;273:23929–23937.
188. Oganesian A, et al. Protein tyrosine phosphatase RQ is a phosphatidylinositol phosphatase that can regulate cell survival and proliferation. *Proc Natl Acad Sci USA.* 2003;100:7563–7568.
189. Nambiar R, McConnell RE, Tyska MJ. Myosin motor function: the ins and outs of actin-based membrane protrusions. *Cell Mol Life Sci.* 2010;67:1239–1254.
190. Sakaguchi H, et al. Dynamic compartmentalization of protein tyrosine phosphatase receptor Q at the proximal end of stereocilia: implication of myosin VI-based transport. *Cell Motil Cytoskeleton.* 2008;65:528–538.
191. Hertzano R, et al. A Myo6 mutation destroys coordination between the myosin heads, revealing new functions of myosin VI in the stereocilia of mammalian inner ear hair cells. *PLoS Genet.* 2008;4:e1000207.
192. Self T, et al. Role of myosin VI in the differentiation of cochlear hair cells. *Dev Biol.* 1999;214:331–341.
193. Zhao H, Williams DE, Shin JB, Brugger B, Gillespie PG. Large membrane domains in hair bundles specify spatially constricted radixin activation. *J Neurosci.* 2012;32:4600–4609.
194. Oshima K, et al. Mechanosensitive hair cell-like cells from embryonic and induced pluripotent stem cells. *Cell.* 2010;141:704–716.
195. Inokuchi J, Kabayama K. Modulation of growth factor receptors in membrane microdomains. *Trends in Glycosci Glycotech.* 2008;20:353–371.

CHAPTER SIX

# Gangliosides in Cancer Cell Signaling

**Sophie Groux-Degroote*, Macarena Rodríguez-Walker†, Justine H. Dewald*, Jose L. Daniotti†, Philippe Delannoy*,[1]**
* University of Lille, Villeneuve d'Ascq, France
† National University of Cordoba, Ciudad Universitaria, Córdoba, Argentina
[1] Corresponding author. E-mail address: philippe.delannoy@univ-lille1.fr

## Contents



## Abstract

At the outer leaflet of the plasma membrane, gangliosides are found with other glycosphingolipids, phospholipids, and cholesterol in glycolipid-enriched microdomains, in which they interact with signaling molecules including receptor tyrosine kinases and signal transducers. The role of gangliosides in the regulation of signal transduction has been reported for many cases and in different cell types. The biosynthesis of gangliosides involves specific enzymes, mainly glycosyltransferases that control together with glycohydrolases, the steady state of gangliosides at the cell surface. Changes in ganglioside composition are therefore correlated with modifications of glycosyltransferases or glycohydrolases expression and result in the deregulation of cellular signals. In several types of cancers, the overexpression of

*Progress in Molecular Biology and Translational Science*, Volume 156
ISSN 1877-1173
http://dx.doi.org/10.1016/bs.pmbts.2017.10.003

disialogangliosides, such as GD3 or GD2 mainly results in the activation of cell signaling, increasing cell proliferation and migration, as well as tumor growth. In this chapter, we summarize our current knowledge of ganglioside biosynthesis, degradation, and of their role in cell signaling regulation in cancers.

## ABBREVIATIONS

| | |
|---|---|
| Cer | ceramide |
| CSC | cancer stem cells |
| EGF | epidermal growth factor |
| EGFR | epidermal growth factor receptor |
| EMT | epithelial-mesenchymal transition |
| ER | endoplasmic reticulum |
| FAK | focal adhesion kinase |
| FGF | fibroblast growth factor |
| GalCer | galactosylceramide |
| GD3S | GD3 synthase |
| GEM | glycolipid-enriched microdomain |
| GlcCer | glucosylceramide |
| GM3S | GM3 synthase |
| GSL | glycosphingolipids |
| GT | glycosyltransferase |
| LacCer | lactosylceramide |
| PDGF | platelet-derived growth factor |
| RTK | receptor tyrosine kinase |
| TF | transcription factor |
| TGF | transforming growth factor |
| VEGF | vascular endothelial growth factor |

## 1. INTRODUCTION

Gangliosides, the sialic acid-containing glycosphingolipids (GSL), are major components of the plasma membrane of eukaryotic cells. They are found in almost all tissues, especially in membranes of the central and peripheral nervous system where they make up to 6% of total lipids.[1] Localized at the outer leaflet of the plasma membrane, gangliosides interact with other GSLs, phospholipids, cholesterol, and transmembrane proteins, forming glycolipid-enriched microdomains (GEMs), also known as lipid rafts. Gangliosides are usually classified in four Series (0-, a-, b-, and c-Series) according to the number of sialic acid residues linked to the lactosylceramide (LacCer) (Fig. 1).[2] Normal extraneural-human tissues mainly express "simple" gangliosides from 0- and a-Series, whereas "complex"

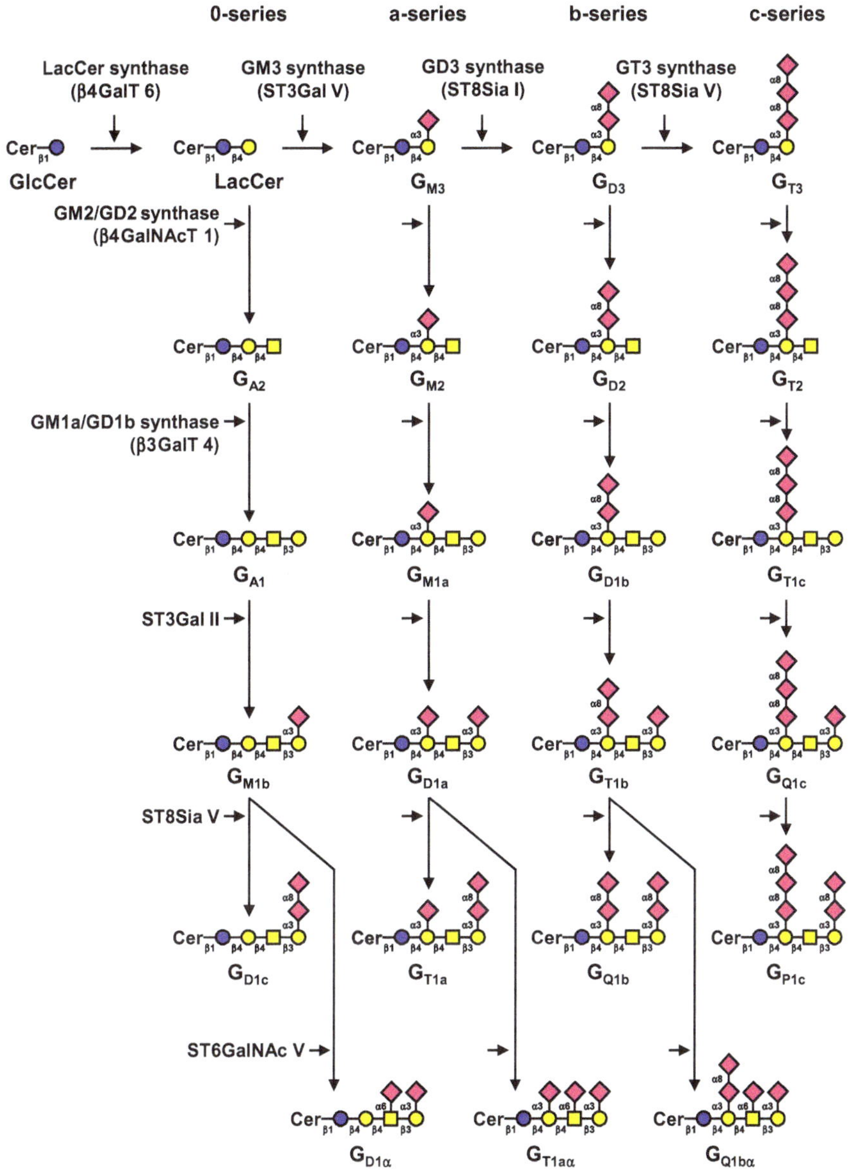


**Fig. 1** Biosynthesis pathway for gangliosides. Gangliosides are synthesized by the stepwise addition of monosaccharides to ceramide (Cer). Cer is the acceptor for UDP-Glc: ceramide β-glucosyltransferase. Extension of GlcCer occurs by the action of UDP-Gal: GlcCer β1,4-galactosyltransferase (β4GalT 6) to make LacCer. The action of ST3 Gal V (GM3 synthase), ST8Sia I (GD3 synthase), and ST8Sia V (GT3 synthase) leads to the biosynthesis of the precursors of a-, b-, and c-Series gangliosides, respectively. The 0-Series gangliosides are directly synthesized from LacCer. Elongation is performed by the sequential action of *N*-acetyl-galactosaminyltransferase (β4GalNAcT 1), galactosyltransferase (β3GalT 4) and sialyltransferases (ST3 Gal II and ST8Sia V). α-Series gangliosides derive from the action of ST6GalNAc V on GM1b, GD1a, or GT1b. Gangliosides nomenclature is that of Svennerholm.[2] ●, Glc; ○, Gal; □, GalNAc; ◆, Neu5Ac.

gangliosides from b- and c-Series are essentially found in developing tissues, during embryogenesis, and mainly restricted to the nervous system of healthy adults.[3] The expression of complex gangliosides has been found to be increased in pathological conditions including several types of cancer, especially those from neuroectoderm origin, such as melanoma, neuroblastoma, glioblastoma, or breast cancer, but also lung cancer, in which they play a key role in invasion and metastasis,[4,5] making complex gangliosides attractive target molecules for cancer immunotherapy.[6–9] The biosynthesis of gangliosides is catalyzed by specific glycosyltransferases (GTs) that control, together with glycohydrolases, the cell surface expression of gangliosides. Changes in ganglioside composition are therefore usually correlated with modifications of the expression of GTs. For example, the increased GD3 synthase (GD3S) expression was reported for many types of cancer cells. Due to their extracellular orientation, gangliosides are involved in cell–cell and cell–matrix interactions. Within lipid rafts, gangliosides also interact with signaling molecules including receptor tyrosine kinases (RTKs) or signal transducers, controlling, and regulating different signaling processes potentially involved in cancer progression.[10,11] Although the role of gangliosides in the regulation of signal transduction has been reported in a variety of cancer cell lines, the molecular mechanisms sustaining these functions are not fully understood. Several examples have shown that direct carbohydrate-to-carbohydrate interaction through glycosynapses, interaction with extracellular matrix components, or crosstalk with integrins or other transmembrane proteins could be involved in the regulation of cell signaling by gangliosides.[12,13] Finally, within lipid rafts, gangliosides are also implicated in CD95 death receptor-mediated signaling and apoptosis, potentially providing new targets for cancer therapy.[14]

## 2. GANGLIOSIDE BIOSYNTHESIS AND DEGRADATION

The first step of ganglioside biosynthesis is ceramide (Cer) formation by endoplasmic reticulum (ER)-resident Cer synthases.[15] Cer is then transported to the Golgi apparatus by vesicular transport or by Cer transfer protein,[16] where it is used by specific GTs, which add monosaccharides in a stepwise manner to produce the glycosidic moiety of the ganglioside. The first glycosylation step corresponds to glucose or galactose residues transferred to the 1-hydroxy group of Cer, yielding the simple GSL glucosylceramide (GlcCer) or galactosylceramide (GalCer), respectively. Human

GalCer synthase is an ER-located transmembrane protein, with its catalytic site facing the lumen of the ER.[17] GalCer is a precursor for 3-sulfated-galactosylceramide (sulfatide) and GM4, both of which are synthesized in the Golgi apparatus, by the action of galactosylceramide 3-O-sulfotransferase or of α2,3-sialyltransferase ST3 Gal V, respectively.[18,19]

Human GlcCer synthase, encoded by *UGCG* gene, is a transmembrane protein localized on the cis-Golgi, with its catalytic site facing the cytosol.[20] Since higher GSL biosynthesis takes place in the lumen of the Golgi apparatus, GlcCer synthesized on the cytosolic side of the membrane has to translocate across the Golgi membrane to be used as a substrate by LacCer synthase (UDP-Gal: GlcCer β1,4-galactosyltransferase, β4GalT 6) and other GTs involved in ganglioside biosynthesis.[21] It was proposed by Halter and coworkers that the essential GlcCer flipping step is accomplished via a complex trafficking itinerary. Four-phosphate adaptor protein 2 (FAPP2), a cytoplasmic lipid transfer protein, transports GlcCer from the cytosolic side of the Golgi apparatus to the ER, then GlcCer has to flip across the ER membrane, which is facilitated by ATP-independent ER flippases, and be delivered to the lumen of the Golgi apparatus by vesicular traffic.[22,23] Eventually, GlcCer could be transported by FAPP2 directly from proximal to distal Golgi apparatus compartments.[24] Almost all complex GSLs are synthesized from GlcCer following additional reactions sequentially catalyzed by Golgi-localized GTs. LacCer synthase first transfers galactose from UDP-Gal to GlcCer to produce LacCer (Galβ1-4Glcβ1-Cer) in the Golgi apparatus. The transfer of the first sialic acid residue to LacCer is achieved by α2,3-sialyltransferase ST3 Gal V (GM3 synthase). GM3 synthase was recently found to be responsible for the synthesis of GM4 (sialylated GalCer), both in vitro and in vivo. The mechanism underlying GM4 expression remains unclear, and it seems that excess amounts of ST3 Gal V are necessary for GM4 synthesis in mammalian cells.[19]

Additional sialic acid residues can be added by sialyltransferases ST8Sia I (GD3S), and ST8Sia V (GT3 synthase), which synthesize GD3 and GT3, respectively. GD3S uses GM3 as a preferential substrate, but it is also able to sialylate GD3 to synthesize GT3,[25] whereas ST8Sia V shows a broader substrate specificity.[26] Although ganglioside biosynthesis pathways have been located to the Golgi apparatus, it was recently demonstrated that both ectopically and endogenously expressed GD3S were active at the plasma membrane, using GM3 and CMP-Neu5Ac as substrates, which could regulate local ganglioside composition.[27]

LacCer, GM3, GD3, and GT3 are the precursors for more complex gangliosides of the 0-, a-, b-, and c-Series, respectively, by sequential glycosylation steps catalyzed by β4GalNAcT 1, β3GalT 4, ST3 Gal II, and ST8Sia V (Fig. 1). In the central nervous system of mammals, a shift in the expression of gangliosides, from GD3 at early stages to GD1a at late developmental stages, is usually observed, which correlates with the transcriptional regulation of ST8Sia I and β4GalNAcT 1 observed during development.[28] Gangliosides GM1, GD1a, GD1b, and GT1b represent more than 90% of gangliosides in the brain of mammals and birds.

After synthesis, gangliosides are transported to the plasma membrane via vesicular transport where they are part of GEMs on the outer leaflet and key effectors in diverse cellular functions.[29–31] They can subsequently be released to the extracellular milieu or undergo endocytosis, and once internalized, they can be recycled back to the plasma membrane directly from endosomes, sorted from endosomes to the Golgi apparatus where they can be reglycosylated, or they can be degraded at the lysosomal level.[32,33]

Catabolism of gangliosides takes place mainly at the luminal surface of lysosomes by the action of glycohydrolases that sequentially cleave off the monosaccharide units from the nonreducing end of the ganglioside glycan chains.[34](Fig. 2). Adequate lysosomal ganglioside catabolism requires the presence of an acidic pH, glycohydrolases and in some cases also of lipid-transfer proteins for degradation of simpler gangliosides.[35] In this sense, it is known that GSLs with four carbohydrate residues or less require the presence of small lipid binding glycoproteins (GM2 activator protein or one of the four saposins A, B, C, and D), which extract glycolipids from membranes and present them to the soluble acid hydrolases.[34,36] Additionally, anionic lipids, such as bis (monoacylglycero)phosphate are key players in ganglioside degradation since they participate in the electrostatic attraction of glycosidases and lipid binding proteins to the ganglioside-containing luminal membrane surfaces. If a defect is inherited in the genes encoding glycohydrolases and lipid-transfer proteins, the substrate is accumulated in the organelle resulting in severe neurodegenerative pathologies called gangliosidosis.[35,37] Eventually, degradation of gangliosides can also occur at the plasma membrane by the action of sialidases (mainly Neu3), β-galactosidase, β-glucosidase, and β-hexosaminidase,[38–40] that can remodel cell surface glycans, and consequently cause rapid changes in critical signal transduction pathways. Interestingly, some of these enzymes are able to hydrolyze the ganglioside substrates present in both the plasma membrane from its own cell or when exposed at the cell surface of neighboring cells, evidencing new regulatory mechanisms of ganglioside expression.

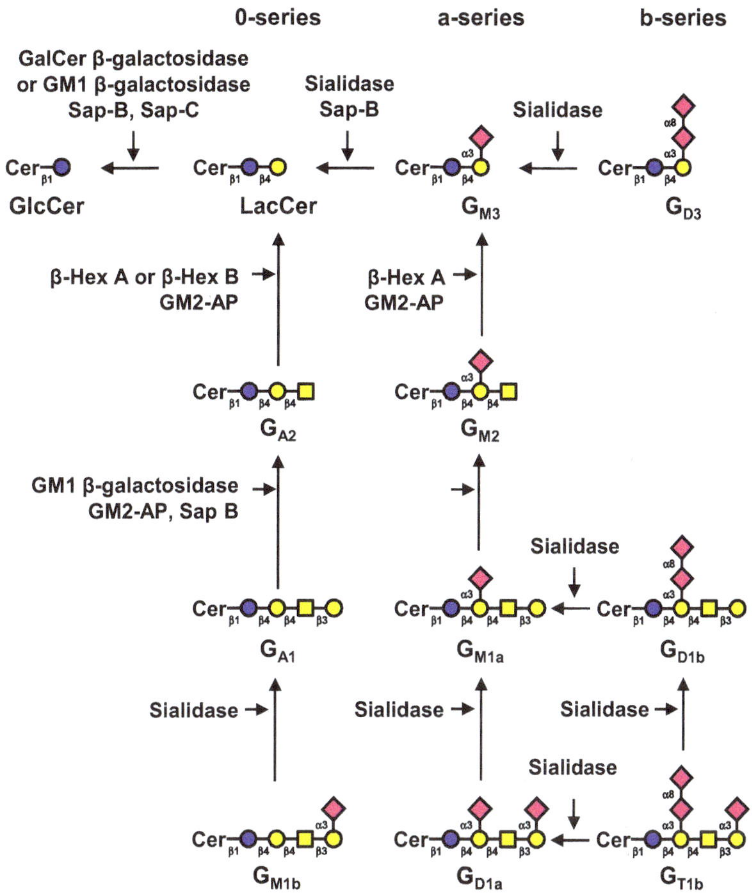


**Fig. 2** Representative reactions of ganglioside catabolism. Enzymes and lipid-transfer proteins abbreviations are as follow: β-Hex A, β-hexosaminidase A; β-Hex B, β-hexosaminidase B; GM2-AP, GM2 activator protein; Sap-B, saposin B; Sap-C, saposin C. The sugar symbols are the same as in Fig. 1.

## 3. CELLULAR AND MOLECULAR BIOLOGY OF GTS AND SIALIDASES

### 3.1 Transcriptional and Cellular Regulation of GT Involved in Ganglioside Expression

The biosynthesis of glycan chains is mostly regulated at the transcriptional level. The tissue specificity of GT genes expression depends on the use of alternative promoters that control tissue specific transcripts, differing in their

5′ untranslated region (5′UTR), and on the involvement of specific transcription factors (TF).[41] However, the identity of TF controlling GT expression and regulation in normal and pathological conditions is largely unknown, and precise mechanisms involved in TF-mediated GT expression are available for only a limited number of genes. For example, the RelB subunit of NF-κB is implicated in the expression of α2,3-sialyltransferase genes *ST3GAL1*, *ST3GAL2*, and *ST3GAL6*, resulting in GD1a and sialylparagloboside overexpression on prostatic cancer cells.[42] Other examples are the well-described interactions of GATA-3 and T-bet TFs that regulate the expression of sialyl-LewisX homing receptors on Th1/Th2 lymphocytes[43] and the activation of the *B3galt4* promoter by the Ets-1 transcription factor on PC12 cells.[44]

Epigenetic modifications are also involved in the regulation of GT genes and several studies have shown a direct link between the methylation status of GT genes and their expression profile.[45] The epigenetic regulation of GT genes involved in ganglioside biosynthesis has recently been shown to be crucial for neural development and neuronal differentiation of neuronal stem cells. In mouse brain, histone acetylation of GT genes contributes to the modifications of ganglioside expression during development.[46] Acetylation of histones H3 and H4 on the GM2/GD2 synthase *B4GALNT1* gene promoter allows the recruitment of SP1 and AP-2, which activate *B4GALNT1* expression and neural differentiation in a primary neuroepithelium culture in response to exogenous GM1 addition.[47] In addition, GM1 itself directly binds to acetylated histones on *B4GALNT1* and neurogenic TF *NeuroD1* promoters in differentiated neurons, suggesting that GM1 is involved in the chromatin complex that promotes neuronal differentiation.[48]

The elevated expression of specific gangliosides described in many types of cancers and involved in tumor aggressiveness suggests that epigenetic regulation of ganglioside GT genes could underlie some pathogenic mechanisms, and future studies would be useful to develop new strategies and find new targets for disease treatment.

The precise Golgi localization and supramolecular organization of glycosylation enzymes is also of critical importance. GTs involved in ganglioside biosynthesis are typical type II membrane-anchored proteins with a gradient distribution in the Golgi apparatus. The first glycosylation steps take place in the *cis*/medial-Golgi and the later in the *trans*-Golgi and *trans*-Golgi network. The relatively short-single transmembrane domain of GTs seems to be a critical signal for Golgi retention,[49] but the luminal and cytoplasmic domains also contain multiple signals involved in GT precise Golgi localization.[50–52] Disulfide-bonded dimerization, glycosylation, and secretion of GTs can also play a role in the processing and function of the different enzymes in the Golgi apparatus.[52] Moreover, ganglioside GTs organize in distinct multienzyme complexes, and β4GalNAcT 1/β3GalT 4, β4GalT 6/ST3 Gal V/ST8Sia I, and ST8Sia I/β4GalNAcT 1 complexes have been characterized.[53–55] Evidence indicates that the β4GalNAcT 1/β3GalT 4 complex concentrates in a sub-Golgi compartment more distal than the one bearing the β4GalT 6/ST3 Gal V/ST8Sia I complex, suggesting that particular GTs organized in distinct complexes and concentrated in different Golgi subcompartments, are involved in the regulation of ganglioside biosynthesis. Regulation of ganglioside biosynthesis can also be linked to the synthesis of different GT isoforms exhibiting different properties or activities. As an example, mouse GM3 synthase (GM3S) has three isoforms (M1-GM3S, M2-GM3S, and M3-GM3S), with different N-terminal cytoplasmic tails. M1-GM3S is localized in the ER, as a result of retrograde transport signals, and has a low GM3 synthesis activity compared with the other isoforms. In contrast, both M2-GM3S and M3-GM3S are localized in the Golgi apparatus, but M2 isoform is rapidly degraded in lysosomes, whereas M3-GM3S is retained in the Golgi apparatus. The existence of different GM3S isoforms having such different properties, that it is probably important in the regulation of GM3 biosynthesis under various pathological and physiological conditions.[56]

### 3.2 Expression and Regulation of Ganglioside GTs in Cancer: The Key Role of GD3S

The expression and regulation of GTs involved in the synthesis of gangliosides can be altered in pathological conditions including cancer. In particular, overexpression of complex gangliosides is observed in malignant cells, such as melanoma, central nervous system tumors, lung and breast cancers due to the increased expression of GD3S gene *ST8SIA1*. The *ST8SIA1* gene consists of five coding exons spanning over 135 kpb,[57] and its 5′UTR have been described in melanoma, glioblastoma, neuroblastoma, and breast cancer cell lines, showing different transcription start sites located upstream the initiation codon on the first exon.[58–61] The analysis of *ST8SIA1* core promoter implicated Tumor Necrosis Factor in GD3S overexpression in estrogen receptor-negative breast cancer cells via the NF-κB pathway, as well as the repressive role of estradiol in estrogen receptor-positive breast cancer cells.[61] Indeed, estradiol inhibits the *ST8SIA1* core promoter activity by preventing NF-κB nuclear translocation in ER-positive MCF-7 breast cancer cell line.[61] Other studies showed that Tumor Necrosis Factor modulates *B4GALNT1* expression in renal carcinoma,[62] and that NF-κB could play an essential role in the transcriptional regulation of the human GD3S expression in Fas-induced Jurkat-T cells.[63] Finally, testosterone was shown to activate ST3 Gal II sialyltransferase expression in prostate cancer cells through epigenetic regulation and NF-κB signaling.[64]

## 3.3 Regulation of Ganglioside Biosynthesis and Epithelial–Mesenchymal Transition

A growing body of evidence indicates that the epithelial–mesenchymal transition (EMT), which is undeniably implicated in cancer progression and metastasis, is likely associated with changes in ganglioside metabolism, and aberrant sialylation of gangliosides plays an important role in cell adhesion and motility in many cancers. Sarkar and coworkers revealed that GD3S could be involved in the EMT that bestows breast cancer cells with increased metastatic potential and cancer stem cell (CSC) properties.[65] Indeed, GD3S knockdown in the mesenchymal MDA-MB-231 cell line led to the expression of the epithelial cadherin and the downregulation of mesenchymal markers. GD3S silencing was also found to reduce the motility of breast cancer cells in vitro and metastasis in mice.[65] Surprisingly, normal restriction to the brain, the expression of ST6GalNAc V, a sialyltransferase implicated in the biosynthesis of α-Series gangliosides, decreases the interactions between breast cancer cells and the human blood–brain barrier.[66] If changes in GSL metabolism can modulate EMT, EMT can conversely affect the expression of genes encoding GTs. Thus, TGF-β-induced EMT is associated with a decreased expression of *B3galt4* gene in normal mouse mammary gland cells[67] and an increased expression of the GD3S in the MCF10A human epithelial breast cell line.[65] Moreover, many TFs involved in EMT have been identified to regulate the expression of genes encoding ganglioside-metabolizing enzymes. For instance, the Smad3/4 complex binds to the *B3galt4* promoter, leading to a decreased expression of the enzyme.[68] Zeb1 enhances the promoter activity of both GM3S and GD3S genes (*St3gal5* and *St8sia1*) by direct binding and furthermore represses the expression of a microRNA from the miR-200 family that targets the 3′UTR of the murine *St3gal5* gene (GM3S).[59,69] Finally, the overexpression of Twist/Snail enhances the expression of GD3S in transformed human mammary epithelial (HMLER) cells, as well as GD2 expression, suggesting a role of EMT in the origin of GD2-positive breast CSCs.[70] GD3S expression has also been shown to be triggered by NF-κB via FOXC2 transcription factor in human triple-negative breast cancer cells.[65] Importantly, GD3S was shown to display higher expression in melanoma cell lines and in estrogen receptor-negative breast cancer tumors, which both have a mesenchymal-like gene signature.[71–73] High expression of GD3S was furthermore associated with poor histopathological grading and a decreased overall survival of patients, suggesting that GD3S overexpression contributes to increase the malignant phenotype of breast cancer cells.[74] In this context, it has been demonstrated that GD3S expression in an estrogen receptor-negative breast cancer cell line induced proliferative properties in the absence of serum or exogenous growth factors, and increased tumor growth.[75] Taken together, these findings show that GD3S plays a key role in the maintenance of a mesenchymal phenotype in cells that have undergone EMT, especially in response to Snail, Twist, and TGF-β1 in several biological systems.

## 3.4 Expression and Regulation of Sialidases

As stated, gangliosides are mono- or multisialylated GSLs, which are synthesized, among other GTs, by a set of specific sialyltransferases in a cell type-dependent manner. The opposite reaction, desialylation, is catalyzed by a family of sialidases that operate at different cellular levels. Sialidases hydrolyze the terminal nonreducing sialic acid linkage in glycolipids and glycoproteins. To date four types of mammalian sialidases have been identified and characterized, designated Neu1, Neu2, Neu3, and Neu4, with each one having different subcellular localization and substrate specificity.[76,77]

Human Neu2 is the only member of the mammalian sialidases that has been crystallized[78] and it shows the typical β-propeller fold already described for bacterial and viral sialidases,[79,80] indicating that despite the low-sequence identity between viral, bacterial, and mammalian sialidases, they have a common folding topology, with active site residues highly conserved among species. Homology models of the remaining three members of the family have been developed based on the crystal structure of Neu2. All of them are assumed to share the typical β-propeller structure organized in six blades, each composed of four antiparallel β-sheets.[81] However, several structural differences in the active sites of these enzymes have been found, and this provides better insight into the differential substrate recognition and activity of each member.

Neu1 is a lysosomal sialidase that hydrolyzes preferentially glycoproteins and oligosaccharides. It is associated with carboxypeptidase protective protein/cathepsin A and β-galactosidase as a complex, with dissociation of the complex leading to sialidase inactivation.[82] Besides its well-known lysosomal catabolic function, Neu1 can be also targeted to the cell surface and assume the previously unrecognized role as a functional modulator of cellular receptors.[83] Neu2 and Neu4 have broad substrate specificity and can act on glycoproteins and oligosaccharides, as well as on glycolipids. Neu2 is a cytosolic sialidase while Neu4 has been found to be present in lysosomes, mitochondria, and the ER.[76]

Among human sialidases, the plasma membrane–associated sialidase Neu3 is by far the most studied. Neu3 is a key glycosidase for ganglioside degradation that preferentially hydrolyzes α2–3 and α2–8 terminal sialyl linkages, but is ineffective on an α2,3 inner sialic acid branch. Consequently, gangliosides GM3, GD3, GD1a, GD1b, and GT1b, but not GM1 and GM2, are substrates for the enzyme.[84] Neu3 has been characterized at the molecular level in various animal species and found to be ubiquitously expressed. However, a remarkable upregulation of Neu3 has been observed in various human cancer cells. Neu3 modifies the cell surface ganglioside composition by shifting polysialylated species to the monosialo derivatives and GM3 to LacCer. Besides, sialidase Neu3 is able to hydrolyze gangliosides present in the cell surface of neighboring cells, supporting its transcatalytic activity[85] and interestingly, it was recently demonstrated that Neu3 can also desialylate *N*-glycans present in glycoproteins.[86] The major subcellular localization of Neu3 sialidase is the plasma membrane, and it is also expressed in other cellular membrane components, such as the endosomal compartments.[87,88] Given that Neu3 is tightly associated to the plasma membrane,[88,89] it is thought to be an important molecule for various cell surface events. Indeed, many of the stimuli triggered by the sialidase involves modulation of transmembrane signaling not only through the modulation of ganglioside expression, but also by direct interaction with signaling molecules, as will be discussed in the following sections. Recently, the participation of Neu3 in endocytosis was also reported, regulating specifically the molecular machinery involved in clathrin-mediated internalization.[87]

## 4. REGULATION OF CELL SIGNALING BY GANGLIOSIDES IN CANCER CELLS

The role of gangliosides as regulators of signal transduction was first established by supplementation with exogenous gangliosides directly added in the culture medium of cancer cells.[90] With the identification and the molecular cloning of ganglioside biosynthetic enzymes, numbers of papers reported ectopic expression or antisense inhibition strategies targeting specific GTs to analyze the role of gangliosides in the regulation of signal transduction. All these approaches have clearly demonstrated that gangliosides are fine regulators of RTKs signaling[91] and that changes in cell membrane ganglioside composition that occur in cancer cells result in different cellular responses.[92,93] RTKs activate various intracellular signaling pathways and regulate cell survival, proliferation, differentiation, migration, and invasion. RTKs are usually activated by the binding of the ligand, inducing receptor dimerization and autophosphorylation of the kinase domain. In cancer cells, RTKs signaling is often overactivated through mutation or chromosomal translocation, leading to constitutive kinase domain activation and upregulation of downstream signaling.[94] At the plasma membrane, RTKs are localized in GEMs with other lipid raft-associated proteins including integrins and tetraspanins. Changes in ganglioside expression modify GEMs molecular composition and structure, leading to the reorganization and/or the exclusion of RTKs from GEMs.[95–97] Several RTKs including receptors for epidermal growth factor (EGF), fibroblast growth factor (FGF), hepatocyte growth factor, platelet-derived growth factor (PDGF) or nerve growth factor were demonstrated to be positively or negatively regulated by gangliosides in cancer cells.[11,93] These different observations show contrasting effects of ganglioside expression on cell signaling, monosialogangliosides mostly downregulating RTKs activation whereas disialogangliosides acting as activators of RTKs signaling pathways (Fig. 3).[13]

## 4.1 Monosialogangliosides Negatively Regulate Cell Signaling

Converging evidence demonstrates that monosialogangliosides, especially GM3, negatively regulate the activity of RTKs in cancer cells.[98] The negative effect of GM3 was first demonstrated for EGF receptor (EGFR) signaling in a variety of cell lines including hepatoma, hepatocellular carcinoma, and neuroblastoma cells.[99–102] In murine hepatoma cells Hca-F25, GM3 expression suppressed cell motility and migration via the inhibition of EGFR phosphorylation and the PI3K/Akt signaling pathway.[99] GM3 inhibits the dimerization of EGFR by preventing the autophosphorylation of the intracellular kinase domain in response to ligand binding.[101] It was shown that GM3 directly interacts with EGFR on a site distinct from the EGF-binding site through direct carbohydrate–carbohydrate interactions between GM3 and a GlcNAc-terminated *N*-glycan on EGFR.[103] It was also shown that the inhibition of the sialidase Neu3 suppresses cancer-cell growth by attenuation of EGFR signaling through GM3 accumulation.[104] In human

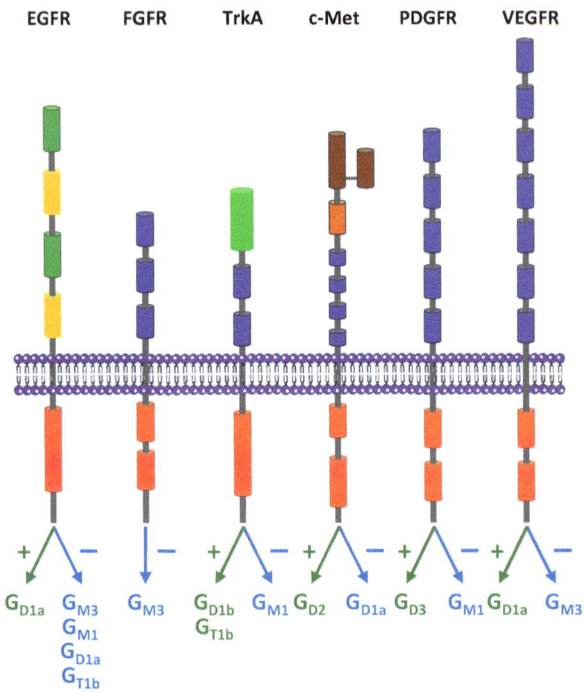


**Fig. 3** Main effects of gangliosides on RTKs signaling. Gangliosides can either inhibit (*red arrows*) or activate (*green arrows*) RTK signaling. FGFR, FGF receptor; PDGFR, PDGF receptor; VEGFR, VEGF receptor. *Adapted from Julien S, Bobowski M, Steenackers A, Le Bourhis X, Delannoy P. How do gangliosides regulate RTKs signaling? Cells. 2013;2:751–767.*[11]

neuroblastoma cells, proliferation and EGFR phosphorylation is not only inhibited by GM3 but also by GM1, GD1a, or GT1b. [102] GM3 is also implicated in the decrease of vascular endothelial growth factor (VEGF) receptor 2 phosphorylation and subsequent inhibition of Akt signaling pathway in human umbilical vein endothelial cells.[105] GM3 decreases VEGF receptor 2 activation by blocking receptor dimerization and VEGF binding through a GM3-specific interaction with the extracellular domain of VEGF receptor 2.[106] In cultured retinal glial cells, GM3 depletion by GlcCer synthase inhibition enhances tyrosine phosphorylation of the FGF receptor, activates PI3K/Akt pathway and increases the interactions of the FGF receptor with integrins.[107,108]

The monosialoganglioside GM1 also negatively regulates RTK signaling. For example, in human glioma cells, GM1 expression by transfection with GM2/GD2 synthase and GM1/GD1b synthase cDNAs, resulted in reduced

PDGF receptor phosphorylation and signaling, due to the exclusion of the receptor from GEMs.[109] The Csk binding protein PAG (phosphoprotein associated with glycosphingolipid-enriched microdomains) regulates PDGF receptor partitioning in caveolae and its association with Src family protein tyrosine kinases by controlling GM1 levels at the plasma membrane.[110] In rat pheochromocytoma cell line PC12, the expression of GM1 by transfection of β3GalT 4 cDNA that converts GM2 into GM1, inhibited nerve growth factor-induced TrkA dimerization and phosphorylation, as well as the downstream pathways.[111]

## 4.2 Tandem Disialogangliosides as Activators of RTKs

Contrasting with the negative effect of monosialogangliosides, more complex gangliosides showing Neu5Acα2–8Neu5Acα2-3 disaccharide motif linked to LacCer such as GD2 or GD3, are considered as positive regulators of RTKs signaling. In PC12 cells, the introduction of the GD3S (*ST8SIA1*) cDNA resulted in the overexpression of GD1b and GT1b. These gangliosides triggered a conformational change of TrkA that formed constitutively active dimers, activating its downstream signal pathways including Erk1/2 and PI3K/Akt, and leading to the enhancement of cell proliferation.[112] Furukawa and coworkers analyzed in depth the implication of GD3 in the malignant properties and cell signaling of human melanoma cells. By stable transfection of the GD3S cDNA into GD3-negative SK-MEL-28 mutant, they showed the enhancement of p130Cas, paxillin, and focal adhesion kinase (FAK) phosphorylation in GD3-positive cells.[113] The adhesion signals via integrins were also enhanced due to the clustering of integrins into GEMs associated with GD3 expression.[114] They also found higher amounts of constitutively activated Yes kinase in GEMs in GD3-positive cells.[115] They suggested that GD3 ganglioside plays a crucial role in the convergence of adhesion and RTKs signals, leading to the synergistic effects of those signals on malignant properties of melanomas.[116] Similarly, in glioma cells and tissues, GD3 ganglioside, PDGF receptor α, and activated kinase Yes colocalized in GEMs, promoting proliferation, invasion and a malignant phenotype.[117] In MDA-MB-231 breast cancer cells, the expression of GD3S induced the accumulation of b- and c-Series gangliosides including GD3, GD2, and GT3.[118,119] Among these complex gangliosides, GD2 was found to be involved in the constitutive activation of c-Met, and the subsequent activation of MEK/Erk and PI3K/Akt signaling pathways, leading to enhanced cell migration, proliferation and tumor growth in severe combined immunodeficiency (SCID) mice. This was shown by competition assays using anti-GD2 mAb that inhibited c-Met phosphorylation, demonstrating the role of the GD2 glycan moiety in c-Met activation.[119] Moreover, silencing of the GM2/GD2 synthase efficiently reduced both GD2 expression and c-Met phosphorylation. Of importance, the GD2-dependent activation of c-Met occurred in the absence of hepatocyte growth factor.[75] Several papers have recently underlined the role of GD2 ganglioside and GD3S in renewal and properties of cancer stem cells. GD2 was identified as a new specific cell surface marker of $CD44^{hi}CD24^{lo}$ breast CSC from human breast cancer cell lines and patient samples, capable of forming mammospheres and initiating tumors.[70] Gene expression analysis revealed that several GT genes involved in GD2 biosynthesis (*ST3GAL5*, *B4GALNT1*, and *ST8SIA1*) are highly expressed in CSC.[70,120] The reduction of GD2 expression by *ST8SIA1* knockdown reduced mammosphere formation and cell motility and completely abrogated tumor formation in vivo, changing the phenotype from CSC to non-CSC.[70,120] Similarly, GD3 and GD3S are also key drivers for glioblastoma stem cells maintenance and tumorigenicity.[121] Finally, it was also demonstrated that the interaction of GD3 with EGFR is responsible for sustaining its expression and downstream signaling to maintain the self-renewal of mouse neural stem cells in vitro.[122]

Disialogangliosides that belong to the a-Series, such as GD1a, were found to have different effects on cell signaling. In mouse osteosarcoma cell variant FBJ-LL cells, GD1a inhibits the hepatocyte growth factor-induced motility and scattering through the suppression of phosphorylation of c-Met.[123,124] On the contrary, in normal human dermal fibroblasts, GD1a promotes ligand-independent EGFR dimerization, enhances EGFR-mediated activation of the MAPK signaling pathway and EGFR phosphorylation is significantly reduced with the knockdown of ST3 Gal II, an enzyme that converts GM1 into GD1a.[125,126]

## 5. GANGLIOSIDES AND APOPTOSIS SIGNALING

Gangliosides, especially the disialoganglioside GD3, are involved in mitochondrial damage and apoptosis. GD3 was first implicated in apoptosis in lymphoid and myeloid tumor cells where the activation of the apoptosis-inducing CD95 death receptor (Fas) induced GD3 synthesis and accumulation in the membrane of mitochondria, leading to disruption of mitochondrial transmembrane potential and apoptosis in a caspase-independent fashion.[127] Apoptosis can be induced by direct exposure to GD3 exogenously added in the medium, or by overexpression of GD3S. GD3 accumulates within mitochondria of cells undergoing apoptosis and inhibition of GD3S expression prevents GD3-induced mitochondrial changes, caspase 9 activation, and apoptosis.[128] Although apoptosis could be induced by GM1 in thymocytes and by GD1b in MCF-7 breast cancer cells, most data available show that GD3 is the major player in programmed cell death.[129,130] Importantly, the association of specific gangliosides, mainly GD3 with cytoskeletal elements, such as the cytoplasmic peripheral membrane protein Ezrin, is required to trigger CD95/Fas-mediated apoptosis.[131] Efficient apoptotic signaling mediated by the CD95 death receptor required its translocation in GEMs. Thus, the structure and composition of GEMs play a major role in the promotion of either survival or cell death.[14] It has been shown that the proapoptotic activity of GD3 is suppressed by sialic acid O-acetylation. In childhood acute lymphoblastic leukemia, sialic acid O-acetylation on GD3 prevents its apoptotic effect and promotes survival of lymphoblasts.[132] Gangliosides bearing 9-O-acetylated sialic acid are expressed at high levels in tumors of neuroectodermal origin, such as melanoma, where they play roles in the growth and migration.[133] In tumor cells, 9-O-acetyl-GD3 could therefore modify the proapoptotic activity of GD3 either by affecting its transport to the mitochondria, or its interactions with mitochondrial components.

The targeting of tumor-associated gangliosides with antibodies also affects signaling pathways and leads to cell death including apoptosis.[134,135] As an example, the antiproliferative and proapoptotic activity of GD2-specific mAb 3F8 has been reported in human melanoma cells.[136] In human neuroblastoma cell lines, treatment with the anti-GD2 mAb 14G2a downregulated the PI3K/Akt/mTOR signaling network, decreasing cell viability.[137] In 2015, Dinutuximab (a chimeric monoclonal antibody to GD2 ganglioside) was approved by the US Food and Drug Administration (FDA) and is currently used in a combination immunotherapeutic regimen for the treatment of children with high-risk neuroblastoma.[138]

## 6. BIOLOGICAL ROLES AND RELEVANCE OF SIALIDASES IN CANCER

During the past few years, experimental evidence pointed out the potential involvement of mammalian sialidases in the occurrence of various kinds of tumors. Changes in sialidase expression levels have been observed in primary tumor samples, as well as in cancer cell lines, with different sialidases having been reported to either promote or revert malignant phenotypes. In fact, it seems like three of the sialidases, Neu1, Neu2, and Neu4 showed a tendency for downregulation, while Neu3 showed marked upregulation during carcinogenesis.[139]

Multiple observations suggest that, in addition to lysosomal catabolism, Neu1 modulates transmembrane signaling through desialylation of surface molecules, presumably after mobilization to the cell surface. In this sense, overexpression of Neu1 leads to suppression of experimental pulmonary metastasis and reversion of malignant phenotype in murine B16 melanoma cells.[140] This was consistent with pioneering observations that metastatic potential is inversely correlated with lysosomal-type sialidase activity in transformed rat 3Y1 fibroblasts.[141] It was proposed that a low level of Neu1 in tumors causes hypersialylation of the carbohydrate portion of glycoproteins, in line with the observation that malignant cells are often hypersialylated. To elucidate the mechanism underlying the modulation of malignant properties by Neu1, the human sialidase gene was overexpressed or silenced in colon cancer HT-29 cells. When Neu1-overexpressing cells were injected transsplenically into mice, in vivo liver metastasis was significantly reduced. Neu1 overexpression suppressed cell migration, invasion and adhesion in vitro, whereas its silencing resulted in the opposite. One of the target molecules of Neu1 was found to be integrin β4, which undergoes desialylation and decreased phosphorylation followed by attenuation of focal adhesion kinase and ERK1/2 downstream pathway.[142] Furthermore, it has been demonstrated that cell surface-residing Neu1 downregulates the proliferation of human aortic smooth-muscle cells and fibroblasts by desialylation of other cell surface receptors that directly propagate mitogenic signals, such as PDGF and insulin-like growth factor-2 receptors. Moreover, fibroblasts derived from patients with sialidosis bearing exclusive deficiency in Neu1 demonstrate higher basal proliferation rates and greater responsiveness to PDGF and insulin-like growth factor-2 than normal fibroblasts.[143] Although, several cancers show a tendency for decreased Neu1 expression, recent reports suggest that this is not a general property of cancer cells. For example, in ovarian cancer tissues of patients, Neu1 was expressed at a higher level than that in adjacent normal tissues, and in OVCAR3 and SKOV3 ovarian carcinoma cells, Neu1 knockdown inhibited proliferation, invasion, arrested cells cycle at G0/G1 phase, and induced apoptosis in vitro.[144] In line with these results, it has been proposed that Neu1 hydrolyzes α-2,3-sialic acid residues of glycosylated receptors, such as EGFR and TrkA on the cell surface to remove steric hindrance and to facilitate growth factor association, subsequent receptor activation, and downstream signaling.[145,146]

As mentioned before, upregulation of Neu3 is observed in various types of tumors including colon,[147] renal,[148] ovarian,[149] prostate,[150,151] and head and neck cancers.[152] In general, its upregulation has been associated with apoptosis suppression and increased cell proliferation, invasion, and migration. In colon cancer HCT-116 cells, Neu3 overexpression increases the level of antiapoptotic molecules, such as Bcl-2, and decreases the amount of proapoptotic molecules, such as caspase-3.[147] In colon cancer DLD-1 cells, this sialidase also regulates cell adhesion by enhancing tyrosine phosphorylation of integrin β4 and promoting adhesion to laminins.[153] In human prostate cancer–cell lines PC-3 and LNCaP, upregulation of Neu3 causes androgen-independent proliferation by enhancing expression of androgen receptor signaling-related molecules.[150] This was consistent with another report in which mice injected with PC-3 M prostate carcinoma cells carrying the Neu3 siRNA developed fewer bone metastases than control mice.[151] In squamous carcinoma HSC-2 and SAS cells, Neu3 promoted cell motility and invasion, accompanied by the increased expression of matrix metallopeptidase 9.[152] Furthermore, it has been recently demonstrated that Neu3 influences the expression of stem cell pluripotency marker genes and Wnt/β-catenin signaling related genes in HT-29 and HCT116 colon cancer cells, suggesting that this sialidase might contribute to maintenance of an undifferentiated and tumorigenic state of cancer cells.[154] On the other hand, silencing of the Neu3 gene in different cancer cell lines significantly inhibited cell growth, decreasing cell proliferation and increasing the propensity to undergo apoptosis.[150,155,156]

The role of Neu3 in tumorigenesis in vivo has been studied in animal models. Neu3 transgenic mice showed increased azoxymethane-induced aberrant crypt foci formation in the colonic mucosa due to suppression of apoptosis,[157] whereas Neu3 knock-out mice were less susceptible than wild-type mice to the colitis-associated colon carcinogenesis induced by azoxymethane and dextran sodium sulfate.[158]

One of the possible molecular mechanisms that explains the changes caused by Neu3 upregulation is that this sialidase stimulates EGFR phosphorylation in response to EGF with a consequent activation of its downstream molecules, ERK1/2 and Akt.[86,156,157,159] As already mentioned, ligand-dependent activation of EGFR is known to be inhibited by ganglioside GM3 through carbohydrate–carbohydrate interactions between GM3 and *N*-glycans in the ectodomain of EGFR.[160] Analysis of glycolipid content in cells overexpressing Neu3 indicated a significant decrease in GM3 substrate and a marked increase in LacCer, suggesting that modulation of gangliosides as a consequence of Neu3 catalytic activity could remove the inhibitory effect exerted by GM3 on EGFR, promoting its activation. Interestingly, it was also proposed that Neu3 might activate EGFR through direct desialylation of the receptor, suggesting that glycoproteins may be also among the physiological substrates of Neu3.[86] Therefore, Neu3 may protect against cell death both by promoting EGFR activation through its association and eventual desialylation, as well as through gangliosides pattern modulation. In addition, Neu3 might exert an influence on the cell surface by interacting directly with other signaling molecules including integrin β4,[153] low-density lipoprotein receptor-related protein 6,[154] Rac-1,[161] Cav-1,[162] Src,[159] and Grb-2.[163] In conclusion, the net effect of Neu3 upregulation in cancer cells seems to be the activation of prosurvival pathways that are used as an advantage for tumor progression.

Much less is known about the involvement of the other two members of the sialidase family, Neu2 and Neu4, in carcinogenesis. In clear contrast to Neu3, some studies suggest that downregulation of Neu2 may contribute to the invasive properties and protection against programmed cell death of tumors. For example, ectopic expression of Neu2 in B16 melanoma cells led to marked suppression of lung metastasis accompanied by decrease in invasiveness and cell motility.[164] Similar results were obtained when the cytosolic sialidase was overexpressed in highly metastatic mouse colon adenocarcinoma NL17 cells.[165] In the same sense, Neu2 expression in myeloid leukemia K562 cells remarkably weakened the antiapoptotic axis Bcr-Abl, leading to an increased susceptibility to apoptosis and a marked reduction in proliferation rate. The molecular link between Neu2 activity and Bcr-Abl signaling pathway is not completely understood, but may rely on the desialylation of some cytosolic glycoconjugates.[166]

Regarding Neu4, the level of this sialidase was found to be markedly decreased in human colon cancers as compared with the noncancerous mucosa.[139,167] Besides, transfection of Neu4 in cultured human colon cancer cells resulted in acceleration of apoptosis and in decreased invasion and motility. The siRNA-mediated Neu4 targeting, on the other hand, caused a significant inhibition of apoptosis and promotion of invasion and motility.[167] It has been proposed that Neu4 plays a physiological role in controlling sialyl-Lewis antigens through desialylation of O-glycans, therefore contributing to invasive properties of cancer.[168] In clear contrast,

Neu4 is upregulated in glioblastoma stem cells and promotes β-catenin signaling, enhancing stem-like malignant cell growth.[169,170]

Overall, it is clear that sialidases are likely to be involved not only in catabolism of glycoconjugates but also in functional modulation of transmembrane signaling through protein interactions with signaling molecules. Moreover, the effects of sialidase expression on malignant phenotype may be due to different mechanisms depending on the cell type and cellular conditions.

## 7. CONCLUSIONS

The last 3 decades have been very fruitful in increasing our knowledge about the metabolism of gangliosides and related lipids, their expression in different physiological and pathological conditions, and very importantly, the cellular signal mechanisms regulated by these sialoglycolipids. It is now clear that gangliosides regulate RTK signaling in cancer cells either by inhibiting or activating receptor dimerization and autophosphorylation. From a general point of view, monosialogangliosides, such as GM3 or GM1 can be considered as negative regulators of RTKs signaling whereas disialogangliosides including GD2, mostly activate RTKs-mediated signal transduction involved in cell proliferation and migration, angiogenesis, and tumor metastasis, making both gangliosides and related glycolipids important predictive biomarkers for clinical outcome of different tumors.[171]

Undoubtedly, gangliosides have reemerged as promising targets for developing anticancer therapies, which includes new and better anticancer antibodies; the development of immunogene therapies, such as antiganglioside-specific chimeric antigen receptor (CAR) to redirect T cells to particular tumor cells;[172] and the design of novel and specific inhibitors of glycolipid synthesis. In this context, the understanding of the mechanisms by which gangliosides regulate cell signaling is of primary importance to identify new targets in cancer therapy.

## ACKNOWLEDGMENTS

This work was supported by the University of Lille, the CNRS and the "comité du Nord de La Ligue contre le Cancer" and by grants (to JLD) from the Universidad Nacional de Córdoba, Consejo Nacional de Investigaciones Científicas y Técnicas (CONICET, Argentina), the Agencia Nacional de Promoción Científica y Tecnológica (Argentina) and the Mizutani Foundation for Glycoscience (Japan).

## REFERENCES


1. Fishman PH, Brady RO. Biosynthesis and function of gangliosides. *Science.* 1976;194:906–915.
2. Svennerholm L. Ganglioside designation. *Adv Exp Med Biol.* 1980;125:11.
3. Yamashita T, Wada R, Sasaki T, Deng C, Bierfreund U, Sandhoff K, Proia RL. A vital role for glycosphingolipid synthesis during development and differentiation. *Proc Natl Acad Sci USA.* 1999;96:9142–9147.
4. Birklé S, Zeng G, Gao L, Yu RK, Aubry J. Role of tumor-associated gangliosides in cancer progression. *Biochimie.* 2003;85:455–463.
5. Bobowski M, Cazet A, Steenackers A, Delannoy P. Role of complex gangliosides in cancer progression. *Carbohydr Chem.* 2012;37:1–20.
6. Furukawa K, Hamamura K, Aixinjueluo W, Furukawa K. Biosignals modulated by tumor-associated carbohydrate antigens: novel targets for cancer therapy. *Ann NY Acad Sci.* 2006;1086:185–198.
7. Krengel U, Bousquet PA. Molecular recognition of gangliosides and their potential for cancer immunotherapies. *Front Immunol.* 2014;5:325.
8. Rabu C, McIntosh R, Jurasova Z, Durrant L. Glycans as targets for therapeutic anti-tumor antibodies. *Future Oncol.* 2012;8:943–960.
9. Suzuki M, Cheung NK. Disialoganglioside GD2 as a therapeutic target for human diseases. *Expert Opin Ther Targets.* 2015;19:349–362.
10. Hakomori S. Carbohydrate-to-carbohydrate interaction, through glycosynapse, as a basis of cell recognition and membrane organization. *Glycoconj J.* 2004;21: 125–137.
11. Julien S, Bobowski M, Steenackers A, Le Bourhis X, Delannoy P. How do gangliosides regulate RTKs signaling? *Cells.* 2013;2:751–767.
12. Harding AS, Hancock JF. Using plasma membrane nanoclusters to build better signaling circuits. *Trends Cell Biol.* 2008;18:364–371.
13. Furukawa K, Ohkawa Y, Yamauchi Y, Hamamura K, Ohmi Y, Furukawa K. Fine tuning of cell signals by glycosylation. *J Biochem.* 2012;151:573–578.
14. Gajate C, Mollinedo F. Lipid rafts and raft-mediated supramolecular entities in the regulation of CD95 death receptor apoptotic signaling. *Apoptosis.* 2015;20:584–606.
15. Stiban J, Tidhar R, Futerman AH. Ceramide synthases: roles in cell physiology and signaling. *Adv Exp Med Biol.* 2010;688:60–71.
16. Hanada K, Kumagai K, Tomishige N, Yamaji T. CERT-mediated trafficking of ceramide. *Biochim Biophys Acta.* 2009;1791:684–691.
17. Sprong H, Kruithof B, Leijendekker R, Slot JW, van Meer G, van der Sluijs P. UDP-galactose: ceramide galactosyltransferase is a class I integral membrane protein of the endoplasmic reticulum. *J Biol Chem.* 1998;273:25880–25888.
18. Honke K, Tsuda M, Hirahara Y, Ishii A, Makita A, Wada Y. Molecular cloning and expression of cDNA encoding human 3′-phosphoadenylylsulfate:galactosyl ceramide 3′-sulfotransferase. *J Biol Chem.* 1997;272:4864–4868.
19. Uemura S, Go S, Shishido F, Inokuchi J. Expression machinery of GM4: the excess amounts of GM3/GM4S synthase (ST3GAL5) are necessary for GM4 synthesis in mammalian cells. *Glycoconj J.* 2014;31:101–108.
20. Marks DL, Wu K, Paul P, Kamisaka Y, Watanabe R, Pagano RE. Oligomerization and topology of the Golgi membrane protein glucosylceramide synthase. *J Biol Chem.* 1999;274:451–456.
21. Lannert H, Bünning C, Jeckel D, Wieland FT. Lactosylceramide is synthesized in the lumen of the Golgi apparatus. *FEBS Lett.* 1994;342:91–96.
22. Halter D, Neumann S, van Dijk SM, Wolthoorn J, de Mazière AM, Vieira OV, Mattjus P, Klumperman J, van Meer G, Sprong H. Pre- and post-Golgi translocation of glucosylceramide in glycosphingolipid synthesis. *J Cell Biol.* 2007;179:101–115.

23. Chalat M, Menon I, Turan Z, Menon AK. Reconstitution of glucosylceramide flip-flop across endoplasmic reticulum: implications for mechanism of glycosphingolipid biosynthesis. *J Biol Chem.* 2012;287:15523–15532.
24. D'Angelo G, Polishchuk E, Di Tullio G, Santoro M, Di Campli A, Godi A, West G, Bielawski J, Chuang CC, van der Spoel AC, Platt FM, Hannun YA, Polishchuk R, Mattjus P, De Matteis MA. Glycosphingolipid synthesis requires FAPP2 transfer of glucosylceramide. *Nature.* 2007;449:62–67.
25. Nakayama J, Fukuda MN, Hirabayashi Y, Kanamori A, Sasaki K, Nishi T, Fukuda M. Expression cloning of a human GT3 synthase. GD3 AND GT3 are synthesized by a single enzyme. *J Biol Chem.* 1996;271:3684–3691.
26. Kim YJ, Kim KS, Do S, Kim CH, Kim SK, Lee YC. Molecular cloning and expression of human alpha2,8-sialyltransferase (hST8Sia V). *Biochem Biophys Res Commun.* 1997;235:327–330.
27. Crespo PM, Demichelis VT, Daniotti JL. Neobiosynthesis of glycosphingolipids by plasma membrane-associated glycosyltransferases. *J Biol Chem.* 2010;285: 29179–29190.
28. Panzetta P, Maccioni HJ, Caputto R. Synthesis of retinal gangliosides during chick embryonic development. *J Neurochem.* 1980;35:100–108.
29. Regina Todeschini A, Hakomori SI. Functional role of glycosphingolipids and gangliosides in control of cell adhesion, motility, and growth, through glycosynaptic microdomains. *Biochim Biophys Acta.* 2008;1780:421–433.
30. Crespo PM, Iglesias-Bartolomé R, Daniotti JL. Ganglioside GD3 traffics from the trans-Golgi network to plasma membrane by a Rab11-independent and brefeldin A-insensitive exocytic pathway. *J Biol Chem.* 2004;279:47610–47618.
31. Daniotti JL, Iglesias-Bartolomé R. Metabolic pathways and intracellular trafficking of gangliosides. *IUBMB Life.* 2011;63:513–520.
32. Tettamanti G. Ganglioside/glycosphingolipid turnover: new concepts. *Glycoconj J.* 2004;20:301–317.
33. Iglesias-Bartolomé R, Trenchi A, Comín R, Moyano AL, Nores GA, Daniotti JL. Differential endocytic trafficking of neuropathy-associated antibodies to GM1 ganglioside and cholera toxin in epithelial and neural cells. *Biochim Biophys Acta.* 2009;1788:2526–2540.
34. Kolter T. Ganglioside biochemistry. *ISRN Biochem.* 2012;2012:506160.
35. Sandhoff K. Metabolic and cellular bases of sphingolipidoses. *Biochem Soc Trans.* 2013;41:1562–1568.
36. Kolter T, Sandhoff K. Lysosomal degradation of membrane lipids. *FEBS Lett.* 2010;584:1700–1712.
37. d'Azzo A, Bonten E. Molecular mechanisms of pathogenesis in a glycosphingolipid and a glycoprotein storage disease. *Biochem Soc Trans.* 2010;38:1453–1457.
38. Aureli M, Samarani M, Murdica V, Mauri L, Loberto N, Bassi R, Prinetti A, Sonnino S. Gangliosides and cell surface ganglioside glycohydrolases in the nervous system. *Adv Neurobiol.* 2014;9:223–244.
39. Prinetti A, Chigorno V, Mauri L, Loberto N, Sonnino S. Modulation of cell functions by glycosphingolipid metabolic remodeling in the plasma membrane. *J Neurochem.* 2007;103(Suppl 1):113–125.
40. Daniotti JL, Vilcaes AA, Rodriguez-Walker M, Ruggiero FM. Metabolism of Glycolipids at the Cell Surface. In: Catala A, ed. *Sphingolipids: Biology, Synthesis and Functions. New York, NY: Nova Science Publishers Inc; 2015:67–90.*
41. Harduin-Lepers A. Comprehensive analysis of sialyltransferases in vertebrate genomes. *Glycobiology Insights.* 2010;2:29–61.
42. Hatano K, Miyamoto Y, Nonomura N, Kaneda Y. Expression of gangliosides, GD1a, and sialyl paragloboside is regulated by NF-κB-dependent transcriptional control of

α2,3-sialyltransferase I, II, and VI in human castration-resistant prostate cancer cells. *Int J Cancer*. 2011;129:1838–1847.

43. Chen GY, Osada H, Santamaria-Babı LF, Kannagi R. Interaction of GATA-3/T-bet transcription factors regulates expression of sialyl Lewis X homing receptors on Th1/Th2 lymphocytes. *Proc Natl Acad Sci USA*. 2006;103:16894–16899.
44. Raimondi LP, Daniotti JL, Maccioni HJ. ETS-1 transcription factor activates the expression of mouse UDP-Gal:GA2/GM2/GD2/GT2 galactosyltransferase gene. *FEBS Lett*. 2004;576:487–491.
45. Antony P, Rose M, Heidenreich A, Knüchel R, Gaisa NT, Dahl E. Epigenetic inactivation of ST6GAL1 in human bladder cancer. *BMC Cancer*. 2014;14:901.
46. Suzuki Y, Yanagisawa M, Ariga T, Yu RK. Histone acetylation-mediated glycosyltransferase gene regulation in mouse brain during development. *J Neurochem*. 2011;116: 874–880.
47. Tsai YT, Yu RK. Epigenetic activation of mouse ganglioside synthase genes: implications for neurogenesis. *J Neurochem*. 2014;128:101–110.
48. Itokazu Y, Tsai YT, Yu RK. Epigenetic regulation of ganglioside expression in neural stem cells and neuronal cells. *Glycoconj J*. 2016. [Epub ahead of print].
49. Munro S. An investigation of the role of transmembrane domains in Golgi protein retention. *EMBO J*. 1995;14:4695–4704.
50. Zerfaoui M, Fukuda M, Langlet C, Mathieu S, Suzuki M, Lombardo D, El-Battari A. The cytosolic and transmembrane domains of the beta 1,6 *N*-acetylglucosaminyltransferase (C2GnT) function as a cis to medial/Golgi-targeting determinant. *Glycobiology*. 2002;12:15–24.
51. Tu L, Tai WC, Chen L, Banfield DK. Signal-mediated dynamic retention of glycosyltransferases in the Golgi. *Science*. 2008;321:404–407.
52. El-Battari A, Prorok M, Angata K, Mathieu S, Zerfaoui M, Ong E, Suzuki M, Lombardo D, Fukuda M. Different glycosyltransferases are differentially processed for secretion, dimerization, and autoglycosylation. *Glycobiology*. 2003;13:941–953.
53. Giraudo CG, Daniotti JL, Maccioni HJ. Physical and functional association of glycolipid *N*-acetyl-galactosaminyl and galactosyl transferases in the Golgi apparatus. *Proc Natl Acad Sci USA*. 2001;98:1625–1630.
54. Giraudo CG, Maccioni HJ. Ganglioside glycosyltransferases organize in distinct multienzyme complexes in CHO-K1 cells. *J Biol Chem*. 2003;278:40262–40271.
55. Bieberich E, MacKinnon S, Silva J, Li DD, Tencomnao T, Irwin L, Kapitonov D, Yu RK. Regulation of ganglioside biosynthesis by enzyme complex formation of glycosyltransferases. *Biochemistry*. 2002;41:11479–11487.
56. Uemura S, Yoshida S, Shishido F, Inokuchi J. The cytoplasmic tail of GM3 synthase defines its subcellular localization, stability, and in vivo activity. *Mol Biol Cell*. 2009;20:3088–3100.
57. Furukawa K, Horie M, Okutomi K, Sugano S, Furukawa K. Isolation and functional analysis of the melanoma specific promoter region of human GD3 synthase gene. *Biochim Biophys Acta*. 2003;1627:71–78.
58. Kang NY, Kim CH, Kim KS, Ko JH, Lee JH, Jeong YK, Lee YC. Expression of the human CMP-NeuAc:GM3 alpha2,8-sialyltransferase (GD3 synthase) gene through the NF-kappaB activation in human melanoma SK-MEL-2 cells. *Biochim Biophys Acta*. 2007;1769:622–630.
59. Dae HM, Kwon HY, Kang NY, Song NR, Kim KS, Kim CH, Lee JH, Lee YC. Isolation and functional analysis of the human glioblastoma-specific promoter region of the human GD3 synthase (hST8Sia I) gene. *Acta Biochim Biophys Sin (Shanghai)*. 2009;41:237–245.
60. Kwon HY, Dae HM, Song NR, Kim KS, Kim CH, Lee YC. Valproic acid induces transcriptional activation of human GD3 synthase (hST8Sia I) in SK-N-BE(2)-C human neuroblastoma cells. *Mol Cells*. 2009;27:113–118.

61. Bobowski M, Vincent A, Steenackers A, Colomb F, Van Seuningen I, Julien S, Delannoy P. Estradiol represses the G(D3) synthase gene ST8SIA1 expression in human breast cancer cells by preventing NFκB binding to ST8SIA1 promoter. *PLoS One.* 2013;8:e62559.
62. Raval G, Biswas S, Rayman P, Biswas K, Sa G, Ghosh S, Thornton M, Hilston C, Das T, Bukowski R, Finke J, Tannenbaum CS. TNF-alpha induction of GM2 expression on renal cell carcinomas promotes T cell dysfunction. *J Immunol.* 2007;178:6642–6652.
63. Kang NY, Kang SK, Lee YC, Choi HJ, Lee YS, Cho SY, Kim YS, Ko JH, Kim CH. Transcriptional regulation of the human GD3 synthase gene expression in Fas-induced Jurkat T cells: a critical role of transcription factor NF-kappaB in regulated expression. *Glycobiology.* 2006;16:375–389.
64. Hatano K, Miyamoto Y, Mori M, Nimura K, Nakai Y, Nonomura N, Kaneda Y. Androgen-regulated transcriptional control of sialyltransferases in prostate cancer cells. *PLoS One.* 2012;7:e31234.
65. Sarkar TR, Battula VL, Werden SJ, Vijay GV, Ramirez-Peña EQ, Taube JH, Chang JT, Miura N, Porter W, Sphyris N, Andreeff M, Mani SA. GD3 synthase regulates epithelial-mesenchymal transition and metastasis in breast cancer. *Oncogene.* 2015;34:2958–2967.
66. Drolez A, Vandenhaute E, Delannoy CP, Dewald JH, Gosselet F, Cecchelli R, Julien S, Dehouck MP, Delannoy P, Mysiorek C. ST6GALNAC5 expression decreases the interactions between breast cancer cells and the human blood–brain barrier. *Int J Mol Sci.* 2016;17:E1309.
67. Guan F, Schaffer L, Handa K, Hakomori SI. Functional role of gangliotetraosylceramide in epithelial-to-mesenchymal transition process induced by hypoxia and by TGF-β. *FASEB J.* 2010;24:4889–4903.
68. Guo J, Song B, Li X, Hε C, Yang G, Yang X, Guan F. Downregulation of gangliotetraosylceramide and β1,3-galactosyltransferase-4 gene expression by Smads during transforming growth factor β-induced epithelial–mesenchymal transition. *Mol Med Rep.* 2015;11:2241–2247.
69. Mathow D, Chessa F, Rabionet M, Kaden S, Jennemann R, Sandhoff R, Gröne HJ, Feuerborn A. Zeb1 affects epithelial cell adhesion by diverting glycosphingolipid metabolism. *EMBO Rep.* 2015;16:321–331.
70. Battula VL, Shi Y, Evans KW, Wang RY, Spaeth EL, Jacamo RO, Guerra R, Sahin AA, Marini FC, Hortobagyi G, Mani SA, Andreeff M. Ganglioside GD2 identifies breast cancer stem cells and promotes tumorigenesis. *J Clin Invest.* 2012;122:2066–2078.
71. Ruan S, Lloyd KO. Glycosylation pathways in the biosynthesis of gangliosides in melanoma and neuroblastoma cells: relative glycosyltransferase levels determine ganglioside patterns. *Cancer Res.* 1992;52:5725–5731.
72. Yamashiro S, Okada M, Haraguchi M, Furukawa K, Lloyd KO, Shiku H, Furukawa K. Expression of alpha 2,8-sialyltransferase (GD3 synthase) gene in human cancer cell lines: high level expression in melanomas and up-regulation in activated T lymphocytes. *Glycoconj J.* 1995;12:894–900.
73. Ruckhäberle E, Rody A, Engels K, Gaetje R, von Minckwitz G, Schiffmann S, Grösch S, Geisslinger G, Holtrich U, Karn T, Kaufmann M. Microarray analysis of altered sphingolipid metabolism reveals prognostic significance of sphingosine kinase 1 in breast cancer. *Breast Cancer Res Treat.* 2008;112:41–52.
74. Ruckhäberle E, Karn T, Rody A, Hanker L, Gätje R, Metzler D, Holtrich U, Kaufmann M. Gene expression of ceramide kinase, galactosyl ceramide synthase and ganglioside GD3 synthase is associated with prognosis in breast cancer. *J Cancer Res Clin Oncol.* 2009;135:1005–1013.
75. Cazet A, Lefebvre J, Adriaenssens E, Julien S, Bobowski M, Grigoriadis A, Tutt A, Tulasne D, Le Bourhis X, Delannoy P. GD3 synthase expression enhances proliferation

and tumor growth of MDA-MB-231 breast cancer cells through c-Met activation. *Mol Cancer Res.* 2010;8:1526–1535.
76. Miyagi T, Yamaguchi K. Mammalian sialidases: physiological and pathological roles in cellular functions. *Glycobiology.* 2012;22:880–896.
77. Monti E, Bonten E, D'Azzo A, Bresciani R, Venerando B, Borsani G, Schauer R, Tettamanti G. Sialidases in vertebrates: a family of enzymes tailored for several cell functions. *Adv Carbohydr Chem Biochem.* 2010;64:403–479.
78. Chavas LM, Tringali C, Fusi P, Venerando B, Tettamanti G, Kato R, Monti E, Wakatsuki S. Crystal structure of the human cytosolic sialidase Neu2. Evidence for the dynamic nature of substrate recognition. *J Biol Chem.* 2005;280:469–475.
79. Varghese JN, Laver WG, Colman PM. Structure of the influenza virus glycoprotein antigen neuraminidase at 2.9 Å resolution. *Nature.* 1983;303:35–40.
80. Crennell SJ, Garman EF, Laver WG, Vimr ER, Taylor GL. Crystal structure of a bacterial sialidase (from Salmonella typhimurium LT2) shows the same fold as an influenza virus neuraminidase. *Proc Natl Acad Sci USA.* 1993;90:9852–9856.
81. Magesh S, Suzuki T, Miyagi T, Ishida H, Kiso M. Homology modeling of human sialidase enzymes NEU1, NEU3 and NEU4 based on the crystal structure of NEU2: hints for the design of selective NEU3 inhibitors. *J Mol Graph Model.* 2006;25:196–207.
82. Pshezhetsky AV, Ashmarina M. Lysosomal multienzyme complex: biochemistry, genetics, and molecular pathophysiology. *Prog Nucleic Acid Res Mol Biol.* 2001;69:81–114.
83. Pshezhetsky AV, Hinek A. Where catabolism meets signalling: neuraminidase 1 as a modulator of cell receptors. *Glycoconj J.* 2011;28:441–452.
84. Miyagi T, Wada T, Iwamatsu A, Hata K, Yoshikawa Y, Tokuyama S, Sawada M. Molecular cloning and characterization of a plasma membrane-associated sialidase specific for gangliosides. *J Biol Chem.* 1999;274:5004–5011.
85. Papini N, Anastasia L, Tringali C, Croci G, Bresciani R, Yamaguchi K, Miyagi T, Preti A, Prinetti A, Prioni S, Sonnino S, Tettamanti G, Venerando B, Monti E. The plasma membrane-associated sialidase MmNEU3 modifies the ganglioside pattern of adjacent cells supporting its involvement in cell-to-cell interactions. *J Biol Chem.* 2004;279:16989–16995.
86. Mozzi A, Forcella M, Riva A, Difrancesco C, Molinari F, Martin V, Papini N, Bernasconi B, Nonnis S, Tedeschi G, Mazzucchelli L, Monti E, Fusi P, Frattini M. NEU3 activity enhances EGFR activation without affecting EGFR expression and acts on its sialylation levels. *Glycobiology.* 2015;25:855–868.
87. Rodriguez-Walker M, Vilcaes AA, Garbarino-Pico E, Daniotti JL. Role of plasma-membrane-bound sialidase NEU3 in clathrin-mediated endocytosis. *Biochem J.* 2015;470:131–144.
88. Zanchetti G, Colombi P, Manzoni M, Anastasia L, Caimi L, Borsani G, Venerando B, Tettamanti G, Preti A, Monti E, Bresciani R. Sialidase NEU3 is a peripheral membrane protein localized on the cell surface and in endosomal structures. *Biochem J.* 2007;408:211–219.
89. Rodriguez-Walker M, Daniotti JL. Human sialidase Neu3 is *S*-acylated and behaves like an integral membrane protein. Sci Rep 2017;7, 4167
90. Bremer EG, Schlessinger J, Hakomori SI. Ganglioside-mediated modulation of cell growth. Specific effects of GM3 on tyrosine phosphorylation of the epidermal growth factor receptor. *J Biol Chem.* 1986;261:2434–2440.
91. Bremer EG, Hakomori SI. Gangliosides as receptor modulators. *Adv Exp Med Biol.* 1984;174:381–394.
92. Hakomori S, Igarashi Y. Functional role of glycosphingolipids in cell recognition and signaling. *J Biochem.* 1995;118:1091–1103.
93. Kaucic K, Liu Y, Ladisch S. Modulation of growth factor signaling by gangliosides: positive or negative? *Methods Enzymol.* 2006;417:168–185.

94. Nelson KN, Peiris MN, Meyer AN, Siari A, Donoghue DJ. Receptor tyrosine kinases: translocation partners in hematopoietic disorders. *Trends Mol Med.* 2017;23:59–79.
95. Lai AZ, Abella JV, Park M. Crosstalk in Met receptor oncogenesis. *Trends Cell Biol.* 2009;19:542–551.
96. Park SY, Yoon SJ, Freire-de-Lima L, Kim JH, Hakomori SI. Control of cell motility by interaction of gangliosides tetraspanins and epidermal growth factor receptor in A431 versus KB epidermoid tumor cells. *Carbohydr Res.* 2009;344:1479–1486.
97. Todeschini AR, Dos Santos JN, Handa K, Hakomori SI. Ganglioside GM2-tetraspanin CD82 complex inhibits met and its cross-talk with integrins providing a basis for control of cell motility through glycosynapse. *J Biol Chem.* 2007;282:8123–8133.
98. Hakomori SI, Handa K. GM3 and cancer. *Glycoconj J.* 2015;32:1–8.
99. Huang X, Li Y, Zhang J, Xu Y, Tian Y, Ma K. Ganglioside GM3 inhibits hepatoma cell motility via down-regulating activity of EGFR and PI3K/Akt signaling pathway. *J Cell Biochem.* 2013;114:1616–1624.
100. Li Y, Huang X, Zhang J, Li Y, Ma K. Synergistic inhibition of cell migration by tetraspanin CD82 and gangliosides occurs via the EGFR or cMet-activated Pl3K/Akt signalling pathway. *Int J Biochem Cell Biol.* 2013;45:2349–2358.
101. Coskun Üul., Grzybek M, Drechsel D, Simons K. Regulation of human EGF receptor by lipids. *Proc Natl Acad Sci USA.* 2011;108:9044–9048.
102. Mirkin BL, Clark SH, Zhang C. Inhibition of human neuroblastoma cell proliferation and EGF receptor phosphorylation by gangliosides GM1 GM3 GD1A and GT1B. *Cell Prolif.* 2002;35:105–115.
103. Kawashima N, Yoon SJ, Itoh K, Nakayama K. Tyrosine kinase activity of epidermal growth factor receptor is regulated by GM3 binding through carbohydrate to carbohydrate interactions. *J Biol Chem.* 2009;284:6147–6155.
104. Yoshinaga A, Kajiya N, Oishi K, Kamada Y, Ikeda A, Chigwechokha PK, Kibe T, Kishida M, Kishida S, Komatsu M, Shiozaki K. NEU3 inhibitory effect of naringin suppresses cancer cell growth by attenuation of EGFR signaling through GM3 ganglioside accumulation. *Eur J Pharmacol.* 2016;782:21–29.
105. Seyfried TN, Mukherjee P. Ganglioside GM3 is antiangiogenic in malignant brain cancer. *J Oncol.* 2010;2010:961243.
106. Chung TW, Kim SJ, Choi HJ, Kim KJ, Kim MJ, Kim SH, Lee HJ, Ko JH, Lee YC, Suzuki A, Kim CH. Ganglioside GM3 inhibits VEGF/VEGFR-2-mediated angiogenesis: direct interaction of GM3 with VEGFR-2. *Glycobiology.* 2009;19:229–239.
107. Meuillet E, Cremel G, Dreyfus H, Hicks D. Differential modulation of basic fibroblast and epidermal growth factor receptor activation by ganglioside GM3 in cultured retinal Müller glia. *Glia.* 1996;17:206–216.
108. Toledo MS, Suzuki E, Handa K, Hakomori S. Effect of ganglioside and tetraspanins in microdomains on interaction of integrins with fibroblast growth factor receptor. *J Biol Chem.* 2005;280:16227–16234.
109. Mitsuda T, Furukawa K, Fukumoto S, Miyazaki H, Urano T, Furukawa K. Overexpression of ganglioside GM1 results in the dispersion of platelet-derived growth factor receptor from glycolipid-enriched microdomains and in the suppression of cell growth signals. *J Biol Chem.* 2002;277:11239–11246.
110. Veracini L, Simon V, Richard V, Schraven B, Horejsi V, Roche S, Benistant C. The Csk-binding protein PAG regulates PDGF-induced Src mitogenic signaling via GM1. *J Cell Biol.* 2008;182:603–614.
111. Nishio M, Fukumoto S, Furukawa K, Ichimura A, Miyazaki H, Kusunoki S, Urano T, Furukawa K. Overexpressed GM1 suppresses nerve growth factor (NGF) signals by modulating the intracellular localization of NGF receptors and membrane fluidity in PC12 cells. *J Biol Chem.* 2004;279:33368–33378.

112. Fukumoto S, Mutoh T, Hasegawa T, Miyazaki H, Okada M, Goto G, Furukawa K, Urano T. GD3 synthase gene expression in PC12 cells results in the continuous activation of TrkA and ERK1/2 and enhanced proliferation. *J Biol Chem.* 2000;275: 5832–5838.
113. Hamamura K, Furukawa K, Hayashi T, Hattori T, Nakano J, Nakashima H, Okuda T, Mizutani H, Hattori H, Ueda M, Urano T, Lloyd KO, Furukawa K. Ganglioside GD3 promotes cell growth and invasion through p130Cas and paxillin in malignant melanoma cells. *Proc Natl Acad Sci USA.* 2005;102:11041–11046.
114. Ohkawa Y, Miyazaki S, Hamamura K, Kambe M, Miyata M, Tajima O, Ohmi Y, Yamauchi Y, Furukawa K, Furukawa K. Ganglioside GD3 enhances adhesion signals and augments malignant properties of melanoma cells by recruiting integrins to glycolipid-enriched microdomains. *J Biol Chem.* 2010;285:27213–27223.
115. Hamamura K, Tsuji M, Hotta H, Ohkawa Y, Takahashi M, Shibuya H, Nakashima H, Yamauchi Y, Hashimoto N, Hattori H, Ueda M, Furukawa K, Furukawa K. Functional activation of Src family kinase yes protein is essential for the enhanced malignant properties of human melanoma cells expressing ganglioside GD3. *J Biol Chem.* 2011;286:18526–18537.
116. Furukawa K, Kambe M, Miyata M, Ohkawa Y, Tajima O, Furukawa K. Ganglioside GD3 induces convergence and synergism of adhesion and hepatocyte growth factor/Met signals in melanomas. *Cancer Sci.* 2014;105:52–63.
117. Ohkawa Y, Momota H, Kato A, Hashimoto N, Tsuda Y, Kotani N, Honke K, Suzumura A, Furukawa K, Ohmi Y, Natsume A, Wakabayashi T, Furukawa K. Ganglioside GD3 enhances invasiveness of gliomas by forming a complex with platelet-derived growth factor receptor α and yes kinase. *J Biol Chem.* 2015;290: 16043–16058.
118. Cazet A, Groux-Degroote S, Teylaert B, Kwon KM, Lehoux S, Slomianny C, Kim CH, Le Bourhis X, Delannoy P. GD3 synthase overexpression enhances proliferation and migration of MDA-MB-231 breast cancer cells. *Biol Chem.* 2009;390: 601–609.
119. Cazet A, Bobowski M, Rombouts Y, Lefebvre J, Steenackers A, Popa I, Guérardel Y, Le Bourhis X, Tulasne D, Delannoy P. The ganglioside GD2 induces the constitutive activation of c-Met in MDA-MB-231 breast cancer cells expressing the GD3 synthase. *Glycobiology.* 2012;22:806–816.
120. Liang YJ, Ding Y, Levery SB, Lobaton M, Handa K, Hakomori SI. Differential expression profiles of glycosphingolipids in human breast cancer stem cells vs cancer non-stem cells. *Proc Natl Acad Sci USA.* 2013;110:4968–4973.
121. Yeh SC, Wang PY, Lou YW, Khoo KH, Hsiao M, Hsu TL, Wong CH. Glycolipid GD3 and GD3 synthase are key drivers for glioblastoma stem cells and tumorigenicity. *Proc Natl Acad Sci USA.* 2016;113:5592–5597.
122. Wang J, Yu RK. Interaction of ganglioside GD3 with an EGF receptor sustains the self-renewal ability of mouse neural stem cells in vitro. *Proc Natl Acad Sci USA.* 2013;110:19137–19142.
123. Hyuga S, Kawasaki N, Hyuga M, Ohta M, Shibayama R, Kawanishi T, Yamagata S, Yamagata T, Hayakawa T. Ganglioside GD1a inhibits HGF-induced motility and scattering of cancer cells through suppression of tyrosine phosphorylation of c-Met. *Int J Cancer.* 2001;94:328–334.
124. Zhang L, Wang Y, Wang L, Cao T, Hyuga S, Sato T, Wu Y, Yamagata S, Yamagata T. Ganglioside GD1a negatively regulates hepatocyte growth factor expression through caveolin-1 at the transcriptional level in murine osteosarcoma cells. *Biochim Biophys Acta.* 2011;1810:759–768.
125. Liu Y, Li R, Ladisch S. Exogenous ganglioside GD1a enhances epidermal growth factor receptor binding and dimerization. *J Biol Chem.* 2004;279:36481–36489.

126. Yang HJ, Jung KY, Kwak DH, Lee SH, Ryu JS, Kim JS, Chang KT, Lee JW, Choo YK. Inhibition of ganglioside GD1a synthesis suppresses the differentiation of human mesenchymal stem cells into osteoblasts. *Dev Growth Differ.* 2011;53:323–332.
127. De Maria R, Lenti L, Malisan F, d'Agostino F, Tomassini B, Zeuner A, Rippo MR, Testi R. Requirement for GD3 ganglioside in CD95- and ceramide-induced apoptosis. *Science.* 1997;277:1652–1655.
128. Rippo MR, Malisan F, Ravagnan L, Tomassini B, Condo I, Costantini P, Susin SA, Rufini A, Todaro M, Kroemer G, Testi R. GD3 ganglioside directly targets mitochondria in a bcl-2-controlled fashion. *FASEB J.* 2000;14:2047–2054.
129. Zhou J, Shao H, Cox NR, Baker HJ, Ewald SJ. Gangliosides enhance apoptosis of thymocytes. *Cell Immunol.* 1998;183:90–98.
130. Ha SH, Lee JM, Kwon KM, Kwak CH, Abekura F, Park JY, Cho SH, Lee K, Chang YC, Lee YC, Choi HJ, Chung TW, Ha KT, Chang HW, Kim CH. Exogenous and endogeneous disialosyl ganglioside GD1b induces apoptosis of MCF-7 human breast cancer cells. *Int J Mol Sci.* 2016;17:E652.
131. Giammarioli AM, Garofalo T, Sorice M, Misasi R, Gambardella L, Gradini R, Fais S, Pavan A, Malorni W. GD3 glycosphingolipid contributes to Fas-mediated apoptosis via association with ezrin cytoskeletal protein. *FEBS Lett.* 2001;506:45–50.
132. Mukherjee K, Chava AK, Mandal C, Dey SN, Kniep B, Chandra S, Mandal C. O-acetylation of GD3 prevents its apoptotic effect and promotes survival of lymphoblasts in childhood acute lymphoblastic leukaemia. *J Cell Biochem.* 2008;105:724–734.
133. Kohla G, Stockfleth E, Schauer R. Gangliosides with O-acetylated sialic acids in tumors of neuroectodermal origin. *Neurochem Res.* 2002;27:583–592.
134. Horwacik I, Rokita H. Targeting of tumor-associated gangliosides with antibodies affects signaling pathways and leads to cell death including apoptosis. *Apoptosis.* 2015;20:679–688.
135. Doronin II, Vishnyakova PA, Kholodenko IV, Ponomarev ED, Ryazantsev DY, Molotkovskaya IM, Kholodenko RV. Ganglioside GD2 in reception and transduction of cell death signal in tumor cells. *BMC Cancer.* 2014;14:295.
136. Tsao CY, Sabbatino F, Cheung NV, Hsu JC, Villani V, Wang X, Ferrone S. Anti-proliferative and pro-apoptotic activity of GD2 ganglioside-specific monoclonal antibody 3F8 in human melanoma cells. *Oncoimmunology.* 2015;4:e1023975.
137. Durbas M, Horwacik I, Boratyn E, Kamycka E, Rokita H. GD2 ganglioside specific antibody treatment downregulates PI3K/Akt/mTOR signaling network in human neuroblastoma cell lines. *Int J Oncol.* 2015;47:1143–1159.
138. Horta ZP, Goldberg JL, Sondel PM. Anti-GD2 mAbs and next-generation mAb-based agents for cancer therapy. *Immunotherapy.* 2016;8:1097–1117.
139. Miyagi T, Wada T, Yamaguchi K, Shiozaki K, Sato I, Kakugawa Y, Yamanami H, Fujiya T. Human sialidase as a cancer marker. *Proteomics.* 2008;8:3303–3311.
140. Kato T, Wang Y, Yamaguchi K, Milner CM, Shineha R, Satomi S, Miyagi T. Overexpression of lysosomal-type sialidase leads to suppression of metastasis associated with reversion of malignant phenotype in murine B16 melanoma cells. *Int J Cancer.* 2001;92:797–804.
141. Miyagi T, Sato K, Hata K, Taniguchi S. Metastatic potential of transformed rat 3Y1 cell lines is inversely correlated with lysosomal-type sialidase activity. *FEBS Lett.* 1994;349:255–259.
142. Uemura T, Shiozaki K, Yamaguchi K, Miyazaki S, Satomi S, Kato K, Sakuraba H, Miyagi T. Contribution of sialidase NEU1 to suppression of metastasis of human colon cancer cells through desialylation of integrin beta4. *Oncogene.* 2009;28: 1218–1229.
143. Hinek A, Bodnaruk TD, Bunda S, Wang Y, Liu K. Neuraminidase-1, a subunit of the cell surface elastin receptor, desialylates and functionally inactivates adjacent receptors

interacting with the mitogenic growth factors PDGF-BB and IGF-2. *Am J Pathol.* 2008;173:1042–1056.
144. Ren LR, Zhang LP, Huang SY, Zhu YF, Li WJ, Fang SY, Shen L, Gao YL. Effects of sialidase NEU1 siRNA on proliferation, apoptosis, and invasion in human ovarian cancer. *Mol Cell Biochem.* 2016;411:213–219.
145. Gilmour AM, Abdulkhalek S, Cheng TS, Alghamdi F, Jayanth P, O'Shea LK, Geen O, Arvizu LA, Szewczuk MR. A novel epidermal growth factor receptor-signaling platform and its targeted translation in pancreatic cancer. *Cell Signal.* 2013;25:2587–2603.
146. Jayanth P, Amith SR, Gee K, Szewczuk MR. Neu1 sialidase and matrix metalloproteinase-9 cross-talk is essential for neurotrophin activation of Trk receptors and cellular signaling. *Cell Signal.* 2010;22:1193–1205.
147. Kakugawa Y, Wada T, Yamaguchi K, Yamanami H, Ouchi K, Sato I, Miyagi T. Up-regulation of plasma membrane-associated ganglioside sialidase (Neu3) in human colon cancer and its involvement in apoptosis suppression. *Proc Natl Acad Sci USA.* 2002;99:10718–10723.
148. Ueno S, Saito S, Wada T, Yamaguchi K, Satoh M, Arai Y, Miyagi T. Plasma membrane-associated sialidase is up-regulated in renal cell carcinoma and promotes interleukin-6-induced apoptosis suppression and cell motility. *J Biol Chem.* 2006;281:7756–7764.
149. Nomura H, Tamada Y, Miyagi T, Suzuki A, Taira M, Suzuki N, Susumu N, Irimura T, Aoki D. Expression of NEU3 (plasma membrane-associated sialidase) in clear cell adenocarcinoma of the ovary: its relationship with T factor of pTNM classification. *Oncol Res.* 2006;16:289–297.
150. Kawamura S, Sato I, Wada T, Yamaguchi K, Li Y, Li D, Zhao X, Ueno S, Aoki H, Tochigi T, Kuwahara M, Kitamura T, Takahashi K, Moriya S, Miyagi T. Plasma membrane-associated sialidase (NEU3) regulates progression of prostate cancer to androgen-independent growth through modulation of androgen receptor signaling. *Cell Death Differ.* 2012;19:170–179.
151. Li X, Zhang L, Shao Y, Liang Z, Shao C, Wang B, Guo B, Li N, Zhao X, Li Y, Xu D. Effects of a human plasma membrane-associated sialidase siRNA on prostate cancer invasion. *Biochem Biophys Res Commun.* 2011;416:270–276.
152. Shiga K, Takahashi K, Sato I, Kato K, Saijo S, Moriya S, Hosono M, Miyagi T. Upregulation of sialidase NEU3 in head and neck squamous cell carcinoma associated with lymph node metastasis. *Cancer Sci.* 2015;106:1544–1553.
153. Kato K, Shiga K, Yamaguchi K, Hata K, Kobayashi T, Miyazaki K, Saijo S, Miyagi T. Plasma-membrane-associated sialidase (NEU3) differentially regulates integrin-mediated cell proliferation through laminin- and fibronectin-derived signalling. *Biochem J.* 2006;394(Pt 3):647–656.
154. Takahashi K, Hosono M, Sato I, Hata K, Wada T, Yamaguchi K, Nitta K, Shima H, Miyagi T. Sialidase NEU3 contributes neoplastic potential on colon cancer cells as a key modulator of gangliosides by regulating Wnt signaling. *Int J Cancer.* 2015;137: 1560–1573.
155. Tringali C, Lupo B, Cirillo F, Papini N, Anastasia L, Lamorte G, Colombi P, Bresciani R, Monti E, Tettamanti G, Venerando B. Silencing of membrane-associated sialidase Neu3 diminishes apoptosis resistance and triggers megakaryocytic differentiation of chronic myeloid leukemic cells K562 through the increase of ganglioside GM3. *Cell Death Differ.* 2009;16:164–174.
156. Wada T, Hata K, Yamaguchi K, Shiozaki K, Koseki K, Moriya S, Miyagi T. A crucial role of plasma membrane-associated sialidase in the survival of human cancer cells. *Oncogene.* 2007;26:2483–2490.
157. Shiozaki K, Yamaguchi K, Sato I, Miyagi T. Plasma membrane-associated sialidase (NEU3) promotes formation of colonic aberrant crypt foci in azoxymethane-treated transgenic mice. *Cancer Sci.* 2009;100:588–594.

158. Yamaguchi K, Shiozaki K, Moriya S, Koseki K, Wada T, Tateno H, Sato I, Asano M, Iwakura Y, Miyagi T. Reduced susceptibility to colitis-associated colon carcinogenesis in mice lacking plasma membrane-associated sialidase. *PLoS One*. 2012;7:e41132.
159. Yamamoto K, Takahashi K, Shiozaki K, Yamaguchi K, Moriya S, Hosono M, Shima H, Miyagi T. Potentiation of epidermal growth factor-mediated oncogenic transformation by sialidase NEU3 leading to Src activation. *PLoS One*. 2015;10:e0120578.
160. Yoon SJ, Nakayama K, Hikita T, Handa K, Hakomori SI. Epidermal growth factor receptor tyrosine kinase is modulated by GM3 interaction with N-linked GlcNAc termini of the receptor. *Proc Natl Acad Sci USA*. 2006;103:18987–18991.
161. Yamaguchi K, Hata K, Wada T, Moriya S, Miyagi T. Epidermal growth factor-induced mobilization of a ganglioside-specific sialidase (NEU3) to membrane ruffles. *Biochem Biophys Res Commun*. 2006;346:484–490.
162. Wang Y, Yamaguchi K, Wada T, Hata K, Zhao X, Fujimoto T, Miyagi T. A close association of the ganglioside-specific sialidase Neu3 with caveolin in membrane microdomains. *J Biol Chem*. 2002;277:26252–26259.
163. Sasaki A, Hata K, Suzuki S, Sawada M, Wada T, Yamaguchi K, Obinata M, Tateno H, Suzuki H, Miyagi T. Overexpression of plasma membrane-associated sialidase attenuates insulin signaling in transgenic mice. *J Biol Chem*. 2003;278:27896–27902.
164. Tokuyama S, Moriya S, Taniguchi S, Yasui A, Miyazaki J, Orikasa S, Miyagi T. Suppression of pulmonary metastasis in murine B16 melanoma cells by transfection of a sialidase cDNA. *Int J Cancer*. 1997;73:410–415.
165. Sawada M, Moriya S, Saito S, Shineha R, Satomi S, Yamori T, Tsuruo T, Kannagi R, Miyagi T. Reduced sialidase expression in highly metastatic variants of mouse colon adenocarcinoma 26 and retardation of their metastatic ability by sialidase overexpression. *Int J Cancer*. 2002;97:180–185.
166. Tringali C, Lupo B, Anastasia L, Papini N, Monti E, Bresciani R, Tettamanti G, Venerando B. Expression of sialidase Neu2 in leukemic K562 cells induces apoptosis by impairing Bcr-Abl/Src kinases signaling. *J Biol Chem*. 2007;282:14364–14372.
167. Yamanami H, Shiozaki K, Wada T, Yamaguchi K, Uemura T, Kakugawa Y, Hujiya T, Miyagi T. Down-regulation of sialidase NEU4 may contribute to invasive properties of human colon cancers. *Cancer Sci*. 2007;98:299–307.
168. Shiozaki K, Yamaguchi K, Takahashi K, Moriya S, Miyagi T. Regulation of sialyl Lewis antigen expression in colon cancer cells by sialidase NEU4. *J Biol Chem*. 2011;286:21052–21061.
169. Tringali C, Cirillo F, Lamorte G, Papini N, Anastasia L, Lupo B, Silvestri I, Tettamanti G, Venerando B. NEU4L sialidase overexpression promotes β-catenin signaling in neuroblastoma cells, enhancing stem-like malignant cell growth. *Int J Cancer*. 2012;131:1768–1778.
170. Silvestri I, Testa F, Zappasodi R, Cairo CW, Zhang Y, Lupo B, Galli R, Di Nicola M, Venerando B, Tringali C. Sialidase NEU4 is involved in glioblastoma stem cell survival. *Cell Death Dis*. 2014;5:e1381.
171. Aloia A, Petrova E, Tomiuk S, Bissels U, Déas O, Saini M, Zickgraf FM, Wagner S, Spaich S, Sütterlin M, Schneeweiss A, Reitberger M, Rüberg S, Gerstmayer B, Agorku D, Knöbel S, Terranegra A, Falleni M, Soldati L, Sprick MR, Trumpp A, Judde JG, Bosio A, Cairo S, Hardt O. The sialyl-glycolipid stage-specific embryonic antigen 4 marks a subpopulation of chemotherapy-resistant breast cancer cells with mesenchymal features. *Breast Cancer Res*. 2015;17:146.
172. Lo AS, Ma Q, Liu DL, Junghans RP. Anti-GD3 chimeric sFv-CD28/T-cell receptor zeta designer T cells for treatment of metastatic melanoma and other neuroectodermal tumors. *Clin Cancer Res*. 2010;16:2769–2780.

CHAPTER SEVEN

# Gangliosides in Diabetic Wound Healing

**Duncan Hieu M. Dam, Amy S. Paller**[1]
Northwestern University, Chicago, IL, United States
[1] Corresponding author. E-mail address: apaller@northwestern.edu

## Contents



## Abstract

An organized series of complicated biological and molecular phenomena is required for normal skin wound healing. These processes depend on normal cellular responses to cytokines, growth factors, and other mediators, such as clotting factors, prostaglandins, free radicals, and nitric oxide. In diabetic ulcers, impaired responses to these molecules lead to abnormalities in vascularization, innervation, matrix reconstruction, and reepithelialization of wounds. keratinocyte migration and proliferation on an extracellular matrix is critical in reepithelialization, but the response to growth factors is blunted in diabetes, including the insulin/IGF-1signaling axis. Ganglioside GM3, a sialylated epidermal glycosphingolipid, has been identified as a key mediator of the inhibition of insulin/IGF-1 signaling in response to factors, such as tumor necrosis factor-alpha (TNF-α) and hyperglycemia. Decreased expression of GM3 and the enzyme required for its synthesis, GM3 synthase (GM3S), leads to increased insulin/IGF-1 receptor signaling and accelerated keratinocyte migration, even in the presence of high glucose levels. GM3 depletion in GM3S knockout diabetic mice and diet-induced diabetic mice treated topically with nanoconstruct-mediated GM3S-targeting gene regulation also accelerates wound healing. These recent observations, coupled

*Progress in Molecular Biology and Translational Science*, Volume 156
ISSN 1877-1173
http://dx.doi.org/10.1016/bs.pmbts.2017.12.006

with evidence that GM3 depletion reverses distal innervation abnormalities in diabetic mice, suggest that GM3-depleting strategies are a promising new approach for human diabetic wounds.

## 1. INTRODUCTION

Disruption of the normal orderly progression of events in wound healing has been observed in wounds with preexisting pathophysiological abnormalities, such as diabetic ulcers.[1,2] This impaired healing has been attributed to several intrinsic factors (e.g., neuropathy, vascular problems leading to ischemia, and signaling abnormalities in several wound cell types), as well as extrinsic factors (e.g., wound infection and trauma).[3,4] Macrovascular[5] and microvascular defects, which often occur in patients with diabetes, result in capillary size reduction and basement membrane thickening, limiting blood flow, and migration of dermal and epidermal cells into wounds.[6,7] The slowed microcirculation is further exacerbated by denervation,[8,9] with sensory nerve abnormalities that limit the patient's intrinsic protection from external trauma[3,4,10–12] and increase the risk of trauma-initiated ulceration.

## 2. GM3 MEDIATES INSULIN RESISTANCE IN DIABETES

Growing evidence suggests that GM3 mediates tumor necrosis factor-alpha (TNF-α) and glucose-induced development of insulin resistance and type 2 diabetes. Gangliosides are sialylated glycosphingolipids that mediate membrane-based growth factor receptor signaling in skin and other organs.[13–15] Monosialylated GM3, a product of lactosylceramide by GM3 synthase (GM3S), is the most abundant ganglioside in skin and precursor for more complex gangliosides. Levels of GM3 and GM3S (also called SAT-I/ST3Gal-V and coded by the *ST3GAL5* gene) are increased in the kidneys, liver, adipose tissue, and muscle of murine models of diabetes.[16–18] Ganglioside levels in the serum of diabetic patients with microvascular complications are alsoincreased,[19] and GM3S expression is higher in the kidneys of diabetics with nephropathy.[20] Knocking out GM3S in mice[21] and treating diabetic mice or rats with a glucosylceramide synthase inhibitor, which depletes GM3 precursors, improves insulin sensitivity,[22–24] ameliorates hepatic steatosis,[23] and suppresses the development of diabetic renal hypertrophy.[17]

## 3. GM3 IS A DRIVER OF IMPAIRED WOUND HEALING IN TYPE 2 DIABETIC MICE

The impaired wound healing of type 2 diabetes is associated with hyperglycemia and chronic, low-grade elevation of TNF-α, which is known to drive insulin resistance. We found that the expression of GM3S and GM3 is increased threefold in diet-induced obese (DIO) and genetic (ob/ob) diabetic mouse skin,[14] and that GM3S and GM3 are three- and >twofold higher, respectively, in human diabetic plantar skin than in age- and site-matched controls.[25] These observations suggested the possibility that regional increases in GM3S/GM3 might contribute to poor wound healing in diabetes. Indeed, DIO GM3S knockout ($GM3S^{-/-}$) diabetic mice have normal wound healing, in contrast to the delayed healing of their DIO wild-type (WT) diabetic littermates,[14] despite comparable obesity and often persistent abnormal glucose tolerance in the $GM3S^{-/-}$ mice (Fig. 1A–C). Furthermore, wound healing in DIO diabetic mice is normalized by nanoconstruct-mediated, skin-specific suppression of GM3S expression, and reduction of GM3 levels in skin.[25] This regional in vivo inhibition of GM3S is achieved by the topical application to the wound area of siRNA spherical nucleic acids (SNA), in which the GM3S-targeting siRNA are densely arrayed on a gold nanoparticle surface, facilitating penetration of epidermis at the wound edge without adverse effects (Fig. 1D–F). In both $GM3S^{-/-}$ and GM3S SNA-treated diabetic mice, GM3 depletion reverses the suppression of insulin growth factor 1 receptor (IGF1R) phosphorylation in epidermis at the wound border,[14,25] suggesting that reversing the suppressed IGF1R activation in diabetic skin is associated with improved healing.

Furthermore, diabetic mouse models with GM3 depletion display improvement in wound healing parameters beyond the keratinocytes (KC), suggesting a multitissue impact of GM3 depletion. An impact on fibroblasts and the vasculature is suggested by the increase in granulation tissue and vasculature in GM3-depleted wounds.[25] The small fiber neuropathy in the mouse footpad is thought to be responsible for neuropathic pain in diabetic mouse models and is representative of the neuropathy associated with acral impaired wound healing in humans.[26] Depletion of

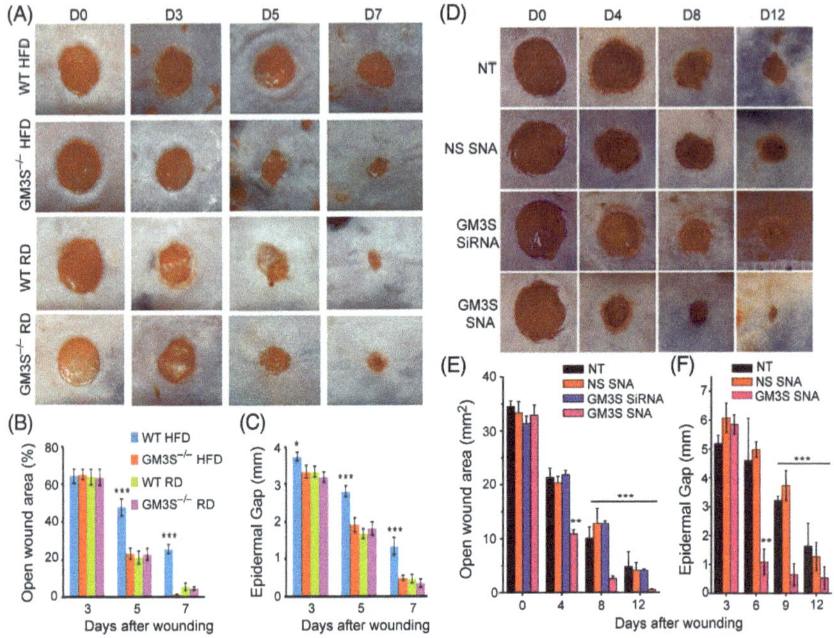


**Fig. 1** Accelerated wound healing in diabetic mice through depletion of GM3. (A–C) Complete deletion of GM3S (*St3gal5*) improved wound healing in diet-induced obese (DIO) GM3S knockout (GM3S$^{-/-}$) diabetic mice, in contrast to the delayed healing of their DIO wild-type (WT) diabetic littermates.[14] (D–F) Wound healing in DIO diabetic mice is normalized by nanoconstruct-mediated, skin-specific suppression of GM3S expression, and reduction of GM3 levels in skin through topical application. *HFD*, High fat diet; *NS*, nonsense; *NT*, nontreated; *RD*, regular diet; *SNA*, spherical nucleic acid. *Part (A–C), reprinted with permission from Elsevier; Part (D–F), reprinted from Randeria PS, Seeger MA, Wang XQ, Wilson H, Shipp D, Mirkin CA, Paller AS. siRNA-based spherical nucleic acids reverse impaired wound healing in diabetic mice by ganglioside GM3 synthase knockdown. Proc Natl Acad Sci USA. 2015;112:5573–5578,*[25] *with permission.*

GM3S and resultant decrease in GM3 expression rescued the denervation in mouse footpad skin, promoted sprouting of small fiber nerves into the epidermal layer, and fully ceased neuropathic nerve pain in DIO mice with biochemical confirmation of diabetes. Reduction of GM3 in dorsal root ganglia derived from diabetic mice also prevented hyperexcitability of calcium influx or calcium toxicity, which contributes to small fiber degeneration in diabetes.[26] These observations suggest a more global role for regional GM3 depletion in accelerating diabetic wound healing and deserve further investigation.

## 4. THE INSULIN/IGF1 SIGNALING AXIS IN DIABETES AND DIABETIC WOUNDS

Insulin resistance is a cardinal feature of type 2 diabetes. Insulin and insulin-like growth factor (IGF-1) bind to both the insulin receptor (IR) and IGF-1 receptor (IGF1R) and trigger activation of the IGF-1/insulin signaling axis (with insulin the major ligand for IR and IGF-1 for IGF1R).[27,28] Impaired IGF-1-/insulin-induced signaling impairs endothelial repair in type 2 diabetes-related atherosclerosis[29] and diabetic retinopathy.[30] Activation of IGF1R by IGF-1 stimulates hypoxia-inducible factor 1a protein synthesis and enhances the expression of vascular endothelial growth factor (VEGF) in human retinal pigment epithelial cells.[30,31] This induction of VEGF by IGF-1 plays a critical role during the development of proliferative diabetic retinopathy. IGF-1 also stabilizes nascent blood vessels by ERK activation.[32] IGF-1 also improves neuromuscular recovery and nerve regeneration after nerve injury by acting primarily on the axon, Schwann cells, and indirectly on neuromuscular junctions.[33] Furthermore, IGF-1 acts during development of the peripheral nervous system, while promoting axon regeneration of motor and sensory axons in a nerve crush model.[34]

Importantly, IGF-1 and, to a lesser extent, insulin are growth factors that stimulate KC migration through activation primarily of IGF1R and IR, respectively, during wound closure.[27,35] Indeed, studies in IR- and IGF1R-null mice have shown that deletion of IGF1R profoundly suppresses KC migration and proliferation, whereas deletion of IR has little effect.[36,37] IGF-1 stimulates lamellipodial protrusion and cell spreading,[28] and promotes KC migration through activation of the phosphatidylinositol-3-kinase and Rac1 signaling pathways.[28,38] Decreased IGF1R phosphorylation has been described at the wound edge epidermis of DIO diabetic mice with impaired healing, suggesting the role of suppressed IGF1R signaling in the impaired diabetic wound healing.[14] In addition, basal KCs from human diabetic foot ulcer skin are known to be deficient in IGF-1 expression.[39] Topical supplementation of IGF-1 to the wound edge of diabetic rats has been shown to accelerate wound closure,[40,41] and topical application of insulin has similarly been shown to encourage diabetic wound closure by enhancing protein

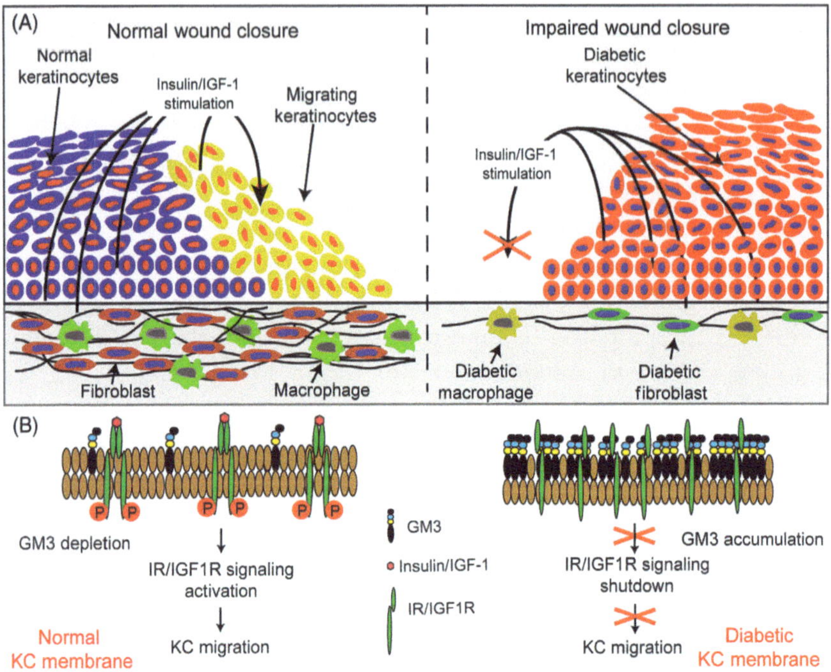


**Fig. 2** Increased keratinocyte GM3 expression is associated with reduction in the insulin/IGF-1 signaling axis and impaired healing in diabetes. (A) Diabetic keratinocytes at the edge of the wound have increased GM3 expression and blunted responses to growth factors, particularly of the insulin/IGF-1 signaling axis. GM3 also affects the responses of fibroblasts, macrophages, vessels, and nerves. The result is an impairment in wound closure. (B) GM3 accumulation in the keratinocyte membrane shuts down IR/IGF1R signaling, leading to reduction in keratinocyte migration and scratch wound reepithelialization.

kinase B(Akt) and extracellular signal-regulated kinases (ERK) signaling pathways (Fig. 2).[42]

## 5. GANGLIOSIDES REGULATE IGF1R SIGNALING AXIS IN CULTURED KCS

In vitro, TNF-α and glucose supplementation increase GM3S and GM3 expression in canonical insulin target cells (adipocytes, myocytes, and hepatocytes),[16,18] and directly suppress IR and IR substrate-1 (IRS-1) tyrosine phosphorylation in cultured adipocytes, decreasing glucose uptake.[16] In normal human epidermal KCs, high ambient glucose (e.g., a 2.5-fold increase to simulate hyperglycemia) and chronic exposure to low-dose TNF-α (e.g., 100 pM for 4 days) slow cell migration and inhibit proliferation.[43,44] Similarly, TNF-α and glucose supplementation elevate expression of GM3S and GM3 in cultured KCs,[14,38] leading to inhibition of bothinsulin- and IGF-1-induced receptor phosphorylation.[14,43,44] As noted, IGF1R activation is thought to play a more important role than IR activation in KCs migration, proliferation, and survival. Indeed, increases in GM3 by pharmacological supplementation, such as exposure to excess glucose or blockade of GM3 metabolism with accumulation of GM3, inhibit IGF-1-induced IGF1R phosphorylation and dramatically slow cell movement to a much greater extent than its effect on IR activation. The mechanism of this inhibition of migration has been shown to involve prevention of lamellipodial protrusion through suppression of Rac1 signaling and activation of RhoA.[38]

The trisialylated epidermal ganglioside GT1b[45] has been shown in scratch assays to inhibit KC migration through a variety of mechanisms, including its attachment to fibronectin,[46] its binding to integrin α5β1 to block the integrin-fibronectin interaction,[46] and its inhibition of the urokinase-type plasminogen activator (uPA)-integrin interaction.[15] Whether GT1b is increased in diabetic skin has not been studied; however, KC migration is slowed despite GT1b depletion when GM3 accumulates from gene knockdown of the enzyme needed to synthesize GT1b from GM3. This finding further suggests that GM3, not GT1b, plays the predominant role in blocking KC migration, including in diabetes.[38]

The ganglioside-specific impact on KC migration in vitro is further confirmed by its reversal in the face of GM3 depletion through treatment with GM3S SNAs or glucosylceramide synthase inhibitors. GM3 depletion leads to ligand-independent and increased ligand-dependent receptor activation, which accelerates KC migration, promotes directional lamellipodial protrusion, and increases cell velocity, displacement and persistence.[38] GM3 depletion activates IGFR and Rac1 signaling[38] and reverses the glucose- and TNF-α-induced suppression of p-IGF1R and p-IR.[14,25] In the presence of glucose or chronic low-dose TNF-α, this increase in KC movement from GM3 depletion further suggests that GM3 is the required intermediate in glucose-induced exacerbation of insulin resistance. Depletion of GM3 has been shown to promote glucose uptake into cells and GLUT1 transport to the membrane of KCs, suggesting that the accelerated migration in the face of GM3 depletion and glucose excess relates to increased availability of this carbon source.[47] GM3 supplementation or overexpression has also been shown to inhibit epidermal growth factor receptor (EGFR) activation both directly and through preventing in the matrix metallopeptidase 9-integrin interaction in squamous carcinoma cells[48] and normal human KCs, leading to inhibition of both cell proliferation and migration.[13] It is possible that GM3-induced inhibition of EGFR activation or increased phosphorylation of EGFR by GM3 depletion also plays a role in the wound healing impairment in diabetes and its reversal, respectively. However, the gene knockdown of IGF1R or inhibition of its activation dramatically suppresses the migration in scratch assays induced by GM3 depletion, and migration ceases with the combination of inhibition of both IGF1R and IR.[38] This finding suggests that EGFR inhibition plays a lesser role in GM3-modulated inhibition of KC migration, but further investigation into the role of GM3-induced EGFR inhibition is warranted.

## 6. CONCLUSIONS

The wound healing process requires a carefully orchestrated series of events. Diabetic ulcers affect a variety of cells and are multifactorial in pathogenesis. Central to poor healing is the dampened response to growth factors, including activators of the insulin/IGF-1 signaling axis. Ganglioside GM3 has been shown to be a mediator of TNF-α and glucose-induced insulin resistance and directly suppresses the cutaneous IGF-1/insulin signaling axis. The reduced IGF-1 levels in diabetic wounds, improved diabetic wound epidermal gap closure with IGF-1 overexpression, and critical role of IGF1R signaling in wound neovascularization all suggest that activating insulin/IGF-1 signaling to accelerate reepithelialization in diabetic wounds can be a novel means of intervention. Furthermore, reduction of GM3 through topically administered gene therapy at the wound edge in diabetic mice reduces their excess cutaneous GM3S and GM3, leading to normalization of wound healing. These findings reinforce a pivotal role for GM3-induced insulin resistance in impairing diabetic reepithelialization and wound closure and suggest GM3-depleting strategies as a novel approach for human diabetic wounds.

## ACKNOWLEDGMENTS

This work was supported by the National Institutes of Health Grants R01AR44619 and R21AR062898 (AP), and the Postgraduate Training in Cutaneous Biology T32 AR060710 (DD).


## REFERENCES


1. Loots MAM, Kenter SB, Au FL, van Galen WJM, Middelkoop E, Bos JD, Mekkes JR. Fibroblasts derived from chronic diabetic ulcers differ in their response to stimulation with EGF, IGF-I, bFGF and PDGF-AB compared to controls. *Eur J Cell Biol.* 2002;81: 153–160.
2. Loots MAM, Lamme EN, Zeegelaar J, Mekkes JR, Bos JD, Middelkoop E. Differences in cellular infiltrate and extracellular matrix of chronic diabetic and venous ulcers versus acute wounds. *J Invest Dermatol.* 1998;111:850–857.
3. Jeffcoate WJ, Harding KG. Diabetic foot ulcers. *Lancet.* 2003;361:1545–1551.
4. Jeffcoate WJ, Price P, Harding KG. Wound heating and treatments for people with diabetic foot ulcers. *Diabetes Metab Res.* 2004;20:S78–S89.
5. Ferrier TM. Comparative study of arterial disease in amputated lower limbs from diabetics and non-diabetics (with special reference to feet arteries). *Med J Aust.* 1967;1:5–11.
6. Dinh T, Veves A. Microcirculation of the diabetic foot. *Curr Pharm Des.* 2005;11: 2301–2309.
7. Dinh TL, Veves A. A review of the mechanisms implicated in the pathogenesis of the diabetic foot. *Int J Low Extrem Wounds.* 2005;4:154–159.
8. LoGerfo FW, Coffman JD. Current concepts. Vascular and microvascular disease of the foot in diabetes. Implications for foot care. *N Engl J Med.* 1984;311:1615–1619.
9. Veves A, Akbari CM, Primavera J, Donaghue VM, Zacharoulis D, Chrzan JS, DeGirolami U, LoGerfo FW, Freeman R. Endothelial dysfunction and the expression of endothelial nitric oxide synthetase in diabetic neuropathy, vascular disease, and foot ulceration. *Diabetes.* 1998;47:457–463.
10. Boulton AJ, Kirsner RS, Vileikyte L. Clinical practice. Neuropathic diabetic foot ulcers. *N Engl J Med.* 2004;351:48–55.
11. Pecoraro RE, Ahroni JH, Boyko EJ, Stensel VL. Chronology and determinants of tissue repair in diabetic lower-extremity ulcers. *Diabetes.* 1991;40:1305–1313.
12. Brem H, Sheehan P, Boulton AJ. Protocol for treatment of diabetic foot ulcers. *Am J Surg.* 2004;187:1S–10S.
13. Wang X, Rahman Z, Sun P, Meuillet E, George D, Bremer EG, Al-Qamari A, Paller AS. Ganglioside modulates ligand binding to the epidermal growth factor receptor. *J Invest Dermatol.* 2001;116:69–76.
14. Wang XQ, Lee S, Wilson H, Seeger M, Iordanov H, Gatla N, Whittington A, Bach D, Lu JY, Paller AS. Ganglioside GM3 depletion reverses impaired wound healing in diabetic mice by activating IGF-1 and insulin receptors. *J Invest Dermatol.* 2014;134: 1446–1455.
15. Wang XQ, Sun P, Paller AS. Gangliosides inhibit urokinase-type plasminogen activator (uPA)-dependent squamous carcinoma cell migration by preventing uPA receptor/alphabeta integrin/epidermal growth factor receptor interactions. *J Invest Dermatol.* 2005;124:839–848.
16. Tagami S, Inokuchi Ji J, Kabayama K, Yoshimura H, Kitamura F, Uemura S, Ogawa C, Ishii A, Saito M, Ohtsuka Y, Sakaue S, Igarashi Y. Ganglioside GM3 participates in the pathological conditions of insulin resistance. *J Biol Chem.* 2002;277:3085–3092.
17. Zador IZ, Deshmukh GD, Kunkel R, Johnson K, Radin NS, Shayman JA. A role for glycosphingolipid accumulation in the renal hypertrophy of streptozotocin-induced diabetes mellitus. *J Clin Invest.* 1993;91:797–803.

18. Memon RA, Holleran WM, Uchida Y, Moser AH, Ichikawa S, Hirabayashi Y, Grunfeld C, Feingold KR. Regulation of glycosphingolipid metabolism in liver during the acute phase response. *J Biol Chem.* 1999;274:19707–19713.
19. Hamed EA, Zakary MM, Abdelal RM, Abdel Moneim EM. Vasculopathy in type 2 diabetes mellitus: role of specific angiogenic modulators. *J Physiol Biochem.* 2011;67: 339–349.
20. Wei D, Tao R, Zhang Y, White MF, Dong XC. Feedback regulation of hepatic gluconeogenesis through modulation of SHP/Nr0b2 gene expression by Sirt1 and FoxO1. *Am J Physiol Endocrinol Metab.* 2011;300:E312–E320.
21. Yamashita T. Glycosphingolipid modification: structural diversity, functional and mechanistic integration of diabetes. *Diabetes Metab J.* 2011;35:309–316.
22. Aerts JM, Ottenhoff R, Powlson AS, Grefhorst A, van Eijk M, Dubbelhuis PF, Aten J, Kuipers F, Serlie MJ, Wennekes T, Sethi JK, O'Rahilly S, Overkleeft HS. Pharmacological inhibition of glucosylceramide synthase enhances insulin sensitivity. *Diabetes.* 2007;56:1341–1349.
23. Zhao H, Przybylska M, Wu IH, Zhang J, Maniatis P, Pacheco J, Piepenhagen P, Copeland D, Arbeeny C, Shayman JA, Aerts JM, Jiang C, Cheng SH, Yew NS. Inhibiting glycosphingolipid synthesis ameliorates hepatic steatosis in obese mice. *Hepatology.* 2009;50: 85–93.
24. Zhao H, Przybylska M, Wu IH, Zhang J, Siegel C, Komarnitsky S, Yew NS, Cheng SH. Inhibiting glycosphingolipid synthesis improves glycemic control and insulin sensitivity in animal models of type 2 diabetes. *Diabetes.* 2007;56:1210–1218.
25. Randeria PS, Seeger MA, Wang XQ, Wilson H, Shipp D, Mirkin CA, Paller AS. siRNA-based spherical nucleic acids reverse impaired wound healing in diabetic mice by ganglioside GM3 synthase knockdown. *Proc Natl Acad Sci USA.* 2015;112:5573–5578.
26. Menichella DM, Jayaraj ND, Wilson HM, Ren D, Flood K, Wang XQ, Shum A, Miller RJ, Paller AS. Ganglioside GM3 synthase depletion reverses neuropathic pain and small fiber neuropathy in diet-induced diabetic mice. *Mol Pain.* 2016;12: doi:10.1177/1744806916666284.
27. Li Y, Fan J, Chen M, Li W, Woodley DT. Transforming growth factor-alpha: a major human serum factor that promotes human keratinocyte migration. *J Invest Dermatol.* 2006;126:2096–2105.
28. Haase I, Evans R, Pofahl R, Watt FM. Regulation of keratinocyte shape, migration and wound epithelialization by IGF-1- and EGF-dependent signalling pathways. *J Cell Sci.* 2003;116:3227–3238.
29. Cubbon RM, Kearney MT, Wheatcroft SB. Endothelial IGF-1 receptor signalling in diabetes and insulin resistance. *Trends Endocrinol Metab.* 2016;27:96–104.
30. Sall JW, Klisovic DD, O'Dorisio MS, Katz SE. Somatostatin inhibits IGF-1 mediated induction of VEGF in human retinal pigment epithelial cells. *Exp Eye Res.* 2004;79:465–476.
31. Fukuda R, Hirota K, Fan F, Jung YD, Ellis LM, Semenza GL. Insulin-like growth factor 1 induces hypoxia-inducible factor 1-mediated vascular endothelial growth factor expression, which is dependent on MAP kinase and phosphatidylinositol 3-kinase signaling in colon cancer cells. *J Biol Chem.* 2002;277:38205–38211.
32. Jacobo SM, Kazlauskas A. Insulin-like growth factor 1 (IGF-1) stabilizes nascent blood vessels. *J Biol Chem.* 2015;290:6349–6360.
33. Apel PJ, Ma J, Callahan M, Northam CN, Alton TB, Sonntag WE, Li Z. Effect of locally delivered IGF-1 on nerve regeneration during aging: an experimental study in rats. *Muscle Nerve.* 2010;41:335–341.
34. Rabinovsky ED. The multifunctional role of IGF-1 in peripheral nerve regeneration. *Neurol Res.* 2004;26:204–210.
35. Ando Y, Jensen PJ. Epidermal growth factor and insulin-like growth factor I enhance keratinocyte migration. *J Invest Dermatol.* 1993;100:633–639.

36. Sadagurski M, Yakar S, Weingarten G, Holzenberger M, Rhodes CJ, Breitkreutz D, Leroith D, Wertheimer E. Insulin-like growth factor 1 receptor signaling regulates skin development and inhibits skin keratinocyte differentiation. *Mol Cell Biol.* 2006;26:2675–2687.
37. Stachelscheid H, Ibrahim H, Koch L, Schmitz A, Tscharntke M, Wunderlich FT, Scott J, Michels C, Wickenhauser C, Haase I, Bruning JC, Niessen CM. Epidermal insulin/IGF-1 signalling control interfollicular morphogenesis and proliferative potential through Rac activation. *Embo J.* 2008;27:2091–2101.
38. Dam DH, Wang XQ, Sheu S, Vijay M, Shipp D, Miller L, Paller AS. Ganglioside GM3 mediates glucose-induced suppression of IGF-1 receptor-Rac1 activation to inhibit keratinocyte motility. *J Invest Dermatol.* 2017;137:440–448.
39. Blakytny R, Jude EB, Martin Gibson J, Boulton AJ, Ferguson MW. Lack of insulin-like growth factor 1 (IGF1) in the basal keratinocyte layer of diabetic skin and diabetic foot ulcers. *J Pathol.* 2000;190:589–594.
40. Bitar MS. Insulin-like growth factor-1 reverses diabetes-induced wound healing impairment in rats. *Horm Metab Res.* 1997;29:383–386.
41. Korolkiewicz RP, Tashima K, Fujita A, Kato S, Takeuchi K. Exogenous insulin-like growth factor (IGF)-1 improves the impaired healing of gastric mucosal lesions in diabetic rats. *Pharmacol Res.* 2000;41:221–229.
42. Lima MH, Caricilli AM, de Abreu LL, Araujo EP, Pelegrinelli FF, Thirone AC, Tsukumo DM, Pessoa AF, dos Santos MF, de Moraes MA, Carvalheira JB, Velloso LA, Saad MJ. Topical insulin accelerates wound healing in diabetes by enhancing the AKT and ERK pathways: a double-blind placebo-controlled clinical trial. *PLoS One.* 2012;7:e36974.
43. Lan CC, Liu IH, Fang AH, Wen CH, Wu CS. Hyperglycaemic conditions decrease cultured keratinocyte mobility: implications for impaired wound healing in patients with diabetes. *Br J Dermatol.* 2008.
44. Spravchikov N, Sizyakov G, Gartsbein M, Accili D, Tennenbaum T, Wertheimer E. Glucose effects on skin keratinocytes: implications for diabetes skin complications. *Diabetes.* 2001;50:1627–1635.
45. Paller AS, Arnsmeier SL, Robinson JK, Bremer EG. Alteration in keratinocyte ganglioside content in basal cell carcinomas. *J Invest Dermatol.* 1992;98:226–232.
46. Paller AS, Arnsmeier SL, Chen JD, Woodley DT. Ganglioside GT1b inhibits keratinocyte adhesion and migration on a fibronectin matrix. *J Invest Dermatol.* 1995;105:237–242.
47. Shehu A, Lu JY, Wilson H, Bach D, Shipp D, Randeria PS, Mirkin CA, Paller AS. Cytoplasmic sequestion of keratinocyte GLUT1 by ganglioside GM3 mediates impaired diabetic wound healing. *J Invest Dermatol.* 2013;133:S249.
48. Wang XQ, Sun P, Paller AS. Ganglioside GM3 inhibits matrix metalloproteinase-9 activation and disrupts its association with integrin. *J Biol Chem.* 2003;278:25591–25599.

CHAPTER EIGHT

# Gangliosides in Nerve Cell Specification

**Yutaka Itokazu**[*,†], **Jing Wang**[*,†], **Robert K. Yu**[*,†,1]
[*] Department of Neuroscience and Regenerative Medicine, Medical College of Georgia, Augusta University, Augusta, GA, United States
[†] Charlie Norwood VA Medical Center, Augusta, GA, United States
[1] Corresponding author. E-mail address: ryu@augusta.edu

## Contents



## Abstract

The central nervous system is generated from progenitor cells that are recognized as neural stem cells (NSCs). NSCs are defined as undifferentiated neural cells that are characterized by the capacity for self-renewal and multipotency. Throughout neural development, NSCs undergo proliferation, migration, and cellular differentiation, and dynamic changes are observed in the composition of carbohydrate-rich molecules, including gangliosides. Gangliosides are sialic acid-containing glycosphingolipids with essential and multifaceted functions in brain development and NSC maintenance, which reflects the complexity of brain development. Our group has pioneered research on the importance of gangliosides for growth factor receptor signaling and epigenetic regulation of ganglioside biosynthesis in NSCs. We found that GD3 is the predominant ganglioside species in NSCs (>80%) and modulates NSC proliferation by interacting with epidermal growth factor receptor signaling. In postnatal brain, GD3 is required for long-term maintenance of NSCs. Deficiency in GD3 leads to developmental and behavioral deficits, such as depression. The synthesis of GD3 is switched to the synthesis of complex, brain-type gangliosides, namely, GM1, GD1a,

*Progress in Molecular Biology and Translational Science*, Volume 156
ISSN 1877-1173
http://dx.doi.org/10.1016/bs.pmbts.2017.12.008

GD1b, and GT1b, resulting in terminal differentiation and loss of "stemness" of NSCs. In this process, GM1 is augmented by a novel GM1-modulated epigenetic gene regulation mechanism of glycosyltransferases at a later differentiation stage. Consequently, our research suggests that stage-specific gangliosides play specific roles in maintaining NSC activities and in cell fate determination.

## 1. INTRODUCTION

Neural stem cells (NSCs) are fundamental cells that are capable of differentiating into various cell types in the nervous system. Gangliosides undergo dramatic qualitative and quantitative developmental changes that are correlated with cellular proliferation, differentiation, and biological functions of nerve cells. In this article, we introduce the expression pattern and potential functions of gangliosides in NSCs and differentiating neuronal cells. We also discovered a novel epigenetic mechanism that regulates ganglioside expression and may play a role in cell fate determination.

In early neural development, neuroepithelial cells (NECs) proliferate by repeating symmetric cell divisions at the wall of the neural tube; as NECs accumulate, the wall gradually becomes thicker. Proliferated NECs that are transformed into bipolar-shaped cells are known as radial glial cells (RGCs). The NEC elongates its fibers and becomes an RGC whose cell body lines the ventricular zone (VZ), and the apical surface meets the ventricles with the radial fibers reaching the pial surface. RGCs lose some of their epithelial properties in favor of certain glial characteristics, but RGCs retain some of the original properties of NECs, such as apicobasal polarity, contacting with the ventricular surface. Previously, RGCs were thought of as specialized glial cells whose major function was to guide migration of newborn neurons.[1–7] Recently, RGCs have been recognized as the precursors of neurons and glia. An RGC generates another RGC and an intermediate progenitor cell (IPC) or immature neuron by asymmetric cell division.[8–10] IPCs remain in the subventricular zone (SVZ) to proliferate and give rise to more neurons. Immature neurons migrate along with radial fibers into the cortical plate, and then become mature neurons. At first, RGCs produce inner-layer neurons that shift and become outer-layer neurons at later stages. RGCs also give rise to the major types of macroglia in the central nervous system (CNS), including oligodendrocytes and ependymal cells, and RGCs can eventually differentiate into astrocytes. Both NECs and RGCs are considered as NSCs.[7,11] A specialized microenvironment, known as the NSC niche, maintains stem cells in a multipotent and undifferentiated state. The NSC niche contains a variety of stem/progenitor cells, such as NECs, RGCs, and IPCs. Altogether, these versatile progenitors cooperate for neurogenesis and gliogenesis in the developing CNS.

During development, neurons and glia are generated in the CNS in a limited time-dependent manner. At early developmental stages, a preplate consisting of the earliest-born neurons and possibly other cell types is formed between the VZ and meninges at the brain surface. The VZ is a highly packed cell layer formed by morphologically homogeneous RGCs, and the SVZ is a second proliferative layer. Newly generated neurons migrate radially outward of the VZ/SVZ and form a new laminar structure. This preplate is subsequently split into the marginal zone and subplate by waves of migrating neurons. During this development, the VZ becomes smaller, and after neurogenesis is completed, the VZ is replaced by an ependymal cell layer. Postnatally, most of the SVZ disappears except along the lateral wall of the lateral ventricles, where it is considered a site for the NSC niche in the adult stage.[11–13] Astrocytes fulfill a range of essential functions, including regulation of neurogenesis, synaptogenesis, and development of the neural circuit.[14–17] Toward the end of cortical development, most RGCs transform into astrocytes. Astrocytic-destined IPCs are reported to divide locally before terminal differentiation into astrocytes in the embryonic and postnatal stages.[18,19] In rodents, on the day of birth [postnatal day (P) 0], most astrocyte precursors are found in the inner half of the cortical width. On P4, the majority of the astrocyte precursors are distributed in the outer half of the cortical width. The pattern of gliogenesis in the early postnatal rat thus shows an inside-out tendency, in analogy to neurogenesis.[20,21] Whether astrocyte IPCs are involved in this astrocytic differentiation in vivo is not well known. Oligodendrocytes, the chief myelin-forming cells in the CNS, are derived from RGCs. The myelin structure provides efficient axon insulation and facilitates saltatory conduction of nerve impulses. Most of the oligodendrocytes present in the adult cortex of mammals are generated after birth.[20,21]

For most mammals, neurogenesis commences during early embryonic stages and is almost complete shortly after birth; it continues to occur at a much slower pace and in a limited manner throughout the entire adult life. In the adult brain of mammals, neurogenesis persists primarily in two germinal zones, the SVZ of the lateral ventricles[22,23] and the subgranular zone in the dentate gyrus of the hippocampus.[24,25] In the adult SVZ, four distinct cell types are present. Type B cells are RGC-like cells and are considered as NSCs. Type B NSCs are slow dividing (duration of cell cycle >15 days) and express glial fibrillary acidic protein (GFAP). Type C cells are transient amplifying cells that are rapidly proliferating (duration of cell cycle about 13 h) and express the transcription factor Mash1. Type A cells are neuronal precursors that have already committed to differentiate into neurons, and these cells express polysialic acid-neural adhesion molecules on the cell surface (duration of cell cycle about 13 h).[26] Ependymal cells are lined on the wall of the ventricle and have multimotile cilia, which are important for controlling the flow of cerebrospinal fluid. Multipotency of the ependymal cells has been reported,[27] although this is still not settled.[22,28,29] Recently, ependymal cells have been shown to be the most quiescent type of NSCs whose cell cycle is strictly regulated and reinitiated under specific circumstances. In certain restricted situations, a subpopulation of ependymal cells may develop into neurons, and these cells are considered as NSCs.[30,31] In the subgranular zone, five types of cells have been described.[32] Type 1 cells are considered quiescent neural progenitors that are RGC-like cells and largely equivalent to Type B NSCs in the SVZ. Type 2 cells express nestin, and this cell type has been classified into two cell populations: Type 2a cells are amplifying neural progenitors that are similar to Type C transient amplifying cells in the SVZ; Type 2b and 3 cells are neuroblasts that express polysialic acid-neural adhesion molecules.[33,34] The other type of cells is mature granule neurons. Recently, it has been reported that Mash1$^+$ cells do not amplify and are therefore not Type 2a amplifying neural progenitors that can directly differentiate into early neuroblasts without mitosis.[35]

With regard to gangliosides, we have shown that GD3 is the predominant ganglioside in NSCs, and it can serve as a convenient cell surface marker of these cells.[36] The interaction of GD3 with epidermal growth factor receptor (EGFR) plays a crucial role in maintaining the self-renewal capacity of NSCs by directing the EGFR through the recycling pathway rather than through the degradative pathway after endocytosis.[37] We have reported that efficient histone acetylation of glycosyltransferase (GT) genes contributes to the developmental alteration of ganglioside expression in mouse brain.[38] Furthermore, we have demonstrated that acetylation of histones H3 and H4 on the *GM2/GD2S* gene promoter leads to recruitment of *trans*-activation factors Sp1 and AP-2 during neuronal differentiation.[39] More recently, we found that nuclear GM1 binds with acetylated histones on the promoters of the *GM2/GD2S* gene as well as on the neurogenic transcription factor, *NeuroD1* gene, in differentiated neurons.[40] Here, we will further discuss the importance of ganglioside regulation in neural development and neuronal differentiation of NSCs.

## 2. GLYCOSPHINGOLIPID EXPRESSION IN NEURAL DEVELOPMENT

During neural development, dramatic and consistent changes in the composition of glycoconjugates, including glycolipids, glycoproteins, and proteoglycans, occur.[41–43] Glycosphingolipids (GSLs) contain one or more monosaccharide residues bound by a glycosidic linkage to the hydrophobic moiety, ceramide. It is known that changes in the expression of glycolipids, including gangliosides that contain one or more sialic acid residues, in the nervous system correlate with neurodevelopmental events.[44–47] For example, in fertilized eggs, the *globo*-series of glycolipids (e.g., stage-specific embryonic antigen, SSEA-3/SSEA-4; Fig. 1) are robustly expressed. As cell division proceeds, the *lacto*-series GSLs (e.g., lactoneotetraosylceramide, nLc4Cer) are expressed at embryonic day (E) 1.5, followed by the *ganglio*-series GSLs in the developing brain (from E7). The lipid portion of GSLs, including gangliosides, is ceramide, which is synthesized primarily in the endoplasmic reticulum (ER) from a sphingosine base and a fatty acid residue. Ceramide is transferred to the Golgi apparatus and glucosylated to glucosylceramide (GlcCer) and catalyzed by glucosyltransferase (GlcT or GlcCer-synthase) on the cytosolic leaflet of the *cis*-Golgi. The expression of GlcCer is abundant in early embryonic brain and decreases from E16.[41] The synthesized GlcCer is transported to the *trans*-Golgi by the phosphoinositol 4-phosphate adaptor protein 2 (FAPP2) and *trans*-located to the luminal side of the Golgi vesicle.[48,49] On the other direction, FAPP2 also mediates GlcCer transportation from the *cis*-Golgi back to the ER.[50] Then, GlcCer is flip-flopped into the lumen of the ER, and GlcCer ultimately reaches the lumen of the Golgi by vesicular transport.[51] At the luminal side of the Golgi, GlcCer is converted to lactosylceramide (LacCer). The sequential action catalyzed by sialyltransferases and galactosyltransferase can elongate the carbohydrate moieties to gangliosides. Another pathway from ceramide is catalyzed by galactosyltransferase III (GalT-III or GalCer-synthase) in the ER to form galactosylceramide (GalCer), sulfatide, and GM4.[52] Each step is catalyzed by a unique, specifically controlled GT (Fig. 1A).

Gangliosides are sialic acid-containing GSLs expressed primarily, but not exclusively, on the outer leaflet of the plasma membrane of cells in all

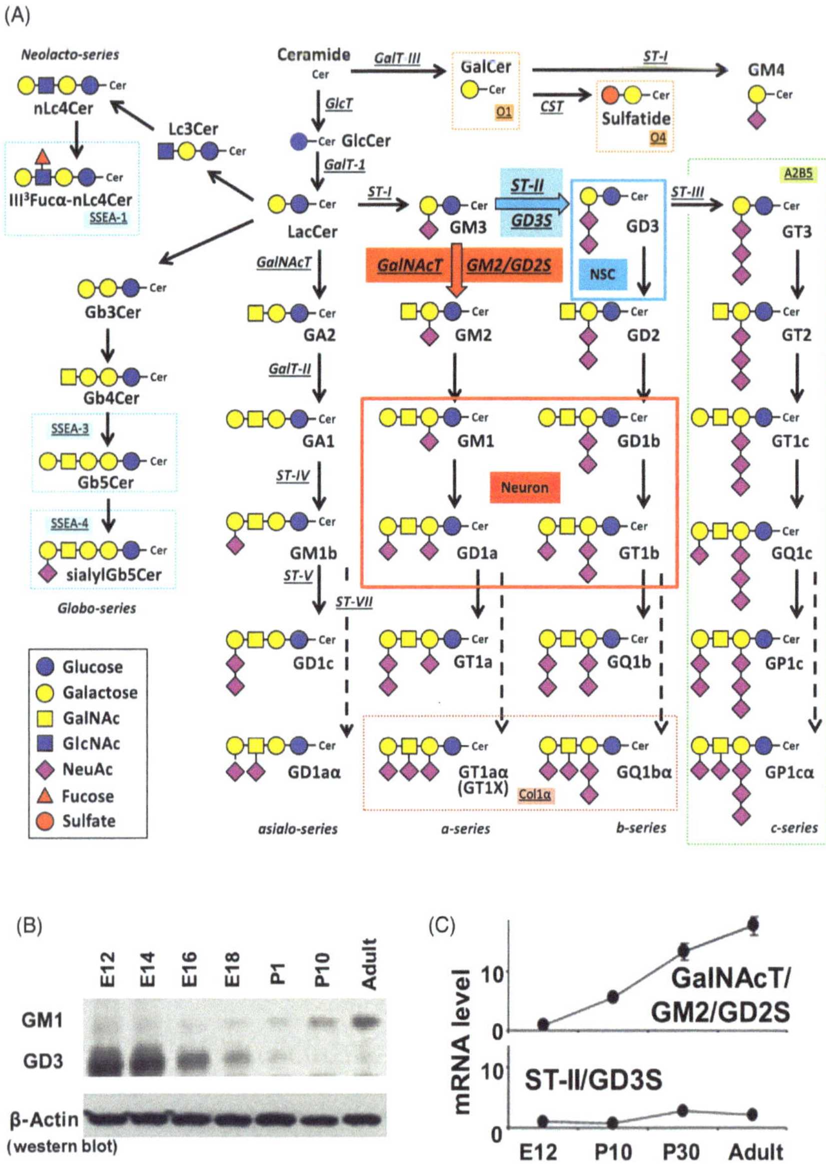


**Fig. 1** Glycosphingolipids and glycosyltransferases in the nervous system. (A) Structures and biosynthetic pathways of glycosphingolipids (GSLs). Cer, Ceramide; CST, cerebroside sulfotransferase (*Gal3st1*, sulfatide synthase); GalNAc-T, *N*-acetylgalactosaminyltransferase I (*B4galnt1*, GA2/GM2/GD2/GT2-synthase); GalT-I, galactosyltransferase I (*B4galt6*, lactosylceramide synthase); GalT-II, galactosyltransferase II (*B3galt4*, GA1/GM1/GD1b/GT1c-synthase); GalT-III, galactosyltransferase III (*Ugt8a*, galactosylceramide synthase); GlcT, glucosyltransferase (*Ugcg*, glucosylceramide synthase); ST-I, sialyltransferase I

vertebrates. In early embryonic rodent brains, the ganglioside pattern is characterized by the expression of a large quantity of simple gangliosides, such as GM3 and GD3. In later developmental stages, more complex gangliosides prevail, particularly GM1, GD1a, GD1b, and GT1b, which account for more than 80% of the total gangliosides.[41,43] Thus, the expression of neural gangliosides changes dramatically during cellular differentiation and brain development (Fig. 1B and C). This unique expression pattern of specific gangliosides can be used for specific cell lineage markers and may reflect the functional roles they play at specific developmental stages. Abundant evidence supports the notion that GSLs, including gangliosides, serve regulatory roles in cellular events, including proliferation and neural differentiation, as exemplified by neuritogenesis, axonogenesis, and synaptogenesis.[41,45,53–57] The functional importance of gangliosides has been evaluated using specific enzyme gene knockout (KO) mice.

## 3. DEFICIENCY OF GLYCOSPHINGOLIPIDS

In recent years, with the advent of contemporary molecular genetics and biology, several mutant lines of genetically modified mice have been established in which the expression of gangliosides and other GSLs has been altered or depleted; this has greatly facilitated elucidating the biological functions of GSLs. For example, GM2/GD2-synthase (GM2/GD2S; *N*-acetylgalactosaminyltransferase, GalNAcT) is one of the key enzymes needed for the synthesis of the major brain-type gangliosides, including GM1, GD1a, GD1b, and GT1b. Mice lacking this enzyme do not express

◀(*St3gal5*, GM3-synthase); ST-II, sialyltransferase II (*St8Sia1*, GD3-synthase); ST-III, sialyltransferase III (*St8Sia3*, GT3-synthase); ST-IV, sialyltransferase IV (*St3gal2*, GM1b/GD1a/GT1b/GQ1c-synthase); ST-V, sialyltransferase V (*St8sia5*, GD1c/GT1a/GQ1b/GP1c-synthase); ST-VII, sialyltransferase VII (*St6galnac6*, GD1aα/GT1aα/GQ1bα/GP1cα-synthase). Official symbols of genes are represented in italics in this figure legend. GD3 is the most abundant ganglioside in rodent NSCs. c-Series gangliosides are A2B5 antigens. GM1, GD1a, GD1b, and GT1b are the most abundant ganglioside species in adult mammalian brain and neurons. Oligodendrocyte markers O1 and O4 are GalCer and sulfatide, respectively. GT1aα and GQ1bα are cholinergic-specific antigens (Chol-1α). (B) Thin-layer chromatography (TLC) and (C) quantitative PCR studies demonstrate the structural complexity of gangliosides that develops with age. Note the dynamic changes from E12 up through the adult for expression of gangliosides (GM1 and GD3) and glycosyltransferases (GalNAcT; GD2/GM2S and ST-II; GD3S), indicative of developmental regulation.

GalNAc-containing gangliosides. Phenotypically, they are developmentally abnormal and exhibit neurological problems, such as axonal degeneration, sensory, motor and behavioral deficits, and other neurological dysfunctions.[58–63] During brain development, gangliosides are assumed to modulate ceramide-induced apoptosis and to maintain cellular survival and differentiation.[53] GM3-synthase (GM3S; sialyltransferase I, ST-I) is a critical enzyme for the synthesis of all complex gangliosides. Mutation of GM3-synthase is associated with human autosomal recessive infantile-onset symptomatic epilepsy syndrome,[64] and an alteration of the *GM2/GD2S* gene has been reported in patients of hereditary spastic paraplegias.[65] These studies clearly demonstrate that deletion of complex gangliosides can be associated with human diseases. A lack of b- and c-series gangliosides results in clear and subtle developmental and behavioral deficits with mice lacking these gangliosides, which exhibited sudden death from audiogenic seizures.[66] Both GM2/GD2S- and GM3S-deficient mice, which lack all gangliosides, die soon after weaning at 3 weeks of age.[67] Taken together, these observations clearly indicate that GSLs, including gangliosides, have important biological functions in the developing nervous system.

## 4. GD3 ON NSC SELF-RENEWAL

b-Series ganglioside GD3 (CD60a) is expressed in the neural tube during early development and can be detected using a GD3-specific monoclonal antibody (mAb R24).[68] Upon closer examination, GD3 is found to be robustly expressed on NECs in the neural tube and in RGCs of the VZ in embryos, and the SVZ of postnatal and adult rodents.[36,69–72] Fig. 2A shows that GD3 is colocalized with nestin at the SVZ. $GD3^+$ cells also express SSEA-1 in the SVZ of mouse brain.[36] GD3 is the predominant ganglioside species in cultured mouse neurospheres, accounting for more than 80% of the total GSLs. In addition to GD3, another b-series ganglioside, GD2, has also been shown to be highly expressed in human NSCs.[73] NSCs are costained with GD3 and SSEA-1, and GD3+ cells also express other NSC markers, such as Sox2, nestin, and Musashi-1. After cell differentiation, GD3 expression is decreased in cultured NSCs. For this reason, GD3 has been proposed to be a useful biomarker for mouse NSCs. Most living cells (80%) were positive for GD3 in E14 striata and in P2 and P10 SVZs. On the contrary, fewer cells (30%) are positive for GD3 in P30 and adult SVZs. Similar strong expression of GD3 is observed in NSCs isolated from embryonic, postnatal, and adult

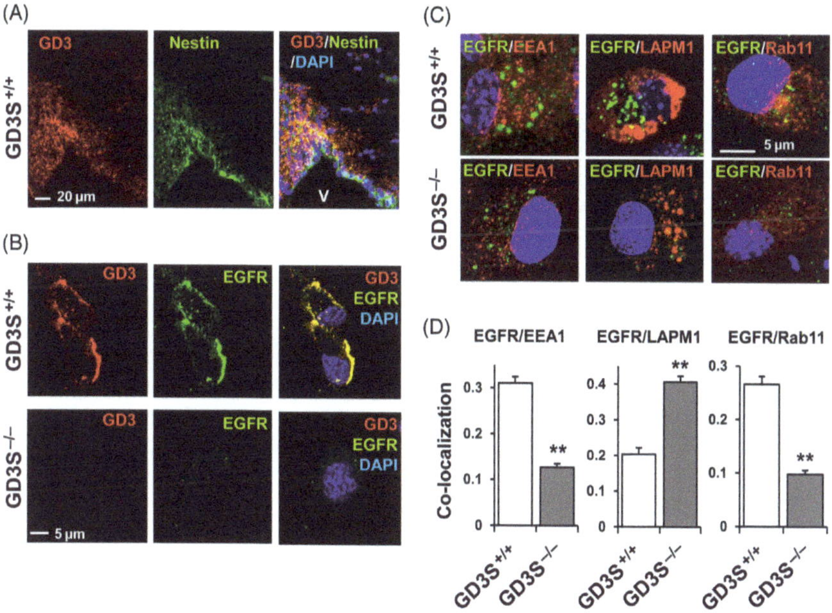


**Fig. 2** Interaction of GD3 with EGFR modulates self-renewal of murine NSCs. (A) GD3 *(red)* is colocalized with nestin *(green)* in SVZ (3-month-old wild-type mouse). (B) Cell surface EGFR expression *(green)* on $GD3^{+/+}$ or $GD3^{-/-}$ NSCs. NSCs were cultured with 20 ng/mL EGF and 20 ng/mL FGF2. A decrease of EGFR expression is detected in GD3S-KO NSCs. The decrease in EGFR expression is accompanied by an accelerated loss of the self-renewal capability in the GD3S-KO NSCs (data not shown). (C) GD3 regulates the endocytosis of EGFR. Colocalization of EGFR *(green)* with EEA1 *(red)*, LAMP1 *(red)*, and Rab11 *(red)* in monolayer NSCs under normal culture conditions (20 ng/mL EGF and 20 ng/mL FGF2). The colocalization coefficiency was measured by Zeiss LSM-enhanced colocalization software. Values are expressed as mean ± SEM; $n = 7$. $^{**}P < 0.01$. EGFR showed significantly increased colocalization with the lysosomal marker LAMP1 in GD3S-KO NSCs. On the contrary, colocalization rates with a marker of recycling endosomes, Rab11 and the early endosomal marker, EEA1, showed decreases in GD3S-KO NSCs. V, Ventricle.

brains. Most $GD3^+$ cells isolated by FACS were also positive for nestin and SSEA-1, but negative for differentiated cell markers. $GD3^+$ cells can generate neurospheres more efficiently than $GD3^-$ cells. The $GD3^+$ cells are capable of differentiating into neuronal and glial cells. Those results clearly suggest that GD3 is a useful cell surface marker for isolating living NSCs, especially from adult brain tissues in which NSCs are less abundant.[36]

To clarify whether GD3 modulates the self-renewal capacity or cell fate determination of NSCs, NSCs from GD3-synthase (GD3S)-KO mice were analyzed. Surprisingly, no significant difference in the proliferation rate and expression of lineage-associated markers was found between GD3S$^{+/+}$ and GD3S-KO NSCs that were cultured in the presence of FGF-2 but in the absence of EGF.[74] GD3S-KO NSCs that were cultured with EGF, however, showed dramatically suppressed proliferation.[37] The expression of nestin and EGFR was also strongly downregulated and the MAPK/ERK pathway signaling was impaired in GD3S-KO NSCs. In addition, EGFR degradation and the reduction of p-EGFR and p-ERK1/2 were correlated in the GD3S-KO NSCs after EGF stimulation. Subsequently, a decrease of the MAPK/ERK proliferation pathway was found in GD3S-KO NSCs. Furthermore, more membrane EGFRs (Fig. 2B) were expressed on the cell surface of GD3S$^{+/+}$ NSCs (53%) than in GD3S-KO NSCs (22%). Likewise, EGFR and GD3 were found to be colocalized in NSCs, and they interacted in the lipid raft region of the cell surface as well as in intracellular vesicles. Interestingly, EGFR was found to exist on nonlipid raft fractions in the GD3S-KO NSCs. Those findings provide convincing evidence that EGFR and GD3 are colocalized in lipid raft microdomains and support the notion that GD3 is essential for the maintenance of the self-renewal capability in NSCs by recruiting EGFR to the microdomains to sustain the EGF-induced downstream signaling.[37]

Endocytosis is a basic cellular process that is used by cells to internalize a variety of molecules. Cells are estimated to internalize, via endocytosis, about half their plasma membrane per hour.[75] GSLs are recognized to undergo endocytosis. Once internalized, GSLs can be (1) recycled to the plasma membrane; (2) sorted to the Golgi complex, in which they can be reglycosylated; or (3) degraded in the lysosome. As EGFR expression was reduced in GD3S-KO NSCs, internalization of its ligand EGF was investigated. It was found that the number of NSCs with internalized biotinylated-fluorescent-EGF was significantly reduced in the absence of GD3. As it is known that EGFR can be recruited to the endosomes for recycling or sorted to lysosomes for degradation, it was found that in the GD3S-KO NSCs, EGFR exhibited increased colocalization with the LAMP 1 (lysosome) as well as diminished colocalization with a marker of recycling endosomes, Rab11 (Fig. 2C and D). The colocalization rate with EEA1 (an early endosome marker) also showed a decrease in the GD3S-KO NSCs. Pursuing the endocytosed EGFR indicated that a large amount of EGFR in GD3S-KO NSCs underwent the endosomal–lysosomal degradative pathway, whereas a greater portion of EGFR was subject to the recycling pathway in the GD3S$^{+/+}$ NSCs than in GD3S$^{-/-}$ NSCs. Hence, the interaction of GD3 and EGFR in NSCs is responsible for sustaining EGFR surface expression and downstream signaling to maintain the self-renewal of the cells.[37]

## 5. GD3 MAINTAINS POSTNATAL NSC POOLS

In postnatal brain, GM1, GD1a, GD1b, and GT1b are strongly expressed. On the contrary, the expression of GD3 is decreased and restricted. As shown in Fig. 2A, the expression of GD3 and nestin is found in identical cells at the SVZ of postnatal mice. In these nestin$^+$ cells, GD3 is the predominant GSL species (>80%). We compared the histology and cellularity of SVZ and DG in 6-month-old wild-type (WT) and GD3S-KO mice.[76] The cellularity at the SVZ and DG of GD3S-KO mice was significantly reduced compared with that of their WT littermates (Fig. 3A). These findings indicated that there are progressive reductions of the cellular pool in the SVZ and DG in adult GD3S-KO mice. To investigate the population of adult NSCs, we counted the nestin and GFAP double-positive cells in the adult brain (Fig. 3B). We found that nestin$^+$/GFAP$^+$ radial glia-like NSCs were reduced in the SVZ of 1-month-old GD3S-KO mice (57.7 ± 16.2% of WT mice). In the 6-month-old SVZ, nestin$^+$/GFAP$^+$ cells were nearly absent from the GD3S-KO brains. Consistent with this observation, the number of nestin$^+$/GFAP$^+$ radial glia-like cells at the DG of GD3S-KO mice was also significantly reduced (45.7 ± 13.6% at 1-month-old mice; 83.2 ± 15.2% reduced at 6-month-old mice) compared with their WT littermates. To evaluate neurogenesis, bromo-deoxy-uridine (BrdU) was administered to 4-month-old mice (i.p.) and imunohistochemical analyses were performed at 4 weeks after BrdU injection (Fig. 3C). Newly generated neurons (NeuN$^+$/BrdU$^+$) in the olfactory bulb were significantly decreased in GD3S-KO mice. The density of NeuN in the granular cell layers was much lower in the GD3S-KO mice than in their WT littermates, which indicated a long-term insufficiency of the supply of new neurons to this region. At the SVZ and DG, the numbers of BrdU-incorporated cells were significantly reduced in the GD3S-KO mice compared with WT mice. Our study clearly demonstrates that GD3 plays a crucial role in the long-term maintenance of NSC populations in postnatal mouse brain. Moreover, impaired neurogenesis in the adult GD3S-KO mice can lead to depression-like behaviors.[76] Thus, our results provide strong evidence linking ganglioside deficiency and neurogenesis defects to behavioral deficits.

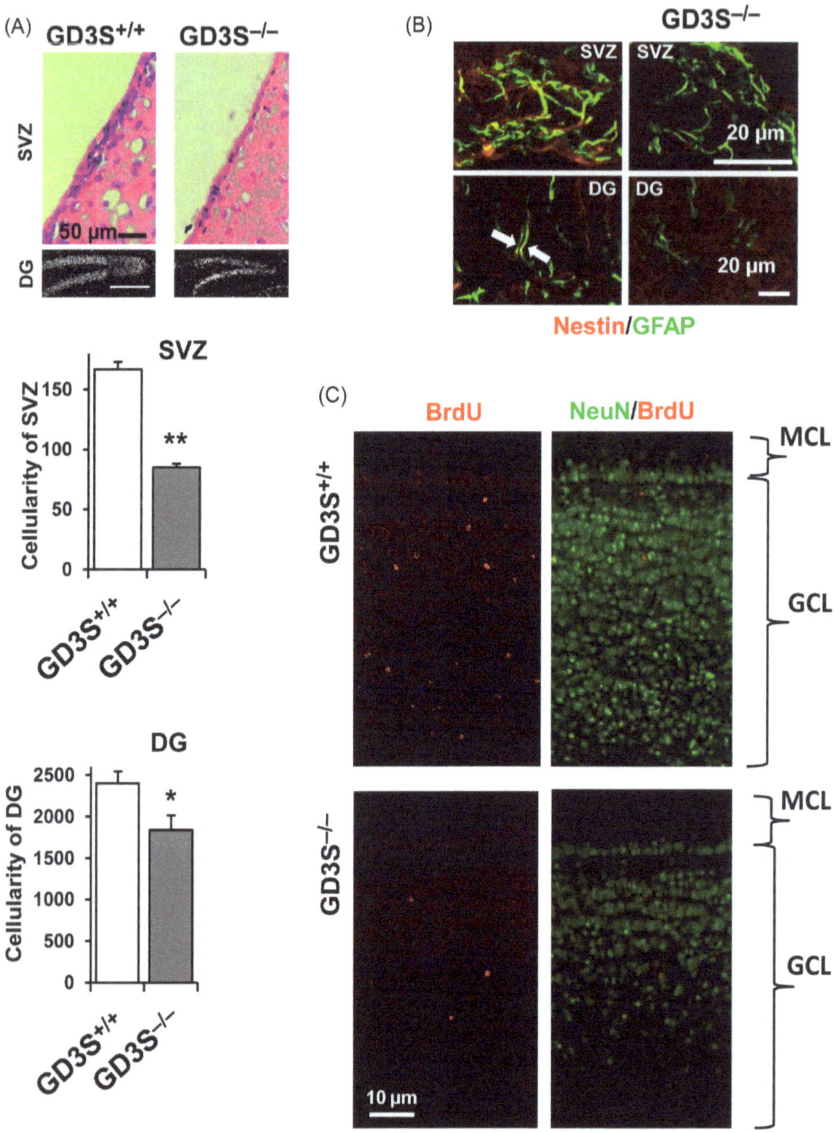


**Fig. 3** NSC pools are reduced in GD3S-KO mouse brain in vivo. (A) Reduced cellularity in the SVZ and DG of adult GD3S-KO mouse brain (6-month-old mice). Hematoxylin and eosin staining was used to quantify the SVZ cellularity, and DAPI staining was used to trace the DG cellularity. Values are expressed as mean ± SEM; $n = 4$. *$P < 0.05$; **$P < 0.01$. (B) Nestin$^+$/GFAP$^+$ radial glia-like NSCs were reduced in the SVZ and DG of GD3S-KO mouse. In the 6-month-old SVZ, nestin$^+$/GFAP$^+$ cells were nearly absent from the GD3S-KO brains (B, right upper panel). Consistent with this observation, the number of nestin$^+$/GFAP$^+$ radial glia-like cells at DG of GD3S-KO mice was also significantly

## 6. A2B5 IN PROGENITOR CELLS

A2B5 antigens, including the c-series gangliosides, are well-known markers for immature progenitor cells in the nervous system and are the first GSL antigens expressed in cells of the glial lineage.[77–80] A2B5 is a mouse mAb originally developed using chicken embryonic retina cells as the immunogen.[77] The antigens recognized by mAb A2B5 have been established as the c-series gangliosides, including GQ1c, GT1c, and GT3,[78,80] and the antibody recognizes the Neu5Acα2-8-Neu5Acα2-8Neu5Acα-or α2,8-trisialosyl (triSia) structure.[81] These c-series gangliosides are phylogenetically conserved and developmentally regulated, and are most abundant in the brain of lower vertebrates, such as fish, and the brain of mammalian embryos, but not adult brains.[82–86] In adult zebrafish brain, c-series gangliosides comprise up to 70% of the total gangliosides.[87] Accordingly, damaged zebrafish brains have high regeneration capacity, and the brain is rich in $A2B5^+$ cells.[88] In proliferating NSCs of mammals, GT3-synthase (GT3S or ST-III), an enzyme for the synthesis for c-series gangliosides (A2B5 antigens), is highly expressed.[89,90] Some GFAP-expressing cells are considered as NSCs as well as astrocytes, and $A2B5^+/GFAP^+$ cells have a lipid composition distinct from mature astrocytes, but they are more similar to stem/progenitor cells.[91] In maturing mammalian brain, the concentration of c-series gangliosides decreases drastically,[46,82–84] and this decrease in the synthesis of c-series gangliosides is compensated for by a pathway shift in favor of the accretion of a- and b-series gangliosides.[41,45,46] Glial-restricted precursors (GRPs) have been recognized by the expression of the A2B5 epitope.[92] It is uncertain, however, whether GRPs exist in vivo. A2B5 antigens in these progenitors have been identified as GT3 and *O*-acetyl GT3.[93] In the developing mouse brain, it has been reported that mAb A2B5 also reacts with four glycoproteins in addition to c-series gangliosides.[81]

◀ reduced (45.7 ± 13.6% at 1-month-old mice; 83.2 ± 15.2% reduced at 6-month-old mice) compared with their WT littermates. (C) Bromo-deoxy-uridine (BrdU) was administered to 4-month-old mice (i.p.) and imunohistochemical analyses were performed at 4 weeks after BrdU injection. Newly generated neurons ($NeuN^+/BrdU^+$) in the olfactory bulb (OB) were significantly decreased in GD3S-KO mice. The density of NeuN in the granular cell layers (GCL) was much lower in the GD3S-KO mice than in their WT littermates, which indicated a long-term insufficiency of the supply of new neurons to this region. MCL, Molecular cell layer.

## 7. CHOLINERGIC-SPECIFIC ANTIGEN (CHOL-1α)

The presence of neuronal gangliosides may also be functionally specific. For example, Chol-1α gangliosides (GT1aα and GQ1bα) are normally minor species in the brain and serve as unique markers of cholinergic neurons.[94,95] They have been shown immunocytochemically in human CNS and may find applications in human neuropathology. The expression of Chol-1α gangliosides in rat brain regions, such as the hippocampus, is developmentally regulated, and their concentrations increase with aging.[96] Ando *et al.*[97] have further shown that treatment of anti-Chol-1α mAb inhibits the release of acetylcholine from synaptosomes. Interestingly, the memory and learning abilities of rats given anti-Chol-1α mAb are remarkably suppressed. On the contrary, addition of Chol-1α gangliosides of the synaptosomal preparation induces high affinity choline uptake into synaptosomes and enhances synthesis of acetylcholine. Thus, Chol-1α gangliosides may participate in maintaining cognitive functions, such as memory and learning. In addition, Chol-1α gangliosides have been shown to alleviate the decreased synaptic functions of aged brains.[98,99] NSCs have been found to express both gangliosides, GT1aα and GQ1bα, in vitro.[41] These findings suggest that Chol-1α antigens may play an important role in cholinergic synaptic transmission and participate in cognitive function.

## 8. EPIGENETIC REGULATION OF NEURONAL DIFFERENTIATION BY GM1

During neuronal differentiation, the concentration of GD3, which is the predominant ganglioside in NSCs, is rapidly decreased. Concomitantly, the levels of GM1, GD1a, GD1b, and GT1b continuously increase in young animals, reaching a plateau during adulthood.[38,41,100] This pattern change follows closely with the upregulation of GM2/GD2S expression.[41] The dramatic changes in the expression profile of gangliosides clearly reflect the biological needs of GalNAc-containing ganglio-series gangliosides at particular stages during brain development. During neuronal development, GM1-expressing cells are considered as neuronal progenitor cells and neurons.[101–103]

Recently, our laboratory has demonstrated for the first time that histone acetylation in chromatin of the 5′-region of the mouse *GM2/GD2S* gene can regulate mRNA expression, and this epigenetic modification is highly correlative to the stage-specific alterations of GM2/GD2S mRNA levels in mouse brain during development.[38,39] In an initial study, we investigated the epigenetic regulation of two key GTs, GM2/GD2S and GD3S, in embryonic, postnatal, and adult mouse brains. Interestingly, the temporal expression patterns of GM2/GD2S and GD3S mRNAs are correlated with histone H3 and H4 acetylation (AcH3/AcH4) of the gene's 5′-flanking region in chromatin.[38] These observations suggest that the 5′-region of the GM2/GD2S and GD3S genes could be targets for epigenetic regulation by histone acetylation. Furthermore, we have demonstrated that acetylation of histones H3 and H4 on the *GM2/GD2S* gene promoter leads to recruitment of the *trans*-activation factors Sp1 and AP-2.

When the cellular histone deacetylase (HDAC) activity is globally inhibited by short-chain fatty acids, such as valproic acid and butyric acid, neuronal differentiation of NSCs is promoted with extensive neurite outgrowth.[38,39] More GM2/GD2S or GD3S mRNA can be detected by the application of valproic acid in the NSC culture, which is triggered due to a loading boost of the transcription factors Ap-2 and Sp1 on the promoter region.[39] Individually knocking down HDAC1 and HDAC2 gene expression increases the levels of AcH3 and AcH4 on the *GM2/GD2S* gene. However, the *GM2/GD2S* gene does not show any corresponding change in the promoter-binding abilities for the transcription factors or mRNA expression level by individually knocking down. Intriguingly, when both HDAC1 and HDAC2 are knocked down, the expression of GM2/GD2S mRNA is upregulated and Ap-2- or Sp1-loading is significantly increased, reflecting the elevated level of histone acetylation on the *GM2/GD2S* gene. This is not the case for the *GD3S* gene, however, which remains unchanged with respect to the levels of mRNA, the binding of transcription factors, or the acetylated histone status. Our results clearly indicate that transcription of *GM2/GD2S* and *GD3S* can be regulated by different HDAC isoforms because double-knockdown by si-HDAC1 and si-HDAC2 leads to *GM2/GD2S* gene *trans*-activation, but not *GD3S*.[39]

It is known that GM1 enhances neurite outgrowth in primary neuronal cultures. It has been shown that GM1 is also present in the nuclear membranes and upregulation of GM1 in the nuclear membrane accompanies the process of neurite outgrowth. This observation prompted Ledeen and coworkers to propose that the elevated GM1 has a modulatory effect on $Ca^{2+}$ homeostasis in the nucleus, which is mediated by a tight association of GM1 with $Na^{+}/Ca^{2+}$exchanger.[104,105] In earlier studies, it has been reported that GM1 induces tubulin mRNA accumulation during GM1-stimulated neurite outgrowth.[106] Interestingly, NSCs cultured with a supplement of GM1 for 7 days also exhibit a significantly enhanced neurogenic effect.[39] Additional GM1, but not GD3, enhanced GM2/GD2S expression of mRNA, whereas the mRNA level of GD3S remained unchanged. The presence of ectopic GM1 renders an enrichment of the acetylated histones on the gene loci of *GM2/GD2S*, but not *GD3S*, which is accompanied by the recruitment of transcription factors AP-2 and Sp1 within the gene promoter region. It is possible that in this enhanced neurogenic process, exogenous GM1 induces NSCs to transcribe more GM2/GD2S mRNA with a higher level of AcH4 on the *GM2/GD2S*'s promoter region where more transcription factors are recruited. On the contrary, the *GD3S* gene does not show significant changes in mRNA expression and AcH4 binding by addition of GM1. This result may represent a potential mechanism accounting for the correlations between the ganglioside pattern shift and epigenetic modifications of ganglioside synthase expression over neuronal differentiation and neural development. In this scenario, GM1 generates a positive feedback loop for NSCs to promote neuronal differentiation and to produce more GM1 and brain-type gangliosides, such as GD1a, GD1b, and GT1b. In this regard, GM1 is considered to play a crucial role in modulating the pathway switch in ganglioside expression in the developing brain.

The nuclear envelope, including the nuclear lamina and nuclear pore complexes, is an important structure to maintain chromatin architecture and cell-specific gene expression.[107] Our confocal microscopy study reveals that nuclear GM1 is colocalized with lamin B1 (a protein of the nuclear lamina) or nucleoporin (a protein of the nuclear pore complex) on the nuclear periphery of induced neurons from NSCs.[40] Coimmunoprecipitation experiments show that nuclear GM1 interacts with AcH3 and AcH4, both active epigenetic marks, but not with H3K27me3, an inactive epigenetic mark. Chromatin immunoprecipitation assays reveal that the promoter regions of the *GM2/GD2S* and *NeuroD1* genes are associated with GM1. In situ hybridization assays have revealed that GM1 and the promoter region of the *GM2/GD2S* gene are in close proximity in the nucleus of a neuron. Despite the low expression of nuclear GM1 in NSCs, GM1 and the *GM2/GD2S* promoter do not colocalize therein. This result suggests that the interaction of GM1 and the *GM2/GD2S* gene promoter occurs in a differentiation stage-specific manner.[40]

## 9. CONCLUSION AND FUTURE STUDIES

Gangliosides are recognized as significant brain lipids that play important functions in cell–cell recognition, adhesion, and signal transduction. Brain gangliosides undergo dramatic qualitative and quantitative developmental changes that are correlated with cellular proliferation, differentiation, and play critical biological functions in the nervous system. Future studies should focus on gaining a better understanding of the molecular basis of the dramatic changes in ganglioside expression and their associated functional roles in neural cell fate determination and the regulatory mechanisms of GTs that account for the GSL pathway switch from simple to complex gangliosides during neural differentiation. Changes in GT expression in NSCs rationalize the dramatic changes of ganglioside expression during differentiation. The synthesis of GD3 is switched to the synthesis of complex gangliosides (GM1, GD1a, GD1b, and GT1b), resulting in terminal differentiation and loss of the "stemness" of NSCs. In the adult, NSCs are enriched in the SVZ of the lateral ventricles and the subgranular layer of the DG in the hippocampus. There is a growing awareness that incorporation of NSCs is associated with not only improved learning and memory, but also in functional recovery from brain injuries. This latter observation also indicates that events impacting adult NSC homeostasis may significantly contribute to the pathogenic mechanisms associated with neurodegenerative diseases. Finally, our preliminary experiments showing that intraventricular infusion of ganglioside GD3 augments NSC pools in the adult mouse brain (unpublished data) suggest a mechanism for regulating adult neurogenesis. Future studies in this regard will contribute greatly to regenerative and reparative biology.

Although recent evidence of GSLs including gangliosides has shed light on their roles in modulating signaling pathways during cellular differentiation and reprograming, mice deficient in some of these molecules show only subtle phenotypic abnormalities compared with the WT animals in early development. Clearly, the biological function of one glycoconjugate can be substituted by another, albeit with less efficiency. However, the aberrant ganglioside expression becomes progressively more serious in the adult stage and pathogenic conditions. The "biological redundancy" can be considered for more important functions of these molecules. Future studies will clarify the cell- and stage-specific function and importance of gangliosides in maintaining NSC activities and in neural differentiation.

Cellular signaling events that mediate lipid metabolism are among the most crucial processes controlling maintenance of brain function at the structural level. Progressive imbalance of cell membrane lipid composition is a physicochemical property altered with normal aging, and further disruptions to these processes are observed in neurodegenerative diseases. Ganglioside expression profiles are associated not only with cell differentiation of stem cells, but also with pathogenic mechanisms of neurodegenerative diseases in CNS, such as Alzheimer's disease, Parkinson disease, Huntington's disease, amyotrophic lateral sclerosis, and multiple sclerosis.[108–111] In addition, the aberrant and elevated expression of gangliosides has also been observed in many types of cancer cells, thereby promoting tumor survival.[112,113] Novel epigenetic regulatory mechanisms for GSLs will contribute to a better understanding of the pathway switch observed during neuronal differentiation of NSCs. In addition, epigenetic regulation of GSLs will provide clues underlying the pathogenic mechanisms, which are useful in developing novel strategies for disease treatment and tissue damage repair. Futures studies will prove extremely fruitful in this regard.

## ACKNOWLEDGMENTS

This work was supported by a VA Merit Review Award (1 I01BX001388 to RKY), NIH grants (RO1 NS100839, NS26994, and NS11853 to RKY), and Mizutani Foundation for Glycoscience (150026 to YI). The authors declare no conflicts of interest.

## REFERENCES

1. Alvarez-Buylla A, Garcia-Verdugo JM, Tramontin AD. A unified hypothesis on the lineage of neural stem cells. *Nat Rev Neurosci.* 2001;2(4):287–293.
2. Fishell G, Kriegstein A. Cortical development: new concepts. *Neuron.* 2005;46(3): 361–362.
3. Fishell G, Kriegstein AR. Neurons from radial glia: the consequences of asymmetric inheritance. *Curr Opin Neurobiol.* 2003;13(1):34–41.
4. Fujita S. The discovery of the matrix cell, the identification of the multipotent neural stem cell and the development of the central nervous system. *Cell Struct Funct.* 2003;28(4):205–228.
5. Gotz M, Huttner WB. The cell biology of neurogenesis. *Nat Rev Mol Cell Biol.* 2005;6(10):777–788.
6. Miller FD, Gauthier AS. Timing is everything: making neurons versus glia in the developing cortex. *Neuron.* 2007;54(3):357–369.
7. Shimojo H, Ohtsuka T, Kageyama R. Dynamic expression of notch signaling genes in neural stem/progenitor cells. *Front Neurosci.* 2011;5:78.
8. Malatesta P, Hartfuss E, Gotz M. Isolation of radial glial cells by fluorescent-activated cell sorting reveals a neuronal lineage. *Development.* 2000;127(24):5253–5263.
9. Miyata T, Kawaguchi A, Okano H, Ogawa M. Asymmetric inheritance of radial glial fibers by cortical neurons. *Neuron.* 2001;31(5):727–741.

10. Noctor SC, Flint AC, Weissman TA, Dammerman RS, Kriegstein AR. Neurons derived from radial glial cells establish radial units in neocortex. *Nature.* 2001;409(6821):714–720.
11. Franco SJ, Muller U. Shaping our minds: stem and progenitor cell diversity in the mammalian neocortex. *Neuron.* 2013;77(1):19–34.
12. Pinto L, Gotz M. Radial glial cell heterogeneity—the source of diverse progeny in the CNS. *Prog Neurobiol.* 2007;83(1):2–23.
13. Qian X, Shen Q, Goderie SK, et al. Timing of CNS cell generation: a programmed sequence of neuron and glial cell production from isolated murine cortical stem cells. *Neuron.* 2000;28(1):69–80.
14. Jayakumar AR, Tong XY, Curtis KM, et al. Decreased astrocytic thrombospondin-1 secretion after chronic ammonia treatment reduces the level of synaptic proteins: in vitro and in vivo studies. *J Neurochem.* 2014;131(3):333–347.
15. Rodnight RB, Gottfried C. Morphological plasticity of rodent astroglia. *J Neurochem.* 2013;124(3):263–275.
16. Sofroniew MV, Vinters HV. Astrocytes: biology and pathology. *Acta Neuropathol.* 2010;119(1):7–35.
17. Tsai HH, Li H, Fuentealba LC, et al. Regional astrocyte allocation regulates CNS synaptogenesis and repair. *Science.* 2012;337(6092):358–362.
18. Hajos F, Woodhams PL, Basco E, Csillag A, Balazs R. Proliferation of astroglia in the embryonic mouse forebrain as revealed by simultaneous immunocytochemistry and autoradiography. *Acta Morphol Acad Sci Hung.* 1981;29(4):361–364.
19. Ichikawa M, Shiga T, Hirata Y. Spatial and temporal pattern of postnatal proliferation of glial cells in the parietal cortex of the rat. *Brain Res.* 1983;285(2):181–187.
20. Kriegstein A, Alvarez-Buylla A. The glial nature of embryonic and adult neural stem cells. *Annu Rev Neurosci.* 2009;32:149–184.
21. Rowitch DH, Kriegstein AR. Developmental genetics of vertebrate glial-cell specification. *Nature.* 2010;468(7321):214–222.
22. Doetsch F, Caille I, Lim DA, Garcia-Verdugo JM, Alvarez-Buylla A. Subventricular zone astrocytes are neural stem cells in the adult mammalian brain. *Cell.* 1999;97(6):703–716.
23. Doetsch F, Garcia-Verdugo JM, Alvarez-Buylla A. Cellular composition and three-dimensional organization of the subventricular germinal zone in the adult mammalian brain. *J Neurosci.* 1997;17(13):5046–5061.
24. Seri B, Garcia-Verdugo JM, McEwen BS, Alvarez-Buylla A. Astrocytes give rise to new neurons in the adult mammalian hippocampus. *J Neurosci.* 2001;21(18):7153–7160.
25. Suhonen JO, Peterson DA, Ray J, Gage FH. Differentiation of adult hippocampus-derived progenitors into olfactory neurons in vivo. *Nature.* 1996;383(6601):624–627.
26. Morshead CM. Adult neural stem cells: attempting to solve the identity crisis. *Dev Neurosci.* 2004;26(2–4):93–100.
27. Johansson CB, Momma S, Clarke DL, Risling M, Lendahl U, Frisen J. Identification of a neural stem cell in the adult mammalian central nervous system. *Cell.* 1999;96(1):25–34.
28. Chiasson BJ, Tropepe V, Morshead CM, van der Kooy D. Adult mammalian forebrain ependymal and subependymal cells demonstrate proliferative potential, but only subependymal cells have neural stem cell characteristics. *J Neurosci.* 1999;19(11):4462–4471.
29. Laywell ED, Rakic P, Kukekov VG, Holland EC, Steindler DA. Identification of a multipotent astrocytic stem cell in the immature and adult mouse brain. *Proc Natl Acad Sci USA.* 2000;97(25):13883–13888.
30. Carlen M, Meletis K, Goritz C, et al. Forebrain ependymal cells are Notch-dependent and generate neuroblasts and astrocytes after stroke. *Nat Neurosci.* 2009;12(3):259–267.

31. Coskun V, Wu H, Blanchi B, et al. CD133+ neural stem cells in the ependyma of mammalian postnatal forebrain. *Proc Natl Acad Sci USA*. 2008;105(3):1026–1031.
32. Filippov V, Kronenberg G, Pivneva T, et al. Subpopulation of nestin-expressing pro genitor cells in the adult murine hippocampus shows electrophysiological and morphological characteristics of astrocytes. *Mol Cell Neurosci*. 2003;23(3):373–382.
33. Encinas JM, Vaahtokari A, Enikolopov G. Fluoxetine targets early progenitor cells in the adult brain. *Proc Natl Acad Sci USA*. 2006;103(21):8233–8238.
34. Steiner B, Klempin F, Wang L, Kott M, Kettenmann H, Kempermann G. Type-2 cells as link between glial and neuronal lineage in adult hippocampal neurogenesis. *Glia*. 2006;54(8):805–814.
35. Lugert S, Vogt M, Tchorz JS, Muller M, Giachino C, Taylor V. Homeostatic neurogenesis in the adult hippocampus does not involve amplification of Ascl1 (high) intermediate progenitors. *Nat Commun*. 2012;3:670.
36. Nakatani Y, Yanagisawa M, Suzuki Y, Yu RK. Characterization of GD3 ganglioside as a novel biomarker of mouse neural stem cells. *Glycobiology*. 2010;20(1):78–86.
37. Wang J, Yu RK. Interaction of ganglioside GD3 with an EGF receptor sustains the self-renewal ability of mouse neural stem cells in vitro. *Proc Natl Acad Sci USA*. 2013;110(47): 19137–19142.
38. Suzuki Y, Yanagisawa M, Ariga T, Yu RK. Histone acetylation-mediated glycosyltransferase gene regulation in mouse brain during development. *J Neurochem*. 2011;116(5): 874–880.
39. Tsai YT, Yu RK. Epigenetic activation of mouse ganglioside synthase genes: implications for neurogenesis. *J Neurochem*. 2014;128(1):101–110.
40. Tsai YT, Itokazu Y, Yu RK. GM1 ganglioside is involved in epigenetic activation loci of neuronal cells. *Neurochem Res*. 2016;41(1–2):107–115.
41. Ngamukote S, Yanagisawa M, Ariga T, Ando S, Yu RK. Developmental changes of glycosphingolipids and expression of glycogenes in mouse brains. *J Neurochem*. 2007;103 (6):2327–2341.
42. Yanagisawa M, Yu RK. The expression and functions of glycoconjugates in neural stem cells. *Glycobiology*. 2007;17(7):57R–74R.
43. Yu RK, Macala LJ, Taki T, Weinfield HM, Yu FS. Developmental changes in ganglioside composition and synthesis in embryonic rat brain. *J Neurochem*. 1988;50(6): 1825–1829.
44. Xu YH, Barnes S, Sun Y, Grabowski GA. Multi-system disorders of glycosphingolipid and ganglioside metabolism. *J Lipid Res*. 2010;51(7):1643–1675.
45. Yu RK, Nakatani Y, Yanagisawa M. The role of glycosphingolipid metabolism in the developing brain. *J Lipid Res*. 2009;50(suppl):S440–S445.
46. Yu RK, Itokazu Y. Glycolipid and glycoprotein expression during neural development. *Adv Neurobiol*. 2014;9:185–222.
47. Itokazu Y, Tsai YT, Yu RK. Epigenetic regulation of ganglioside expression in neural stem cells and neuronal cells. *Glycoconj J*. 2017;34(6):749–756.
48. D'Angelo G, Polishchuk E, Di Tullio G, et al. Glycosphingolipid synthesis requires FAPP2 transfer of glucosylceramide. *Nature*. 2007;449(7158):62–67.
49. D'Angelo G, Uemura T, Chuang CC, et al. Vesicular and non-vesicular transport feed distinct glycosylation pathways in the Golgi. *Nature*. 2013;501(7465):116–120.
50. Halter D, Neumann S, van Dijk SM, et al. Pre- and post-Golgi translocation of glucosylceramide in glycosphingolipid synthesis. *J Cell Biol*. 2007;179(1):101–115.
51. Chalat M, Menon I, Turan Z, Menon AK. Reconstitution of glucosylceramide flip-flop across endoplasmic reticulum: implications for mechanism of glycosphingolipid biosynthesis. *J Biol Chem*. 2012;287(19):15523–15532.
52. Yu RK, Tsai YT, Ariga T. Functional roles of gangliosides in neurodevelopment: an overview of recent advances. *Neurochem Res*. 2012;37(6):1230–1244.

53. Bieberich E, MacKinnon S, Silva J, Yu RK. Regulation of apoptosis during neuronal differentiation by ceramide and b-series complex gangliosides. *J Biol Chem.* 2001;276 (48):44396–44404.
54. Fang Y, Wu G, Xie X, Lu ZH, Ledeen RW. Endogenous GM1 ganglioside of the plasma membrane promotes neuritogenesis by two mechanisms. *Neurochem Res.* 2000;25(7): 931–940.
55. Wu G, Fang Y, Lu ZH, Ledeen RW. Induction of axon-like and dendrite-like processes in neuroblastoma cells. *J Neurocytol.* 1998;27(1):1–14.
56. Wu G, Lu ZH, Xie X, Li L, Ledeen RW. Mutant NG108-15 cells (NG-CR72) deficient in GM1 synthase respond aberrantly to axonogenic stimuli and are vulnerable to calcium-induced apoptosis: they are rescued with LIGA-20. *J Neurochem.* 2001;76(3):690–702.
57. Yu RK, Bieberich E, Xia T, Zeng G. Regulation of ganglioside biosynthesis in the nervous system. *J Lipid Res.* 2004;45(5):783–793.
58. Furukawa K, Aixinjueluo W, Kasama T, et al. Disruption of GM2/GD2 synthase gene resulted in overt expression of 9-*O*-acetyl GD3 irrespective of Tis21. *J Neurochem.* 2008;105(3):1057–1066.
59. Sheikh KA, Sun J, Liu Y, et al. Mice lacking complex gangliosides develop Wallerian degeneration and myelination defects. *Proc Natl Acad Sci USA.* 1999;96(13):7532–7537.
60. Sugiura Y, Furukawa K, Tajima O, Mii S, Honda T. Sensory nerve-dominant nerve degeneration and remodeling in the mutant mice lacking complex gangliosides. *Neuroscience.* 2005;135(4):1167–1178.
61. Susuki K, Baba H, Tohyama K, et al. Gangliosides contribute to stability of paranodal junctions and ion channel clusters in myelinated nerve fibers. *Glia.* 2007;55(7):746–757.
62. Takamiya K, Yamamoto A, Furukawa K, et al. Mice with disrupted GM2/GD2 synthase gene lack complex gangliosides but exhibit only subtle defects in their nervous system. *Proc Natl Acad Sci USA.* 1996;93(20):10662–10667.
63. Wu G, Lu ZH, Kulkarni N, Amin R, Ledeen RW. Mice lacking major brain gangliosides develop parkinsonism. *Neurochem Res.* 2011;36(9):1706–1714.
64. Simpson MA, Cross H, Proukakis C, et al. Infantile-onset symptomatic epilepsy syndrome caused by a homozygous loss-of-function mutation of GM3 synthase. *Nat Genet.* 2004;36(11):1225–1229.
65. Boukhris A, Schule R, Loureiro JL, et al. Alteration of ganglioside biosynthesis responsible for complex hereditary spastic paraplegia. *Am J Hum Genet.* 2013;93(1):118–123.
66. Kawai H, Allende ML, Wada R, et al. Mice expressing only monosialoganglioside GM3 exhibit lethal audiogenic seizures. *J Biol Chem.* 2001;276(10):6885–6888.
67. Yamashita T, Wada R, Sasaki T, et al. A vital role for glycosphingolipid synthesis during development and differentiation. *Proc Natl Acad Sci USA.* 1999;96(16):9142–9147.
68. Rosner H, al-Aqtum M, Rahmann H. Gangliosides and neuronal differentiation. *Neurochem Int.* 1992;20(3):339–351.
69. Bannerman PG, Oliver TM, Xu Z, Shieh A, Pleasure DE. Effects of FGF-1 and FGF-2 on GD3 immunoreactive spinal neuroepithelial cells. *J Neurosci Res.* 1996;45(5):549–557.
70. Cammer W, Zhang H. Ganglioside GD3 in radial glia and astrocytes in situ in brains of young and adult mice. *J Neurosci Res.* 1996;46(1):18–23.
71. Cammer W, Zhang H. Carbonic anhydrase II in microglia in forebrains of neonatal rats. *J Neuroimmunol.* 1996;67(2):131–136.
72. Goldman JE, Hirano M, Yu RK, Seyfried TN. GD3 ganglioside is a glycolipid characteristic of immature neuroectodermal cells. *J Neuroimmunol.* 1984;7(2–3):179–192.
73. Yanagisawa M, Yoshimura S, Yu RK. Expression of GD2 and GD3 gangliosides in human embryonic neural stem cells. *ASN Neuro.* 2011;3(2). pii: e00054.
74. Yu RK, Yanagisawa M. Glycosignaling in neural stem cells: involvement of glycoconjugates in signal transduction modulating the neural stem cell fate. *J Neurochem.* 2007;103(suppl 1):39–46.

75. Steinman RM, Mellman IS, Muller WA, Cohn ZA. Endocytosis and the recycling of plasma membrane. *J Cell Biol.* 1983;96(1):1–27.
76. Wang J, Cheng A, Wakade C, Yu RK. Ganglioside GD3 is required for neurogenesis and long-term maintenance of neural stem cells in the postnatal mouse brain. *J Neurosci.* 2014;34(41):13790–13800.
77. Eisenbarth GS, Walsh FS, Nirenberg M. Monoclonal antibody to a plasma membrane antigen of neurons. *Proc Natl Acad Sci USA.* 1979;76(10):4913–4917.
78. Kasai N, Yu RK. The monoclonal antibody A2B5 is specific to ganglioside GQ1c. *Brain Res.* 1983;277(1):155–158.
79. Raff MC, Abney ER, Cohen J, Lindsay R, Noble M. Two types of astrocytes in cultures of developing rat white matter: differences in morphology, surface gangliosides, and growth characteristics. *J Neurosci.* 1983;3(6):1289–1300.
80. Saito M, Kitamura H, Sugiyama K. The specificity of monoclonal antibody A2B5 to c-series gangliosides. *J Neurochem.* 2001;78(1):64–74.
81. Inoko E, Nishiura Y, Tanaka H, et al. Developmental stage-dependent expression of an alpha2,8-trisialic acid unit on glycoproteins in mouse brain. *Glycobiology.* 2010;20(7): 916–928.
82. Ando S, Yu RK. Isolation and characterization of two isomers of brain tetrasialogangliosides. *J Biol Chem.* 1979;254(23):12224–12229.
83. Freischutz B, Saito M, Rahmann H, Yu RK. Activities of five different sialyltransferases in fish and rat brains. *J Neurochem.* 1994;62(5):1965–1973.
84. Freischutz B, Saito M, Rahmann H, Yu RK. Characterization of sialyltransferase-IV activity and its involvement in the c-pathway of brain ganglioside metabolism. *J Neurochem.* 1995;64(1):385–393.
85. Rosner H, Greis C, Henke-Fahle S. Developmental expression in embryonic rat and chicken brain of a polysialoganglioside-antigen reacting with the monoclonal antibody Q 211. *Brain Res.* 1988;470(2):161–171.
86. Yu RK, Ando S. Structures of some new complex gangliosides of fish brain. *Adv Exp Med Biol.* 1980;125:33–45.
87. Viljetic B, Labak I, Majic S, Stambuk A, Heffer M. Distribution of mono-, di- and trisialo gangliosides in the brain of Actinopterygian fishes. *Biochim Biophys Acta.* 2012;1820(9):1437–1443.
88. Kishimoto N, Shimizu K, Sawamoto K. Neuronal regeneration in a zebrafish model of adult brain injury. *Dis Model Mech.* 2012;5(2):200–209.
89. Itokazu Y, Yu RK. Amyloid beta-peptide 1-42 modulates the proliferation of mouse neural stem cells: upregulation of fucosyltransferase IX and notch signaling. *Mol Neurobiol.* 2014;50:186–196.
90. Koon NA, Itokazu Y, Yu RK. Ganglioside-dependent neural stem cell proliferation in Alzheimer's disease model mice. *ASN Neuro.* 2015;7(6). pii: 1759091415618916.
91. Itokazu Y, Tajima N, Kerosuo L, et al. A2B5+/GFAP+ cells of rat spinal cord share a similar lipid profile with progenitor cells: a comparative lipidomic study. *Neurochem Res.* 2016;41(7):1527–1544.
92. Rao MS, Subbarao V. Effect of dexamethasone on ciprofibrate-induced cell proliferation and peroxisome proliferation. *Fundam Appl Toxicol.* 1997;35(1):78–83.
93. Farrer RG, Quarles RH. GT3 and its *O*-acetylated derivative are the principal A2B5-reactive gangliosides in cultured O2A lineage cells and are down-regulated along with *O*-acetyl GD3 during differentiation to oligodendrocytes. *J Neurosci Res.* 1999;57 (3):371–380.
94. Ando S, Hirabayashi Y, Kon K, Inagaki F, Tate S, Whittaker VP. A trisialoganglioside containing a sialyl alpha 2-6 *N*-acetylgalactosamine residue is a cholinergic-specific antigen, Chol-1 alpha. *J Biochem.* 1992;111(3):287–290.

95. Hirabayashi Y, Nakao T, Irie F, Whittaker VP, Kon K, Ando S. Structural characterization of a novel cholinergic neuron-specific ganglioside in bovine brain. *J Biol Chem.* 1992;267(18):12973–12978.
96. Derrington EA, Borroni E. The developmental expression of the cholinergic-specific antigen Chol-1 in the central and peripheral nervous system of the rat. *Brain Res Dev Brain Res.* 1990;52(1–2):131–140.
97. Ando S, Tanaka Y, Kobayashi S, et al. Synaptic function of cholinergic-specific Chol-1alpha ganglioside. *Neurochem Res.* 2004;29(4):857–867.
98. Ando S, Tanaka Y, Waki H, Kon K, Iwamoto M, Fukui F. Gangliosides and sialylcholesterol as modulators of synaptic functions. *Ann N Y Acad Sci.* 1998;845:232–239.
99. Ando S. Glycoconjugate changes in aging and age-related diseases. *Adv Neurobiol.* 2014;9:415–447.
100. Hirschberg K, Zisling R, van Echten-Deckert G, Futerman AH. Ganglioside synthesis during the development of neuronal polarity. Major changes occur during axonogenesis and axon elongation, but not during dendrite growth or synaptogenesis. *J Biol Chem.* 1996;271(25):14876–14882.
101. Androutsellis-Theotokis A, Walbridge S, Park DM, Lonser RR, McKay RD. Cholera toxin regulates a signaling pathway critical for the expansion of neural stem cell cultures from the fetal and adult rodent brains. *PLoS One.* 2010;5(5):e10841.
102. Liour SS, Dinkins MB, Su CY, Yu RK. Spatiotemporal expression of GM1 in murine medial pallial neural progenitor cells. *J Comp Neurol.* 2005;491(4):330–338.
103. Maric D, Maric I, Chang YH, Barker JL. Prospective cell sorting of embryonic rat neural stem cells and neuronal and glial progenitors reveals selective effects of basic fibroblast growth factor and epidermal growth factor on self-renewal and differentiation. *J Neurosci.* 2003;23(1):240–251.
104. Ledeen RW, Wu G, Lu ZH, Kozireski-Chuback D, Fang Y. The role of GM1 and other gangliosides in neuronal differentiation. Overview and new finding. *Ann N Y Acad Sci.* 1998;845:161–175.
105. Ledeen RW, Wu G. The multi-tasked life of GM1 ganglioside, a true factotum of nature. *Trends Biochem Sci.* 2015;40(7):407–418.
106. Rybak S, Ginzburg I, Yavin E. Gangliosides stimulate neurite outgrowth and induce tubulin mRNA accumulation in neural cells. *Biochem Biophys Res Commun.* 1983;116(3):974–980.
107. Talamas JA, Capelson M. Nuclear envelope and genome interactions in cell fate. *Fron Genet.* 2015;6:95.
108. Yu RK, Ueno K, Glaser GH, Tourtellotte WW. Lipid and protein alterations of spinal cord and cord myelin of multiple sclerosis. *J Neurochem.* 1982;39(2):464–477.
109. Yu RK, Ledeen RW, Eng LF. Ganglioside abnormalities in multiple sclerosis. *J Neurochem.* 1974;23(1):169–174.
110. Ariga T. The pathogenic role of ganglioside metabolism in Alzheimer's disease-cholinergic neuron-specific gangliosides and neurogenesis. *Mol Neurobiol.* 2017;54(1):623–638.
111. Ariga T. Pathogenic role of ganglioside metabolism in neurodegenerative diseases. *J Neurosci Res.* 2014;92(10):1227–1242.
112. Ladisch S, Liu Y. Dynamic aspects of neural tumor gangliosides. *Adv Neurobiol.* 2014;9:501–515.
113. Groux-Degroote S, Guerardel Y, Delannoy P. Gangliosides: structures, biosynthesis, analysis, and roles in cancer. *Chembiochem.* 2017;18(13):1146–1154.

CHAPTER NINE

# Gangliosides in Inflammation and Neurodegeneration

**Koichi Furukawa*,†,1, Yuhsuke Ohmi*, Orie Tajima*, Yuki Ohkawa*, Yuji Kondo†, Ji Shuting†, Noboru Hashimoto†, Keiko Furukawa***

* Department of Biomedical Sciences, Chubu University College of Life and Health Sciences, Kasugai, Japan
† Department of Biochemistry II, Nagoya University Graduate School of Medicine, Nagoya, Japan
[1] Corresponding author. E-mail address: koichi@isc.chubu.ac.jp

## Contents



*Progress in Molecular Biology and Translational Science,* Volume 156
ISSN 1877-1173
http://dx.doi.org/10.1016/bs.pmbts.2018.01.009

## Abstract

Gangliosides play roles in the regulation of cell signaling that are mediated via membrane microdomains, lipid rafts. In this review, functions of gangliosides in the maintenance of nervous systems with a focus on regulation of inflammation and neurodegeneration are addressed. During analyses of various ganglioside-lacking mutant mice, we demonstrated that nervous tissues exhibited inflammatory reactions and subsequent neurodegeneration. Among inflammation-related genes, factors of the complement system showed up-regulation with aging. Analyses of architectures and compositions of lipid rafts in nervous tissues from these mutant mice revealed that dysfunctions of complement regulatory proteins based on disrupted lipid rafts were main factors to induce the inflammatory reactions resulting in neurodegeneration. Ganglioside changes in development and senescence, and implication of them in the integrity of cell membranes and cellular phenotypes in physiological and pathological conditions including Alzheimer disease have been summarized. Novel directions to further analyze mechanisms for ganglioside functions in membrane microdomains have been also addressed.

## 1. INTRODUCTION

Gangliosides are sialic acid-containing glycosphingolipids, and are distributed in almost all tissues and cells of vertebrates. High level expression of gangliosides in brain has long suggested that they function in the development and regulation of nervous systems.[1] Most striking characteristics of them are that they consist of hydrophilic sugar moiety and hydrophobic lipid portion.[2] They are mainly expressed on the cell membrane, particularly at the outer layer of double lipid-layer structures.[3] Therefore, it has been one of puzzling issues how gangliosides are involved in the regulation of signaling for cell differentiation, activation, and malignant transformation.[4] In this chapter, ganglioside profiles during inflammation and neurodegeneration have been summarized, and effects of ganglioside changes on the integrity of cell membranes and cellular phenotypes have been summarized with focus on inflammation and resulting neurodegeneration. Finally, the puzzling issue about modes of actions of gangliosides in cells has been addressed.

## 1.1 Phenotypes of Knockout Mice and Gradual Severity of Inflammation

To investigate functional roles of glycosphingolipids in our bodies, we isolated cDNAs for ganglioside GM2/GD2 synthase,[5] GD3 synthase,[6] GM1/GD1b/GA1 synthase,[7] GM3 synthase,[8] Gb3 synthase,[9] and many other glycosyltransferases.[10–13] We also established KO mice lines of these glycosyltransferase genes, such as GM2/GD2 synthase,[14] GD3 synthase,[15] Gb3 synthase,[16] lactosylceramide synthase,[17] and double KO of GM2/GD2 synthase and GD3 synthase genes.[18] Synthetic pathways of main glycosphingolipids and deleted structures in the individual KO mouse lines are shown in Fig. 1. Generally, abnormal phenotypes found in these KO mice were much milder than expected probably because remaining glycolipid structures in the individual KO mice can compensate defective functions of primarily existing gangliosides[14] as shown in Table 1. However, intriguing findings commonly observed among these glycosyltransferase gene KO mice are inflammatory reactions mainly found in central nervous system (CNS).[19] During the study

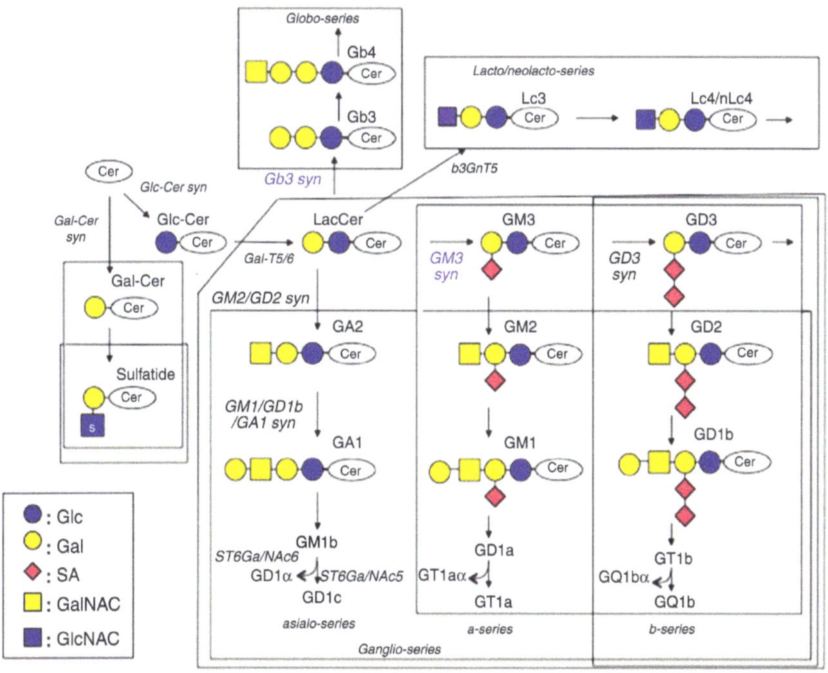


**Fig. 1** Biosynthetic pathways of glycosphingolipids. Deleted structures in individual knockout mice are encircled.

**Table 1** Major Phenotypes of Individual Knockout Mice With Focus on the Inflammatory Reaction and Neurodegeneration.

| *KO Genes* | GM3 Synthase | GD3 Synthase | GM2/GD2 Synthase |
|---|---|---|---|
| *Deleted structures* | Ganglio-series (a-, b-, c-) | b-series | All complex gangliosides |
| *Remaining structures* | Asialo-series | a-series and asialo-series | GM3, GD3 |
| | Gal-Cer and sulfatides | Gal-Cer and sulfatides | Gal-Cer and sulfatides |
| | Neutral glycolipids | Neutral glycolipids | Neutral glycolipids |
| *Outcome (inflammation and degeneration)* | No inflammation | Mild inflammation | Moderate inflammation |
| | No apparent disorders | Mild disorders | Progressive disorders (neurodegeneration) |

Table summarizes ganglioside and glycosphingolipid expression in each knockout mouse and their inflammatory and neurodegenerative outcomes.

on phenotypes of double KO (DKO) mice of GM2/GD2 synthase and GD3 synthase, biochemical and morphological changes indicating inflammatory reactions were detected not only in DKO, but also in single gene KO mice, more or less[20] (Table 1).

## 1.2 Inflammatory Reactions and Gangliosides

Changes in ganglioside profiles upon various inflammatory reactions have been observed.[21] For example, changes in gangliosides on glial cells in multiple sclerosis,[22,23] or by growth factors[24] were reported. Moreover, ganglioside GD3 was induced when brain was stimulated by inflammatory conditions in rat[25,26] and mouse.[27] GD3 was also induced on activated T cells via stimulation of T cell receptor (TCR), IL-2 receptor, or by various mitogens.[28,29] Functional T cells, in turn, expressed ganglioside GD2.[30] This might reflect the fact that HTLV-1-infected T cells expressed GD2 via viral p40tax expression.[31] Thus, changes in ganglioside expression of immune cells during various inflammatory reactions have been frequently observed.[32]

## 1.3 Aging and Gangliosides

There have been many reports on the compositional changes in gangliosides in brain tissues along with aging.[33,34] As known well, in the early developmental stage of mice at E12–14 days, simple gangliosides GM3 and GD3 are major species in brain tissues. During differentiation stage with neuritogenesis and synaptic formation, that is, after E16, mature type gangliosides such as GM1, GD1a, GD1b, and GT1b increased and become major species in brain tissues.[35,36] On the other hand, ganglioside profiles in brain scarcely alter during adult life, while total amounts gradually decrease. Then, gradual changes in some areas emerge with aging as reported.[33] However, no comprehensive and precise changes in ganglioside profiles in CNS, particularly in individual regions have been studied with some exceptions.[37–39]

## 1.4 Neurodegeneration Exhibited in Ganglioside Synthase Knockout Mice

It can be easily imagined that continued inflammatory conditions induce destructive changes in nervous tissues, that is, neurodegeneration.[40] We need to investigate which inflammatory cytokines, chemokines, and growth factors affect intact features of neuron, glia, vascular systems, and immune systems such as microglia, monocyte-macrophage, and T cell subsets. KO mice of GM3 synthase and those of GD3 synthase did not show apparent morphological changes,[15] suggesting defects of ganglio-series gangliosides or those of b-series gangliosides do not induce apparent neurodegeneration.[15] However, KO of GM2/GD2 synthase resulted in progressive neurodegeneration, suggesting that complex gangliosides are essential for the maintenance of healthy architecture and functions of nervous tissues.[41] Major gangliosides in brain, GD1a and GT1b might play central roles in these functions as shown by series of knockout of ST3Gals,[37,42] while it was hard to be concluded due to their diverse substrate specificity toward glycolipids and glycoproteins. DKO mice showed much intense destruction of brain tissues in cerebella and spinal cords,[19,43] suggesting that there is a quantitative threshold for neuroinflammation to determine whether it can induce actual degeneration of nervous tissues.

## 1.5 Membrane Microdomains and Gangliosides

Various extracellular stimuli are transduced into cells through membrane receptors and surrounding molecules, that form molecular complexes named lipid rafts or glycolipid-enriched microdomain (GEM)/rafts[44,45] (Fig. 2). A number of studies with cell lines and KO mice suggest that ganglioside alteration largely affects the nature of lipid rafts, and thereby control cell signaling and finally cell phenotypes.[46] Structures of individual gangliosides

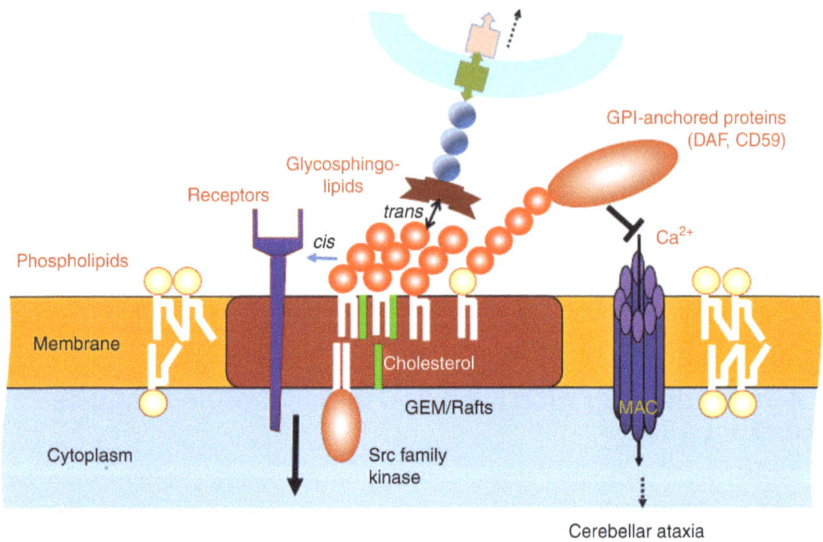


**Fig. 2** Glycosphingolipids play roles in the maintenance of the integrity of nervous tissues. Glycosphingolipids interact with membrane proteins or outward molecules by cis- or trans-reactions, respectively.

have large influence on the quality of regulatory functions of lipid rafts, and eventually directions of cell fates. Therefore, relatively simple examination for the intactness of lipid rafts has been performed by determining intracellular localization of raft-residing molecules, glycosphingolipids, and other membrane molecules that are frequently recruited to lipid rafts depending on cellular situation.[47–49] Antibody staining of these molecules has been also an effective approach to determine their colocalization on the same microdomains. Furthermore, verification of physical association of these molecules on the membrane of living cells seems to be essential and prerequisite to investigate roles of gangliosides and their cooperating molecules on the cell surface.[50]

## 2. ABNORMAL PHENOTYPES OF KNOCKOUT MICE AND GRADUAL SEVERITY OF INFLAMMATION

Recently, we reported novel roles of gangliosides in the regulation of neuroinflammation, and of complement system.[51] During analysis of DKO phenotypes and their mechanisms, we compared complement activation, inflammatory reaction, and disruption of GEM/rafts among various KO mice of ganglioside synthases, that is, GD3 synthase KO, GM2/GD2 synthase KO, DKO as well as wild type (WT) mice. We found that gene expression of complement-related genes was generally up-regulated, and astrocyte proliferation and microglia infiltration occurred with gradual severity depending on degree of defects in ganglioside compositions. Correspondingly, genes of inflammatory cytokines underwent up-regulation with aging. Molecular mechanisms for these inflammatory reactions due to defects of gangliosides were examined with focus on alteration of GEM/rafts.[20]

## 2.1 Neuroinflammation Depending on the Degree of Defects in Ganglioside Expression

Gangliosides have been considered to be involved in the development, differentiation, and function of the nervous system in vertebrates.[52] However, it has been demonstrated that gangliosides mainly play roles in the maintenance and repair of nervous tissues based on abnormal phenotypes observed in gene-engineered animals.[53,54] Generally, mice disrupted of ganglioside synthase genes exhibited neurodegeneration in the nervous systems.[13,14,17,55,56] GM2/GD2 synthase KO mice exhibited progressive neurodegenerative changes with aging, while they showed subtle neurological dysfunctions at birth.[14] In contrast, DKO of GM2/GD2 synthase and GD3 synthase showed definite neurodegeneration even at early stage of life with high severity.[18,57] Although KO of GlcCer synthase resulted in embryonic lethality,[58] conditional KO mice of the enzyme, in which GlcCer synthase was disrupted after birth, also showed neurodegenerative changes.[59] All these results suggested that lack of gangliosides causes defects in maintenance of integrity of nervous tissues, and induces neurodegeneration, while mechanisms for neurodegeneration due to ganglioside defects remained to be clarified.

Among various features of inflammation, abnormal proliferation of astrocytes and accumulation of microglia were markedly and characteristically formed in the cerebellum of ganglioside-lacking mice.[19] Moreover, inflammatory reactions were confirmed by immunohistostaining of astrocytes with antibody for glial fibrillary acidic protein (GFAP), and microglia with antibody reactive with F4/80. GFAP+ cells definitely increased in DKO mice at 15weeks after birth and further increased with aging. In 50-week-old mice, GFAP+ cells increased in the GD3 synthase KO mice and GM2/GD2 synthase KO mice also. Microglia were also accumulated in the DKO mice at 15 weeks after birth, and the accumulation increased with aging.

This microglia accumulation was also found in GD3 synthase KO and GM2/GD2 synthase KO mice at 50 weeks after birth.[19]

As for inflammatory cytokines, we measured mRNA expression levels of inflammatory cytokines in the cerebella with q-RT-PCR. Expression of *IL-1β* and *TNFα* genes was up-regulated in the DKO mice (Fig. 3). Expression levels of these genes tended to increase with aging in the DKO mice, but expression levels of cytokine genes were not so different between the WT and single gene KO mice. These results suggested that the cerebella of DKO mice underwent serious inflammation leading to neurodegeneration, but those of single gene KO mice underwent milder inflammation or gliosis than of DKO mice.

## 2.2 Roles of the Complement System in the Inflammatory Reactions

In the analyses of gene expression profiles between the DKO and WT mice, we found that complement-related genes were generally up-regulated in DKO cerebella. These results suggested that there is extensive consumption of complement components based on complement activation.[60] Indeed, there were deposits of a complement component, C1q in cerebellum tissues in DKO, but also in GM2/GD2 synthase KO. Furthermore, complement activation and subsequent inflammatory reactions were observed in DKO mice, probably due to the disruption of GEM/rafts[44,61] (Fig. 4). Gene expression and protein deposits of the complement system increased in the cerebella, depending on degree of ganglioside-deficiency.[51]

To examine whether complement activation in cerebella of DKO mice exacerbated with aging, we performed q-RT-PCR to clarify chronological

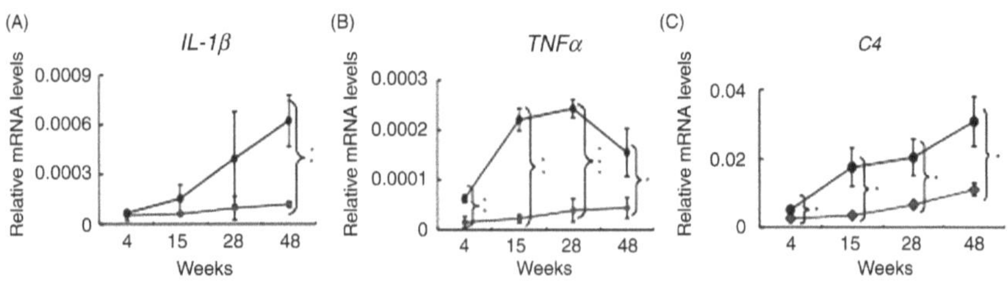


**Fig. 3** Progressive inflammatory reactions with aging as detected in the mRNA expression levels of inflammatory cytokines and complement-related proteins. Results of RT-qPCR were shown in a chronological manner. RNA was extracted from cerebella of DKO mice (GM2/GD2 synthase and GD3 synthase)as shown by –◆–, and by -●- for WT samples.

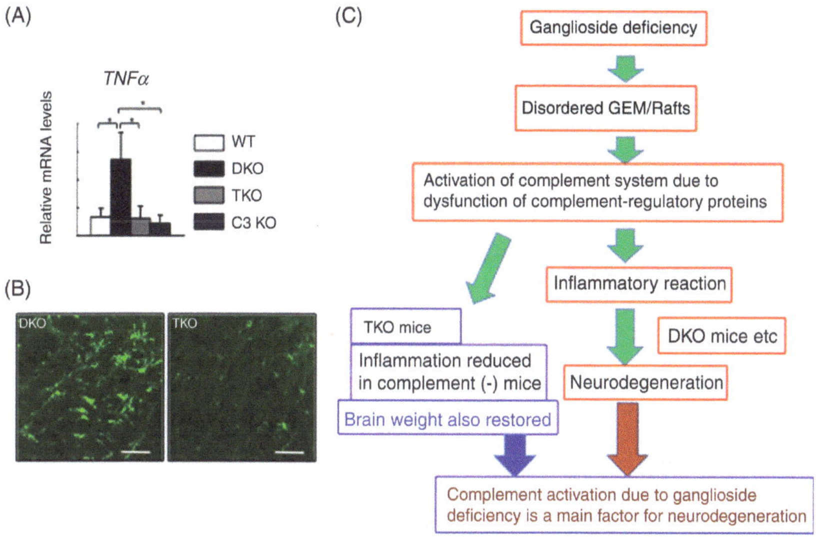


**Fig. 4** A schematic diagram of proposed molecular events leading to inflammation and neurodegeneration in the CNS of ganglioside-deficient mice. Defects in ganglioside composition induce unstable architecture and dysfunction of GEM/rafts, leading to complement activation and inflammatory reaction. Prolonged inflammation results in neurodegeneration. (A) Restoration of TNF-α mRNA level in cerebellum of triple KO mice (GM2/GD2 synthase, GD3 synthase, C3). (B) Reduced microglial assembly in TKO (*right*) compared with DKO (*left*). (C) Flow of brain pathology from ganglioside deficiency to neurodegeneration via inflammatory reactions.

changes in the expression levels of *C1qα*, *C3*, *C4*, using mRNAs from the cerebella of 15- and 50-week-old mice.[19] *C1qα* expression levels clearly increased in the DKO mice between 15 and 50 weeks, and differences between the DKO and the WT became definite with aging. In GM2/GD2 synthase KO mice, expression levels of complement genes were moderately up-regulated.

To clarify roles of complement activation in the inflammatory reactions and subsequent neurodegeneration, triple KO (TKO) mice of GM2/GD2 synthase, GD3 synthase, and C3 genes were generated.[19] Comprehensive analysis of TKO revealed that complement activation is essentially involved in the complement deposits, increased secretion of inflammatory cytokines, and neurodegeneration as shown by prevention of brain weight loss in these mice compared to the DKO mice (Fig. 4).

As for neurological disorders caused by antibodies reactive with gangliosides, that is, Guillain-Barré syndrome and Miller Fisher syndrome, essential involvement of complement system has been reported.[62,63] Inhibitors of factors in the complement system have been tested in disease control and showed fairly good therapeutic effects in mouse models[64,65] and patients.[66]

# 3. INFLAMMATORY REACTIONS AND GANGLIOSIDES

## 3.1 Gliosis and Gangliosides

Glia cells such as astrocytes respond to various damages and stimulation, and exhibit proliferation as previously reported.[25,67,68] Among altered expression of gangliosides on glia cells, increased expression of GD3 with reactive gliosis has been documented in a variety of neurodegenerative disorders and in microglia reactive states.[67]

## 3.2 Morphological Changes of Glia in Single Knockout Mice

In GM2/GD2 synthase KO mice, lacking all complex gangliosides, marked activation and proliferation of astrocytes were detected along with aging.[55] Enlarged and thick processes of astrocytes with densely packed glial filaments were extended among the neuropils and around brain blood vessels, suggesting glial responses to compensate defects of nervous tissues and vascular systems probably including brain blood barrier.[55] Even in GD3 synthase KO mice, mild astrocytosis and microglia assembly were found,[19] suggesting that these glial responses are universal phenomena in ganglioside-deficient nervous tissues.

## 3.3 Ganglioside Expression on Immune Cells

Ganglioside expression on immune cells has been studied by many researchers. Some population of T lymphocytes expressed ganglioside GD3, and its expression increased with immune stimulation.[28] As for GD2, cytotoxic T cells expressed high levels of GD2,[30] corresponding with up-regulation of GM2/GD2 synthase gene in stimulated murine thymocytes.[69] Lack of GM2/GD2 synthase resulted in reduced reaction to IL-2 stimulation and attenuated signaling in murine lymphocytes.[70] Consequently, general activation of T lymphocytes results in expression of GD3 and GD3 synthase gene,[29] while marked activation of T cells, such as TCR-mediated activation[71] brings about expression of GD2.

## 3.4 Gangliosides in Leukemias and Lymphomas

Malignant transformed cells from lymphocytic lineages frequently express characteristic gangliosides. T-cell acute lymphoblastic leukemia cells expressed GD3.[72–74] We also reported GD3 expression in T-ALL cells.[31] In T-lymphotropic virus type I (HTLV-I)-infected cells expressed GD2, and its expression was induced by HTLV-I p40tax protein.[31,32] Recently, expression of 9-O-acetyl GD3 in pre-B-ALL was reported.[75] All these results suggest that expression of unique gangliosides on leukemia cells might be reflection of those gangliosides in normal lymphocytes in particular differentiation and/or activation status.

## 4. AGING, NEURODEGENERATION, AND GANGLIOSIDES

### 4.1 Ganglioside Changes in Central Nervous System With Aging

There have been some reports on the changes in ganglioside profiles during development till senescence in mice[63,76] and rats.[63,77] Yu et al. also reported that ganglioside profiles in brain tissues alter with aging.[78] As for human brain gangliosides, there have been reports on the alteration of ganglioside profiles in human brains with aging.[79–81] Generally, ganglioside contents gradually decreased,[79] and a-series gangliosides tended to decrease mainly in frontal cortex.[80] In turn, b-series gangliosides reduced with aging in cerebellum.[80] The content of membrane lipids diminished continuously up to 90 years of age, when a marked diminution in level of gangliosides and cerebrosides occurred.[81] In a study of 118 subjects, age 20–100 years, gangliosides showed an almost constant concentration between 20 and 70 years of age.[82] But ganglioside pattern showed continuous change with aging with decreasing proportions of GM1 and GD1a. Similar findings to these prior studies were detected using the latest analytic technologies (Orbitrap mass spectrometry[83]). However, no detailed functional relevance of ganglioside profiles with pathological changes has been investigated.

### 4.2 Gangliosides in Alzheimer Disease

In various neurodegenerative diseases, particularly in Alzheimer's disease (AD), an important factor involved in the death of neurons is local inflammation.[84,85] As reported previously,[86] all components involved in the classical complement pathway were detected in nervous tissues, and this pathway has been reported to be activated in AD as shown by fibrillar β amyloid peptide[85] or neurofibrillary tangles.[87] These findings suggest that complement activation and subsequent inflammation are main mechanisms leading to brain damages in AD.

In the last few decades, it has been demonstrated that the complement system is involved in various neurodegenerative disorders such as AD.[88] C1q and other components were deposited in amyloid plaques and neurofibrill tangles,[86,87] resulting in complement activation. mRNA levels of complement components were largely increased at the diseased regions.[89] Furthermore, an inhibitor of C1q attenuated AD pathology,[90] suggesting that C1q exerts a detrimental effect on neuronal integrity, and activation of complement systems plays an essential role in neuroinflammation and subsequent neurodegeneration.[59,84] Intriguingly, activation of the complement system detected in DKO mice appears to play similar roles to those reported in AD.

In addition, the complement system has been considered to play a positive and a negative role[91] in disease control. For example, functions in development,[92,93] and neurogenesis in adults[94] were reported. Roles in the protection of disease-susceptible mutant mice like Tg/Aβ have been also reported.[95] In particular, physiological roles of complements such as the elimination of unnecessary degraded cell components and the refinement of the inflammatory reaction[92] seem to be interesting.

As for roles of gangliosides in AD, roles of GM1 as a triggering seed of Aβ peptide aggregation on the plasma membrane were previously reported.[96,97] Mechanisms of sequential accumulation of Aβ were demonstrated.[98,99] AD model mice were generated in the genetic background of GD3 synthase KO,[100] or GM2/GD2 synthase KO.[101] Milder phenotypes were found in the former case,[100] and worse phenotypes were observed in the latter case.[101] In turn, different changes in ganglioside compositions were found among various transgenic mouse models of AD expressing human amyloid precursor protein,[102] suggesting that these transgenic models would represent valuable tools to further investigate the role of altered ganglioside metabolism in the pathogenesis of AD.

Recently, it was reported that *B4GALNT1* expression enhanced Aβ processing, suggesting that complex gangliosides are involved in the promotion of AD pathology.[103]

## 5. MEMBRANE MICRODOMAINS AND GANGLIOSIDES

Glycosphingolipids have been considered to be enriched in GEM/rafts[104] as shown in Fig. 2. The maximum ganglioside density in GEM/rafts was observed in fully differentiated neurons.[105] A majority of gangliosides are actually localized in lipid rafts, while they appear to disperse from lipid rafts under particular conditions.[106] Marked changes in gangliosides and/or other related molecules such as caveolin-1 may affect intracellular localization of glycosphingolipids due to unknown mechanisms.[107]

## 5.1 Correct Ganglioside Composition is Essential for the Integrity of Glycolipid-Enriched Microdomain/Rafts

Surprisingly, GPI-anchored proteins and GEM/Raft markers dispersed from GEM/rafts in the cerebella of ganglioside-deficient mice. We reported that the GEM/rafts of the cerebella in DKO mice underwent severe disorders by deficiency of gangliosides.[19,20,108] Therefore, we performed immunoblotting of GPI-anchored proteins, that is, decay-accelerating factor (DAF), Thy1.2, NCAM, and GEM/raft markers (flotillin-1 and caveolin-1) to examine whether those molecules were altered in the floating patterns inside/outside of GEM/rafts in various ganglioside-deficient mice.

Flotillin-1 and caveolin-1 largely dispersed from the GEM/rafts in all mutant mice. DAF and NCAM markedly shifted in DKO mice, and DAF tended to move to non-GEM/rafts in GM2/GD2 synthase KO mice. However, Thy1.2 scarcely moved in all mutant mice. The total protein levels of those five molecules were not different among all mouse lines. These results suggested that impairment of GEM/rafts was induced more prominently in the DKO mice compared with single KO mice.

## 5.2 Disruption of Glycolipid-Enriched Microdomain/Rafts Occurred Depending On the Degree of Defects in Ganglioside Composition in Knockout Mice: Influences on Other Membrane Lipids

To investigate the distribution of glycolipids and other lipids in GEM/rafts and non-GEM/rafts in cerebella of ganglioside-deficient mice, we performed thin layer chromatography of acidic glycolipids, neutral glycolipids, phospholipids, and cholesterol using lipids of GEM/rafts and non-GEM/rafts fractions.[20] The majority of gangliosides in cerebella were localized in GEM/raft fractions of all mouse lines including WT, but some of GD1b and GM3 were also localized in non-GEM/raft fractions. GalCer was exclusively localized in GEM/raft fraction in all mouse lines. Phosphatidylethanolamine and phosphatidylcholine were localized in both GEM/raft and non-GEM/raft fractions in all mouse lines. GalCer and phospholipids in GEM/rafts fraction tended to decrease in GM2/GD2 synthase KO and DKO mice compared with other mouse lines. Sphingomyelin and cholesterol were enriched in GEM/rafts, and tended to decrease in GM2/GD2 synthase KO and DKO mice.

## 5.3 Linkages Between Complement Activation and Disrupted Lipid Rafts

As expected from the results of DKO,[19] single gene KO mice also showed dispersion of some of GPI-anchored proteins as GEM/rafts marker proteins. These results suggested that disorder of GEM/rafts in brain tissues might cause a dysfunction of GPI-anchored proteins. As described earlier, amounts of GalCer, phospholipids, and cholesterol in GEM/rafts tended to decrease in GM2/GD2 synthase KO and DKO mice in the distribution analysis of glycolipids and other lipids in GEM/rafts and non-GEM/rafts. These results suggested that changes in ganglioside compositions affected localization of not only proteins such as GPI-anchored proteins and raft markers, but also other lipids.

One of the most important factors affecting complement activation and resulting neuronal damages are complement-regulatory proteins. Indeed, it is well known that expression levels of CD59 are largely suppressed in the diseased areas of AD.[109] DAF is also a critical molecule for maintaining the integrity of tissues.[110] Many of these complement-regulatory proteins are GPI-anchored proteins, and are enriched in GEM/raft fractions.[111] Therefore, disruption of GEM/rafts should have induced dislocation of GPI-anchored proteins and dysfunction of complement regulatory molecules, leading to activation of complement system and inflammation.

## 5.4 Opposite Functions Between Monosialyl and Disialyl Gangliosides

To analyze the effects of individual gangliosides on transduced signals in neuronal cells, a rat pheochromocytoma cell line, PC12 has been used.[112] GM1-overexpressing PC12 cells showed low sensitivity to NGF, showing poor neurite extension and suppressed activation of TrkA/Ras/ERK1/2.[113] In turn, GD3-over-expressing PC12 cells showed increased phosphorylation levels of TrkA and ERK1/2 even before NGF treatment.[114] Similar contrastive effects of the expression of GM1 synthase gene and GD3 synthase gene have been demonstrated in various cancer cell lines.[115,116] All these results were summarized as opposite effects between disialylated gangliosides and monosialylated gangliosides on malignant properties of cancer cells.[117]

## 5.5 Implication of Ceramides in the Intracellular Localization of Gangliosides

There have been studies on ceramide compositions in brain gangliosides, indicating differential expression of long-chain base and fatty acids depending on the developmental stages and anatomical sites.[118] In particular, constant increase of C20:1 long-chain base during development was emphasized.[119] A majority of glycosphingolipids including gangliosides have been considered to be enriched in GEM/rafts as its original nomenclature. However, some of glycosphingolipids appear to be located outside of GEM/rafts. Over expression of a raft marker, caveolin-1 in a melanoma cell line resulted in disordered distribution of ganglioside GD3, that is, some GD3 dispersed from GEM/rafts in caveolin-1-transfectant cells.[107] Mass spectrometry (MS) analysis revealed that GD3 located outside of GEM/rafts contained unsaturated forms of fatty acids. Furthermore, when GM1 synthase was stably expressed in a melanoma cell line, gangliosides confined in GEM/rafts in the parent cell lines dispersed into non-GEM/rafts.[116] Comparison of ceramide structures between gangliosides in GEM/rafts versus those in non-GEM/rafts revealed that gangliosides in non-GEM/rafts contained relatively higher amounts of unsaturated fatty acids.[116] Collectively, ceramide structures largely affect intracellular localization of gangliosides. These events might affect roles of glycolipids in the interaction with glycolipid-recognizing molecules and resulting effects on cell phenotypes.

In addition to intracellular localization of gangliosides, ceramide structures, particularly those of long chain base might be region-specific in the brain.[120] Namely, age-dependent accumulation of C20-ganglioside molecular species in the molecular layer of dentate gyrus of mice has been reported, suggesting distinct function of a specific structure.

# 6. CONCLUDING REMARKS

Changes in ganglioside compositions caused disruption of GEM/rafts, and also alterations in contents of cholesterol and phospholipids. Consequently, the intensities of inflammatory reactions might determine the degree of neuronal damage and neurodegeneration in KO mice. In addition to the effects of altered GEM/rafts on GPI-anchored proteins, other functional proteins expressed on the cell membrane such as neurotrophic factor receptors[53] and integrins[121] might be also affected by disordered

GEM/rafts. The fact that deficiency of both DAF and CD59 did not show marked neurological disorders[122] suggests that not only complement activation due to dysfunction of complement-regulatory proteins, but dysfunction of other membrane proteins due to abnormal GEM/rafts and altered composition of membrane lipids might be also responsible for neurodegeneration in these mutant mice.

Here, predominant roles of glycosphingolipids to regulate the modes of GEM/rafts by interacting with individual ligand proteins should be important, and their mechanisms are now under investigation. Simons et al. reported three phases of raft formation.[123] Important roles of gangliosides in these three phases of GEM/rafts formation need to be clarified. As we described earlier, quantity and/or quality of GEM/rafts should be decreased in ganglioside-deficient mice. Consequently, there should be interaction between glycosphingolipids and their ligand proteins in the vicinity of cell membrane under physiological and pathological situations. Identification of interacting molecules with gangliosides has progressed mainly in the field of tumor biology.[124,125] Enzyme-mediated activation of radical sources-mass spectrometry approach seems to be a very efficient way to detect intramolecular interaction in living cells.[126]

In the recognition of glycosphingolipids by glycolipid-interacting molecules, not only carbohydrate moiety but also lipid moiety, for example, ceramide structures have become more important. In the binding of globotetraosylceramide (Gb4) with LPS receptor, TLR4-MD2, ceramide structure in Gb4 largely affected their binding, that is, only saturated forms of ceramide-containing Gb4 could be bound by MD2 protein.[127] A molecular docking model suggested that unsaturated fatty acid-containing Gb4 was hard to be incorporated in the gap site of MD2. The mechanisms of those interactions are now under investigation with novel approaches including high-resolution single molecule imaging.[128]

## REFERENCES


1. Skaper SD, Leon A, Toffano G. Ganglioside function in the development and repair of the nervous system. From basic science to clinical application. *Mol Neurobiol.* 1989;3:173–199.
2. Wiegandt H. Glycolipids. In: Wiegandt H, ed. *Gangliosides. New York: Elsevier; 1985:199–260.*
3. Yu RK, Tsai YT, Ariga T, Yanagisawa M. Structures. biosynthesis, and functions of gangliosides-an overview. *J Oleo Sci.* 2011;60:537–544.
4. Hakomori SI. Structure and function of glycosphingolipids and sphingolipids: recollections and future trends. *Biochim Biophys Acta.* 2008;1780:325–346.
5. Nagata Y, Yamashiro S, Yodoi J, Lloyd KO, Shiku H, Furukawa K. Expression cloning of beta 1,4 N-acetylgalactosaminyltransferase cDNAs that determine the expression of GM2 and GD2 gangliosides. *J Biol Chem.* 1992;267:12082–12089.

6. Haraguchi M, Yamashiro S, Yamamoto A, et al. Isolation of GD3 synthase gene by expression cloning of GM3 alpha-2,8-sialyltransferase cDNA using anti-GD2 monoclonal antibody. *Proc Natl Acad Sci USA*. 1994;91:10455–10459.
7. Miyazaki H, Fukumoto S, Okada M, Hasegawa T, Furukawa K. Expression cloning of rat cDNA encoding UDP-galactose:GD2 beta1,3-galactosyltransferase that determines the expression of GD1b/GM1/GA1. *J Biol Chem*. 1997;272:24794–24799.
8. Fukumoto S, Miyazaki H, Goto G, Urano T, Furukawa K, Furukawa K. Expression cloning of mouse cDNA of CMP-NeuAc:Lactosylceramide alpha2,3-sialyltransferase, an enzyme that initiates the synthesis of gangliosides. *J Biol Chem*. 1999;274:9271–9276.
9. Kojima Y, Fukumoto S, Furukawa K, et al. Molecular cloning of globotriaosyl- ceramide/CD77 synthase, a glycosyltransferase that initiates the synthesis of globo series glycosphingolipids. *J Biol Chem*. 2000;275:15152–15156.
10. Okajima T, Fukumoto S, Ito H, et al. Molecular cloning of brain-specific GD1alpha synthase (ST6GalNAc V) containing CAG/Glutamine repeats. *J Biol Chem*. 1999;274: 30557–30562.
11. Okajima T, Fukumoto S, Furukawa K, Urano T. Molecular basis for the progeroid variant of Ehlers-Danlos syndrome. Identification and characterization of two mutations in galactosyltransferase I gene. *J Biol Chem*. 1999;274:28841–28844.
12. Okajima T, Yoshida K, Kondo T, Furukawa K. Human homolog of Caenorhabditis elegans sqv-3 gene is galactosyltransferase I involved in the biosynthesis of the glycosaminoglycan-protein linkage region of proteoglycans. *J Biol Chem*. 1999;274: 22915–22918.
13. Okajima T, Fukumoto S, Miyazaki H, et al. Molecular cloning of a novel alpha2,3-sialyltransferase (ST3 Gal VI) that sialylates type II lactosamine structures on glycoproteins and glycolipids. *J Biol Chem*. 1999;274:11479–11486.
14. Takamiya K, Yamamoto A, Furukawa K, et al. Mice with disrupted GM2/GD2 synthase gene lack complex gangliosides but exhibit only subtle defects in their nervous system. *Proc Natl Acad Sci USA*. 1996;93:10662–10667.
15. Okada M, Itoh Mi M, Haraguchi M, et al. b-series Ganglioside deficiency exhibits no definite changes in the neurogenesis and the sensitivity to Fas-mediated apoptosis but impairs regeneration of the lesioned hypoglossal nerve. *J Biol Chem*. 2002;277: 1633–1636.
16. Okuda T, Tokuda N, Numata S, et al. Targeted disruption of Gb3/CD77 synthase gene resulted in the complete deletion of globo-series glycosphingolipids and loss of sensitivity to verotoxins. *J Biol Chem*. 2006;281:10230–10235.
17. Tokuda N, Numata S, Li X, et al. β4GalT6 is involved in the synthesis of lactosylceramide with less intensity than β4GalT5. *Glycobiology*. 2013;23:1175–1183.
18. Inoue M, Fujii Y, Furukawa K, et al. Refractory skin injury in the complex knock-out mice expressing only GM3 ganglioside. *J Biol Chem*. 2002;277:29881–29888.
19. Ohmi Y, Tajima O, Ohkawa Y, et al. Gangliosides play pivotal roles in the regulation of complement systems and in the maintenance of integrity in nerve tissues. *Proc Natl Acad Sci USA*. 2009;106:22405–22410.
20. Ohmi Y, Tajima O, Ohkawa Y, et al. Gangliosides are essential in the protection of inflammation and neurodegeneration via maintenance of lipid rafts: elucidation by a series of ganglioside-deficient mutant mice. *J Neurochem*. 2011;116:926–935.
21. Furukawa K, Ohmi Y, Kondo Y, Ohkawa Y, Tajima O, Furukawa K. Regulatory function of glycosphingolipids in the inflammation and degeneration. *Arch Biochem Biophys*. 2015;571:58–65.
22. Marconi S, De Toni L, Lovato L, et al. Expression of gangliosides on glial and neuronal cells in normal and pathological adult human brain. *J Neuroimmunol*. 2005;170:115–121.
23. Sela BA, Offner H, Konat G, Lev-Ram V, Cohen O, Cohen IR. Immunological expression of gangliosides in multiple sclerosis and in a demyelinating model disease in rabbits. *Adv Exp Med Biol*. 1984;174:441–453.

24. Pilkington GJ, Dunan JR, Rogers JP, Clarke TM, Knott JC. Growth factor modulation of surface ganglioside expression in cloned neoplastic glia. *Neurosci Lett.* 1993;149:1–5.
25. Kawai K, Kuroda S, Watarai S, Takahashi H, Ikuta F. Occurrence of GD3 ganglioside in reactive astrocytes–an immunocytochemical study in the ratbrain. *Neurosci Lett.* 1994;174:225–227.
26. Reynolds R, Wilkin GP. Cellular reaction to an acute demyelinating/remyelinating lesion of the ratbrain stem: localisation of GD3 ganglioside immunoreactivity. *J Neurosci Res.* 1993;36:405–422.
27. Cammer W, Zhang H. Ganglioside GD3 in radial glia and astrocytes in situ in brains of young and adult mice. *J Neurosci Res.* 1996;46:18–23.
28. Welte K, Miller G, Chapman PB, et al. Stimulation of T lymphocyte proliferation by monoclonal antibodies against GD3 ganglioside. *J Immunol.* 1987;139:1763–1771.
29. Yamashiro S, Okada M, Haraguchi M, et al. Expression of alpha 2,8-sialyltransferase (GD3 synthase) gene in human cancer cell lines: high level expression in melanomas and up-regulation in activated T lymphocytes. *Glycoconj J.* 1995;12:894–900.
30. Yoshimura A, Takamiya K, Kato I, Nakayama E, Shiku H, Furukawa K. GD2 ganglioside-specific monoclonal antibody reacts with murine cytotoxic T lymphocytes reactive with FBL-3N erythroleukaemia. *Scand J Immunol.* 1994;40:557–563.
31. Furukawa K, Akagi T, Nagata Y, et al. GD2 ganglioside on human T-lymphotropic virus type I-infected T cells: possible activation of beta-1,4-N-acetylgalactosaminyl- transferase gene by p40tax. *Proc Natl Acad Sci USA.* 1993;90:1972–1976.
32. Okada M, Furukawa K, Yamashiro S, et al. High expression of ganglioside alpha-2,8-sialyltransferase (GD3 synthase) gene in adult T-cell leukemia cells unrelated to the gene expression of human T-lymphotropic virus type I. *Cancer Res.* 1996;56:2844–2848.
33. Schengrund CL. Gangliosides: glycosphingolipids essential for normal neural development and function. *Trends Biochem Sci.* 2015;40:397–406.
34. Skaper SD1. Leon A, Toffano G. Ganglioside function in the development and repair of the nervous system. From basic science to clinical application. *Mol Neurobiol.* 1989;3:173–199.
35. Ngamukote S1. Yanagisawa M, Ariga T, Ando S, Yu RK. Developmental changes of glycosphingolipids and expression of glycogenes in mouse brains. *J Neurochem.* 2007;103:2327–2341.
36. Yu RK1. Macala LJ, Taki T, Weinfield HM, Yu FS. Developmental changes in ganglioside composition and synthesis in embryonic rat brain. *J Neurochem.* 1988;50: 1825–1829.
37. Suzaki K. The pattern of mammalian brain gangliosides. 3. Regional and developmental differences. *J Neurochem.* 1965;12:969–979.
38. Vrbaski SR. Cerebellum gangliosides in postnatal rats. *J Neurochem.* 1980;34: 1032–1033.
39. Trindade VM, Perry ML, Bernard EA. Gangliosides and sialoglycoproteins in hypothalamus of normal, postnatal, and pre- and postnatal protein undernourished rats. *J Neurol Sci.* 1992;107:93–97.
40. Bjartmar C, Trapp BD. Axonal and neuronal degeneration in multiple sclerosis: mechanisms and functional consequences. *Curr Opin Neurol.* 2001;14:271–278.
41. Furukawa K, Ohmi Y, Ohkawa Y, Tajima O, Furukawa K. Glycosphingolipids in the regulation of the nervous system. *Adv Neurobiol.* 2014;9:307–320.
42. Sturgill ER, Aoki K, Lopez PH, et al. Biosynthesis of the major brain gangliosides GD1a and GT1b. *Glycobiology.* 2012;22:1289–1301.
43. Ohmi Y, Ohkawa Y, Tajima O, Sugiura Y, Furukawa K, Furukawa K. Ganglioside deficiency causes inflammation and neurodegeneration via the activation of complement system in the spinal cord. *J Neuroinflammation.* 2014;11:61. 10.1186/1742-2094-11-61.

44. Simons K, Ikonen E. Functional rafts in cell membranes. *Nature.* 1997;387:569–572.
45. Hakomori S, Yamamura S, Handa AK. Signal transduction through glyco (sphingo) lipids. Introduction and recent studies on glyco(sphingo)lipid-enriched microdomains. *Ann N Y Acad Sci.* 1998;845:1–10.
46. Furukawa K, Ohmi Y, Ohkawa Y, et al. Roles of glycosphingolipids in the regulation of the membrane organization and cell signaling in lipid rafts. In: Catala A, ed. *Membrane Organization and Lipid Rafts in the Cell and Artificial Membranes. London: Nova Science Publishers; 2016:129–146.*
47. Hamamura K, Tsuji M, Hotta H, et al. Functional activation of Src family kinase yes protein is essential for the enhanced malignant properties of human melanoma cells expressing ganglioside GD3. *J Biol Chem.* 2011;286:18526–18537.
48. Ohkawa Y, Miyazaki S, Hamamura K, et al. Ganglioside GD3 enhances adhesion signals and augments malignant properties of melanoma cells by recruiting integrins to glycolipid-enriched microdomains. *J Biol Chem.* 2010;285:27213–27223.
49. Hakomori SI. Glycosynaptic microdomains controlling tumor cell phenotype through alteration of cell growth, adhesion, and motility. *FEBS Lett.* 2010;584:1901–1906.
50. Kotani N, Gu J, Isaji T, Udaka K, Taniguchi N, Honke K. Biochemical visualization of cell surface molecular clustering in living cells. *Proc Natl Acad Sci USA.* 2008;105:7405–7409.
51. Furukawa K, Ohmi Y, Ohkawa Y, et al. Regulatory mechanisms of nervous systems with glycosphingolipids. *Neurochem Res.* 2011;36:1578–1586.
52. Schengrund CL. The roles of gangliosides in neural differentiation and repair: a perspective. *Brain Res Bull.* 1990;24:131–141.
53. Furukawa K, Tajima O, Okuda T, Tokuda N, Furukawa K. Knockout mice and glycolipids. In: Kamerling JP, Boons GJ, Lee YC, Suzuki A, Taniguchi N, Voragen AGJ, eds. *Comprehensive Glycoscience. From Chemistry to Systems Biology. Oxford: Elsevier; 2007:149–157.*
54. Kittaka D, Itoh M, Ohmi Y, et al. Impaired hypoglossal nerve regeneration in complex ganglioside-lacking mutant mice: down-regulation of neurotrophic factors and receptors as possible mechanisms. *Glycobiology.* 2008;18:509–516.
55. Sugiura Y, Furukawa K, Tajima O, Mii S, Honda T, Furukawa K. Sensory nerve-dominant nerve degeneration and remodeling in the mutant mice lacking complex gangliosides. *Neuroscience.* 2005;135:1167–1178.
56. Sheikh KA, Sun J, Liu Y, et al. Mice lacking complex gangliosides develop Wallerian degeneration and myelination defects. *Proc Natl Acad Sci USA.* 1999;96:7532–7537.
57. Tajima O, Egashira N, Ohmi Y, et al. Reduced motor and sensory functions and emotional response in GM3-only mice: emergence from early stage of life and exacerbation with aging. *Behav Brain Res.* 2009;198:74–82.
58. Yamashita T, Wada R, Sasaki T, et al. A vital role for glycosphingolipid synthesis during development and differentiation. *Proc Natl Acad Sci USA.* 1999;96:9142–9147.
59. Jennemann R, Sandhoff R, Wang S, et al. Cell-specific deletion of glucosyl- ceramide synthase in brain leads to severe neural defects after birth. *Proc Natl Acad Sci USA.* 2005;102:12459–12464.
60. Truedsson L, Bengtsson AA, Sturfelt G. Complement deficiencies and systemic lupus erythematosus. *Autoimmunity.* 2007;40:560–566.
61. Michel V, Bakovic M. Lipid rafts in health and disease. *Biol Cell.* 2007;99:129–140.
62. McGonigal R, Rowan EG, Greenshields KN, et al. Anti-GD1a antibodies activate complement and calpain to injure distal motor nodes of Ranvier in mice. *Brain.* 2010;133:1944–1960.
63. McGonigal R, Cunningham ME, Yao D, et al. C1q-targeted inhibition of the classical complement pathway prevents injury in a novel mouse model of acute motor axonal neuropathy. *Acta Neuropathol Commun.* 2016;4:23. 10.1186/s40478-016-0291-x.

64. Willison HJ, Halstead SK, Beveridge E, et al. The role of complement and complement regulators in mediating motor nerve terminal injury in murine models of Guillain-Barré syndrome. *J Neuroimmunol.* 2008;201–202:172–182.
65. Halstead SK, Zitman FM, Humphreys PD, et al. Eculizumab prevents anti-ganglioside antibody-mediated neuropathy in a murine model. *Brain.* 2008;131:1197–1208.
66. Davidson AI, Halstead SK, Goodfellow JA, et al. Inhibition of complement in Guillain-Barré syndrome: the ICA-GBS study. *J Peripher Nerv Syst.* 2017;22:4–12.
67. Fredman P, Lekman A. Glycosphingolipids as potential diagnostic markers and/or antigens in neurological disorders. *Neurochem Res.* 1997;22:1071–1083.
68. Lekman A, Fredman PA. New procedure for determining ganglioside GD3 a potential glial cell activation marker in cerebrospinal fluid. *Neurochem Int.* 1998;33:103–108.
69. Takamiya K, Yamamoto A, Yamashiro S, et al. T cell receptor-mediated stimulation of mouse thymocytes induces up-regulation of the GM2/GD2 synthase gene. *FEBS Lett.* 1995;358:79–83.
70. Zhao J, Furukawa K, Fukumoto S, et al. Attenuation of interleukin 2 signal in the spleen cells of complex ganglioside-lacking mice. *J Biol Chem.* 1999;274:13744–13747.
71. Villanueva-Cabello TM, Mollicone R, Cruz-Muñoz ME, López-Guerrero DV, Martínez-Duncker I. Activation of human naïve Th cells increases surface expression of GD3 and induces neoexpression of GD2 that colocalize with TCR clusters. *Glycobiology.* 2015;25:1454–1464.
72. Merritt WD, Casper JT, Lauer SJ, Reaman GH. Expression of GD3 ganglioside in childhood T-cell lymphoblastic malignancies. *Cancer Res.* 1987;47:1724–1730.
73. Merritt WD, Sztein MB, Taylor B, Reaman GH. Immunoreactivity of leukemic lymphoblasts of T-cell and B-cell precursor origin with monoclonal anti-GD3 and anti-GM3 antibodies. *Leukemia.* 1991;5:1087–1091.
74. Siddiqui B, Buehler J, DeGregorio MW, Macher BA. Differential expression of ganglioside GD3 by human leukocytes and leukemia cells. *Cancer Res.* 1984;44:5262–5265.
75. Parameswaran R, Lim M, Arutyunyan A, et al. O-acetylated N-acetylneuraminic acid as a novel target for therapy in human pre-B acute lymphoblastic leukemia. *J Exp Med.* 2013;210:805–819.
76. Ohsawa T. Changes of mouse brain gangliosides during aging from young adult until senescence. *Mech Ageing Dev.* 1989;50:169–177.
77. Hilbig R, Lauke G, Rahmann H. Brain gangliosides during the life span (embryogenesis to senescence) of the rat. *Dev Neurosci.* 1983-1984;6:260–270.
78. Yu RK, Nakatani Y, Yanagisawa M. The role of glycosphingolipid metabolism in the developing brain. *J Lipid Res.* 2009;50:440–445.
79. Segler-Stahl K, Webster JC, Brunngraber EG. Changes in the concentration and composition of human brain gangliosides with aging. *Gerontology.* 1983;29:161–168.
80. Svennerholm L, Boström K, Helander CG, Jungbjer B. Membrane lipids in the aging human brain. *J Neurochem.* 1991;56:2051–2059.
81. Kracun I, Rosner H, Drnovsek V, et al. Gangliosides in the human brain development and aging. *Neurochem Int.* 1992;20:421–431.
82. Svennerholm L, Boström K, Jungbjer B, Olsson L. Membrane lipids of adult human brain: lipid composition of frontal and temporal lobe in subjects of age 20 to 100 years. *J Neurochem.* 1994;63:1802–1811.
83. Sarbu M, Dehelean L, Munteanu CV, Vukelić Ž. Zamfir AD. Assessment of ganglioside age-related and topographic specificity in human brain by Orbitrap mass spectrometry. *Anal Biochem.* 2017;521:40–54.
84. McGeer EG, McGeer PL. The importance of inflammatory mechanisms in Alzheimer disease. *Exp Gerontol.* 1998;33:371–378.
85. Rogers J, Cooper NR, Webster S, et al. Complement activation by beta-amyloid in Alzheimer disease. *Proc Natl Acad Sci USA.* 1992;89:10016–10020.

86. Terai K, Walker DG, McGeer EG, McGeer PL. Neurons express proteins of the classical complement pathway in Alzheimer disease. *Brain Res.* 1997;769:385–390.
87. Shen Y, Lue L, Yang L, et al. Complement activation by neurofibrillary tangles in Alzheimer's disease. *Neurosci Lett.* 2001;305:165–168.
88. Gasque P, Dean YD, McGreal EP, VanBeek J, Morgan BP. Complement components of the innate immune system in health and disease in the CNS. *Immuno-pharmacology.* 2000;49:171–186.
89. Yasojima K, Schwab C, McGeer EG, McGeer PL. Up-regulated production and activation of the complement system in Alzheimer's disease brain. *Am J Pathol.* 1999;154:927–936.
90. Farkas I, Takahashi M, Fukuda A, et al. Complement C5a receptor-mediated signaling may be involved in neurodegeneration in Alzheimer's disease. *J Immunol.* 2003;170: 5764–5771.
91. Tenner AJ. Complement in Alzheimer's disease: opportunities for modulating protective and pathogenic events. *Neurobiol Aging.* 2001;22:849–861.
92. Stevens B, Allen NJ, Vazquez LE, et al. The classical complement cascade mediates CNS synapse elimination. *Cell.* 2007;131:1164–1178.
93. Bénard M, Gonzalez BJ, Schouft MT, et al. Characterization of C3a and C5a receptors in rat cerebellar granule neurons during maturation. Neuroprotective effect of C5a against apoptotic cell death. *J Biol Chem.* 2004;279:43487–43496.
94. Rahpeymai Y, Hietala MA, Wilhelmsson U, et al. Complement: a novel factor in basal and ischemia-induced neurogenesis. *EMBO J.* 2006;25:1364–1374.
95. Wyss-Coray T, Yan F, Lin AH, et al. Prominent neurodegeneration and increased plaque formation in complement- inhibited Alzheimer's mice. *Proc Natl Acad Sci USA.* 2002;99:10837–10842.
96. Yanagisawa K. Role of gangliosides in Alzheimer's disease. *Biochim Biophys Acta.* 2007;1768:1943–1951.
97. Ariga T, McDonald MP, Yu RK. Role of ganglioside metabolism in the pathogenesis of Alzheimer's disease. *J Lipid Res.* 2008;49:1157–1175.
98. Matsuzaki K, Kato K, Yanagisawa K. A beta polymerization through interaction with membrane gangliosides. *Biochim Biophys Acta.* 2010;1801:868–877.
99. Utsumi M, Yamaguchi Y, Sasakawa H, Yamamoto N, Yanagisawa K, Kato K. Up-and-down topological mode of amyloid beta-peptide lying on hydrophilic/hydrophobic interface of ganglioside clusters. *Glycoconj J.* 2009;26:999–1006.
100. Bernardo A, Harrison FE, McCord M, et al. Elimination of GD3 synthase improves memory and reduces amyloid-beta plaque load in transgenic mice. *Neurobiol Aging.* 2009;30:1777–1791.
101. Oikawa N, Yamaguchi H, Ogino K, et al. Gangliosides determine the amyloid pathology of Alzheimer's disease. *Neuroreport.* 2009;20:1043–1046.
102. Barrier L, Ingrand S, Damjanac M, Rioux Bilan A, Hugon J, Page G. Genotype-related changes of ganglioside composition in brain regions of transgenic mouse models of Alzheimer's disease. *Neurobiol Aging.* 2007;28:1863–1872.
103. Yamaguchi T, Yamauchi Y, Furukawa K, et al. Expression of B4GALNT1, an essential glycosyltransferase for the synthesis of complex gangliosides, suppresses BACE1 degradation and modulates APP processing. *Sci Rep.* 2016;6:34505. 10.1038/srep34505.
104. Hakomori SI. Cell adhesion/recognition and signal transduction through glyco- sphingolipid microdomain. *Glycoconj J.* 2000;17:143–151.
105. Prinetti A, Chigorno V, Prioni S, et al. Changes in the lipid turnover, composition, and organization, as sphingolipid-enriched membrane domains, in rat cerebellar granule cells developing in vitro. *J Biol Chem.* 2001;276:21136–21145.
106. Harder T, Simons K. Caveolae, DIGs, and the dynamics of sphingolipid-cholesterol microdomains. *Curr Opin Cell Biol.* 1997;9:534–542.

107. Nakashima H, Hamamura K, Houjou T, et al. Overexpression of caveolin-1 in a human melanoma cell line results in dispersion of ganglioside GD3 from lipid rafts and alteration of leading edges, leading to attenuation of malignant properties. *Cancer Sci.* 2007;98:512–520.
108. Ohmi Y, Ohkawa Y, Yamauchi Y, Tajima O, Furukawa K, Furukawa K. Essential roles of gangliosides in the formation and maintenance of membrane microdomains in brain tissues. *Neurochem Res.* 2012;37:1185–1191.
109. Yang LB, Li R, Meri S, Rogers J, Shen Y. Deficiency of complement defense protein CD59 may contribute to neurodegeneration in Alzheimer's disease. *J Neurosci.* 2000;20:7505–7509.
110. Mason JC, Yarwood H, Sugars K, Morgan BP, Davies KA, Haskard DO. Induction of decay-accelerating factor by cytokines or the membrane-attack complex protects vascular endothelial cells against complement deposition. *Blood.* 1999;94:1673–1682.
111. Kinoshita T, Fujita M, Maeda Y. Biosynthesis, remodelling and functions of mammalian GPI-anchored proteins: recent progress. *J Biochem.* 2008;144:287–294.
112. Greene LA, Tischler AS. Establishment of a noradrenergic clonal line of rat adrenal pheochromocytoma cells which respond to nerve growth factor. *Proc Natl Acad Sci USA.* 1976;73:2424–2428.
113. Nishio M, Fukumoto S, Furukawa K, et al. Overexpressed GM1 suppresses nerve growth factor (NGF) signals by modulating the intracellular localization of NGF receptors and membrane fluidity in PC12 cells. *J Biol Chem.* 2004;279:33368–33378.
114. Fukumoto S, Mutoh T, Hasegawa T, et al. GD3 synthase gene expression in PC12 cells results in the continuous activation of TrkA and ERK1/2 and enhanced proliferation. *J Biol Chem.* 2000;275:5832–5838.
115. Hamamura K, Furukawa K, Hayashi T, et al. Ganglioside GD3 promotes cell growth and invasion through p130Cas and paxillin in malignant melanoma cells. *Proc Natl Acad Sci USA.* 2005;102:11041–11046.
116. Dong Y, Ikeda K, Hamamura K, et al. GM1/GD1b/GA1 synthase expression results in the reduced cancer phenotypes with modulation of composition and raft-localization of gangliosides in a melanoma cell line. *Cancer Sci.* 2010;101:2039–2047.
117. Furukawa K, Ohkawa Y, Yamauchi Y, Hamamura K, Ohmi Y, Furukawa K. Fine tuning of cell signals by glycosylation. *J Biochem.* 2012;151:573–578.
118. Palestini P, Masserini M, Fiorilli A, Calappi E, Tettamanti G. Age-related changes in the ceramide composition of the major gangliosides present in rat brain subcellular fractions enriched in plasma membranes of neuronal and myelin origin. *J Neurochem.* 1993;61:955–960.
119. Valsecchi M, Palestini P, Chigorno V, Sonnino S, Tettamanti G. Changes in the ganglioside long-chain base composition of rat cerebellar granule cells during differentiation and aging in culture. *J Neurochem.* 1993;60:193–196.
120. Sugiura Y, Shimma S, Konishi Y, Yamada MK, Setou M. Imaging mass spectrometry technology and application on ganglioside study; visualization of age-dependent accumulation of C20-ganglioside molecular species in the mouse hippocampus. *PLoS One.* 2008;3:e3232.
121. Toledo MS, Suzuki E, Handa K, Hakomori S. Effect of ganglioside and tetraspanins in microdomains on interaction of integrins with fibroblast growth factor receptor. *J Biol Chem.* 2005;280:16227–16234.
122. Lin F, Salant DJ, Meyerson H, Emancipator S, Morgan BP, Medof ME. Respective roles of decay-accelerating factor and CD59 in circumventing glomerular injury in acute nephrotoxic serum nephritis. *J Immunol.* 2004;172:2636–2642.
123. Simons K, Gerl MJ. Revitalizing membrane rafts: new tools and insights. *Nat Rev Mol Cell Biol.* 2010;11:688–699.

124. Kaneko K, Ohkawa Y, Hashimoto N, et al. Neogenin, defined as a GD3-associated molecule by enzyme-mediated activation of radical sources, confers malignant properties via intracytoplasmic domain in melanoma cells. *J Biol Chem.* 2016;291: 16630–16643.
125. Ohkawa Y, Momota H, Kato A, et al. Ganglioside GD3 enhances invasiveness of gliomas by forming a complex with platelet-derived growth factor receptor α and Yes kinase. *J Biol Chem.* 2015;290:16043–516048.
126. Honke K, Kotani N. Identification of cell-surface molecular interactions under living conditions by using the enzyme-mediated activation of radical sources (EMARS) method. *Sensors (Basel).* 2012;12:16037–16045.
127. Kondo Y, Ikeda K, Tokuda N, et al. TLR4-MD-2 complex is negatively regulated by an endogenous ligand, globotetraosylceramide. *Proc Natl Acad Sci USA.* 2013;110: 4714–4719.
128. Komura N, Suzuki KG, Ando H, et al. Raft-based interactions of gangliosides with a GPI-anchored receptor. *Nat Chem Biol.* 2016;12:402–410.

CHAPTER TEN

# Glycan Chains of Gangliosides: Functional Ligands for Tissue Lectins (Siglecs/Galectins)

**Robert W. Ledeen*[,1], Jürgen Kopitz**, José Abad-Rodríguez[†], Hans-Joachim Gabius[‡,1]**

[*] Department of Pharmacology, Physiology & Neuroscience, Rutgers New Jersey Medical School, Newark, NJ, United States
[**] Ruprecht-Karls-University Heidelberg, Heidelberg, Germany
[†] Hospital Nacional de Parapléjicos (SESCAM), Toledo, Spain
[‡] Ludwig-Maximilians-University Munich Munich, Germany
[1] Corresponding authors. E-mail address: ledeenro@njms.rutgers.edu, gabius@tiph.vetmed.uni-muenchen.de

## Contents



## Abstract

Molecular signals on the cell surface are responsible for adhesion and communication. Of relevance in this respect, their chemical properties endow carbohydrates with the capacity to store a maximum of information in a minimum of space. One way to present glycans on the cell surface is their covalent conjugation to a ceramide anchor. Among the resulting glycosphingolipids, gangliosides are special due to the presence of at least one sialic acid in the glycan chains. Their spatial accessibility and the dynamic regulation of their profile are factors that argue in favor of a role of glycans of gangliosides as ligands (counterreceptors) for carbohydrate-binding proteins (lectins). Indeed, as discovered first for a bacterial toxin, tissue lectins bind gangliosides

*Progress in Molecular Biology and Translational Science*, Volume 156
ISSN 1877-1173
http://dx.doi.org/10.1016/bs.pmbts.2017.12.004

and mediate contact formation (*trans*) and signaling (*cis*). While siglecs have a preference for higher sialylated glycans, certain galectins also target the monosialylated pentasaccharide of ganglioside GM1. Enzymatic interconversion of ganglioside glycans by sialidase action, relevant for neuroblastoma cell differentiation and growth control in vitro, for axonogenesis and axon regeneration, as well as for proper communication between effector and regulatory T cells, changes lectin-binding affinity profoundly. The GD1a-to-GM1 "editing" is recognized by such lectins, for example, myelin-associated glycoprotein (siglec-4) losing affinity and galectin-1 gaining reactivity, and then translated into postbinding signaling. Orchestrations of loss/gain of affinity, of ganglioside/lectin expression, and of lectin presence in a network offer ample opportunities for fine-tuning. Thus glycans of gangliosides such as GD1a and GM1 are functional counterreceptors by a pairing with tissue lectins, an emerging aspect of ganglioside and lectin functionality.

## 1. INTRODUCTION

Initially described as "sugary coating of cells"[1] and thus abundantly found on the external but not the inner side of the plasma membrane,[2] glycan chains of cellular glycoconjugates obviously reside at a prominent site. Here, they may well fulfill a versatile role as biochemical signals. This assumption explains the large interest in the carbohydrates on cell surfaces. But already the first step to investigate glycans, that is, their structural analysis, has turned out to be an extraordinarily demanding endeavor, because "glycoconjugates are much more complex, variegated, and difficult to study than proteins or nucleic acids."[3] This huge variability, however, enables them to store a maximum of information in a minimum of space. This high-density coding capacity guarantees to place a wide panel of molecular messages onto the cell surface. That oligosaccharides harbor a distinct biological meaning is at the heart of the concept of the sugar code.

Like nucleotides and amino acids, monosaccharides are viewed as "letters" of an alphabet of life. The formation of an oligomer can be likened to writing a message, and its information is solely defined by the sequence in nucleic acids and proteins. The mode of the linkage between the "letters" is uniform in these biopolymers. What makes establishing the unsurpassed structural variability of glycans relative to nucleic acids and proteins possible, that is, permutations of (i) anomeric status, (ii) linkage position, (iii) ring size, (iv) site-specific introduction of branch points, and (v) hydroxyl substituents, is the chemical basis of the sugar code.[4–8] Having to define these parameters at each glycosidic linkage and for each unit explains why the study of glycans is so difficult and so attractive. In summary, complex carbohydrates offer a biochemical platform for high-density coding at spatially readily accessible sites of cells.

The discovery of a wide range of diverse oligosaccharides and their presence at sites suited for a role in cellular communication became strong arguments to postulate their involvement in recognition events.[9–17] The intricate enzymatic machinery for glycan biosynthesis, its well-ordered organization as assembly line, especially in the Golgi apparatus, and the impact of various tunable factors such as substrate availability on the product panel ensure to realize the potential of the glycome for dynamic diversity, metaphorically said to be as characteristic as a fingerprint for cells.[18–26] As scaffolds for glycan presentation, Nature uses proteins and lipids, in the latter case especially the ceramide anchor for neutral and negatively charged glycans in man and animals (sphingolipids).[25,27–36] When a glycan chain containing one or more sialic acid residue(s) is attached to the sphingolipid, then the resulting glycolipid is classified as ganglioside.

This term has its origin in the detection of abundant presence of these compounds in the central nervous system, and here mostly in ganglia cells.[37,38] In his words, E. Klenk introduced this term as follows: "Da sie im Zentralnervensystem vorwiegend oder sogar ausschließlich in den Ganglienzellen lokalisiert sind,[37] schlage ich mit Rücksicht auf ihre Glykosidnatur für sie den Namen Ganglioside vor" [Since they are predominantly, or even exclusively, localized in the ganglion cells in the central nervous system, I suggest, with respect to their glycosidation, the name Ganglioside].[38] The glycans of gangliosides, as respective chains of glycoproteins do, equip cell surfaces with signals. This article examines the hypothesis that sugar-encoded information of gangliosides is translated into cellular effects by proteins. Following a brief introduction of structural aspects, the next section describes initial evidence that glycans of gangliosides are ligands for receptors, which "read" these messages.

## 2. GLYCANS OF GANGLIOSIDES: LIGANDS IN BIORECOGNITION

The origin of the term ganglioside, that is, the reference to their abundance in "Ganglienzellen" (neurons) given in detail above[38] and the analogy to J.L.W. Thudichum's cerebrosides found in the white matter of brain that E. Klenk has deliberately drawn,[27] highlights that the nervous system is especially rich in this type of glycoconjugate. Examples of ganglioside glycans are given in Fig. 1, from rather simple (first row) to more

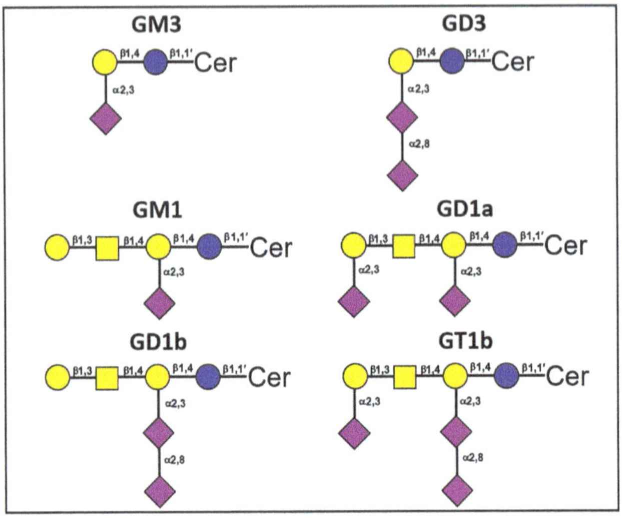


**Fig. 1** Structures of the two simple gangliosides GM3 and GD3 as well as of the four gangliosides with penta- to heptasaccharide chains, which are the most abundant in mammalian brain. Cer, Ceramide; ●, Glc; ○, Gal; ■, GalNAc; ◆, NeuAc.

complex structures (bottom). Biosynthetically, the mono- (M) and disialylated (D) forms are products of stepwise adding one or two sialic acids to lactosylceramide, termed GM3 and GD3, respectively (Fig. 1, first row). In adult human brain, the predominant representatives of this group of glycosphingolipids are the four gangliosides, as shown in Fig. 1 (central and bottom rows). Albeit at (much) lower levels, these and further gangliosides are found, too, in other cell types than neurons and beyond the nervous system. Therefore, they extend a cell's portfolio of glycans beyond those of glycoproteins, and they make neural cells the first choice to initiate work to unveil their physiological significance. In fact, gangliosides add upto about 10% of the total lipid content of the neuronal plasma membrane.[39] Insights obtained by working on gangliosides in the nervous system can thus set a valuable precedent for analogy considerations, as we will illustrate below when discussing the pivotal role of a ganglioside in the communication between effector and regulatory T cells. In this sense, the answers to the question "in what way do these high concentrations and structural complexities serve the spatial needs of the neuron?"[40] will have relevance beyond neurobiology.

Historically, the starting point for routes of in vitro investigations on ganglioside functions was the study of glycosphingolipid storage disorders that result from deficiency in a certain hydrolytic enzyme.[32,33] When degradation is slowed, then the concentration of distinct gangliosides will abnormally be raised, a condition seen to cause neuropathological consequences. Respective electron microscopical monitoring related the pathological excess of gangliosides characteristic for GM1- and GM2-gangliosidoses to triggering ectopic neurites.[41–44] If the glycan structure plays a role in this process, then ganglioside distribution and profile are expected to change in space and time in development. Naturally, the relative quantities among gangliosides indeed are not constant. Striking variations in the ganglioside profile and relative abundance have, for example, been seen in chicken brain and cultures of embryonic neuronal and glial cells.[45,46] To test the emerging idea of gangliosides being molecular switches for cellular events, local concentrations were raised by adding them to cultures of primary neuronal (dorsal root ganglion) and neuroblastoma cells. Confirming the interpretation of the observations on secondary neurites and other aberrant structures made in gangliosidoses, neuritogenesis was triggered.[47,48] Ganglioside GM1, which carries a pentasaccharide shown in Fig. 1,[49] was among the active compounds.[50] Its presence promoted neurite extension in embryonic dorsal root ganglia as well as sympathetic and ciliary ganglia,[51] as it does in murine (N-2A) neuroblastoma cells,[52] resulting in dendrite-like processes in these cells.[53]

In addition to triggering such a complex cellular response, purified gangliosides were discovered to act as matrix for cell adhesion.[54] Embryonic chicken neural retina cells bound efficiently to surface-presented gangliosides, as shown in Fig. 1, especially GT1b, whereas controls with lactosylceramide, sulfatide, or glycerophosphatides were inactive.[55] Murine neuroblastoma cells (N115, S20Y), too, adhered to a ganglioside-presenting plastic surface, especially in the case of GM1.[56] Obviously, certain sugar headgroups of a ceramide anchor have capacity to interact with cell surface determinants in a structurally specific manner. Independent of cell adhesion as read-out, measuring cell association of the glycan was made possible by chemically constructing a labeled marker exposing the ganglioside's glycan as ligand in search of receptors. Such chemically synthesized glycoproteins termed neoglycoproteins had already found application in targeting the conjugate to distinct cells (first hepatocytes), characterizing receptor presence and specificity on cells and histochemical localization of sugar receptors in tissue sections.[57–61] Used as probes to detect binding to cells, the lysoganglioside part of a reporter conjugate with serum albumin as carrier was specifically bound by rat oligodendroglial membranes (ganglioside GT1b)[62,63] and also by human peripheral blood and tumor cells (ganglioside GM1), the latter useful for differential diagnosis by processing tissue sections.[64–66] Altogether, these studies provided clear evidence for a recognition event dependent on the glycan structure of the ganglioside.

When analyzing contact formation between cells, this body of evidence was further extended. The retino-tectal system is a model to test the hypothesis of ganglioside involvement in intercellular recognition. Adhesion of chicken dorsal or ventral retinal cells to appropriate tectal regions is central for specifying retino-tectal projections.[67–69] Interplay of protease-insensitive (ganglioside?) and protease-sensitive epitopes had been suggested to underlie the specificity of the process.[68] Mechanistically, the given lines of evidence converge to view ganglioside glycans as ligand for proteins on cells.

Remarkably, this type of ganglioside-specific binding by a protein was not unknown. Bacterial $AB_5$ toxins, especially the cholera toxin (Ctx) due to its functional pairing with ganglioside GM1, were known to be glycan-specific receptors.[70–73] Ganglioside GD1b, too, has affinity for bacterial toxins,[74] and, discovered later, the glycans, whose structures are illustrated in Fig. 1, are also recognition sites for viral adhesins[75–77]: the pentasaccharide of ganglioside GM1, for example, for simian virus 40 (SV40),[78] the GD1a hexasaccharide for adenovirus (type 37), noro- and sapelovirus adhesins,[79–82] and its GD1b isomer for a BK polyomavirus protein.[82] The productive recognition of a ganglioside's glycan by bacterial toxins, together with the given evidence from cellular neurobiology, shaped the concept that a protein-carbohydrate recognition is responsible for aspects of ganglioside functionality.

Interestingly, in the same period when carbohydrate-mediated adhesion in the retino-tectal system, neuritogenic effects of gangliosides, and their nature as counterreceptors for bacterial toxins were studied, extracts of embryonic chicken brain, spinal cord, and retina, as well as murine neuroblastoma (N-18) cells, were revealed to harbor a special activity: carbohydrate (β-galactoside)-inhibitable agglutination of erythrocytes.[83–85] Using an affinity resin that presents terminal galactose residues, as is the case in cellular glycans, a protein was purified from extracts of the electric organ of *Electrophorus electricus*.[83] Its capacity to bridge cells via glycan binding fits well into the reasoning that "after the genetic code was deciphered, the next important code to solve was the one for cellular recognition",[86] this one presumably a sugar-based code. This finding guides us to the protein side of glycan-dependent recognition, that is, lectins.

## 3. LECTINS: GLYCAN-SPECIFIC RECEPTORS IN BIORECOGNITION

The hemagglutination assay was the classical means to track down activity for erythrocyte crosslinking, first by S. W. Mitchell in the venom of the rattlesnake in 1860,[87] then by H. Stillmark, who studied seeds of *Ricinus communis* in his thesis and called the active principle "ricin".[88] As reviewed in detail elsewhere,[89,90] the ability of these activities in plant extracts to distinguish between the different histo-blood group ABH(0) epitopes on erythrocytes led W. C. Boyd to coin a term that rests on the analogy of reading a message to be selective.[91,92] It is lectin, "from the Latin *lego*, to choose or pick out" or to read.[93] The lectins share ability to "read" the blood group-specific signals with serum antibodies so that they were useful in transfusion medicine for blood-group typing. As happens when working with antibodies, agglutination of erythrocytes is precluded by the presence of a hapten, in this case a cognate carbohydrate. This inhibitory effect of sugars on lectin-dependent crosslinking, first seen for the glucose/mannose-specific concanavalin A in 1936,[94] contributed to the identification of blood-group epitopes as glycans and the structural characterization of these determinants.[95,96] Lectins and ABH(0) blood group-specific antibodies are glycan-binding proteins. Currently, lectins are defined as carbohydrate-binding (glyco)proteins separate from antibodies, enzymes that process the bound glycan and sensor/transport proteins for free mono- to oligosaccharides.[97–99] The occurrence of more than a dozen protein folds among lectins[100,101] and of sequence divergence from an ancestral gene leading to large families in most cases[102–108] definitely excludes concerns that the complexity seen on the level of the glycome is not matched on the lectin side. Within a family, the diversification encompasses the carbohydrate recognition domain (CRD) and the topological mode of its presentation, furthermore presence of single nucleotide polymorphisms for individual proteins. As illustrated in Fig. 2, CRDs can be presented in various structural contexts (please see legend for detailed descriptions and functional implications). The spatial interplay between these features of endogenous lectins with those of cognate glycans, often arranged in clusters,[109] likely is a key factor to make their functional pairing possible and specific.[110] The resulting association of a certain cellular receptor (lectin) and its counterreceptor (glycan on distinct glycoconjugate)[101] is the starting point for the lectin-dependent physiological process, for example, adhesion or signaling (for an overview on triggered activities, please see Table 5 in Ref. 101). The noted surface localization of gangliosides, along

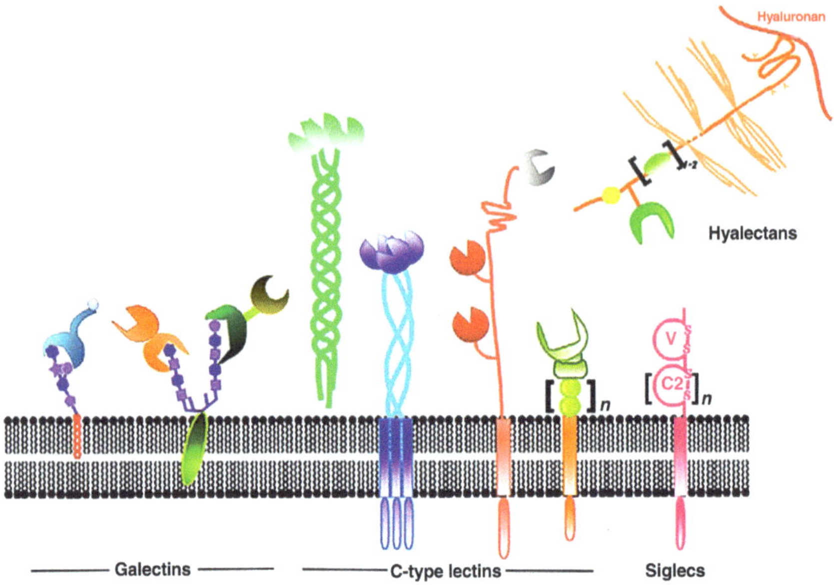


**Fig. 2** Illustration of the strategic ways how carbohydrate recognition domains (CRDs) in animal and human lectins are positioned to reach optimal ligand selection (e.g. to separate self from nonself glycan profiles in innate immunity) and topological complementarity. From left to right, the CRDs display in the three subtypes within the galectin family (chimeric, proto-type. and tandem-repeat-type arrangements binding to a ganglioside or a branched complex-type N-glycan without or with terminal α2,3-sialylation), the presentation of CRDs (C-type or fibrinogen-like domain) in serum and surfactant collectins or ficolins connected to their collagenous stalks and the noncovalent association of lectin sites in transmembrane C-type lectins by α-helical coiled-coil stalks (e. g. asialoglycoprotein and Kupffer cell receptors, the scavenger receptor C-type lectin, CD23, DC-SIGN, or DC-SIGNR) are given. Similar to tandem-repeat-type galectins the C-type family of lectins also has a branch of members with this design, that is, immulectins-1, -2 and -3. Next, the tandem-repeat display in the mannose-specific macrophage receptor (also found on dendritic cells, hepatic endothelial cells, kidney mesangial cells, retinal pigment epithelial cells, and tracheal smooth muscle cells) and the related C-type lectin Endo180 with eight domains as well as in the cation-independent P-type lectin with 15 domains is presented. Capacity for sugar binding is confined to only few domains as depicted. The occurrence of lectin activity for GalNAc-4-SO4-bearing pituitary glycoprotein hormones in the cysteine-rich domain, a member of the β-trefoil fold family with one (QxW)3 domain in the N-terminal section of the macrophage mannose receptor (amino acids 8–128), which is linked via a fibronectin-type-II-repeat-containing module to the tandem-repeat section, is also included into the schematic drawing for these lectins with more than one type of CRD per protein chain. Moving further to the right side, the association of a distal CRD in selectins (attached to an epidermal-growth-factor (EGF)-like domain and two to nine complement-binding consensus repeats) or in the siglec subfamily of I-type lectins using 1–16 C2-set immunoglobulin-like units as spacer equivalents to let the (V-set) CRD reach out to contact ligands and to modulate

with evidence of their active roles (listed above up to binding a toxin) let us anticipate that glycans of gangliosides are counterreceptors for tissue lectins.

## 4. FUNCTIONAL PAIRING OF GANGLIOSIDES WITH TISSUE LECTINS

The presence of a sialic acid residue in a glycan can either constitute a major contact point or preclude lectin binding by blocking access to an essential site.[76,104,110,112–115] Intriguingly, the enzymatic removal of a sialic acid from a glycan chain can thus be a switch off/on signal for different types of lectins. The remodeling of the glycan chain of a ganglioside by a sialidase, as shown in Fig. 3 (top), can therefore change its functional meaning fundamentally into both directions at the same time. To turn this attractive concept of a strategic interplay into cellular reality the presence of sialic acid-specific lectins will be required. The mentioned evidence for binding of labeled probes, which expose glycans of gangliosides, to membranes and the protease sensitivity of this property of cells are indications for their occurrence. As seen in Fig. 2 (right), a lectin family is known that has a domain with explicit specificity to sialic acid. Endowed with a V-set immunoglobulin (Ig)-like domain of this property as common feature (Fig. 2, right), the *s*ialic acid-binding *i*mmuno*g*lobulin-like *lec*tin*s* (siglecs) share an Arg residue on this domain's F-strand to build the salt bridge to the carboxylate of the sialic acid. Within this family, the diversification of glycan specificity in different siglecs enabled them to completely cover the three linkage types of sialic acid in glycans, that is, α2,3, α2,6, and α2,8, and to reach affinity to sialylated epitopes of chains of glycoproteins and gangliosides.[104,116–118]

capacity to serve in *cis*- or *trans*-interactions on the cell surface is shown. The force-dependent alterations of the topological arrangement of the two distal domains in selectins accounts for catch bonds of selectins, a canonical *i*mmunoreceptor *t*yrosine-based *i*nhibitory *m*otif (ITIM) together with a putative tyrosine-based motif is frequently present in the intracellular portion of siglecs. C2-Set domains linked to fibronectin-type-III repeats establish the extracellular section of the I-type lectins L1 and neural cell adhesion molecule (NCAM). In the matrix, the modular proteoglycans (hyalectans/lecticans: aggrecan, brevican, neurocan, and versican) interact (i) with hyaluronan (and also link protein) via the link-protein-type modules of the N-terminal G1 domain (and an Ig-like module), (ii) with receptors binding to the glycosaminoglycan chains in the central region, and (iii) with carbohydrates or proteins (fibulins-1 and -2, and tenascin-R) via the C-type lectin-like domain flanked by EGF-like and complement-binding consensus repeat modules. *(For details, please see Gabius H-J. Animal and human lectins. In: Gabius H-J, ed. The Sugar Code. Fundamentals of Glycosciences. Weinheim, Germany: Wiley-VCH; 2009:317–328,[111] with permission.)*

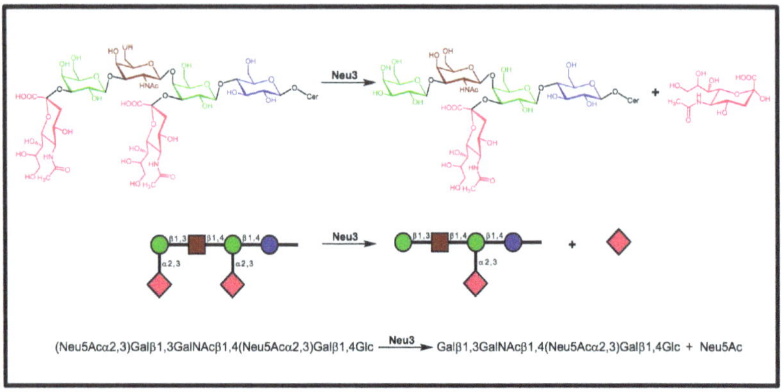


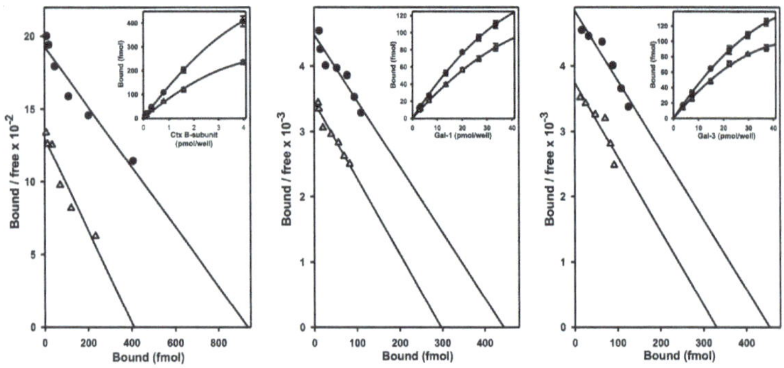


| Type of lectin | Inhibitor presence | $K_D$ [nM] | $B_{max}$ x 10$^4$ |
|---|---|---|---|
| Gal-1 | - | 980 (+/-21) | 260 (+/-39) |
| | + | 850 (+/-19) | 180 (+/-31) |
| Gal-3 | - | 940 (+/-25) | 270 (+/-37) |
| | + | 880 (+/-23) | 200 (+/-30) |
| Ctx B-subunit | - | 49 (+/-5) | 560 (+/-68) |
| | + | 32 (+/-4) | 250 (+/-45) |

**Fig. 3** Enzymatic glycan remodeling by the cell surface ganglioside sialidase (neuraminidase 3, Neu3, top). Cleavage of the α2,3-glycosidic bond at the chain terminus removes one sialic acid residue from the glycan chain, turning ganglioside GD1a into GM1 (please see Fig. 1, middle). Assessment of carbohydrate-inhibitable human neuroblastoma cell (SK-M-MC) surface binding of $^{125}$I-labeled Ctx B-subunit (left panel), Gal-1 (center), and Gal-3 (right panel) by Scatchard analysis (inset: binding curve). Experiments were performed in parallel with cells cultured in the absence (•) and in the presence of 250 μM 2-deoxy-2,3-dehydro-N-acetylneuraminic acid (NeuAc2en) (Δ), each point representing the mean of the three independent measurements. Summary table of dissociation constants ($K_D$) and number of bound proteins per cell at saturation ($B_{max}$) is given (bottom).

## 4.1 Siglecs as Ganglioside Receptors

Siglec-1 (sialoadhesin) has a total of 17 Ig-like domains, which position the distal V-set module far away from the membrane surface, ready for *trans*-contacts, as shown in Fig. 2 (right). Ganglioside GM3 (shown in Fig. 1) on viral (HIV) surfaces is a critical determinant for virus capture by monocyte-derived dendritic cells and ensuing *trans*-infection of lymphocytes.[119–122] Gangliosides such as GM3, GD1a, GD1b, and GT1b are siglec-1 ligands and mediate attachment of cells to a matrix, when this lectin is present on the cell surface.[123,124] Highly sialylated ganglioside chains (with two or more sialic acid moieties and the α2,8-linkage between two sialic acids) are bound well by other siglecs, that is, siglecs-5, -7, -8, -9 and -10.[125] The extensive contact pattern of siglec-7's V-set module with the GT1b heptasaccharide defined in the crystal structure of the complex underlines the involvement of intricate structural recognition beyond the sialic acid.[126] Because this siglec connects target cell binding via a ganglioside such as GD3 to outside-in signaling, activating the modulatory role of the intracellular canonical *i*mmunoreceptor *t*yrosine-based *i*nhibitory *m*otif (ITIM), its functional pairing with tumor cell surface counterreceptors downregulates the cytolytic activity of natural killer (NK) cells,[127] an example for functional consequences beyond binding. The same holds true for a siglec found in the nervous system and also referred to as myelin-associated glycoprotein (MAG).

Combination of a bridging function with a negative effector mechanism, as just described, is seen for siglec-4 (MAG). Its binding to *trans*-counterreceptors contributes to axon-myelin stability and has a negative influence on axon regeneration.[128–130] Particularly the gangliosides GT1b and GD1a are functional binding partners,[123,131–133] corroborating the evidence for a respective receptor obtained by measuring specific binding of the probe with lysoganglioside GT1b in central nervous system myelin.[62] Blocking this pairing to reverse siglec-4-mediated inhibition of axon outgrowth can be accomplished by the presence of a high-affinity glycan-derived synthetic inhibitor.[134,135]

An alternative route toward the same aim is to remove siglec-4's docking site by enzymatic desialylation. A respective enzyme is present on the cell surface, and it is regulated in response to cell signaling. Hereby, higher sialylated gangliosides are turned into GM1 and GD3 into lactosylceramide.[136] The cell surface ganglioside sialidase (neuraminidase 3, Neu3) activity is subject to a remarkable degree of modulation: by studying murine or human neuroblastoma and rat cerebellar granule cells, its activity was found to increase with cell density.[137–139] Glycan remodeling, according to these observations, is linked to growth control, and this observation prompted the question on the identity of "the ganglioside (or glycoprotein?) substrate(s), GM3 or other, that are transducing the action of the sialidase".[139] These results also pointed to ligand properties of the products of Neu3 activity to a class of lectins different from siglecs. In functional terms, the shift from higher sialylated gangliosides such as GD1a to GM1 and the loss of GM3 caused by this enzyme, as well as the association of its cell surface presence to expression of differentiation markers and to growth control in human neuroblastoma (SK-M-MC) cells,[140] are as intriguing as a $Ca^{2+}$ influx-dependent stimulation of neurite outgrowth in three mouse/rat neuroblastoma lines after treatment with a bacterial (*Clostridium perfringens*) sialidase, which is a response blocked by an antibody against ganglioside GM1.[141] When exposing the human neuroblastoma cells to bacterial (*Vibrio cholerae*) sialidase, the same effects were triggered as typical for endogenous Neu3 activity, that is, slowing down proliferation and increasing intracellular acetylcholinesterase.[140]

In sum, neuraminidase activity in normal and malignant neural cells lets counterreceptors for the siglecs vanish and ganglioside GM1 takes their place. Should that be bioactive as ligand, then receptors different from siglecs will be engaged in the molecular rendezvous. The documented developmental regulation, presence in the nervous system, and ability to bridge cells, as well as their specificity to β-galactoside cores, point to *ga*lactoside-binding *lectins* (galectins) as candidates.[83,105,142,143]

## 4.2 Galectins as Ganglioside Receptors

Galectins are a family of β-sandwich proteins that occur in mammals in three structural modes for CRD presentation, that is, as homodimer, as single CRD with an N-terminal tail, and as covalently linked combination of two different CRDs (Fig. 2, left). Its broad expression profile and the ease to obtain relatively large quantities engender that homodimeric galectin-1 (Gal-1) was a popular study object. Initial experiments with glycolipids, ganglioside mixtures, and bead-adsorbed lysoganglioside GM1 revealed binding of mammalian Gal-1 to glycan chains of sphingolipids.[144–146] This member of the galectin family is a homodimer suited for crosslinking ligands (Fig. 2, left). Studies on force-driven dissociation of its complexes with glycans[147] and on bridging of glycodendrimersomes using the wild-type and covalently linked variant proteins[148,149] revealed properties in line with an engagement of the homodimer in *cis*-crosslinking on the cell surface and transient *trans*-bridging between surfaces. Thus, the galectin is structurally suited for initiating contact (*trans*) and signaling (*cis*), hereby eliciting post-binding effects, if a counterreceptor is available for functional pairing, and enzymatic desialylation can make this happen.

As initially noted by measuring hemagglutination activity in extracts of murine (N-18) neuroblastoma cells,[83] Gal-1 is expressed also by the human SK-N-MC neuroblastoma line.[150] Notably, its presence on the cell surface increases in concert with the Neu3 activity.[151] Accessibility of GM1 glycans on the cell surface could be ascertained by the GM1-specific Ctx B-subunit, and human Gal-1 used as probe for detecting counterreceptors, too, led to specific binding (Fig. 3). The extent of binding was decreased by blocking Neu3 activity (Fig. 3), and completely abolished by cognate glycan.[150] Both experiments indicate genuine involvement of GM1, when the endogenous lectin interacts with the neuroblastoma cells. This recognition necessarily required carbohydrate-binding capacity of the lectin (its E71Q mutant failed to bind to cells), was drastically reduced by inhibiting glucosylceramide synthesis and thus ganglioside presence, and was sensitive to perturbing the integrity of microdomains.[152] As cell binding to GM1-presenting beads was impaired by antibody against Gal-1,[150] incorporation of ganglioside GM1 conferred ligand property for Gal-1 binding to lipid monolayers.[153] Assessment of the electron density distribution revealed an orientation of the lectin with its long axis in the surface plane, as expected for counterreceptor crosslinking[153] When combined, these observations make a convincing case for a glycan reprogramming on the level of gangliosides as one part of a molecular switch. On the protein side, Gal-1 fulfills the requirements to be the missing link between the neuraminidase activity and the cell growth control. This line of reasoning was incentive to test what happens when the lectin is added to these cells.

When cultures were treated accordingly, cell proliferation was reduced (Fig. 4). Thus this lectin is a functional ganglioside GM1 receptor in these cells.[151] Other homodimeric members of this lectin family, that is, galectins-2 and -7, too, exhibit this GM1-dependent activity.[154,155] On human T cells, these three galectins induce apoptosis but with differing profiles of caspase activation, arguing against mechanistic redundancy among homodimeric galectins.[156] Turning to normal neuronal cells, GM1–Gal-1 pairing is of physiological significance. The galectin serves as active link to GM1 increases by neuraminidase action in axon-like neuritogenesis in murine (NG108-15) neuroblastoma cells by an autocrine mechanism of action.[157] Absence of GM1 in mutant cells and neurons of ganglio-series KO mice caused marked retardation.[157] Immunohistochemical detection of this (and other) galectin

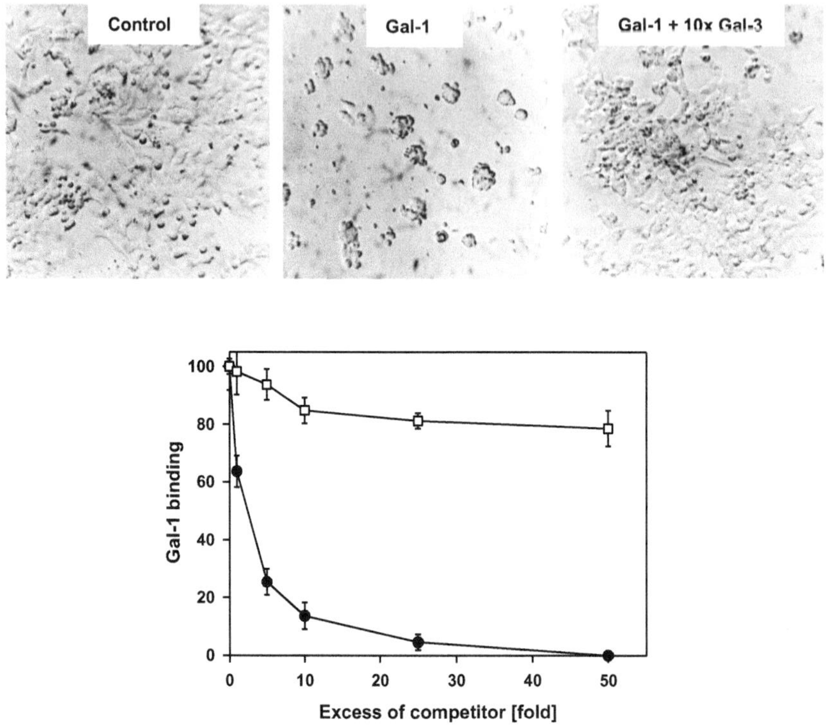


**Fig. 4** Effect of Gal-1 on growth of human neuroblastoma (SK-N-MC) cells in the absence and in the presence of a 10-fold quantity of Gal-3 (top). Cells were plated at a density of $10^4$ cells/well, allowed to attach for 16 h and then treated for 48 h with lectin-free or lectin-containing solution (Gal-1 at 125 μg/ml) prior to recording the microphotographs. Competition experiments of cell surface binding of $^{125}$I-labeled Gal-1 were performed with unlabeled full-length Gal-3 (●) and with proteolytically truncated Gal-3 (□) (bottom). Cells had been grown for five days to reach confluency before the mixture was added. The results are given as means ± SD based on four independent measurements.

(s) has been intensely studied and still is, recently, for example, detecting Gal-1 in interneurons that may secrete the lectin during pathfinding in the murine hippocampus.[158–162] Its signaling led to a $Ca^{2+}$-influx by opening TRPC5 channels, shown in Fig. 5, which bears resemblance to previously delineated responses to Ctx B-subunit binding.[163] The involvement of $\alpha_5\beta_1$-integrin, a Gal-1 counterreceptor in anoikis induction in carcinoma cells,[164–166] leads to autophosphorylation of focal adhesion kinase and activation of phospholipase Cγ and phosphoinositide 3-kinase, when applying Ctx B-subunit and Gal-1 as elicitors.[157,163] Beyond neurons, this

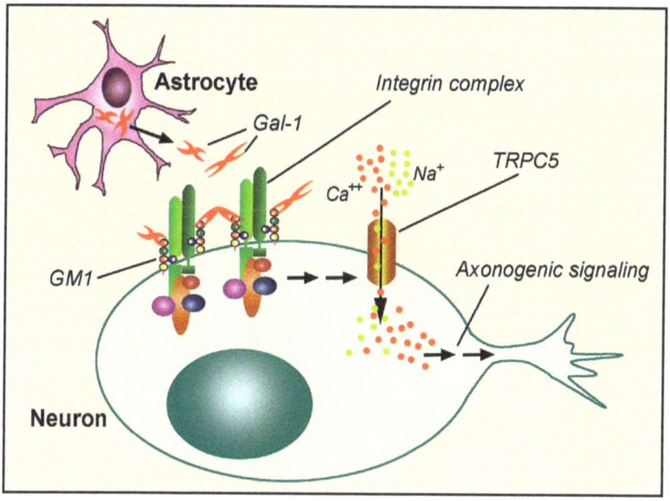


**Fig. 5** Schematic illustration of the downstream effects of functional pairing of Gal-1 with the pentasaccharide of ganglioside GM1 on murine neuroblastoma (NG108-15) cells leading to axon-like neurite outgrowth. The enzymatic conversion from GD1a to GM1 (please see Fig. 3 (top)) and the availability of Gal-1 enable *cis*-cross-linking of GM1-integrin complexes by the lectin homodimers. This lattice formation induces postbinding signaling. It involves autophosphorylation of focal adhesion kinase and activation of phospholipase Cγ and phosphoinositide 3-kinase and induces a transient increase in intracellular $Ca^{2+}$ level via opening of TRPC5 channels. *(For details, please see Wu G, Lu ZH, André S, Gabius H-J, Ledeen RW. Functional interplay between ganglioside GM1 and cross-linking galectin-1 induces axon-like neuritogenesis via integrin-based signaling and TRPC5-dependent* $Ca^{2+}$ *influx. J. Neurochem. 2016;136:550–563.*[157]*)*

orchestration of GD1a-to-GM1 conversion and Gal-1 presence for functional pairing has been observed in the communication between regulatory and effector T cells in autoimmune disorders.[167–169]

The presented evidence underlines the relevance of shifts in the profile of gangliosides by a neuraminidase (endogenous and exogenous) for a galectin. Axon regeneration is driven by Neu3 sorting and activity, which are blocked in the central nervous system.[170] Ectopic overexpression of the enzyme[171] or using bacterial sialidase toward the same end,[172,173] however, restored this capacity, removing the counterreceptor for siglec-4 (GD1a) and making the functional pairing of GM1 with Gal-1 possible. In addition to an effector route via Gal-1, GM1's glycan can associate with N-glycans of the tyrosine kinase-type nerve growth factor receptor TrkA to enhance its autophosphorylation[174] and signaling via modulating actin assembly.[175] The molecular basis of the initial contact is similar to the binding of GM3's glycan to GlcNAc-terminated

N-glycans of the epidermal growth factor receptor[176] and highlights the physiological significance of carbohydrate-carbohydrate recognition.[177] Both effector routes involving glycan binding using GM1, clearly a multifunctional molecule,[178] are illustrated in Fig. 6. In addition, the binding of GM1 to laminin-1 via β-sandwich-type laminin G-like domains, another endogenous GM1 receptor,[179–181] can promote neurite outgrowth via initiating formation of a focal microdomain rich in TrkA and $\beta_1$-integrin, leading to signaling via the Src tyrosine kinase Lyn and also Akt and MAPK.[182]

Having herewith documented the capacity of Gal-1 as functional GM1 receptor in growth control and axonogenesis, a particular property of galectins is considered next. In contrast to siglecs with their common combination of V- and C2 set Ig-like domains, mammalian galectins have three structural types of CRD presentation, as illustrated in Fig. 2 (left). Measurements with the SK-N-MC cells were instrumental to discover (i) functional antagonism between galectins and (ii) variability in receptor activity for ganglioside GM1.

Galectin-3 (Gal-3, also known as MAC-2 antigen[183,184]) is a family member with three functional sections, that is, the CRD, a region with nontriple helical collagen-like repeats and an N-terminal peptide with two sites for serine phosphorylation.[185] Its affinity to the neuroblastoma cell surface resembles that of Gal-1 and binding is reduced by Neu3 inhibition (Fig. 3). Stepwise truncation of the repeat section revealed a threshold required to maintain this level of affinity, complete truncation severely impairing binding capacity.[186] Despite cell binding full-length Gal-3 does not trigger growth control.[151] When present together with Gal-1 in the medium, it competitively interferes with binding of Gal-1 to cells (Fig. 4, bottom) and thereby attenuates its negative effect on cell growth (Fig. 4, top). In comparison, the proteolytically truncated form was only weakly inhibitory (Fig. 4, bottom). Functional antagonism between the two galectins by competition for the counterreceptor has also been detected for N-glycosylation of an integrin in carcinoma growth regulation by a tumor suppressor.[187]

The third type of structural design of galectins is the tandem-repeat-type architecture with a heterodimer display, a covalent connection of two CRDs by a linker (Fig. 2, left). Galectin-4 (Gal-4) is expected to have cross-linking ability, sequence changes relative to Gal-1 made it uncertain whether GM1 binding is maintained. This galectin binds moderately to GM1, weakly to GD1a, and strongly to glycosphingolipids with terminal 3-O-sulfated galactose.[188,189] When associating to the neuroblastoma cell surface, Ctx B-subunit presence did not reduce binding, as shown in Fig. 7, and no effect on proliferation was seen.[190] In contrast, as noted above, the toxin effectively

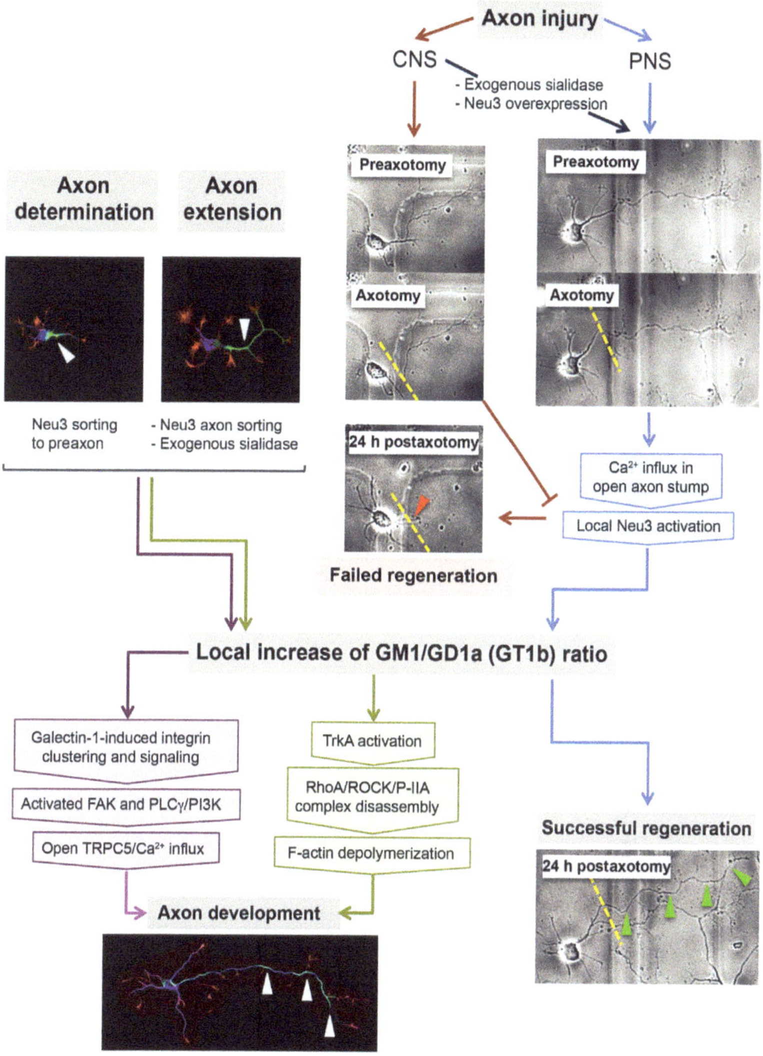


**Fig. 6** Neu3-dependent stimulation of axon development and regeneration. This enzyme's sorting to single neurites of undifferentiated neurons is a means to specify axonal fate and to induce axon extension. Axon regeneration after injury can be installed by exogenous (bacterial) sialidase or ectopic Neu3 expression in central nervous system neurons or, in part, by a $Ca^{2+}$ influx-dependent local Neu3 activation on peripheral nervous system axons. The increase in the GM1/GD1a ratio is sensed by Gal-1 and turned into axonogenesis (please see Fig. 5) and initiates potent signaling by TrkA and small GTPase RhoA (ROCK/profillin-IIA) complex disassembly required for F-actin depolymerization in axon development and in regeneration.

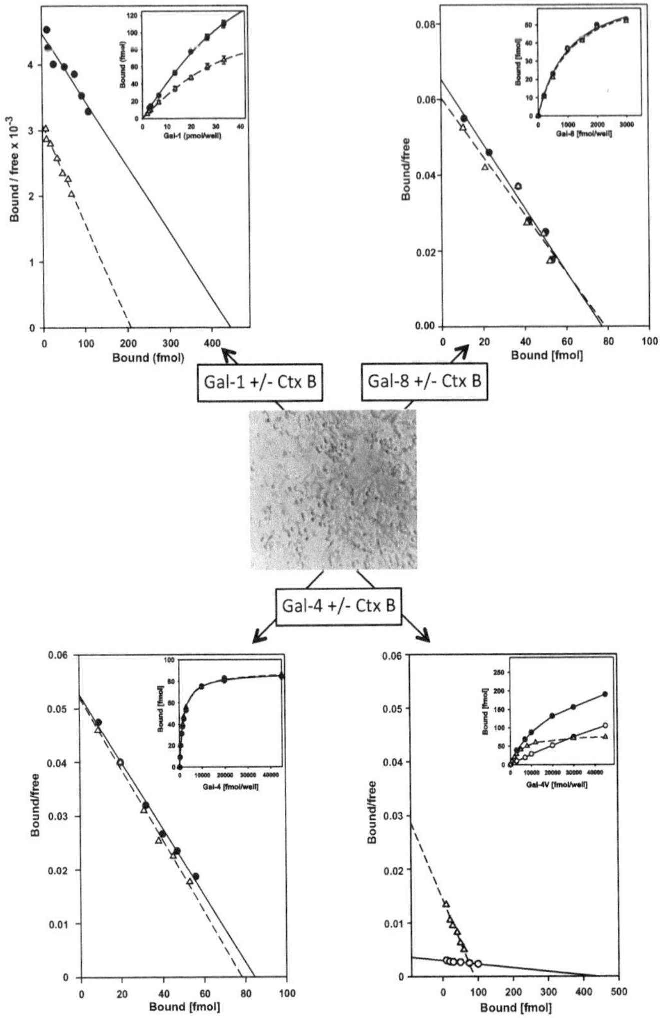


| Type of lectin | $K_D$ (nM) | $B_{max}$ x $10^4$ |
|---|---|---|
| **Gal-4** | **128** (+/-16) | **45** (+/-5) |
| **Gal-4V** | | |
| Ctx B-sensitive | **147** (+/-22) | **258** (+/-29) |
| Ctx B-insensitive | **63** (+/-12) | **57** (+/-7) |
| **Gal-8** | **118** (+/-12) | **46** (+/-8) |

**Fig. 7** Binding of natural and engineered galectins to human neuroblastoma cells and the effect of presence of the Ctx B-subunit. Cells were cultured for five days to reach confluency (microphotograph in center) and then used for binding assays, as described in the legend for the bottom part of Fig. 4. The GM1-specific bacterial toxin Ctx B-subunit

decreased extent of Gal-1 binding (Fig. 7). Interestingly, a length reduction of the 42-amino-acid linker by 26 amino acids generated a Gal-4 variant (Gal-4V), whose cell binding acquired sensitivity to Ctx B-subunit presence.[190] The two sets of binding sites are clearly distinguishable (Fig. 7). This result documents that the CRDs determine not only counterreceptor selection but also the mode of their presentation. As observed for wild-type Gal-4 and Ctx B-subunit, the variant did not affect cell growth.[190] As this alteration affected the hydrodynamic behavior by counterintuitively lowering the translational diffusion constant,[191] the relative positioning of the two CRDs appears to be altered, with an impact on target selection in *cis*-binding of cell surface counterreceptors. Testing galectin-8 (Gal-8) revealed more similarity to siglecs-1 and -4 than to Gal-4. In detail, glycosphingolipids with terminal 3-sialylation (GM3, GD1a)/O-sulfation are strongly reactive with Gal-8 via its N-terminal domain.[192] 3-O-Sulfated lactose is a potent inhibitor of binding of this galectin in tissue sections.[193] This lectin has also been reported to have moderate affinity for the GM1 pentasaccharide in a fluorescence anisotropy assay.[194] On neuroblastoma cell surfaces, Gal-8 bound to the cell surface and elicited moderate growth retardation.[195] Similar to Gal-4, Ctx B-subunit presence did not interfere with Gal-8 binding (Fig. 7).

The lessons emerging from these studies with siglecs and galectins are that (i) glycans of certain gangliosides such as GM1, GM3, or GD1a are functional counterreceptors for these tissue lectins and (ii) divergence in the galectin family in terms of structure/topology establishes functional antagonism on the level of competition for counterreceptor binding and variability in target selection. It is therefore necessary to profile the network of lectins for each family in detail, when more than one member of the family is present in a cell/tissue. This is, for example, the case in tumors,[196–198] in inflamed tissue during pathogenesis of autoimmunological disorders,[199,200] and during development. Monitoring a set of seven galectins covers this aspect in chicken embryogenesis,[201,202] a model for studying galectin involvement since the discovery of high agglutination activity in embryonic

(25 μg) was added 1 h prior to the $^{125}$I-labeled tandem-repeat-type Gal-4 (and its variant Gal-4V), and Gal-8, simultaneously in the case of Gal-1. Experiments were performed in parallel in the absence (•) and in the presence of Ctx B-subunit (Δ) and Scatchard plots were obtained (inset: binding curve), each representing the mean ± SD of three independent experiments. Ctx B-subunit-sensitive binding is given in the graph by data using an open circle. Summary table of $K_D/B_{max}$-values is given (bottom); please see corresponding table in Fig. 3 for comparison.

pectoral muscle (87.5 agglutination units/mg of protein).[83] The structural characterization of the receptor side enabled to proceed from binding studies to defining the details of the molecular fit with ganglioside headgroups. As glycans (a carbohydrate sequence) are endowed with the capacity to form different shapes ("keys"; to draw on E. Fischer's famous analogy[203]), the question is whether different conformers of glycan of a ganglioside such as GM1's pentasaccharide are bioactive.

## 5. STRUCTURAL PAIRING BY CONFORMER SELECTION

The chemical properties of carbohydrates enable structural complexity in chains (first dimension of the sugar code) and in branched oligosaccharides (second dimension of the sugar code). In the third dimension, the energetic profile of conformations becomes structured like a hilly countryside, with (low-energy) valleys of energetically privileged conformers.[204] Metaphorically, "the carbohydrate moves in solution through a bunch of shapes", a glycan generating a "bunch of keys".[205] "It would not be a big surprise if different conformers of a single oligosaccharide interacted with different . . . receptors . . . under different physiological conditions. It is this interplay between proteins and different conformers that likely allows a single carbohydrate structure . . . to be used in many different ways"[3], and GM1's pentasaccharide exemplifies this convincingly.

The analysis of receptors of this pentasaccharide has revealed that the three low-energy conformers are all selected for binding. As illustrated in Fig. 8, Ctx B-subunit and Gal-3 (and also Gal-1 in solution) accommodate different shapes of the same glycan.[206–208] The SV40 capsid protein VP1 prefers the conformer bound by the bacterial toxin.[209,210] Interestingly, factor H, an attenuator of the alternative pathway of complement activation protecting healthy self against damage,[211,212] selects a third shape of GM1's pentasaccharide.[213] This is shown in Fig. 8. Shape change of the glycan under special constraints, for example, in microdomanis, may thus impose shifts in the conformational equilibrium that let the affinity to a receptor become locally adaptable. Ganglioside exposure and the orientation of a glycan chain are factors that contribute to target recognition by antiganglioside GD1a-specific monoclonal antibodies.[214] Changes in the composition of the membrane[215] or the length of the acyl chain in gangliosides[216] also belong to the panel of potential switches. Lack of ganglioside reactivity to a receptor may thus result from conformational masking by the

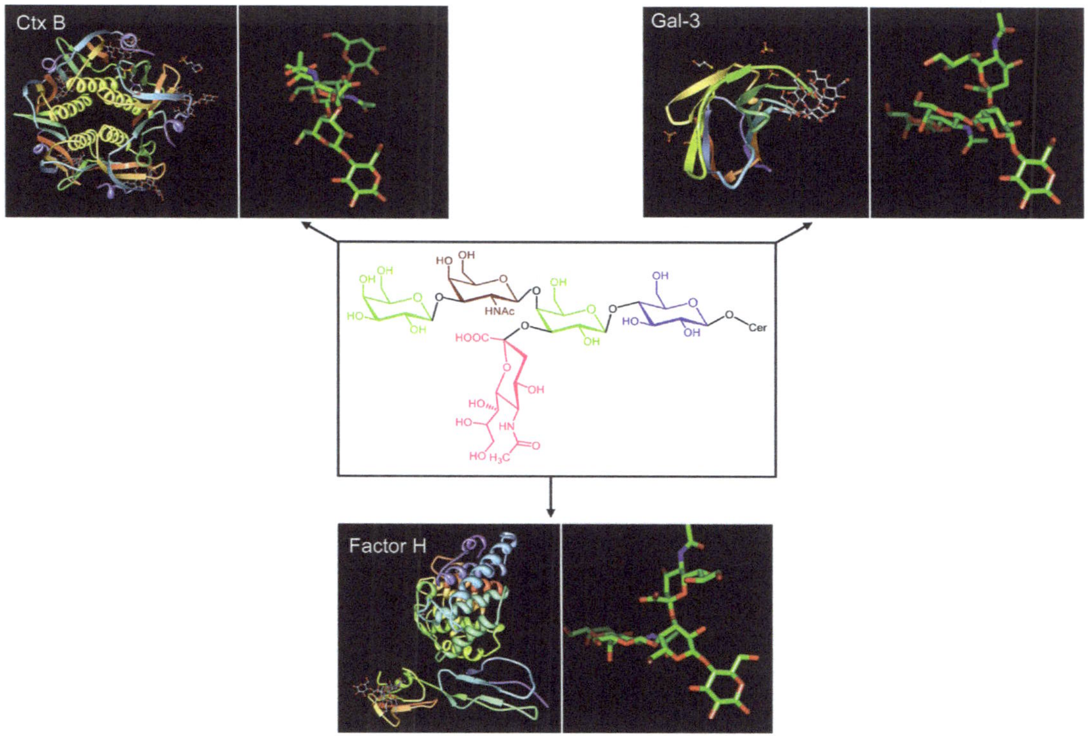


**Fig. 8** Illustration of the complex of the pentasaccharide of ganglioside GM1 (center) with three different receptors and of the pentasaccharide's bound-state conformer detected by crystallography. The cases of the Ctx B-subunit (3CHB; left), the carbohydrate recognition domain (CRD) of human Gal-3 (3AYC; top) and human factor H (4ZH1; right) are given.

microenvironment, an attractive assumption on the significance of the third dimension of the sugar code.

The insight that emerges from these studies is that ganglioside glycans are bioactive by the recognition of their shape through a complementary receptor surface so that different receptors select different conformers. Synthetic inhibitors without conformational flexibility would thus hit conformer binding, with differential reactivity among bacterial/viral receptors, galectins, and factor H. The sialic acid residue is always part of the interaction. A negative charge at the right position is crucial. This holds true, too, when the sugar part of a glycosphingolipid is reduced to a single unit with 3-O-sulfation. This is the case for sulfatides, providing a case of a glycan ligand of minimal size sufficient for lectin binding.

## 6. BEYOND GANGLIOSIDES AS LECTIN LIGANDS: THE CASE STUDY OF SULFATIDES

This term for 3-O-sulfogalactosylceramide, isolated as first sulfoglycolipid from human brain, was coined by J.L.W. Thudichum in 1884, starting the elucidation of an array of its activities at different sites.[217,218] One functional aspect concerns its role in directed transport of glycoproteins together with the tandem-repeat-type Gal-4 mentioned above. Gal-4 can bind to two different ligands, the sulfatide on the microdomain surface and N-glycans of the cargo, illustrated in Fig. 9. This cross-linking of two glycan epitopes, with sulfatide binding to Gal-4's N-terminal domain, is a means of routing cargo from the *trans*-Golgi network to the apical side.[219,220] Cargo loading and transport in microdomains benefit from the ability of Gal-4 to connect glycoproteins with a high density of N-acetyllactosamine termini on N-glycans, the cargo, with recognition of sulfatides with long-chain acyls (up to C26) in the microdomains.[219–222]

This glycosphingolipid–Gal-4 pairing is relevant not only for apical transport but also for segmentation of presence of distinct cargo in the axonal membrane, required for proper axon growth.[223] As shown in Fig. 9, bivalent Gal-4 bridges sulfatide with N-glycans of the axon growth-related neural cell adhesion molecule (NCAM) L1. When sulfatide generation is impaired by treatment with chlorate, overall extent of presence of the glycoprotein L1 and its clustered arrangement is lowered (Fig. 9). Reduction of Gal-4 presence retards axon growth as a consequence.[223]

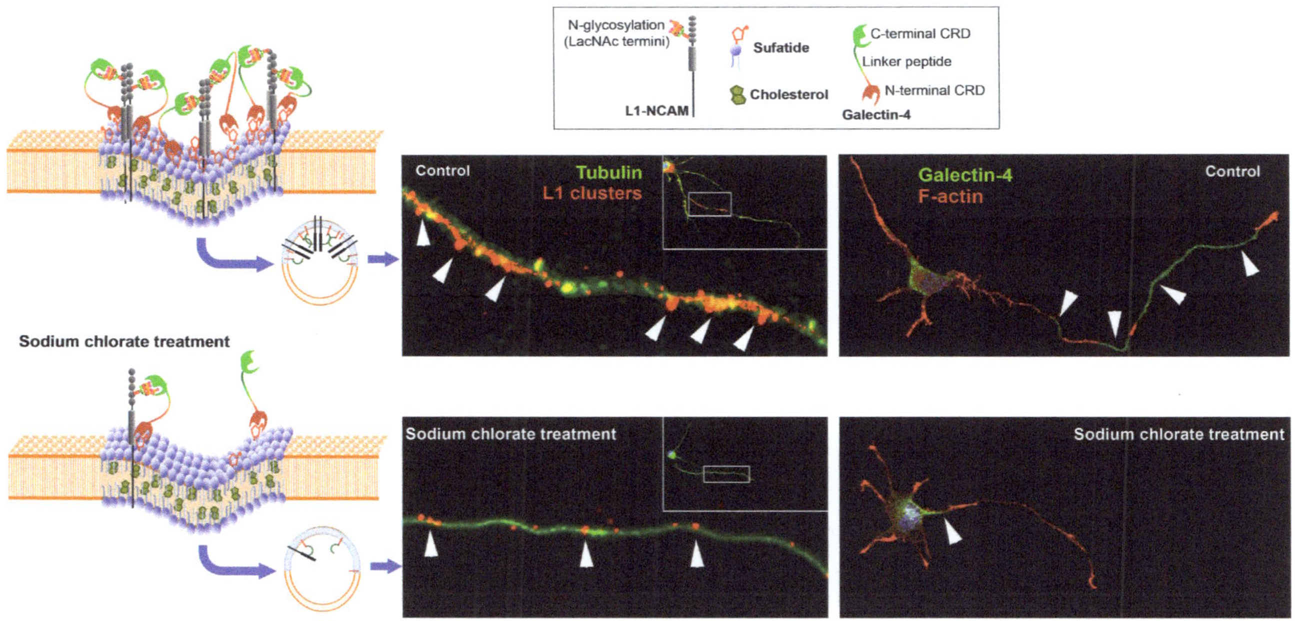


**Fig. 9** The tandem-repeat-type Gal-4 can structurally organize microdomains for polarized transport of glycoproteins by binding to two different types of counterreceptors (top), that is, sulfatide and N-glycans of the cargo glycoproteins, as illustrated by schemes (left) and microphotographs (right), here of NCAM L1 and segmented Gal-4 localization in a cultured hippocampal neuron. Blocking sulfatide synthesis by chlorate impairs glycoprotein routing and also axon growth (bottom). *(For details, please see Velasco S, Díez-Revuelta N, Hernández-Iglesias T, et al. Neuronal galectin-4 is required for axon growth and for the organization of axonal membrane L1 delivery and clustering. J Neurochem. 2013;125:49–62.[223])*

Thus, similar to Gal-4's contribution to maintain epithelial polarity, the lectin, expressed by olfactory, cortical, and dorsal root ganglia neurons,[223–226] accomplishes polarized transport to neuronal membranes, an intracellular function of this galectin in concert with sulfatides. As presence of a sulfate group can take the place of a sialic acid in the 3-O-position of galactose in sphingolipids, turning GM3 into SM3 and GD1a into SB1a, and gangliosides are known to have both substituents,[227] the resulting strong reactivities of Gal-4[188] and Gal-8[192] to these sulfate-bearing glycosphingolipids encourage further analysis, considering the possibility of functional bridging in microdomain assembly.

## 7. CONCLUDING REMARKS

Like the sphinx at the gate of Thebes with its famous riddle, the sphingolipids (and here the gangliosides) pose intriguing questions to those wishing to move from their detection to understanding their functions (of course, a much less dangerous endeavor than trying pass the sphinx, which strangled the ignorant, gangliosides just captivating attention and vigor). The lipid anchor of gangliosides predestines the glycan headgroup for molecular recognition on a membrane's surface. One route to turn the glycan-encoded information into respective cellular activities is via functional pairing with lectins. The underlying structural interplay between cognate binding partners guarantees that gangliosides differing in a single carbohydrate unit can be distinguished as different signals, as the study of ligand properties of GD1a/GM1 exemplified. Matching presence of a receptor (here Gal-1) thus gives a functional meaning to spatiotemporally tightly regulated changes in the ganglioside profile (here the enzymatic processing of GD1a to generate GM1).

Interestingly, the first steps on the ways to delineate ganglioside and galectin functionality, that is, to detect protease-insensitive determinants involved in formation of cell contacts and hemagglutinating activity with potential to be engaged in embryogenesis, have been taken around the same time, about 40 years ago, more than 140 years after the discovery of cerebrosides and sphingomyelin in brain.[228] As the molecular architecture of tissue lectins facilitates ligand crosslinking in *cis*/*trans*-configurations, the profile of cellular responses to the lectin-glycosphingolipid pairing covers a broad panel of activities, from eliciting signaling (*cis*) to adhesion (*trans*). Understanding the impacts of modular design/fine-specificity within lectin networks and density/topological presentation of gangliosides in membranes (at the cell surface or its interior such as nuclear membranes) is as challenging for current and future investigators as solving the riddle of the sphinx was for Oedipus, and that he did is encouragement and inspiration on the way to solving the mysteries of glycosphingolipids.

## ACKNOWLEDGMENTS

We are indebted to Drs. B. Friday and A. Leddoz for inspiring and insightful discussions.

## REFERENCES


1. Bennett HS. Morphological aspects of extracellular polysaccharides. *J Histochem Cytochem*. 1963;11:14–23.
2. Rambourg A, Neutra M, Leblond CP. Presence of a "cell coat" rich in carbohydrate at the surface of cells in the rat. *Anat Rec*. 1966;154:41–71.
3. Roseman S. Reflections on glycobiology. *J Biol Chem* 2001;276:41527–41542.
4. Laine RA. The information-storing potential of the sugar code. In: Gabius H-J, Gabius S, eds. *Glycosciences: Status and Perspectives. London/Weinheim: Chapman & Hall; 1997:1–14.*
5. Rüdiger H, Gabius H-J. The biochemical basis and coding capacity of the sugar code. In: Gabius H-J, ed. *The Sugar Code. Fundamentals of Glycosciences. Weinheim, Germany: Wiley-VCH; 2009:3–13.*
6. Corfield AP, Berry M. Glycan variation and evolution in the eukaryotes. *Trends Biochem Sci*. 2015;40:351–359.
7. Gabius H-J. The magic of the sugar code. *Trends Biochem Sci*. 2015;40:341.
8. Gabius H-J, Roth J. An introduction to the sugar code. *Histochem Cell Biol*. 2017;147:111–117.
9. Roseman S. The synthesis of complex carbohydrates by multiglycosyltransferase systems and their potential function in intercellular adhesion. *Chem Phys Lipids*. 1970;5:270–297.
10. Winterburn PJ, Phelps CF. The significance of glycosylated proteins. *Nature*. 1972;236:147–151.
11. Sharon N. *Complex Carbohydrates. Their Chemistry, Biosynthesis, and Functions*. Reading, MA, USA: Addison-Wesley Publ. Co.; 1975.
12. Cook GMW. Glycobiology of the cell surface: the emergence of sugars as an important feature of the cell periphery. *Glycobiology*. 1995;5:449–461.
13. Gahmberg CG, Tolvanen M. Why mammalian cell surface proteins are glycoproteins. *Trends Biochem Sci*. 1996;21:308–311.
14. Reuter G, Gabius H-J. Eukaryotic glycosylation: whim of nature or multipurpose tool? *Cell Mol Life Sci*. 1999;55:368–422.
15. Gabius H-J, André S, Kaltner H, Siebert H-C. The sugar code: functional lectinomics. *Biochim Biophys Acta*. 2002;1572:165–177.
16. Gabius H-J, Kaltner H, Kopitz J, André S. The glycobiology of the CD system: a dictionary for translating marker designations into glycan/lectin structure and function. *Trends Biochem Sci*. 2015;40:360–376.
17. Schnaar RL. Glycobiology simplified: diverse roles of glycan recognition in inflammation. *J Leukoc Biol*. 2016;99:825–838.
18. Roth J. Subcellular organization of glycosylation in mammalian cells. *Biochim Biophys Acta*. 1987;906:405–436.
19. Brockhausen I, Schachter H. Glycosyltransferases involved in N- and O-glycan biosynthesis. In: Gabius H-J, Gabius S, eds. *Glycosciences: Status and Perspectives. London/Weinheim: Chapman & Hall; 1997:79–113.*

20. Pavelka M. Topology of glycosylation – a histochemist's view. In: Gabius H-J, Gabius S, eds. *Glycosciences: Status and Perspectives. London/Weinheim: Chapman & Hall; 1997:115–120.*
21. Buddecke E. Proteoglycans. In: Gabius H-J, ed. *The Sugar Code. Fundamentals of Glycosciences. Weinheim, Germany: Wiley-VCH; 2009:199–216.*
22. Patsos G, Corfield A. *O*-Glycosylation: structural diversity and function. In: Gabius H-J, ed. *The Sugar Code. Fundamentals of Glycosciences. Weinheim, Germany: Wiley-VCH; 2009:111–137.*
23. Zuber C, Roth J. *N*-Glycosylation. In: Gabius H-J, ed. *The Sugar Code. Fundamentals of Glycosciences. Weinheim, Germany: Wiley-VCH; 2009:87–110.*
24. Hennet T, Cabalzar J. Congenital disorders of glycosylation: a concise chart of glycocalyx dysfunction. *Trends Biochem Sci.* 2015;40:377–384.
25. Schengrund C-L. Gangliosides: glycosphingolipids essential for normal neural development and function. *Trends Biochem Sci.* 2015;40:397–406.
26. Corfield A. Eukaryotic protein glycosylation: a primer for histochemists and cell biologists. *Histochem Cell Biol.* 2017;147:119–147.
27. Klenk E. On the discovery and chemistry of neuraminic acid and gangliosides. *Chem Phys Lipids.* 1970;5:193–197.
28. Wiegandt H. The gangliosides. In: Agranoff BW, Aprison MH, eds. *Advances in Neurochemistry. New York: Plenum; 1982:149–223.*
29. Ledeen RW, Yu RK, Rapport MM, Suzuki K, eds. *Ganglioside Structure, Function and Biomedical Potential. New York: Plenum; 1984.*
30. Stults CLM, Sweeley CC, Macher BA. Glycosphingolipids: structure, biological source, and properties. *Methods Enzymol.* 1989;179:167–214.
31. Sonnino S, Mauri L, Chigorno V, Prinetti A. Gangliosides as components of lipid membrane domains. *Glycobiology.* 2007;17:1R–13R.
32. Kopitz J. Glycolipids. In: Gabius H-J, ed. *The Sugar Code. Fundamentals of Glycosciences. Weinheim, Germany: Wiley-VCH; 2009:177–198.*
33. Ledeen RW, Wu G. Neurobiology meets glycosciences. In: Gabius H-J, ed. *The Sugar Code. Fundamentals of Glycosciences. Weinheim, Germany: Wiley-VCH; 2009:495–516.*
34. Merrill Jr AH. Sphingolipid and glycosphingolipid metabolic pathways in the era of sphingolipidomics. *Chem Rev.* 2011;111:6387–6422.
35. Schnaar RL, Gerardy-Schahn R, Hildebrandt H. Sialic acids in the brain: gangliosides and polysialic acid in nervous system development, stability, disease, and regeneration. *Physiol Rev.* 2014;94:461–518.
36. Kopitz J. Lipid glycosylation: a primer for histochemists and cell biologists. *Histochem Cell Biol.* 2017;147:175–198.
37. Klenk E, Langerbeins H. Über die Verteilung der Neuraminsäure im Gehirn. *Hoppe-Seylers Z Physiol Chem.* 1941;270:185–193.
38. Klenk E. Über die Ganglioside, eine neue Gruppe von zuckerhaltigen Gehirnlipoiden. *Hoppe-Seylers Z Physiol Chem.* 1942;273:76–86.
39. Ledeen RW. Ganglioside structures and distribution: are they localized at the nerve ending? *J Supramol Struct.* 1978;8:1–17.
40. Ledeen RW. Biology of gangliosides: neuritogenic and neuronotrophic properties. *J Neurosci Res.* 1984;12:147–159.
41. Purpura DP, Suzuki K. Distortion of neuronal geometry and formation of aberrant synapses in neuronal storage disease. *Brain Res.* 1976;116:1–21.
42. Purpura DP, Baker HJ. Meganeurites and other aberrant processes of neurons in feline GM1-gangliosidosis: a Golgi study. *Brain Res.* 1978;143:13–26.
43. Walkley SU, Wurzelmann S, Purpura DP. Ultrastructure of neurites and meganeurites of cortical pyramidal neurons in feline gangliosidosis as revealed by the combined Golgi-EM technique. *Brain Res.* 1981;211:393–398.

44. Byrne MC, Ledeen RW. Regional variation of brain gangliosides in feline GM1 gangliosidosis. *Exp Neurol.* 1983;81:210–225.
45. Dreyfus H, Urban PF, Edel-Harth S, Mandel P. Developmental patterns of gangliosides and of phospholipids in chick retina and brain. *J Neurochem.* 1975;25:245–250.
46. Dreyfus H, Louis JC, Harth S, Mandel P. Gangliosides in cultured neurons. *Neuroscience.* 1980;5:1647–1655.
47. Morgan JI, Seifert W. Growth factors and gangliosides: a possible new perspective in neuronal growth control. *J Supramol Struct.* 1979;10:111–124.
48. Roisen FJ, Bartfeld H, Nagele R, Yorke G. Ganglioside stimulation of axonal sprouting in vitro. *Science.* 1981;214:577–578.
49. Kuhn R, Wiegandt H. Die Konstitution der Ganglio-N-Tetraose und des Gangliosids G1. *Chem Ber.* 1963;96:866–880.
50. Byrne MC, Ledeen RW, Roisen FJ, Yorke G, Sclafani JR. Ganglioside-induced neuritogenesis: verification that gangliosides are the active agents, and comparison of molecular species. *J Neurochem.* 1983;41:1214–1222.
51. Skaper SD, Varon S. Ganglioside GM1 overcomes serum inhibition of neuritic outgrowth. *Int J Dev Neurosci.* 1985;3:187–198.
52. Facci L, Leon A, Toffano G, Sonnino S, Ghidoni R, Tettamanti G. Promotion of neuritogenesis in mouse neuroblastoma cells by exogenous gangliosides. Relationship between the effect and the cell association of ganglioside GM1. *J Neurochem.* 1984;42:299–305.
53. Wu G, Fang Y, Lu ZH, Ledeen RW. Induction of axon-like and dendrite-like processes in neuroblastoma cells. *J Neurocytol.* 1998;27:1–14.
54. Schnaar RL. Glycosphingolipids in cell surface recognition. *Glycobiology.* 1991;1:477–485.
55. Blackburn CC, Swank-Hill P, Schnaar RL. Gangliosides support neural retina cell adhesion. *J Biol Chem.* 1986;261:2873–2881.
56. Fueshko SM, Schengrund CL. Murine neuroblastoma cells express ganglioside binding sites on their cell surface. *J Neurochem.* 1990;54:1791–1797.
57. Rogers JC, Kornfeld S. Hepatic uptake of proteins coupled to fetuin glycopeptide. *Biochem Biophys Res Commun.* 1971;45:622–629.
58. Krantz MJ, Holtzman NA, Stowell CP, Lee YC. Attachment of thioglycosides to proteins: enhancement of liver membrane binding. *Biochemistry.* 1976;15:3963–3968.
59. Stowell CP, Lee YC. Neoglycoproteins: the preparation and application of synthetic glycoproteins. *Adv Carbohydr Chem Biochem.* 1980;37:225–281.
60. Gabius H-J, Bodanowitz S, Schauer A. Endogenous sugar-binding proteins in human breast tissue and benign and malignant breast lesions. *Cancer.* 1988;61:1125–1131.
61. Gabius H-J, Kohnke B, Hellmann T, Dimitri T, Bardosi A. Comparative histochemical and biochemical analysis of endogenous receptors for glycoproteins in human and pig nerve. *J Neurochem.* 1988;51:756–763.
62. Tiemeyer M, Yasuda Y, Schnaar RL. Ganglioside-specific binding protein on rat brain membranes. *J Biol Chem.* 1989;264:1671–1681.
63. Tiemeyer M, Swank-Hill P, Schnaar RL. A membrane receptor for gangliosides is associated with central nervous system myelin. *J Biol Chem.* 1990;265:11990–11999.
64. Gabius S, Kayser K, Hellmann KP, Ciesiolka T, Trittin A, Gabius H-J. Carrier-immobilized derivatized lysoganglioside GM1 is a ligand for specific binding sites in various human tumor cell types and peripheral blood lymphocytes and monocytes. *Biochem Biophys Res Commun.* 1990;169:239–244.
65. Hassid S, Salmon I, Bovin NV, Kiss R, Gabius H-J, Danguy A. Histochemical expression of binding sites for labelled hyaluronic acid and carrier-immobilized synthetic (histo-blood group trisaccharides) or biochemically purified (ganglioside $GM_1$) glycoligands in nasal polyps and other human lesions including neoplasms. *Histol Histopathol.* 1996;11:985–992.

66. Kayser K, Böhm G, Blum S, et al. Glyco- and immunohistochemical refinement of the differential diagnosis between mesothelioma and metastatic carcinoma and survival analysis of patients. *J Pathol.* 2001;193:175–180.
67. Barbera AJ, Marchase RB, Roth S. Adhesive recognition and retinotectal specificity. *Proc Natl Acad Sci U S A.* 1973;70:2482–2486.
68. Barbera AJ. Adhesive recognition between developing retinal cells and the optic tecta of the chick embryo. *Dev Biol.* 1975;46:167–191.
69. Marchase RB. Biochemical investigations of retinotectal adhesive specificity. *J Cell Biol.* 1977;75:237–257.
70. van Heyningen WE, Carpenter CC, Pierce NF, Greenough 3rd WB. Deactivation of cholera toxin by ganglioside. *J Infect Dis.* 1971;124:415–418.
71. Cuatrecasas P. Interaction of *Vibrio cholerae* enterotoxin with cell membranes. *Biochemistry.* 1973;12:3547–3558.
72. Eidels L, Proia RL, Hart DA. Membrane receptors for bacterial toxins. *Microbiol Rev.* 1983;47:596–620.
73. Schengrund CL, Ringler NJ. Binding of *Vibrio cholera* toxin and the heat-labile enterotoxin of *Escherichia coli* to GM1, derivatives of GM1, and nonlipid oligosaccharide polyvalent ligands. *J Biol Chem.* 1989;264:13233–13237.
74. Fukuta S, Magnani JL, Twiddy EM, Holmes RK, Ginsburg V. Comparison of the carbohydrate-binding specificities of cholera toxin and *Escherichia coli* heat-labile enterotoxins LTh-I, LT-IIa, and LT-IIb. *Infect Immun.* 1988;56:1748–1753.
75. Zeng F-Y, Gabius H-J. Sialic acid-binding proteins: characterization, biological functions and application. *Z Naturforsch.* 1992;47c:641–653.
76. Lehmann F, Tiralongo E, Tiralongo J. Sialic acid-specific lectins: occurrence, specificity and function. *Cell Mol Life Sci.* 2006;63:1331–1354.
77. Holgersson S, Gustafsson A, Gaunitz S. Bacterial and viral lectins. In: Gabius H-J, ed. *The Sugar Code. Fundamentals of Glycosciences. Weinheim, Germany: Wiley-VCH; 2009:279–300.*
78. Tsai B, Gilbert JM, Stehle T, Lencer W, Benjamin TL, Rapoport TA. Gangliosides are receptors for murine polyoma virus and SV40. *EMBO J.* 2003;22:4346–4355.
79. Taube S, Perry JW, Yetming K, et al. Ganglioside-linked terminal sialic acid moieties on murine macrophages function as attachment receptors for murine noroviruses. *J Virol.* 2009;83:4092–4101.
80. Nilsson EC, Storm RJ, Bauer J, et al. The GD1a glycan is a cellular receptor for adenoviruses causing epidemic keratoconjunctivitis. *Nat Med.* 2011;17:105–109.
81. Kim DS, Son KY, Koo KM, et al. Porcine sapelovirus uses α2,3-linked sialic acid on GD1a ganglioside as a receptor. *J Virol.* 2016;90:4067–4077.
82. Low JA, Magnuson B, Tsai B, Imperiale MJ. Identification of gangliosides GD1b and GT1b as receptors for BK virus. *J Virol.* 2006;80:1361–1366.
83. Teichberg VI, Silman I, Beitsch DD, Resheff G. A β-D-galactoside binding protein from electric organ tissue of *Electrophorus electricus. Proc Natl Acad Sci U S A.* 1975;72:1383–1387.
84. Eisenbarth GS, Riffolo RRJ, Walsh FS, Nirenberg M. Lactose-sensitive lectin of chick retina and spinal cord. *Biochem Biophys Res Commun.* 1978;83:1246–1252.
85. Kobiler D, Beyer EC, Barondes SH. Developmentally regulated lectins from chick muscle, brain and liver have similar chemical and immunological properties. *Dev Biol.* 1978;64:265–272.
86. Barondes SH. Galectins: a personal review. *Trends Glycosci Glycotechnol.* 1997;9:1–7.
87. Mitchell SW. Researches upon the venom of the rattlesnake. *Smithsonian Contributions to Knowledge.* vol. XII. 1860;89–90.
88. Stillmark H. *Ueber Ricin, ein giftiges Ferment aus den Samen von Ricinus comm. L. und einigen anderen Euphorbiaceen.* Dorpat: Schnakenburg's Buchdruckerei; 1888.

89. Kocourek J. Historical background. In: Liener IE, Sharon N, Goldstein IJ, eds. *The Lectins. Properties, Functions and Applications in Biology and Medicine. New York: Academic Press;1986:1–32.*
90. Rüdiger H, Gabius H-J. The history of lectinology. In: Gabius H-J, ed. *The Sugar Code. Fundamentals of Glycosciences. Weinheim, Germany: Wiley-VCH; 2009:261–268.*
91. Boyd WC. The proteins of immune reactions. In: Neurath H, Bailey K, eds. *The Proteins. New York: Academic Press; 1954:756–844.*
92. Boyd WC. The lectins: their present status. *Vox Sang.* 1963;8:1–32.
93. Boyd WC, Shapleigh E. Specific precipitating activity of plant agglutinins (lectins). *Science.* 1954;119:419.
94. Sumner JB, Howell SF. The identification of a hemagglutinin of the jack bean with concanavalin A. *J Bacteriol.* 1936;32:227–237.
95. Watkins WM, Morgan WTJ. Neutralisation of the anti-H agglutinin in eel serum by simple sugars. *Nature.* 1952;169:825–826.
96. Watkins WM. A half century of blood-group antigen research: some personal recollections. *Trends Glycosci Glycotechnol.* 1999;11:391–411.
97. Barondes SH. Bifunctional properties of lectins: lectins redefined. *Trends Biochem Sci.* 1988;13:480–482.
98. Gabius H-J, André S, Jiménez-Barbero J, Romero A, Solís D. From lectin structure to functional glycomics: principles of the sugar code. *Trends Biochem Sci.* 2011;36:298–313.
99. André S, Kaltner H, Manning JC, Murphy PV, Gabius H-J. Lectins: getting familiar with translators of the sugar code. *Molecules.* 2015;20:1788–1823.
100. Solís D, Bovin NV, Davis AP, et al. A guide into glycosciences: how chemistry, biochemistry and biology cooperate to crack the sugar code. *Biochim Biophys Acta.* 2015;1850:186–235.
101. Manning JC, Romero A, Habermann FA, Garcia Caballero G, Kaltner H, Gabius H-J. Lectins: a primer for histochemists and cell biologists. *Histochem Cell Biol.* 2017;147:199–222.
102. Gabius H-J. Detection and functions of mammalian lectins – with emphasis on membrane lectins. *Biochim Biophys Acta.* 1991;1071:1–18.
103. Gabius H-J. Animal lectins. *Eur J Biochem.* 1997;243:543–576.
104. Angata T, Brinkman-van der Linden ECM. I-type lectins. *Biochim Biophys Acta.* 2002;1572:294–316.
105. Cooper DNW. Galectinomics: finding themes in complexity. *Biochim Biophys Acta.* 2002;1572:209–231.
106. Kilpatrick DC. Animal lectins: a historical introduction and overview. *Biochim Biophys Acta.* 2002;1572:187–197.
107. Gready JN, Zelensky AN. Routes in lectin evolution: case study on the C-type lectin-like domains. In: Gabius H-J, ed. *The Sugar Code. Fundamentals of Glycosciences. Weinheim, Germany: Wiley-VCH; 2009:329–346.*
108. Mayer S, Raulf MK, Lepenies B. C-type lectins: their network and roles in pathogen recognition and immunity. *Histochem Cell Biol.* 2017;147:223–237.
109. Roy R, Murphy PV, Gabius H-J. Multivalent carbohydrate-lectin interactions: how synthetic chemistry enables insights into nanometric recognition. *Molecules.* 2016;21:629.
110. Gabius H-J, Manning JC, Kopitz J, André S, Kaltner H. Sweet complementarity: the functional pairing of glycans with lectins. *Cell Mol Life Sci.* 2016;73:1989–2016.
111. Gabius H-J. Animal and human lectins. In: Gabius H-J, ed. *The Sugar Code. Fundamentals of Glycosciences. Weinheim, Germany: Wiley-VCH; 2009:317–328.*
112. Schauer R. Sialic acids and their role as biological masks. *Trends Biochem Sci.* 1985;10:357–360.
113. Reuter G, Gabius H-J. Sialic acids. Structure, analysis, metabolism, and recognition. *Biol Chem Hoppe-Seyler.* 1996;377:325–342.

114. Schwartz-Albiez R. Inflammation and glycosciences. In: Gabius H-J, ed. *The Sugar Code. Fundamentals of Glycosciences. Weinheim, Germany: Wiley-VCH; 2009:447–467.*
115. Bhide GP, Colley KJ. Sialylation of N-glycans: mechanism, cellular compartmentalization and function. *Histochem Cell Biol.* 2017;147:149–174.
116. Lopez PH, Schnaar RL. Gangliosides in cell recognition and membrane protein regulation. *Curr Opin Struct Biol.* 2009;19:549–557.
117. Linnartz-Gerlach B, Mathews M, Neumann H. Sensing the neuronal glycocalyx by glial sialic acid binding immunoglobulin-like lectins. *Neuroscience.* 2014;275:113–124.
118. Macauley MS, Crocker PR, Paulson JC. Siglec-mediated regulation of immune cell function in disease. *Nat Rev Immunol.* 2014;14:653–666.
119. Puryear WB, Yu X, Ramirez NP, Reinhard BM, Gummuluru S. HIV-1 incorporation of host-cell-derived glycosphingolipid GM3 allows for capture by mature dendritic cells. *Proc Natl Acad Sci U S A.* 2012;109:7475–7480.
120. Izquierdo-Useros N, Lorizate M, Puertas MC, et al. Siglec-1 is a novel dendritic cell receptor that mediates HIV-1 trans-infection through recognition of viral membrane gangliosides. *PLoS Biol.* 2012;10:e1001448.
121. Izquierdo-Useros N, Lorizate M, McLaren PJ, Telenti A, Krausslich HG, Martinez-Picado J. HIV-1 capture and transmission by dendritic cells: the role of viral glycolipids and the cellular receptor Siglec-1. *PLoS Pathog.* 2014;10:e1004146.
122. Erikson E, Wratil PR, Frank M, et al. Mouse siglec-1 mediates trans-infection of surface-bound murine leukemia virus in a sialic acid N-acyl side chain-dependent manner. *J Biol Chem.* 2015;290:27345–27359.
123. Collins BE, Yang LJ, Mukhopadhyay G, et al. Sialic acid specificity of myelin-associated glycoprotein binding. *J Biol Chem.* 1997;272:1248–1255.
124. Collins BE, Ito H, Sawada N, Ishida H, Kiso M, Schnaar RL. Enhanced binding of the neural siglecs, myelin-associated glycoprotein and Schwann cell myelin protein, to Chol-1 (α-series) gangliosides and novel sulfated Chol-1 analogs. *J Biol Chem.* 1999;274:37637–37643.
125. Rapoport E, Mikhalyov I, Zhang J, Crocker P, Bovin N. Ganglioside binding pattern of CD33-related siglecs. *Bioorg Med Chem Lett.* 2003;13:675–678.
126. Attrill H, Imamura A, Sharma RS, Kiso M, Crocker PR, van Aalten DM. Siglec-7 undergoes a major conformational change when complexed with the α(2,8)-disialylganglioside GT1b. *J Biol Chem.* 2006;281:32774–32783.
127. Nicoll G, Avril T, Lock K, Furukawa K, Bovin N, Crocker PR. Ganglioside GD3 expression on target cells can modulate NK cell cytotoxicity via siglec-7-dependent and -independent mechanisms. *Eur J Immunol.* 2003;33:1642–1648.
128. Schnaar RL, Lopez PH. Myelin-associated glycoprotein and its axonal receptors. *J Neurosci Res.* 2009;87:3267–3276.
129. Schnaar RL. Brain gangliosides in axon-myelin stability and axon regeneration. *FEBS Lett.* 2010;584:1741–1747.
130. Higuero AM, Díez-Revuelta N, Abad-Rodríguez J. The sugar code in neuronal physiology. *Histochem Cell Biol.* 2017;147:257–267.
131. Vyas AA, Patel HV, Fromholt SE, et al. Gangliosides are functional nerve cell ligands for myelin-associated glycoprotein (MAG), an inhibitor of nerve regeneration. *Proc Natl Acad Sci U S A.* 2002;99:8412–8417.
132. Schnaar RL. Glycolipid-mediated cell-cell recognition in inflammation and nerve regeneration. *Arch Biochem Biophys.* 2004;426:163–172.
133. Pronker MF, Lemstra S, Snijder J, et al. Structural basis of myelin-associated glycoprotein adhesion and signalling. *Nat Commun.* 2016;7:13584.
134. Vyas AA, Blixt O, Paulson JC, Schnaar RL. Potent glycan inhibitors of myelin-associated glycoprotein enhance axon outgrowth in vitro. *J Biol Chem.* 2005;280:16305–16310.

135. Schwardt O, Gathje H, Vedani A, et al. Examination of the biological role of the α(2,6)-linked sialic acid in gangliosides binding to the myelin-associated glycoprotein (MAG). *J Med Chem.* 2009;52:989–1004.
136. Kopitz J, von Reitzenstein C, Sinz K, Cantz M. Selective ganglioside desialylation in the plasma membrane of human neuroblastoma cells. *Glycobiology.* 1996;6:367–376.
137. Schengrund CL, Repman MA. Density-dependent changes in gangliosides and sialidase activity of murine neuroblastoma cells. *J Neurochem.* 1982;39:940–947.
138. Riboni L, Prinetti A, Pitto M, Tettamanti G. Patterns of endogenous gangliosides and metabolic processing of exogenous gangliosides in cerebellar granule cells during differentiation in culture. *Neurochem Res.* 1990;15:1175–1183.
139. Kopitz J, von Reitzenstein C, Mühl C, Cantz M. Role of plasma membrane ganglioside sialidase of human neuroblastoma cells in growth control and differentiation. *Biochem Biophys Res Commun.* 1994;199:1188–1193.
140. Kopitz J, Mühl C, Ehemann V, Lehmann C, Cantz M. Effects of cell surface ganglioside sialidase inhibition on growth control and differentiation of human neuroblastoma cells. *Eur J Cell Biol.* 1997;73:1–7.
141. Wu G, Ledeen RW. Stimulation of neurite outgrowth in neuroblastoma cells by neuraminidase: putative role of GM1 ganglioside in differentiation. *J Neurochem.* 1991;56:95–104.
142. Harrison FL, Chesterton CJ. Factors mediating cell-cell recognition and adhesion. Galaptins, a recently discovered class of bridging molecules. *FEBS Lett.* 1980;122:157–165.
143. Kaltner H, Toegel S, Garcia Caballero G, Manning JC, Ledeen RW, Gabius H-J. Galectins: their network and roles in immunity/tumor growth control. *Histochem Cell Biol.* 2017;147:239–256.
144. Caron M, Joubert-Caron R, Cartier JR, Chadli A, Bladier D. Study of lectin-ganglioside interactions by high-performance liquid affinity chromatography. *J Chromatography.* 1993;646:327–333.
145. Mahanthappa NK, Cooper DNW, Barondes SH, Schwarting GA. Rat olfactory neurons can utilize the endogenous lectin, L-14, in a novel adhesion mechanism. *Development.* 1994;120:1373–1384.
146. Kaltner H, Lips KS, Reuter G, Lippert S, Sinowatz F, Gabius H-J. Quantitation and histochemical localization of galectin-1 and galectin-1-reactive glycoconjugates in fetal development of bovine organs. *Histol Histopathol.* 1997;12:945–960.
147. Dettmann W, Grandbois M, André S, et al. Differences in zero-force and force-driven kinetics of ligand dissociation from β-galactoside-specific proteins (plant and animal lectins, immunoglobulin G) monitored by plasmon resonance and dynamic single molecule force microscopy. *Arch Biochem Biophys.* 2000;383:157–170.
148. Percec V, Leowanawat P, Sun HJ, et al. Modular synthesis of amphiphilic Janus glycodendrimers and their self-assembly into glycodendrimersomes and other complex architectures with bioactivity to biomedically relevant lectins. *J Am Chem Soc.* 2013;135:9055–9077.
149. Zhang S, Moussodia R-O, Murzeau C, et al. Dissecting molecular aspects of cell interactions using glycodendrimersomes with programmable glycan presentation and engineered human lectins. *Angew Chem Int Ed.* 2015;54:4036–4040.
150. Kopitz J, von Reitzenstein C, Burchert M, Cantz M, Gabius H-J. Galectin-1 is a major receptor for ganglioside GM1, a product of the growth-controlling activity of a cell surface ganglioside sialidase, on human neuroblastoma cells in culture. *J Biol Chem.* 1998;273:11205–11211.
151. Kopitz J, von Reitzenstein C, André S, et al. Negative regulation of neuroblastoma cell growth by carbohydrate-dependent surface binding of galectin-1 and functional divergence from galectin-3. *J Biol Chem.* 2001;276:35917–35923.

152. Kopitz J, Bergmann M, Gabius H-J. How adhesion/growth-regulatory galectins-1 and -3 attain cell specificity: case study defining their target on neuroblastoma cells (SK-N-MC) and marked affinity regulation by affecting microdomain organization of the membrane. *IUBMB Life.* 2010;62:624–628.
153. Majewski J, André S, Jones E, Chi E, Gabius H-J. X-ray reflectivity and grazing incidence diffraction studies of interaction between human adhesion/growth-regulatory galectin-1 and DPPE:GM1 lipid monolayer at the air/water interface. *Biochemistry (Moscow).* 2015;80:943–956.
154. Kopitz J, André S, von Reitzenstein C, et al. Homodimeric galectin-7 (p53-induced gene 1) is a negative growth regulator for human neuroblastoma cells. *Oncogene.* 2003;22:6277–6288.
155. André S, Kaltner H, Lensch M, et al. Determination of structural and functional overlap/divergence of five proto-type galectins by analysis of the growth-regulatory interaction with ganglioside GM1 *insilico* and *invitro* on human neuroblastoma cells. *Int J Cancer.* 2005;114:46–57.
156. Sturm A, Lensch M, André S, et al. Human galectin-2: novel inducer of T cell apoptosis with distinct profile of caspase activation. *J Immunol.* 2004;173:3825–3837.
157. Wu G, Lu ZH, André S, Gabius H-J, Ledeen RW. Functional interplay between ganglioside GM1 and cross-linking galectin-1 induces axon-like neuritogenesis via integrin-based signaling and TRPC5-dependent $Ca^{2+}$ influx. *J Neurochem.* 2016;136:550–563.
158. Regan LJ, Dodd J, Barondes SH, Jessell TM. Selective expression of endogenous lactose-binding lectins and lactoseries glycoconjugates in subsets of rat sensory neurons. *Proc Natl Acad Sci U S A.* 1986;83:2248–2252.
159. Joubert R, Kuchler S, Zanetta J-P, et al. Immunohistochemical localization of a β-galactoside-binding lectin in rat central nervous system. I. Light- and electron-microscopical studies on developing cerebral cortex and corpus callosum. *Dev Neurosci.* 1989;11:397–413.
160. Gabius H-J, Bardosi A. Regional differences in the distribution of endogenous receptors for carbohydrate constituents of cellular glycoconjugates, especially lectins, in cortex, hippocampus, basal ganglia and thalamus of adult human brain. *Histochemistry.* 1990;93:581–592.
161. Gabius H-J, Wosgien B, Hendrys M, Bardosi A. Lectin localization in human nerve by biochemically defined lectin-binding glycoproteins, neoglycoprotein and lectin-specific antibody. *Histochemistry.* 1991;95:269–277.
162. Kajitani K, Kobayakawa Y, Nomaru H, Kadoya T, Horie H, Nakabeppu Y. Characterization of galectin-1-positive cells in the mouse hippocampus. *Neuroreport.* 2014;25:171–176.
163. Wu G, Lu ZH, Obukhov AG, Nowycky MC, Ledeen RW. Induction of calcium influx through TRPC5 channels by cross-linking of GM1 ganglioside associated with $\alpha_5\beta_1$-integrin initiates neurite outgrowth. *J Neurosci.* 2007;27:7447–7458.
164. Fischer C, Sanchez-Ruderisch H, Welzel M, et al. Galectin-1 interacts with the $\alpha_5\beta_1$ fibronectin receptor to restrict carcinoma cell growth via induction of p21 and p27. *J Biol Chem.* 2005;280:37266–37277.
165. Sanchez-Ruderisch H, Detjen KM, Welzel M, et al. Galectin-1 sensitizes carcinoma cells to anoikis via the fibronectin receptor $\alpha_5\beta_1$-integrin. *Cell Death Differ.* 2011;18:806–816.
166. Amano M, Eriksson H, Manning JC, et al. Tumour suppressor $p16^{INK4a}$: anoikis-favouring decrease in N/O-glycan/cell surface sialylation by down-regulation of enzymes in sialic acid biosynthesis in tandem in a pancreatic carcinoma model. *FEBS J.* 2012;279:4062–4080.
167. Wang J, Lu ZH, Gabius H-J, Rohowsky-Kochan C, Ledeen RW, Wu G. Cross-linking of GM1 ganglioside by galectin-1 mediates regulatory T cell activity involving TRPC5

channel activation: possible role in suppressing experimental autoimmune encephalomyelitis. *J Immunol.* 2009;182:4036–4045.
168. Wu G, Lu ZH, Gabius H-J, Ledeen RW, Bleich D. Ganglioside GM1 deficiency in effector T cells from NOD mice induces resistance to regulatory T cell suppression. *Diabetes.* 2011;60:2341–2349.
169. Ledeen RW, Wu G, André S, et al. Beyond glycoproteins as galectin counterreceptors: tumor/effector T cell growth control via ganglioside GM1. *Ann N Y Acad Sci.* 2012;1253:206–221.
170. Kappagantula S, Andrews MR, Cheah M, Abad-Rodríguez J, Dotti CG, Fawcett JW. Neu3 sialidase-mediated ganglioside conversion is necessary for axon regeneration and is blocked in CNS axons. *J Neurosci.* 2014;34:2477–2492.
171. Abad-Rodríguez J, Piddini E, Hasegawa T, Miyagi T, Dotti CG. Plasma membrane ganglioside sialidase regulates axonal growth and regeneration in hippocampal neurons in culture. *J Neurosci.* 2001;21:8387–8395.
172. Yang LJ, Lorenzini I, Vajn K, Mountney A, Schramm LP, Schnaar RL. Sialidase enhances spinal axon outgrowth in vivo. *Proc Natl Acad Sci U S A.* 2006;103:11057–11062.
173. Mountney A, Zahner MR, Lorenzini I, Oudega M, Schramm LP, Schnaar RL. Sialidase enhances recovery from spinal cord contusion injury. *Proc Natl Acad Sci U S A.* 2010;107:11561–11566.
174. Mutoh T, Tokuda A, Miyadai T, Hamaguchi M, Fujiki N. Ganglioside GM1 binds to the Trk protein and regulates receptor function. *Proc Natl Acad Sci U S A.* 1995;92:5087–5091.
175. Da Silva JS, Hasegawa T, Miyagi T, Dotti CG, Abad-Rodríguez J. Asymmetric membrane ganglioside sialidase activity specifies axonal fate. *Nat Neurosci.* 2005;8:606–615.
176. Kawashima N, Yoon SJ, Itoh K, Nakayama K. Tyrosine kinase activity of epidermal growth factor receptor is regulated by GM3 binding through carbohydrate to carbohydrate interactions. *J Biol Chem.* 2009;284:6147–6155.
177. Bucior I, Burger MM, Fernàndez-Busquets X. Carbohydrate–carbohydrate interactions. In: Gabius H-J, ed. *The Sugar Code. Fundamentals of Glycosciences. Weinheim, Germany: Wiley-VCH; 2009:347–362.*
178. Ledeen RW, Wu G. The multi-tasked life of GM1 ganglioside, a true factotum of nature. *Trends Biochem Sci.* 2015;40:407–418.
179. Hohenester E, Tisi D, Talts JF, Timpl R. The crystal structure of a laminin G-like module reveals the molecular basis of α-dystroglycan binding to laminins, perlecan, and agrin. *Mol Cell.* 1999;4:783–792.
180. Tisi D, Talts JF, Timpl R, Hohenester E. Structure of the C-terminal laminin G-like domain pair of the laminin α2 chain harbouring binding sites for α-dystroglycan and heparin. *EMBO J.* 2000;19:1432–1440.
181. Timpl R, Tisi D, Talts JF, Andac Z, Sasaki T, Hohenester E. Structure and function of laminin LG modules. *Matrix Biol.* 2000;19:309–317.
182. Ichikawa N, Iwabuchi K, Kurihara H, et al. Binding of laminin-1 to monosialoganglioside GM1 in lipid rafts is crucial for neurite outgrowth. *J Cell Sci.* 2009;122:289–299.
183. Ho MK, Springer TA. Mac-2, a novel 32,000 Mr mouse macrophage subpopulation-specific antigen defined by monoclonal antibodies. *J Immunol.* 1982;128:1221–1228.
184. Cherayil BJ, Weiner SJ, Pillai S. The Mac-2 antigen is a galactose-specific lectin that binds IgE. *J Exp Med.* 1989;170:1959–1972.
185. Hughes RC. Mac-2: a versatile galactose-binding protein of mammalian tissues. *Glycobiology.* 1994;4:5–12.
186. Kopitz J, Vértesy S, André S, Fiedler S, Schnölzer M, Gabius H-J. Human chimera-type galectin-3: defining the critical tail length for high-affinity glycoprotein/cell surface binding and functional competition with galectin-1 in neuroblastoma cell growth regulation. *Biochimie.* 2014;104:90–99.

187. Sanchez-Ruderisch H, Fischer C, Detjen KM, et al. Tumor suppressor p16[INK4a]: downregulation of galectin-3, an endogenous competitor of the pro-anoikis effector galectin-1, in a pancreatic carcinoma model. *FEBS J.* 2010;277:3552–3563.
188. Ideo H, Seko A, Yamashita K. Galectin-4 binds to sulfated glycosphingolipids and carcinoembryonic antigen in patches on the cell surface of human colon adenocarcinoma cells. *J Biol Chem.* 2005;280:4730–4737.
189. Ideo H, Seko A, Yamashita K. Recognition mechanism of galectin-4 for cholesterol 3-sulfate. *J Biol Chem.* 2007;282:21081–21089.
190. Kopitz J, Ballikaya S, André S, Gabius H-J. Ganglioside GM1/galectin-dependent growth regulation in human neuroblastoma cells: special properties of bivalent galectin-4 and significance of linker length for ligand selection. *Neurochem Res.* 2012;37:1267–1276.
191. Göhler A, André S, Kaltner H, Sauer M, Gabius H-J, Doose S. Hydrodynamic properties of human adhesion/growth-regulatory galectins studied by fluorescence correlation spectroscopy. *Biophys J.* 2010;98:3044–3053.
192. Ideo H, Seko A, Ishizuka I, Yamashita K. The N-terminal carbohydrate recognition domain of galectin-8 recognizes specific glycosphingolipids with high affinity. *Glycobiology.* 2003;13:713–723.
193. Roy R, Cao Y, Kaltner H, et al. Teaming up synthetic chemistry and histochemistry for activity screening in galectin-directed inhibitor design. *Histochem Cell Biol.* 2017;147:285–301.
194. Carlsson S, Oberg CT, Carlsson MC, et al. Affinity of galectin-8 and its carbohydrate recognition domains for ligands in solution and at the cell surface. *Glycobiology.* 2007;17:663–676.
195. Ruiz FM, Scholz BA, Buzamet E, et al. Natural single amino acid polymorphism (F19Y) in human galectin-8: detection of structural alterations and increased growth-regulatory activity on tumor cells. *FEBS J.* 2014;281:1446–1464.
196. Camby I, Belot N, Rorive S, et al. Galectins are differentially expressed in supratentorial pilocytic astrocytomas, astrocytomas, anaplastic astrocytomas and glioblastomas, and significantly modulate tumor astrocyte migration. *Brain Pathol.* 2001;11: 12–26.
197. Dawson H, André S, Karamitopoulou E, Zlobec I, Gabius H-J. The growing galectin network in colon cancer and clinical relevance of cytoplasmic galectin-3 reactivity. *Anticancer Res.* 2013;33:3053–3059.
198. Katzenmaier E-M, André S, Kopitz J, Gabius H-J. Impact of sodium butyrate on the network of adhesion/growth-regulatory galectins in human colon cancer *in vitro*. *Anticancer Res.* 2014;34:5429–5438.
199. Toegel S, Bieder D, André S, et al. Human osteoarthritic knee cartilage: fingerprinting of adhesion/growth-regulatory galectins *in vitro* and *in situ* indicates differential upregulation in severe degeneration. *Histochem Cell Biol.* 2014;142:373–388.
200. Weinmann D, Schlangen K, André S, et al. Galectin-3 induces a pro-degradative/inflammatory gene signature in human chondrocytes, teaming up with galectin-1 in osteoarthritis pathogenesis. *Sci Rep.* 2016;6:39112.
201. García Caballero G, Kaltner H, Michalak M, et al. Chicken GRIFIN: a homodimeric member of the galectin network with canonical properties and a unique expression profile. *Biochimie.* 2016;128–129:34–47.
202. Kaltner H, García Caballero G, Sinowatz F, et al. Galectin-related protein: an integral member of the network of chicken galectins. 2. From expression profiling to its immunocyto- and histochemical localization and application as tool for ligand detection. *Biochim Biophys Acta.* 2016;1860:2298–2312.
203. Fischer E. Einfluss der Configuration auf die Wirkung der Enzyme. *Ber Dt Chem Ges.* 1894;27:2985–2993.

204. von der Lieth C-W, Siebert H-C, Kozár T, et al. Lectin ligands: new insights into their conformations and their dynamic behavior and the discovery of conformer selection by lectins. *Acta Anat.* 1998;161:91–109.
205. Hardy BJ. The glycosidic linkage flexibility and time-scale similarity hypotheses. *J Mol Struct.* 1997;395-396:187–200.
206. Merritt EA, Sarfaty S, van den Akker F, L'Hoir C, Martial JA, Hol WG. Crystal structure of cholera toxin B-pentamer bound to receptor GM1 pentasaccharide. *Protein Sci.* 1994;3:166–175.
207. Siebert H-C, André S, Lu S-Y, et al. Unique conformer selection of human growth-regulatory lectin galectin-1 for ganglioside GM1 versus bacterial toxins. *Biochemistry.* 2003;42:14762–14773.
208. Bian CF, Zhang Y, Sun H, Li DF, Wang DC. Structural basis for distinct binding properties of the human galectins to Thomsen-Friedenreich antigen. *PLoS ONE.* 2011;6:e25007.
209. Neu U, Woellner K, Gauglitz G, Stehle T. Structural basis of GM1 ganglioside recognition by simian virus 40. *Proc Natl Acad Sci U S A.* 2008;105:5219–5224.
210. Neu U, Bauer J, Stehle T. Viruses and sialic acids: rules of engagement. *Curr Opin Struct Biol.* 2011;21:610–618.
211. Pouw RB, Vredevoogd DW, Kuijpers TW, Wouters D. Of mice and men: the factor H protein family and complement regulation. *Mol Immunol.* 2015;67:12–20.
212. Parente R, Clark SJ, Inforzato A, Day AJ. Complement factor H in host defense and immune evasion. *Cell Mol Life Sci.* 2017;74:1605–1624.
213. Blaum BS, Frank M, Walker RC, Neu U, Stehle T. Complement factor H and simian virus 40 bind the GM1 ganglioside in distinct conformations. *Glycobiology.* 2016;26:532–539.
214. Lopez PH, Zhang G, Bianchet MA, Schnaar RL, Sheikh KA. Structural requirements of anti-GD1a antibodies determine their target specificity. *Brain.* 2008;131:1926–1939.
215. Rog T, Vattulainen I. Cholesterol, sphingolipids, and glycolipids: what do we know about their role in raft-like membranes? *Chem Phys Lipids.* 2014;184:82–104.
216. Manna M, Javanainen M, Monne HM, Gabius H-J, Rog T, Vattulainen I. Long-chain GM1 gangliosides alter transmembrane domain registration through interdigitation. *Biochim Biophys Acta.* 2017;1859:870–878.
217. Eckhardt M. The role and metabolism of sulfatide in the nervous system. *Mol Neurobiol.* 2008;37:93–103.
218. Takahashi T, Suzuki T. Role of sulfatide in normal and pathological cells and tissues. *J Lipid Res.* 2012;53:1437–1450.
219. Delacour D, Gouyer V, Zanetta J-P, et al. Galectin-4 and sulfatides in apical membrane trafficking in enterocyte-like cells. *J Cell Biol.* 2005;169:491–501.
220. Danielsen EM, Hansen GH. Lipid raft organization and function in brush borders of epithelial cells. *Mol Membr Biol.* 2006;23:71–79.
221. Stechly L, Morelle W, Dessein AF, et al. Galectin-4-regulated delivery of glycoproteins to the brush border membrane of enterocyte-like cells. *Traffic.* 2009;10:438–450.
222. Morelle W, Stechly L, André S, et al. Glycosylation pattern of brush border-associated glycoproteins in enterocyte-like cells: involvement of complex-type *N*-glycans in apical trafficking. *Biol Chem.* 2009;390:529–544.
223. Velasco S, Díez-Revuelta N, Hernández-Iglesias T, et al. Neuronal galectin-4 is required for axon growth and for the organization of axonal membrane L1 delivery and clustering. *J Neurochem.* 2013;125:49–62.
224. Storan MJ, Magnaldo T, Biol-N'Garagba MC, Zick Y, Key B. Expression and putative role of lactoseries carbohydrates present on NCAM in the rat primary olfactory pathway. *J Comp Neurol.* 2004;475:289–302.

225. Wei Q, Eviatar-Ribak T, Miskimins WK, Miskimins R. Galectin-4 is involved in p27-mediated activation of the myelin basic protein promoter. *J Neurochem.* 2007;101:1214–1223.
226. Stancic M, Slijepcevic D, Nomden A, et al. Galectin-4, a novel neuronal regulator of myelination. *Glia.* 2012;60:919–935.
227. Wiegandt H. The chemical constitution of gangliosides of the vertebrate nervous system. *Behav Brain Res.* 1995;66:85–97.
228. Thudichum JLW. Researches on the chemical constitution of the brain. *Report of the Medical Offices of the Privy Council and Local Government Board, London. 1874.*

CHAPTER ELEVEN

# Protein Toxins That Utilize Gangliosides as Host Receptors

**Madison Zuverink, Joseph T. Barbieri**[1]
Medical College of Wisconsin, Milwaukee, WI, United States
[1] Corresponding author. E-mail address: jtb01@mcw.edu

## Contents



## Abstract

Subsets of protein toxins utilize gangliosides as host receptors. Gangliosides are preferred receptors due to their extracellular localization on the eukaryotic cell

*Progress in Molecular Biology and Translational Science*, Volume 156
ISSN 1877-1173
http://dx.doi.org/10.1016/bs.pmbts.2017.11.010

and due to their essential nature in host physiology. Glycosphingolipids, including gangliosides, are mediators of signal transduction within and between eukaryotic cells. Protein toxins possess **AB** structure-function organization, where the **A** domain encodes a catalytic function for the posttranslational modification of a host macromolecule, including proteins and nucleic acids, and a **B** domain, which encodes host receptor recognition, including proteins and glycosphingolipids, alone or in combination. Protein toxins use similar strategies to bind glycans by pockets and loops, generally employing hydrogen bonding and aromatic stacking to stabilize interactions with sugars. In some cases, glycan binding facilitates uptake, while in other cases, cross-linking or a second receptor is necessary to stimulate entry. The affinity that protein toxins have for host glycans is necessary for tissue targeting, but not always sufficient to cause disease. In addition to affinity for binding the glycan, the lipid moiety also plays an important role in productive uptake and tissue tropism. Upon endocytosis, the protein toxin must escape to another intracellular compartment or into cytosol to modify a host substrate, modulating host signaling, often resulting in cytotoxic or apoptotic events in the cell, and a unique morbidity for the organism. The study of protein toxins that utilize gangliosides as host receptors has illuminated numerous eukaryotic cellular processes, identified the basis for developing interventions to prevent disease through vaccines and control bacterial diseases through therapies. In addition, subsets of these protein toxins have been utilized as therapeutic agents to treat numerous human inflictions.

## 1. PROTEIN TOXIN AB STRUCTURE-FUNCTION ORGANIZATION

**AB** protein toxins are classified by domain organization (Fig. 1). **A** is the catalytic domain, which posttranslationally modifies a host protein(s), changing host physiology and often causing death or intoxication.[1] **B** is the binding domain, which binds a host receptor(s) and may also contain a translocation domain to facilitate delivery of **A** into the host cytosol.[1] Inactivation of either the **A** or **B** domains results in a nontoxic protein. For single-chain **AB** toxins, **A** and **B** domains are synthesized within a single protein, while for **AB5** toxins, **A** and **B** domains are synthesized independently and assembled into an oligomer composed of one molecule of **A** bound to a **B** pentamer. The **A** domains of **AB5** toxins comprise an **A1** catalytic domain and an **A2** linker domain. The alpha-helical **A2** linker domain inserts into the **B** pentamer through noncovalent interactions to link the **A1** subdomain with the **B** pentamer to produce the **AB5** toxin.[2]

Single-chain **AB** toxins often contain an interchain disulfide bond, while **AB5** toxins often contain an intrachain disulfide bond between the **A1-** and **A2-**subdomains. In each case, reduction is usually required during the

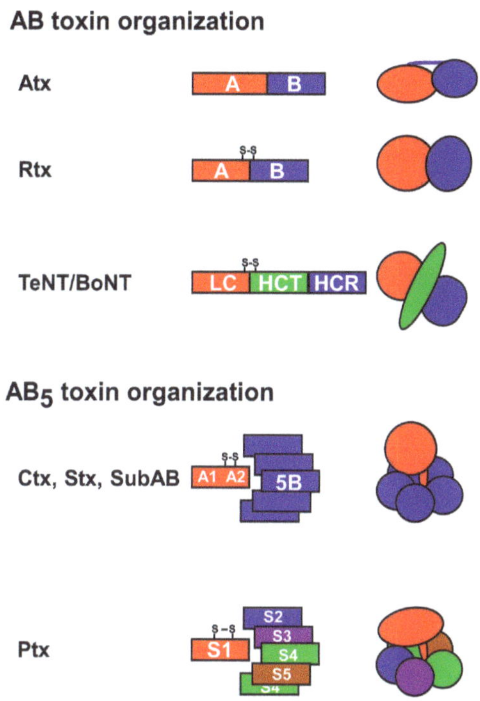


**Fig. 1** **AB** toxin schematic. **AB** protein toxins contain two functional properties: a catalytically active **A** domain and a receptor binding **B** domain. **AB** protein toxins include alpha-toxin (Atx), ricin toxin (Rtx), and clostridial tetanus and botulinum neurotoxins (TeNT/BoNT). Alpha-toxin does not contain an interchain disulfide, but contains a ganglioside binding loop between the **A** and **B** domains. TeNT/BoNT contain a **B** domain, which includes a receptor binding domain (HCR) and a translocation domain (HCT). HCT facilitates delivery of **A** domain, also known as the light chain (LC). **AB5** protein toxins include four families: cholera toxin (Ctx), Shiga toxin (Stx), and subtilase cytotoxin (SubAB) contain an **A** domain that can be divided into **A1**- and **A2**-subdomains, and five identical **B** binding domains. Pertussis toxin (Ptx) contains nonidentical **B** binding components consisting of four homologous subunits S2:S3:S4:S5 in an 1:1:2:1 molecular organization.

delivery of the **A** domain into the host cell cytosol and expression of cellular toxicity.[3–7] **AB** toxins are typically synthesized in an inactive form and require proteolytic processing to be converted into an active form. Single-chain **AB** toxins, such as the clostridial botulinum neurotoxins (BoNTs) and tetanus neurotoxin (TeNT), cleave a trypsin-sensitive site located between **A** and **B** retaining the interchain disulfide.[8] **AB** toxins, such as diphtheria toxin and anthrax toxin, are also proteolyzed by a host protease, furin, for inter-domain cleavage or **B** subunit assembly, respectively.[9,10] Most **AB5** toxins also contain a protease sensitive site, which is proteolyzed by a bacterial- or host-protease.[10,11] The four **AB5** family members; cholera toxin (Ctx), Shiga toxin (Stx), pertussis toxin (Ptx), and subtilase cytotoxin (SubAB), are proteolyzed within the **A** domain, between the **A1** catalytic and **A2** linker-subdomains, which are linked by a disulfide bond.[12–14]

## 2. PROTEIN TOXIN HOST RECEPTORS

**AB** toxins exploit host receptors to coordinate host cell binding and delivery of **A** into the host cell cytosol. Members of the BoNT family have dual host receptors, often a glycolipid (ganglioside) and a protein receptor, while TeNT utilizes two gangliosides as receptors.[15–20] Ctx binds a single class of glycolipid (GM1a) while other members of the Ctx family, the heat labile *E. coli* enterotoxins (LTx's) LT-I and LT-IIa-c, bind multiple gangliosides species with varied affinities.[21,22] Other toxins, such as ricin and Ptx are promiscuous, and bind specific carbohydrate moieties on glycolipids and glycoproteins.[23,24]

The specificity and affinity of protein toxins for host receptors is often assessed by in vitro biochemical and cell biological techniques, including solid-phase assays, surface plasmon resonance, receptor loading, and receptor depletion. Additional in vivo gene knockout studies confirm receptor specificity and tissue tropism. Ultimately, protein toxin action requires a receptor, a productive pathway to traffic the **A** domain to the target substrate, and the presence of the target substrate. This review focuses on protein toxins that utilize a subset of glycosphingolipids, gangliosides, and related carbohydrate moieties as host cell receptors.

## 3. PROTEIN TOXIN ENTRY INTO CELLS

Protein toxins enter cells by multiple pathways. Most **AB** toxins bind a cell surface host receptor and enter via receptor-mediated endocytosis. Early endosomal entry involves clathrin-dependent and clathrin-independent pathways, such as bulk uptake, caveolae, and micropinocytosis. **AB5** toxins Ctx, LTx, Stx, and SubAB are associated with both clathrin-dependent and clathrin-independent pathways depending on the host cell type.[25–28] Following uptake, the catalytic domain escapes the early endosome upon endosome acidification or retrograde traffics to the endoplasmic reticulum

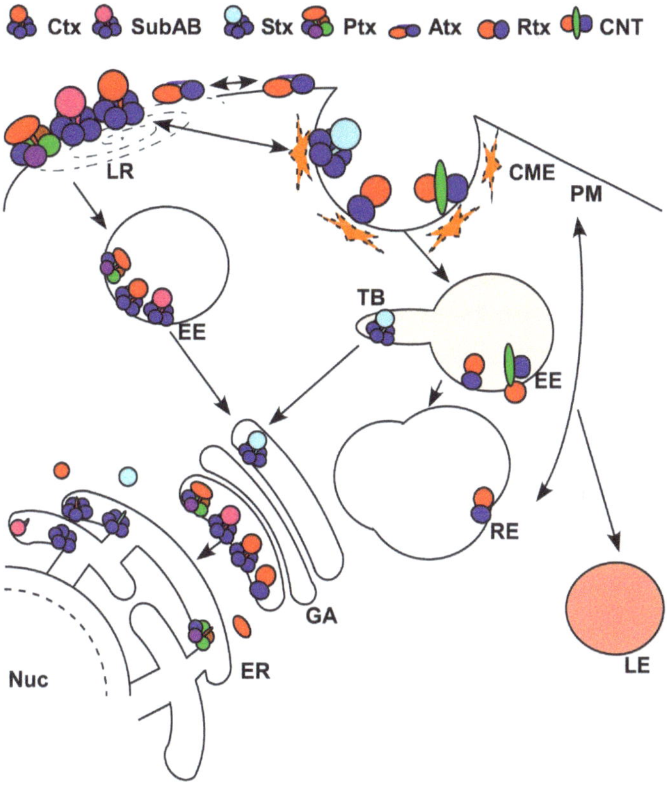


**Fig. 2** Endocytosis of **AB** protein toxins. **AB5** protein toxins (Ctx, SubAB, Stx, and Ptx) enter by clathrin-mediated endocytosis (CME) and clathrin-independent lipid-raft (LR) pathways. Stx enters endosomal tubules (TB) and retrograde traffics with GPP130 through a bypass pathway toward the Golgi apparatus (GA). Other **AB5** protein toxins enter a retrograde trafficking pathway toward the endoplasmic reticulum (ER). The **A** domain of **AB5** protein toxins, excluding SubAB, retrograde translocate into the cytosol to modify a host substrate. SubAB modifies a resident ER protein. Atx associates with the plasma membrane (PM) phospholipids or lipid raft glycosphingolipids. Rtx and clostridial neurotoxins (CNT) enter early endosomes (EE) (in neurons); Rtx can retrograde traffic toward the ER or recycle back to the cell surface. CNTs escape acidifying endosomes (synaptic vesicles for BoNTs and endosomes in inhibitory neurons for TeNT) into the cytosol to cleave SNARE proteins. Other protein toxins (not shown) enter degradative pathways in late endosomes (LE).

(ER) where the **A** domain enters the cytosol to modify a cytosolic host substrate, leading to host cell intoxication or apoptosis[29] (Fig. 2).

Other **AB** toxins enter a retrograde endosomal pathway and traffic to the Golgi and the ER. For **AB5** toxin family members, Ctx/LTx, Stx, Ptx, and SubAB, the **B** pentamer encodes a binding domain, but lacks a translocation domain.[2,29–32] Once these **AB5** toxins retrograde traffic to the ER, the **A** domain dissociates from the **B** pentamer and interacts with a chaperone, such as protein disulfide isomerase (PDI) to reduce the intrachain disulfide.[33–37] Following dissociation and release, the **A1** subdomain usurps the unfolded protein response (UPR) pathway to escape through the Sec61 translocon.[38] For Ctx family members, upon retrograde translocation into the cytosol, via the Sec61 retrieval system, the **A1** subdomain binds cytosolic chaperones, including Arf,[39,40] and refolds before trafficking to ADP-ribosylate the stimulatory heterotrimeric G-protein alpha subunit ($G\alpha_s$). ADP-ribosylation locks $G\alpha_s$ in the GTP bound form to constitutively activate host adenylate cyclase and stimulate cAMP, uncoupling host signaling.[40] An exception within the **AB5** family is SubAB which remains in the ER lumen to modify a chaperone protein involved in UPR.[41,42] The Ctx family members contain a KDEL, RDEL, or HDEL sequence at the C terminus of the **A** domain, which has been associated with retention in the ER by the KDEL receptor. The **B** pentamer also independently retrograde traffics to the ER, which indicates the KDEL sequence(s) optimize retrograde trafficking, but are neither necessary nor sufficient to deliver **AB5** toxins to the ER.[43,44]

## 4. GLYCOLIPIDS

Eukaryotic cells are enveloped by a carbohydrate-rich coating, the glycocalyx. The glycocalyx contains glycoproteins and glycosphingolipids anchored in the outer leaflet of the plasma membrane.[45] Glycans face the exterior milieu and mediate cell–cell interactions via binding complementary molecules on other membranes or regulating activity by forming a buffer zone between host cell and extracellular microbes. Pathogens utilize a capsular glycocalyx coating to avoid host detection and immune cell assault and have evolved to utilize the host cell glycocalyx as receptors, by binding these molecules with adhesins, hemagglutinins, or by releasing protein toxins.[46]

Although glycosphingolipids refer to sphingosine-based lipids containing glycosidic bonds to simple carbohydrate molecules, the lipids are further classified into series based on a core structure.[47] The lipid portion of the molecule, ceramide, contains a sphingoid base connected to a fatty acid through an amide bond.[48] The ceramide portion of the ganglioside can be composed of sphingonine, sphingosine, or phytosphingosine and fatty acids

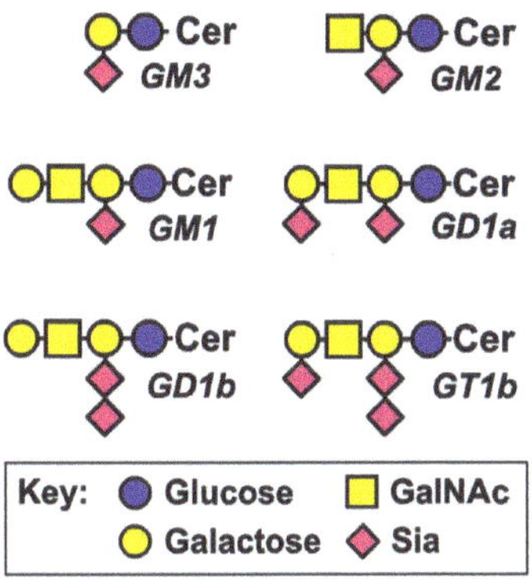


**Fig. 3** Simple and complex gangliosides. Gangliosides are glycosphingolipids which contain a ceramide and oligosaccharides that are complexed with one or more sialic acid (ceramide not depicted). Simple gangliosides are depicted on the top row and more complex gangliosides are depicted on bottom row. Gangliosides are enriched in neurons.

can vary in chain length from ~16 to >24 carbons, which is associated with differences in trafficking.[48,49] Glycosphingolipids are abundantly found in the gray matter of the brain, with 10%–12% of brain sphingolipids being composed of gangliosides.[50] Ganglio-series gangliosides have a common core of Galβ1-3GalNAcβ1-4Galβ1-4GlcβCer, while another common glycolipid species, the globosides, have Galα1-4Galβ1-4GlcβCer (Gb3) as a core[45] (Fig. 3).

While gangliosides are not required for cell survival, gangliosides are necessary for survival in whole-organisms.[51] The hydrophilic, negatively charged headgroups of gangliosides spread on the cell surface and exist in varied conformations at each branch points. Gangliosides are negatively charged due to decoration with 9-carbon sugars, sialic acids, off the core sugars.[50] In humans, sialic acids are composed of *N*-acetylneuraminic acid (Neu5Ac) while other mammals also synthesize *N*-glycolylneuraminic acid (Neu5Gc).[52] In addition to variation in ceramide composition, the number of sialic acid groups connected to the core sugars also vary. Gangliosides contain 75% of the brain's sialic acid, primarily GM1a, GD1a, GD1b, and GT1b (Fig. 3).[53] Gangliosides like other glycosphingolipids are restricted to the outer leaflet of the cell membrane, with the ceramide and first core sugar glucose engaged in this lipid layer.[51] Gangliosides are primarily found in the extracellular plasma membrane, but are also detected in intracellular organelles. The Golgi is responsible for the diversity of cell surface glycans.[54] Neither gangliosides nor globosides are uniformly distributed in the plasma membrane, but cluster in lateral microdomains called detergent-insoluble glycosphingolipid-enriched domains (DIGs) which self-associate due to biophysical properties of long, unsaturated carbon chains.[55] DIGs are known as lipid rafts in living cells and concentrate glycosylphosphatidylinositol-(GPI) anchored proteins, cholesterol, and gangliosides.[56] Lipid rafts are often coupled to signaling complexes, which contain receptor tyrosine kinases (RTKs), G-protein–coupled receptors (GPCRs), and downstream partners, such as mitogen-activated protein (MAP) kinases and heterotrimeric G-proteins (Gαβγ). The binding of protein toxins to glycosphingolipids is prerequisite, but coupling of the lipid species to signaling domains for productive uptake affects tissue susceptibility.

Complex gangliosides, GD1a and GT1b, are ligands for myelin-associated glycoprotein (MAG), promoting axon stability and neurite inhibition.[57] Altered surface expression or storage of gangliosides is associated with abnormal cell–cell contact inhibition and various diseases.[58] Mice lacking some complex gangliosides, GM2 or GD2 synthase knockouts, develop hearing loss and abnormal motor pathology with age.[59] Double knockouts in these aforementioned glycan synthase systems (*St3gal5* or *St8Sia1/B4galnt1*) die shortly after birth; therefore, the necessity for ganglioside synthesis is commonly exploited by pathogens and toxins that bind carbohydrates.[60–62]

## 5. PROTEIN TOXINS UTILIZE GANGLIOSIDES AS HOST RECEPTORS

### 5.1 Clostridial Neurotoxins of *Clostridum botulinum* and *Clostridum tetani*

The clostridial neurotoxins (CNTs) of *Clostridum botulinum* and *tetani* target motor neuron presynaptic termini in the periphery and inhibitory interneurons in the central nervous system, respectively.[8] CNTs comprise seven serologically distinct BoNTs (A–G) and TeNT.[63] CNTs are synthesized as single-chain **AB** toxins cleaved into a dichain; an **A** catalytic light chain (LC) connected to a **B** heavy chain (HC) containing an N-terminal translocation domain and a C-terminal receptor binding domain.[64] The LCs of CNTs are zinc-metalloproteases, which cleave neuron-specific isoforms of SNARE proteins involved in vesicle fusion at the plasma membrane.[65,66] Following cleavage, the tertiary complex cannot form between vesicle and plasma membrane-associated SNARES, and fusion of neurotransmitter-containing vesicles to the presynaptic membrane is inhibited, resulting in paralysis.[67,68]

Toxin trafficking determines the type of paralysis, as both BoNT/B and TeNT cleave the host SNARE at the same scissile bond.[66,69] Flaccid paralysis is characterized by inhibition of acetylcholine release at the neuromuscular junction, with the motorneuron not signaling the downstream muscle to contract.[70] For TeNT, signaling is uncoupled between an interneuron and motorneuron when release of inhibitory glycine into the synaptic cleft is inhibited. This results in spastic paralysis as the motor neuron is persistently releasing acetylcholine, causing muscle contraction.[70] CNTs bind complex gangliosides, which are enriched in the outer leaflet of neurons (Table 1).[71]

TeNT, when isolated from brain and cultured neuronal cell membranes, was found bound to a glycolipid, later identified as GT1b.[72,73] TeNT has two ganglioside binding pockets in the receptor binding domain.[74] Toxin binding two gangliosides is necessary and sufficient for entry into neuronal and nonneuronal cells.[20] Although protein receptors for the toxin have been proposed, none are necessary for intoxication.[75–77] TeNT has been observed to colocalize with signaling endosome proteins, such as the growth factor receptors, which like gangliosides, are enriched on lipid rafts and DIGs.[78] Recent studies indicate that the receptor binding domain pivots with respect to the N-terminal domains of full-length TeNT in response to a change in pH.[79] This may influence the orientation of the translocation domain with respect to channel formation, as BoNT–receptor interactions regulate channel formation.[80,81] BoNTs, excluding BoNT/C, bind a ganglioside and a protein receptor sequentially; upon depolarization of the neuronal membrane, fusion of synaptic vesicle exposes the lumenal domains of a synaptic vesicle protein receptors.[82] Interactions with both the ganglioside and SV2 appear to be sequential for entry to neurons.[83] The overall structure of BoNT/C aligns well with other BoNTs serotypes, except at the C terminus where there are two unique ganglioside binding pockets, which appear sufficient for the entry of BoNT/C into neurons,[84] precluding the use of a host protein receptor.

## 5.2 Cholera Toxin Family of AB5 Toxins (Ctx/LTx) of *Vibrio cholerae* and Enterotoxigenic *Escherichia coli*

The cholera toxin family is comprised of Ctx produced by *Vibrio cholerae* and the related heat-labile enterotoxins (LTx), LT-I, LT-IIa, LT-IIb, and LTIIc, produced by enterotoxigenic *Escherichia coli* (ETEC).[2] *V. cholerae* and ETEC are often found in water and food contaminated with fecal matter. Ctx/LTx target intestinal epithelial cells to cause diarrheal disease, with LTx family members generally eliciting pathology of shorter duration and milder illness

**Table 1** AB Protein Toxin Receptors, Action, and Disease.

| Organisms | Toxins | Receptors | Substrates | Enzyme Actions | Diseases |
|---|---|---|---|---|---|
| *C. tetani* | TeNT | GT1b | VAMP | $Zn^{2+}$ metalloprotease | Spastic paralysis |
| *C. botulinum* | BoNT/A | GT1b, SV2C | SNAP-25 | $Zn^{2+}$ Metalloprotease | Flaccid paralysis |
| *C. perfringens* | Atx | GM1a | Phosphatidyl choline | Phospholipase C | Gas gangrene |
| *V. cholerae* | Ctx | GM1a | $G\alpha_s$ | ADP-ribosyl transferase | Diarrheal disease |
| ETEC | LT-I | GM1a, LPS, A/B blood | $G\alpha_s$ | ADP-ribosyl transferase | Diarrheal disease |
| | LT-IIa | GD1b>GD1a>GM1a | | | |
| | LT-IIb | GD1a>GD1b | | | |
| | LT-IIc | GM1a, long acyl | | | |
| *B. pertussis* | Ptx | Glycolipids, glycoproteins; galactose, lactose, GD1a | $G\alpha_i$ | ADP-ribosyl transferase | Whooping cough |
| *R. communis* | Rtx | Glycolipids, glycoproteins; galactose | 28S rRNA | RNA N-glycosidase | Pulmonary edema, seizures |
| *S. dysentariae* | Stx | Gb3 | 28S rRNA | RNA N-glycosidase | Hemorrhagic colitis, HUS |
| STEC | Stx1 | Gb3 | 28S rRNA | RNA N-glycosidase | Hemorrhagic colitis, HUS |
| | Stx2 | Gb3, Gb4 | 28S rRNA | | |
| STEC | SubAB | Glycolipids, glycoproteins; Neu5Gc | BiP | Subtilase-like serine protease | HUS |

ETEC, Enterotoxigenic *Escherichia coli*; HUS, hemolytic uremic syndrome; STEC, Shiga toxin-producing *Escherichia coli*.

than Ctx.[85] Ctx and LTx cause distention in the rabbit ileal loop model, previously a gold standard for diagnosing this disease.[86] Cholera outbreaks are endemic in many developing countries, while LT-related illnesses are associated with travelers' diarrhea in developing countries and more recently, livestock disease in developed countries.[86]

Ctx and LT-I invade the intestinal epithelia by binding with high affinity to the ganglioside GM1a and retrograde traffic to the ER.[21,87] Ctx/LTx **A1** catalytic domain is an ADP-ribosyltransferase that ADP-ribosylates Arg177 of the stimulatory α-subunit of the heterotrimeric G-protein, $G\alpha_s$, following translocation into the cytosol.[88–91] Addition of the negatively charged ADP-ribosyl group constitutively activates $G\alpha_s$ by preventing hydrolysis of GTP. $G\alpha_s$-GTP binds to host adenylate cyclase, which increases cyclic AMP (cAMP) production by the cyclase and modulates ion channels to release ions into the extracellular space.[92] Persistent chloride ion release and efflux of water out of the intestinal epithelial cells and into the intestine results diarrhea, electrolyte imbalances, and severe dehydration.[93]

Ctx- and LT-I- **B** subunits share ~80% identity and both bind GM1a; however, LT-I also binds LPS and A/B glycolipid antigens on human erythrocytes.[21,94–97] The **B** subunits are less conserved between Ctx and LT-II family members with ~14% shared identity. LT-IIa predominantly binds GD1b, GD1a, and GT1b, and to a lesser extent GM1a and other gangliosides, while LT-IIb predominantly binds only GD1a and GD1b.[22,98] The newest member LT-IIc has been observed to bind GM1 species with long acyl chains.[99] Ctx/LTx bind up to 5 glycans subunits; however, for CTx, binding of one GM1a molecule is sufficient, though less efficient for retrograde trafficking and cell intoxication.[43]

## 5.3 Stx of *Shigella dysenteriae* and Stx1/2

Stx of *S. dysenteriae* and Stx1/2 of Shiga toxin-producing *E. coli* Stx, produced by *S. dysenteriae*, was identified by Kiyoshi Shiga as the causative agent of dysentery.[100] Almost a century later, the toxins now renamed as Shiga-like toxins 1 and 2 (Stx1 and Stx2), were identified in Shiga toxin-producing *E. coli* (STEC). Stx and Stx1 differ by a single amino acid, while Stx1 and Stx2 share ~55% identity.[30] Both *E. coli* Stx1 and Stx2 produce immunologically distinct subtypes. Stx-producing bacteria cause the infection *Shigella*, which is contracted by ingestion of contaminated foods.[101]

Stxs are **AB5** toxins, the **A** subunit contains N-glycosidase activity responsible for removal of an adenine residue from 28S rRNA of the 60S ribosome subunit.[102] Modification of the rRNA inhibits protein synthesis as the ribosome no longer interacts with elongation factor 1.[103,104] Stalled synthesis induces ribotoxin stress and the UPR, resulting in apoptosis.[105] Stx translocates through intestinal epithelial cells to target the endothelial cell layer, and in serious cases intoxicates the kidneys.[101] Following modification of the rRNA substrate, the localized infection presents as hemorrhagic diarrhea and if Stx reaches the kidneys, hemolytic uremic syndrome (HUS) occurs as a result of low platelet count, erythrocyte lysis, and kidney failure.[100] The five identical **B** pentamers of Stx bind the globoside Gb3, Stx1, and Stx2 weakly bind Gb4 (GalNAcβ1-3Galα1-4Galβ1-4GlcβCer), while subtype Stx2e preferentially binds globoside Gb4.[106] Cells and mice lacking Gb3 are insensitive to Stx, while mice with Fabry's disease containing excess Gb3 throughout tissues also possess increased resistance to Stx.[107] Gb3, like other glycosphingolipids, clusters in DIGs or lipid rafts in cells.[108] The presence of cholesterol in microdomains enhances binding and entry of the Stx and Stx1, as pretreatment with cholesterol-binding agents, such as methylβcyclodextrin (MβCD) reduces Golgi-localization and cytotoxicity.[109] Stx2 is associated with lipid rafts in vitro and clinically associated with development of HUS; Stx2 does not interact with the host *trans*-Golgi cycling resident protein, GPP130.[106,110,111]

Shiga toxins are endocytosed by clathrin-dependent and clathrin-independent pathways, depending on the cell type.[32,112] In HeLa cells, clathrin-independent tubular invaginations contribute to sorting of Stx **B** subunits, which bind GPP130 to facilitate retrograde traffic.[113,114]

## 5.4 Ricin of *Ricinus communis*

Ricin (Rtx) is a toxigenic lectin produced by seeds of the castor oil plant *Ricinus communis*. Exposure to Rtx by inhalation or injection is more lethal than digestion, as the toxin is sensitive to bile acids.[115] The ricin **A** subunit, like Stx, is a type II ribosome-inactivating protein, which inhibits protein synthesis by depurination of the rRNA catalyzed by N-glycosidase activity.[102,116] Stx- and Rtx-**A** subunit share <50% identity, but residues involved in adenine abstraction are conserved.[103] Cytotoxic effects of include cell lysis through blebbing and DNA degradation.[27] The route of exposure determines the symptoms; intravenous injection in rodents results in hypoglycemia, inflammation, and ketosis, while inhalation results in fluid accumulation in the lungs and eventual respiratory failure.[117] Like other single-chain **AB** toxins, Rtx **A** subunit catalytic domain connects to the **B** subunit binding domain by a peptide linker. The **B** domain contains two homologous lectin-binding sites specific for galactose.[117–119] Rtx undergoes posttranslational glycosylation on both chains, which remain associated following host cleavage by an interchain disulfide bridge.[4,23]

Rtx binds glycolipids (including gangliosides) and glycoproteins containing terminal galactose (or N-acetylgalactosamine) in β1-4 linkage. In early endosomal sorting, Rtx enters a recycling pathway and traffics back to the cell surface as well as to lysosomes.[120] Uptake of ricin occurs by two pathways: ricin **B** subunit binds galactose-containing host glycolipid and glycoproteins and conversely, glycosylated residues of ricin toxin are bound by a host mannose receptor, which cycles between *trans*-Golgi and late endosomes.[121,122] While galactose-binding by the **B** subunit results in robust uptake of Rtx, the latter mannose-dependent uptake by the mannose receptor is responsible for the cytotoxic effects by delivering RTx to a productive *trans*-Golgi pathway.[121]

## 5.5 SubAB of STEC

Recently, a new AB5 toxin member, subtilase cytotoxin (SubAB) was found to cause HUS-like illness in STEC strain O113:H21. SubAB binds nonhuman Neu5Gc as a receptor and is more cytotoxic than Stx in in vitro cell culture assays.[2] SubAB is also unique, as the **A** subunit is not required to translocate into the cytosol, since the target substrate is located within the lumen of the ER. The **A** subunit contains subtilase-like serine protease activity and cleaves the ER-resident protein BiP.[42] BiP contributes to the UPR as a chaperone for folding nascent proteins to be degraded by the proteasome.[123] BiP also forms a seal within the Sec61 pore; proteolysis of BiP destroys the integrity of this barrier leading to ER permeability.[123] Humans, in contrast to the great apes, have acquired mutations in the biosynthetic pathways to synthesize and convert Neu5Ac into Neu5Gc.[52] With loss of Neu5Gc, Neu5Ac becomes the only sialic acid synthesized. Intriguingly, humans are susceptible to SubAB due to incorporation of dietary Neu5Gc into cellular glycolipids and proteins via meat and dairy consumption.[124]

While SubAB specifically recognizes the sialic acid Neu5Gc, which contains an additional hydroxyl substituent relative to Neu5Ac, many human pathogens have evolved to exploit Neu5Ac as a receptor. For example, the newest member of the enterotoxins, LT-IIc was isolated from avians, some of which produce Neu5Gc, but has similar affinity to both sialic acid species.[99,125] Incorporation of dietary sialic acids and pathogens that bind multiple sialic acid species has potential to expand the host range for zoonotic diseases.

## 5.6 Ptx of *Bordetella pertussis*

Ptx is a major virulence factor produced by *B. pertussis*, which is transmitted by inhalation of aerosolized droplets transmitted early in the infectious state of an infected individual.[126] Although Ptx alone does not cause whooping cough, entry into pulmonary epithelial and endothelial cells leads to inflammation and white blood cell recruitment.[127] Ptx alteration of host signaling also leads to hyperinsulinemia and pulmonary hypertension.[127,128]

Ptx is an **AB5** family member. Like Ctx, the **A** subunit known as S1, ADP-ribosylates the inhibitory heterotrimeric G-protein ($G\alpha_i$) at a C-terminal cysteine, causing dysregulation of cAMP signaling and ion efflux.[129] The **B** pentamer is comprised of four heterologous subunits known as S2, S3, S4, and S5, in a 1:1:2:1 ratio, respectively.[130] Heterogeneity of the **B** pentamer subunits, including the two carbohydrate binding subunits, S2 and S3, allows for Ptx to recognize multiple glycans with varied affinities.[131,132] As a result, Ptx associates with glycolipids and glycoproteins, such as the serum protein fetuin, which is decorated with N-linked glycans.[24] **B** pentamer binding to host receptors alone can induce T cell proliferation and agglutinate erythrocytes.[133] S2 and S3 are 77% identical and each contain a sialic acid (Sia) binding site and an N-terminal carbohydrate binding site.[130] In a glycan array screen, Ptx-bound α2-6–linked Sia in N-glycans, α2-3–linked Sia in O-glycans, and complex gangliosides.[134,135] To discriminate glycans bound by the N-terminal region, dimers lacking the Sia domain were assembled with S4 for stability and bound asialo N-linked glycans.[134]

Although not unique to Ptx, characterization of glycans bound by the **B** subunits of toxins varies depending on the format of the assay. For example, Ptx binds gangliosides, such as GD1a on liposomes, but unlike Ctx, does not detectably bind lysate-derived gangliosides separated by thin layer chromatography.[135,136] Additional considerations for interpretation of toxin affinity for glycan substrates are the orientation of the glycan, flat, or curved surfaces, as well as the linker composition, ceramide, or acyl species of the glycolipids.

## 5.7 Alpha-toxin of *Clostridium perfringens*

*C. perfringens* produces many virulence factors including alpha-toxin, a virulence factor for pathogenesis. Alpha-toxin is the causative agent of gas gangrene, which is characterized by edema, myonecrosis, and gas. Inflammation, resulting from alpha-toxin action leads to septic shock and organ failure.[137,138] Alpha-toxin is also associated with Crohn's Disease, where the toxin targets intestinal epithelial cells and resident macrophages.[139]

Alpha-toxin is a single-chain **AB** toxin, which contains an **A** subunit with zinc-metalloprotease-dependent phospholipase C catalytic activity, which hydrolyzes phospholipids into diacylglycerol (DAG) and ceramide. The **A** subunit is connected to a binding domain that contains both hemolytic activity and a C2-like polycystin-1, lipoxygenase, alpha-toxin (PLAT) domain.[137] In the presence of calcium, the **B** domain binds and partially inserts into the phospholipid membrane.[140] There is also a ganglioside binding site, specific for GM1a, located on a loop between the **A** and **B** subunits.[141] The generation of DAG, a common eukaryotic second messenger, results in uncoupled signaling and release of cytokines.[142]

## 6. MOLECULAR INTERACTIONS OF PROTEIN TOXINS WITH SUGARS

The **AB5** family members Ctx and SubAB contain five sugar binding sites, one per **B** subunit, while Ptx contains up to two functional binding sites on each of the S2 and S3 subunits, and Stx contains up to 3 binding sites per subunit.[106,130,143,144] Each **AB5** family member **B** subunit contains a homologous fold at their C terminus comprised of five or six antiparallel beta strands, which form a barrel topped by a short alpha-helix (Fig. 4A) with and without sugar. This oligonucleotide and oligosaccharide binding domain is known as oligomer binding fold (OBF).[145] Ptx S2 and S3 subunits contain an additional N-terminal domain with homology toward C-type eukaryotic lectins such as rat mannose binding protein.[134]

### 6.1 Cholera Toxin Family of AB5 Toxins (Ctx/LTx) of *Vibrio cholerae and Enterotoxigenic Escherichia coli*

Ctx and LT utilize a pocket formed by loops to bind GM1a; in Ctx these loops are found between β1 and α1, β4 and α2, β5 and β6 on one monomer and the loop between β2 and β3 on the other monomer.[144] The pocket is oriented toward the glycolipid receptor at the membrane. For Ctx, the terminal sugars of GM1a, galactose and sialic acid, predominantly interact with the B subunit.[2,144] The galactose forms hydrogen bonds via oxygen with the nitrogen atoms in the **B** subunit residues Asn90 and Lys91.[143] A hydrophobic interaction between galactose and the indole ring of Trp88 is also observed (Fig. 4B). The sialic acid of GM1a forms hydrogen bonds with the carbon backbone of residues Glu11 and His13, and a hydrophobic interaction with Tyr12.[94] Gly33 lies in a loop between β2 and β3 of the

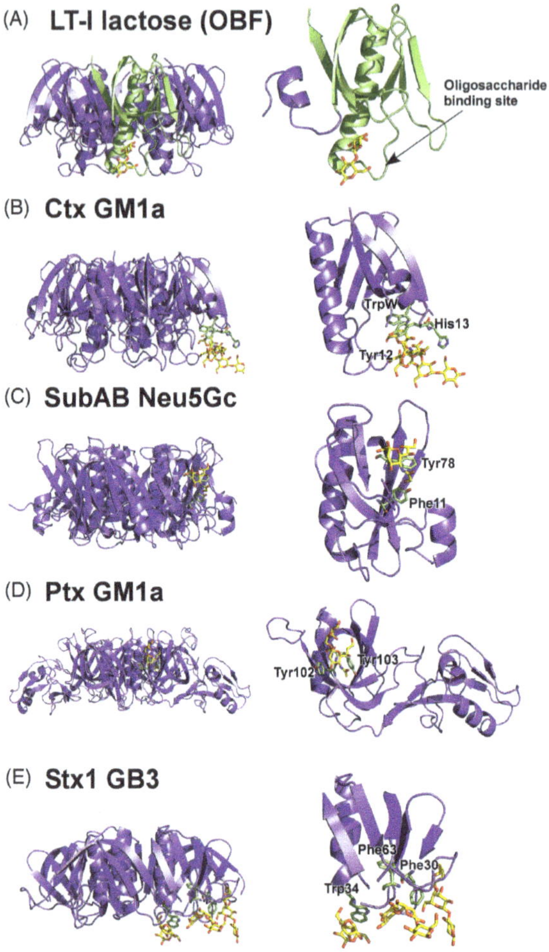


**Fig. 4** **AB5** toxin–aromatic sugar interactions. Crystal structures of the **B** pentamers of **AB5** toxins LT-I, Ctx, SubAB, Ptx, and Stx1 are shown in *blue* on the left of each subpart. Sugar ligands (carbohydrates, gangliosides, or globosides) are shown in *yellow*. Aromatic residues in the **B** domain are shown in *green* and numbered on the right. Polar groups, hydrophobic interactions, and interdomain interactions are left out for clarity. Selected oxygen atoms are shown in *red* and nitrogen atoms shown in *blue*. (A) LT-1 **B** pentamer (PDB: LTT1) contains an oligosaccharide binding fold (OBF) shown in *green* and is bound to lactose. The oligosaccharide binding site is shown magnified on the right. (B) Ctx **B** pentamer (PDB: 3CHB) shown in *blue* with one subunit bound to GM1a. A single subunit is shown on right with aromatic residues involved in binding. (C) SubAB **B** pentamer (PDB: 3DWP) is bound to Neu5Gc, with magnification of this interaction shown on right. (D) Ptx **B** pentamer (PDB: 1PTO) is shown with S3 bound to GM1a. S2 and S3 interactions with GM1a are similar; S3-GM1a is shown on the right. (E) Stx1 **B** pentamer (PDB: 1BOS) is shown bound to three Gb3 molecules. One **B** subunit interacts with three Gb3 molecules shown on the right.

adjacent monomer and forms the pocket edge with one hydrogen bond to galactose.[94,144] In addition to GM1a, LT binds asialo-GM1 and GD1b, derivatives that lack and gain an additional sialic acid, respectively; the substitution of His13 for Arg13 in LT is suspected to accommodate this promiscuity.[146] LPS binding and erythrocyte group A or B antigens by LT have been mapped to residues not involved in binding GM1a.[147]

## 6.2 SubAB of STEC

Unlike Ctx and Stx, SubAB binds the nonhuman derived sialic Neu5Gc in a shallow binding pocket located on outer face of the **B** pentamer.[41] The SubAB pocket resembles the sialic acid pocket of Ptx B subunits S2, S3, and S5, where S2 and S3 bind sialic acid Neu5Ac in a similar orientation, but unique interactions are involved in binding the different sialic species.[41] The chemical structure of Neu5Gc differs from human-derived Neu5Ac by a single hydroxyl group.[52] However, the presence of this hydroxyl group allows SubAB pocket residues Tyr78 and Met10 to form hydrogen bonds with sialic acid, where Neu5Gc stacks with Phe11 of SubAB, forming a van der Waals interaction (Fig. 4C).[2,41] The affinity of SubAB for Neu5Gc is unique and specific; serum from chimpanzees containing glycolipids and proteins decorated with Neu5Gc, but not human serum that predominantly contains Neu5Ac, can compete with SubAB binding of immobilized the Neu5Gc glycan.[124]

## 6.3 Ptx of *Bordetella pertussis*

Ptx binds glycoproteins through **B** subunits, S2 and S3, both of which contain two unique carbohydrate domains. Analysis of a crystal structure incubated with glycolipids identified electron density for Neu5Acα2–6 Gal, but not galactose within the binding pocket[130,131] (Fig. 4D). Sialic acid forms hydrogen bonds with Ser104, and hydrophobic interactions with Tyr102 and Tyr103, on the sides of the pentamer.[130,131] It was previously observed that deletion of Tyr102 and Tyr103 in S2 and S3, or substitution of Ser104 with a charged group reduced biological activity of Ptx.[148]

## 6.4 Stx of *Shigella dysenteriae*

Stx1 binds the Gb3 analog in up to three binding sites per **B** subunit. Gb3 binding pockets are located near the bottom of the **B** pentamer to allow Stx1 to bind the membrane face.[149] Two binding sites are formed between subunits, while one is formed near the membrane. The first site is a pocket formed on a single monomer by loops β2-β3, β5-β6, and strands β3 and β4; hydrogen bonding is observed between both galactose sugars and the **B** subunit residues Thr21, Glu28, Gly60, and Asp17.[149] Additionally, hydrophobic interactions, including a ring stacking interaction with Phe30 are observed with Gb3.[106,149] The second Gb3 binding site is located between monomers and is comprised of loops β2-β3 from site one, β4-α, and β5-β6 (Fig. 4E).[106,109] This site was originally predicted based on known crystal structures of other **AB5** family members and appears to be highly occupied. The binding pocket contacts Gb3 through hydrogen bonding the two galactose sugars with the Asp16, Asn32, Arg33, Asn55, and Phe63 while many hydrophobic interactions are observed between the galactose sugars, glucose interacts with Asn55.[149] At binding site 3, Gb3 is bound within the pentamer cleft by β4–α of the second site, and β2-β3 loop at the first site. Asp18, Trp34, and Asn35 hydrogen bond with the first galactose while the residue Trp34 located within and on an adjacent monomer stack with both the first and second galactose moiety.[106,149]

## 6.5 Rtx of *Ricinus communis*

The Rtx crystal structure exhibits two noncooperative lectin binding domains containing a β-trefoil fold separated by ~3.5 nm.[119] Rtx, binds glycolipids or glycoproteins, and the distance between lectin binding domains allows the **B** subunit to sterically accommodate host receptors.[150] The two domains are composed of three lobes referred to as 1(α, β, and γ) and 2(α, β, and γ), with 1α and 2γ confirmed to bind galactose.[23,118,151,152] Each lobe contains a tripeptide kink and an aromatic residue; the motif Gln-X-Trp.[118] For ricin, mutagenesis, docking analysis, and comparison to other lectin-binding proteins predicts two interactions upon receptor binding: ring stacking of the galactose moiety against an aromatic residue and hydrogen bonding between pocket residues and oxygen groups on the sugar.[150]

## 6.6 Clostridial Neurotoxins of *Clostridum botulinum* and *Clostridum tetani*

Alignment of the clostridial toxins alpha-toxin, TeNT, and most BoNTs serotypes, reveals a conserved GM1a binding motif comprised of ~30 residues (H . . . SXWY . . . G).[19,141] TeNT and BoNTs contain this motif within the receptor binding domain, while alpha-toxin contains this sequence on an exposed loop between the A and B subunits.[64,137] For all toxins, mutagenesis of the aromatic consensus residues tryptophan and tyrosine abrogates binding to the glycolipid receptor.

Tetanus toxin binds the membrane utilizing two pockets. Mutation of Trp1289Ala (W pocket) and Arg1226Leu (R pocket) abrogates ganglioside binding in solid phase and cell binding assays.[20,74] Mutational analysis of single pockets in TeNT discriminated glycan preference bound by each pocket; the R pocket binding b-series and W pocket binding a-series gangliosides.[74] The crystal structure reveals that the R pocket binds a GT1b analog in a shallow pocket primarily through the sialic acid moiety. The bound sialic acid forms hydrogen bonds through oxygens to charged and polar residues Asp1147, Asp1214, Asn1216, and Tyr1229.[74] A salt bridge is also formed between the sialic acid and Arg1226.[74,153] The W pocket, which is also conserved in many BoNTs binds a lactose moiety by stacking with the indole ring of Trp1289 and forming hydrogen bonds between the sugar's hydroxyl group and the charged side chains of His1272, Asp1222, and the amide of Thr1270.[74,153] The recently solved crystal structure of full length TeNT includes GD1a within the W pocket which stabilized the crystal.[79] A similar interaction is observed in the W pocket of BoNT/A incubated with a GT1b analog; analogous hydrogen bonding to His1253, Glu1203, and Ser1264 and hydrophobic interactions with the indole ring of Trp1266.[154]

Alpha-toxin does not bind to sialic acids alone or require complex gangliosides, but is specific to GM1a.[141] Molecular docking simulation predicts that alpha-toxin and GM1a interaction is dependent upon hydrophobic ring stacking and hydrogen bonding of Trp84 in alpha toxin with the sialic acid, while Tyr85 hydrogen bonds with the galactosamine portion of the glycolipid.[155] The binding of GM1a is not necessary for endocytosis of alpha-toxin, but is hypothesized to tether the toxin, before binding to phospholipids via C-terminal aromatic residues.[140] Binding of alpha-toxin to GM1a is necessary to induce signaling through TrkA, a major step in pathogenesis.[142]

## 7. PROTEIN TOXIN-MEDIATED TRAFFICKING AND SIGNALING

Gangliosides located within lipid rafts are coupled with signaling complexes, which include RTKs and GPCRs that modulate downstream signaling upon ligand binding. Interaction between ganglioside and receptor modulates activity through glycan–glycan or glycan–protein interactions; such interactions include hydrogen bonding that interferes with normal ligand–receptor interactions.[156,157] For example, how GM3 modulates fibroblast growth factor receptor or epidermal growth factor receptor may be due to reducing receptor–ligand binding affinity or by exclusion of receptors from lipid rafts that leads to dampened signal.[158,159] Another ganglioside, GM1a, stimulates neurotrophin receptors such TrkA like the endogenous ligand, nerve growth factor, and leads to receptor activation.[160] The studies that elucidated gangliosides as ligands for TrkA found brain slices stimulated with high concentrations of whole-ganglioside, but not the components, such as sialic acids, glycans, or ceramides, induced autophosphorylation of TrkA leading to downstream signaling through MAP kinase pathway.[160] In addition to being ligands for growth factor receptors, gangliosides are also ligands for MAG, inhibiting neuritogenesis and other differentiation processes.[57]

Some protein toxins signal through gangliosides at the membrane to recruit members of endocytic pathways. TeNT receptor binding domain colocalizes with growth factor receptors $p75^{NTR}$ and TrkA in a well-characterized neurotrophic retrograde pathway toward the soma in peripheral motor neurons.[78] TeNT receptor binding domain also stimulates activation of the TrKA receptor in cultured primary neurons.[161] Upon addition of TeNT, dose-dependent TrkA autophosphorylation and activation of phospholipase C leads to activation of MAP kinase ERK1/2 and other neuroprotective pathways.[162,163] TeNT activation of TrkA may mimic the binding of nerve growth factor and induce uptake into signaling endosomes toward the retrograde traffic.

Alpha-toxin is associated with lipid rafts. The binding of phospholipids by a C-terminal domain and hydrolysis causes negative membrane curvature and induces endocytosis.[138] Alpha-toxin also binds ganglioside GM1a through a conserved ganglioside motif. This secondary binding of GM1a activates TrkA, causing a phosphorylation cascade of MAP kinase substrates.[142] Unlike TeNT, alpha-toxin signaling is associated with release of inflammatory chemokines and reactive oxygen species, which is responsible for the severity of gas gangrene symptoms. This signaling is mediated through GM1a, as treatment of cells with neuramidase to cleave the terminal sialic acid or an inhibitor of ganglioside synthesis, PPMP, dose-dependently reduces cytokine release upon exposure to alpha-toxin.[141] Mutation of Trp84-Tyr85 attenuates GM1a binding, activation of TrkA signaling, and release of IL-8.[141] Once alpha-toxin binds the membrane, the phospholipase activity forms DAG which leads to alterations in the membrane that may induce vasculature permeability leading to edema.[138]

## 8. AB5 TOXINS AND GANGLIOSIDE CROSS-LINKING

Cross-linking gangliosides can induce endocytosis through receptor activation and can enrich lipid rafts for TrkA.[164] Ctx clusters ganglioside GM1a by binding five molecules concurrently, leading to more efficient receptor-mediated internalization, and in neurons, the activation of TrkA.[43,165] Stx binds up to 15 molecules of Gb3 at the plasma membrane, which activates the RTK Syk, which phosphorylates clathrin heavy chain.[149,166] This leads to recruitment of adaptor protein 2 and clathrin-coated vesicles to the membrane, and may be responsible for the clathrin-dependent entry of Stx in some cell lines.[167] In some cases, fatty acyl chain saturation and chain length are as important as the carbohydrate moiety for toxin binding, entry, and trafficking into productive pathways. One study utilized lipid probes of varying chemical species and acyl lengths to demonstrate that the ceramide portion of GM1a alters lipid trafficking within the cell;[49] unsaturated ceramide chains retrograde trafficked from plasma membrane to the ER more efficiently than long, saturated ceramides, which were not localized with raft-associated proteins.[49] In contrast, Ctx **B** pentamer did not alter lipid localization, which may explain how toxins exploit lipids trafficking to different compartments.

## 9. AB5 TOXINS AND LIPID TRAFFICKING

Lipid properties directly affect raft composition, which affect ganglioside dynamics by retaining certain lipids and GPI-anchored protein species.[168] Ceramide composition and fatty acid acyl length also affects partitioning properties, cargo sorting, and further increase heterogeneity of the plasma membrane.[49] For Stx1 and Stx2, in vitro binding affinity of Gb3 ganglioside was increased for shorter-acyl chains compared to longer-acyl chains.[169,170] Gangliosides with shorter-acyl chains partition differently within a cell membrane, leading to alterations in ligand accessibility.[171] Chain length may play a physiological role in the association of Stx with lipid rafts in cells, which are also required for clathrin-independent Stx uptake.[166,167] The disruption of Gb3 localization to lipid rafts protects cells against Stx intoxication, but does not affect binding of the toxin.[172] LT-IIb binds nonraft associated GD1a in cultured T84 human colon carcinoma cells; LT-IIb binds, but does not enter a productive retrograde pathway unlike Ctx which binds GM1a.[173] These studies may resolve how host glycolipids as ubiquitous as gangliosides serve to bind and target protein toxins to certain tissues, causing the hallmarks of disease.

## REFERENCES

1. Smyth CJ. In: Joseph EA, Michael RP, eds. The Comprehensive Sourcebook of Bacterial Protein Toxins. San Diego:Academic Press; 2006:1047. Clinical Infectious Diseases. 2006;43(5):669
2. Beddoe T, Paton AW, Le Nours J, Rossjohn J, Paton JC. Structure, biological functions and applications of the AB5 toxins. *Trends Biochem Sci.* 2010;35(7):411–418.
3. Garred Øsah, Dubinina E, Polesskaya A, Olsnes S, Kozlov J, Sandvig K. Role of the disulfide bond in Shiga toxin A-chain for toxin entry into cells. *J Biol Chem.* 1997;272 (17):11414–11419.
4. Lappi DA, Kapmeyer W, Beglau JM, Kaplan NO. The disulfide bond connecting the chains of ricin. *Proc Natl Acad Sci USA.* 1978;75(3):1096–1100.
5. Schiavo G, Papini E, Genna G, Montecucco C. An intact interchain disulfide bond is required for the neurotoxicity of tetanus toxin. *Infect Immun.* 1990;58(12):4136–4141.
6. Tomasi M, Battistini A, Araco A, Roda LG, D'Agnolo G. The role of the reactive disulfide bond in the interaction of cholera-toxin functional regions. *Eur J Biochem.* 1979;93(3):621–627.
7. Zuverink M, Chen C, Przedpelski A, Blum FC, Barbieri JT. A heterologous reporter defines the role of the tetanus toxin interchain disulfide in light-chain translocation. *Infect Immun.* 2015;83(7):2714–2724.
8. Montecucco C, Schiavo G. Mechanism of action of tetanus and botulinum neurotoxins. *Mol Microbiol.* 1994;13(1):1–8.
9. Klimpel KR, Molloy SS, Thomas G, Leppla SH. Anthrax toxin protective antigen is activated by a cell surface protease with the sequence specificity and catalytic properties of furin. *Proc Natl Acad Sci USA.* 1992;89(21):10277–10281.
10. Gordon VM, Klimpel KR, Arora N, Henderson MA, Leppla SH. Proteolytic activation of bacterial toxins by eukaryotic cells is performed by furin and by additional cellular proteases. *Infect Immun.* 1995;63(1):82–87.
11. Gordon VM, Leppla SH. Proteolytic activation of bacterial toxins: role of bacterial and host cell proteases. *Infect Immun.* 1994;62(2):333–340.
12. Lencer WI, Constable C, Moe S, et al. Proteolytic activation of cholera toxin and *Escherichia coli* labile toxin by entry into host epithelial cells: signal transduction by a protease-resistant toxin variant. *J Biol Chem.* 1997;272(24):15562–15568.
13. Lea N, Lord JM, Roberts LM. Proteolytic cleavage of the A subunit is essential for maximal cytotoxicity of *Escherichia coli* 0157:H7 Shiga-like toxin-1. *Microbiology.* 1999;145(5):999–1004.
14. Burns DL, Manclark CR. Role of cysteine 41 of the A subunit of pertussis toxin. *J Biol Chem.* 1989;264(1):564–568.
15. Kozaki S, Kamata Y, Watarai S, Nishiki T-i. Mochida S. Ganglioside GT1b as a complementary receptor component for Clostridium botulinumneurotoxins. *Microb Pathog.* 1998;25(2):91–99.
16. Mahrhold S, Bergstrom T, Stern D, Dorner BG, Astot C, Rummel A. Only the complex N559-glycan in the synaptic vesicle glycoprotein 2C mediates high affinity binding to botulinum neurotoxin serotype A1. *Biochem J.* 2016;473(17): 2645–2654.

17. Mahrhold S, Rummel A, Bigalke H, Davletov B, Binz T. The synaptic vesicle protein 2C mediates the uptake of botulinum neurotoxin A into phrenic nerves. *FEBS Lett.* 2006;580(8):2011–2014.
18. Peng L, Berntsson RPA, Tepp WH, et al. Botulinum neurotoxin D-C uses synaptotagmin I and II as receptors, and human synaptotagmin II is not an effective receptor for type B, D-C and G toxins. *J Cell Sci.* 2012;125(13):3233–3242.
19. Rummel A, Hafner K, Mahrhold S, et al. Botulinum neurotoxins C, E and F bind gangliosides via a conserved binding site prior to stimulation-dependent uptake with botulinum neurotoxin F utilising the three isoforms of SV2 as second receptor. *J Neurochem.* 2009;110(6):1942–1954.
20. Chen C, Fu Z, Kim JJ, Barbieri JT, Baldwin MR. Gangliosides as high affinity receptors for tetanus neurotoxin. *J Biol Chem.* 2009;284(39):26569–26577.
21. Kuziemko GM, Stroh M, Stevens RC. Cholera toxin binding affinity and specificity for gangliosides determined by surface plasmon resonance. *Biochemistry.* 1996;35(20): 6375–6384.
22. Fukuta S, Magnani JL, Twiddy EM, Holmes RK, Ginsburg V. Comparison of the carbohydrate-binding specificities of cholera toxin and *Escherichia coli* heat-labile enterotoxins LTh-I, LT-IIa, and LT-IIb. *Infect Immun.* 1988;56(7):1748–1753.
23. Frankel A, Tagge E, Chandler J, Burbage C, Willingham M. Double-site ricin B chain mutants retain galactose binding. *Protein Eng.* 1996;9(4):371–379.
24. Witvliet MH, Burns DL, Brennan MJ, Poolman JT, Manclark CR. Binding of pertussis toxin to eucaryotic cells and glycoproteins. *Infect Immun.* 1989;57(11):3324–3330.
25. Chong DC, Paton JC, Thorpe CM, Paton AW. Clathrin-dependent trafficking of subtilase cytotoxin, a novel AB5 toxin that targets the endoplasmic reticulum chaperone BiP. *Cell Microbiol.* 2008;10(3):795–806.
26. Nagasawa S, Ogura K, Tsutsuki H, et al. Uptake of Shiga-toxigenic *Escherichia coli* SubAB by HeLa cells requires an actin– and lipid raft -dependent pathway. *Cell Microbiol.* 2014;16(10):1582–1601.
27. Sandvig K, van Deurs B. Endocytosis, intracellular transport, and cytotoxic action of Shiga toxin and ricin. *Physiol Rev.* 1996;76(4):949–966.
28. Torgersen ML, Skretting G, van Deurs B, Sandvig K. Internalization of cholera toxin by different endocytic mechanisms. *J Cell Sci.* 2001;114(20):3737–3747.
29. Falnes PO, Sandvig K. Penetration of protein toxins into cells. *Curr Opin Cell Biol.* 2000;12(4):407–413.
30. Lencer WI, Saslowsky D. Raft trafficking of AB5 subunit bacterial toxins. *Biochimica et Biophysica Acta.* 2005;1746(3):314–321.
31. Plaut RD, Carbonetti NH. Retrograde transport of pertussis toxin in the mammalian cell. *Cell Microbiol.* 2008;10(5):1130–1139.
32. Sandvig K, Grimmer S, Lauvrak SU, et al. Pathways followed by ricin and Shiga toxin into cells. *Histochem Cell Biol.* 2002;117(2):131–141.
33. Yu M, Haslam DB. Shiga toxin is transported from the endoplasmic reticulum following interaction with the luminal chaperone HEDJ/ERdj3. *Infect Immun.* 2005;73(4): 2524–2532.
34. Bellisola G, Fracasso G, Ippoliti R, et al. Reductive activation of ricin and ricin A-chain immunotoxins by protein disulfide isomerase and thioredoxin reductase. *Biochem Pharmacol.* 2004;67(9):1721–1731.
35. Orlandi PA. Protein-disulfide isomerase-mediated reduction of the A subunit of cholera toxin in a human intestinal cell line. *J Biol Chem.* 1997;272(7):4591–4599.
36. Spooner RA, Watson PD, Marsden CJ, et al. Protein disulphide-isomerase reduces ricin to its A and B chains in the endoplasmic reticulum. *Biochem J.* 2004;383(Pt 2): 285–293.

37. Tsai B, Rodighiero C, Lencer WI, Rapoport TA. Protein Disulfide isomerase acts as a redox-dependent chaperone to unfold cholera toxin. *Cell.* 2001;104(6):937–948.
38. Schmitz A, Herrgen H, Winkeler A, Herzog V. Cholera toxin is exported from microsomes by the Sec61p complex. *J Cell Biol.* 2000;148(6):1203–1212.
39. Kahn RA, Gilman AG. Purification of a protein cofactor required for ADP-ribosylation of the stimulatory regulatory component of adenylate cyclase by cholera toxin. *J Biol Chem.* 1984;259(10):6228–6234.
40. Bobak DA, Bliziotes MM, Noda M, Tsai SC, Adamik R, Moss J. Mechanism of activation of cholera toxin by ADP-ribosylation factor (ARF): both low- and high-affinity interactions of ARF with guanine nucleotides promote toxin activation. *Biochemistry.* 1990;29 (4):855–861.
41. Le Nours J, Paton AW, Byres E, et al. Structural basis of subtilase cytotoxin SubAB assembly. *J Biol Chem.* 2013;288(38):27505–27516.
42. Paton AW, Beddoe T, Thorpe CM, et al. AB5 subtilase cytotoxin inactivates the endoplasmic reticulum chaperone BiP. *Nature.* 2006;443(7111):548–552.
43. Jobling MG, Yang Z, Kam WR, Lencer WI, Holmes RK. A single native ganglioside GM1-binding site is sufficient for cholera toxin to bind to cells and complete the intoxication pathway. *MBio.* 2012;3(6).
44. Lencer WI, Constable C, Moe S, et al. Targeting of cholera toxin and *Escherichiacoli* heat labile toxin in polarized epithelia: role of COOH-terminal KDEL. *JCellBiol.* 1995;131 (4):951–962.
45. Varki A. *Essentialsofglycobiology.* 2nd ed Cold Spring Harbor, N.Y: Cold Spring Harbor Laboratory Press; 2009.
46. Kato K, Ishiwa A. The role of carbohydrates in infection strategies of enteric pathogens. *Trop Med Health.* 2015;43(1):41–52.
47. Lingwood CA. Glycosphingolipid functions. *Cold Spring Harb Perspect Biol.* 2011;3(7): a004788.
48. Yu RK. Gangliosides: structure and analysis. In: Ledeen RW, Yu RK, Rapport MM, Suzuki K, eds. *Ganglioside Structure, Function, and Biomedical Potential. Boston, MA: Springer US;1984:39–53.*
49. Chinnapen DJF, Hsieh W-T, te Welscher YM, et al. Lipid-sorting by ceramide structure from plasma membrane to ER for the cholera toxin receptor ganglioside GM1. *Dev Cell.* 2012;23(3):573–586.
50. Kolter T. Ganglioside biochemistry. *ISRNBiochem.* 2012;2012:506160.
51. Yu RK, Tsai Y-T, Ariga T, Yanagisawa M. Structures, biosynthesis, and functions of gangliosides—an overview. *J Oleo Sci.* 2011;60(10):537–544.
52. Varki A. Loss of *N*-glycolylneuraminic acid in humans: mechanisms, consequences, and implications for hominid evolution. *AmJ Phys Anthro.* 2001;116(S33):54–69.
53. Schnaar RL, Gerardy-Schahn R, Hildebrandt H. Sialic acids in the brain: gangliosides and polysialic acid in nervous system development, stability, disease, and regeneration. *Physiol Rev.* 2014;94(2):461–518.
54. Daniotti JL, Iglesias-Bartolome R. Metabolic pathways and intracellular trafficking of gangliosides. *IUBMBLife.* 2011;63(7):513–520.
55. Ikonen E. Roles of lipid rafts in membrane transport. *Curr Opin Cell Biol.* 2001;13 (4):470–477.
56. Simons K, Ikonen E. Functional rafts in cell membranes. *Nature.* 1997;387(6633):569–572.
57. Vyas AA, Patel HV, Fromholt SE, et al. Gangliosides are functional nerve cell ligands for myelin-associated glycoprotein (MAG), an inhibitor of nerve regeneration. *Proc Natl Acad Sci USA.* 2002;99(12):8412–8417.
58. Posse de Chaves E, Sipione S. Sphingolipids and gangliosides of the nervous system in membrane function and dysfunction. *FEBSLett.* 2010;584(9):1748–1759.

59. Yoshikawa M, Go S, Takasaki K, et al. Mice lacking ganglioside GM3 synthase exhibit complete hearing loss due to selective degeneration of the organ of Corti. *Proc Natl Acad Sci USA.* 2009;106(23):9483–9488.
60. Yu RK, Tsai Y-T, Ariga T. Functional roles of gangliosides in neurodevelopment:an overview of recent advances. *Neurochem Res.* 2012;37(6):1230–1244.
61. Kawai H, Allende ML, Wada R, et al. Mice expressing only monosialoganglioside GM3 exhibit lethal audiogenic seizures. *J Biol Chem.* 2001;276(10):6885–6888.
62. Yamashita T, Wu YP, Sandhoff R, et al. Interruption of ganglioside synthesis produces central nervous system degeneration and altered axon-glial interactions. *Proc Natl Acad Sci USA.* 2005;102(8):2725–2730.
63. Rossetto O, Pirazzini M, Montecucco C. Botulinum neurotoxins: genetic, structural and mechanistic insights. *Nat Rev Micro.* 2014;12(8):535–549.
64. Lacy DB, Tepp W, Cohen AC, DasGupta BR, Stevens RC. Crystal structure of botulinum neurotoxin type A and implications for toxicity. *Nat Struct Mol Biol.* 1998;5 (10):898–902.
65. Blasi J, Chapman ER, Link E, et al. Botulinum neurotoxin A selectively cleaves the synaptic protein SNAP-25. *Nature.* 1993;365(6442):160–163.
66. Schiavo G, Benfenati F, Poulain B, et al. Tetanus and botulinum-B neurotoxins block neurotransmitter release by proteolytic cleavage of synaptobrevin. *Nature.* 1992;359 (6398):832–835.
67. Neale EA, Bowers LM, Jia M, Bateman KE, Williamson LC. Botulinum neurotoxin a blocks synaptic vesicle exocytosis but not endocytosis at the nerve terminal. *J Cell Biol.* 1999;147(6):1249–1260.
68. Südhof TC, Rizo J. Synaptic vesicle exocytosis. *Cold Spring Harb Perspect Biol.* 2011;3 (12):a005637.
69. Chen S, Hall C, Barbieri JT. Substrate recognition of VAMP-2 by botulinum neurotoxin B and tetanus neurotoxin. *J Biol Chem.* 2008;283(30):21153–21159.
70. Popoff MR, Poulain B. Bacterial toxins and the nervous system: neurotoxins and multipotential toxins interacting with neuronal cells. *Toxins.* 2010;2(4):683–737.
71. Schengrund C-L, DasGupta BR, Ringler NJ. Binding of botulinum and tetanus neurotoxins to ganglioside GT1b and derivatives thereof. *J Neurochem.* 1991;57(3): 1024–1032.
72. Walton KM, Sandberg K, Rogers TB, Schnaar RL. Complex ganglioside expression and tetanus toxin binding by PC12 pheochromocytoma cells. *J Biol Chem.* 1988;263 (4):2055–2063.
73. Rogers TB, Snyder SH. High affinity binding of tetanus toxin to mammalian brain membranes. *J Biol Chem.* 1981;256(5):2402–2407.
74. Chen C, Baldwin MR, Barbieri JT. Molecular basis for tetanus toxin coreceptor interactions. *Biochemistry.* 2008;47(27):7179–7186.
75. Herreros J, Ng T, Schiavo G. Lipid rafts act as specialized domains for tetanus toxin binding and internalization into neurons. *Mol Biol Cell.* 2001;12(10):2947–2960.
76. Yeh FL, Dong M, Yao J, et al. SV2 mediates entry of tetanus neurotoxin into central neurons. *PLoS Pathog.* 2010;6(11):e1001207.
77. Blum FC, Chen C, Kroken AR, Barbieri JT. Tetanus toxin and botulinum toxin a utilize unique mechanisms to enter neurons of the central nervous system. *Infect Immun.* 2012;80(5):1662–1669.
78. Lalli G, Schiavo G. Analysis of retrograde transport in motor neurons reveals common endocytic carriers for tetanus toxin and neurotrophin receptor p75NTR. *J Cell Biol.* 2002;156(2):233–239.
79. Masuyer G, Conrad J, Stenmark P. The structure of the tetanus toxin reveals pH-mediated domain dynamics. *EMBO Rep.* 2017;18(8):1306–1317.

80. Fischer A, Mushrush DJ, Lacy DB, Montal M. Botulinum neurotoxin devoid of receptor binding domain translocates active protease. *PLoS Pathog.* 2008;4(12):e1000245.
81. Fischer A, Montal M. Molecular dissection of botulinum neurotoxin reveals interdomain chaperone function. *Toxicon.* 2013;75:101–107.
82. Verderio C, Rossetto O, Grumelli C, Frassoni C, Montecucco C, Matteoli M. Entering neurons: botulinum toxins and synaptic vesicle recycling. *EMBORep.* 2006;7(10):995–999.
83. Kroken AR, Blum FC, Zuverink M, Barbieri JT. Entry of botulinum neurotoxin subtypes A1 and A2 into neurons. *Infect Immun.* 2016.
84. Karalewitz AP, Kroken AR, Fu Z, Baldwin MR, Kim JJ, Barbieri JT. Identification of a unique ganglioside binding loop within botulinum neurotoxins C and D-SA. *Biochemistry.* 2010;49(37):8117–8126.
85. Spangler BD. Structure and function of cholera toxin and the related *Escherichiacoli* heat-labile enterotoxin. *Microbiol Rev.* 1992;56(4):622–647.
86. Sack RB. The discovery of cholera—like enterotoxins produced by *Escherichia coli* causing secretory diarrhoea in humans. *IndianJ Med Res.* 2011;133(2):171–178.
87. Backstrom M, Shahabi V, Johansson S, et al. Structural basis for differential receptor binding of cholera and *Escherichia coli* heat-labile toxins: influence of heterologous amino acid substitutions in the cholera B-subunit. *Mol Microbiol.* 1997;24(3): 489–497.
88. Lai C-Y, Cancedda F, Duffy LK. ADP-ribosyl transferase activity of cholera toxin polypeptide A1 and the effect of limited trypsinolysis. *Biochem Biophys Res Commun.* 1981;102(3):1021–1027.
89. Northup JK, Sternweis PC, Smigel MD, Schleifer LS, Ross EM, Gilman AG. Purification of the regulatory component of adenylate cyclase. *Proc Natl Acad Sci USA.* 1980;77(11):6516–6520.
90. Cassel D, Pfeuffer T. Mechanism of cholera toxin action: covalent modification of the guanyl nucleotide-binding protein of the adenylate cyclase system. *Proc Natl Acad Sci USA.* 1978;75(6):2669–2673.
91. Teter K, Allyn RL, Jobling MG, Holmes RK. Transfer of the cholera toxin A1 polypeptide from the endoplasmic reticulum to the cytosol is a rapid process facilitated by the endoplasmic reticulum-associated degradation pathway. *Infect Immun.* 2002;70(11): 6166–6171.
92. Kassis S, Hagmann J, Fishman PH, Chang PP, Moss J. Mechanism of action of cholera toxin on intact cells. Generation of A1 peptide and activation of adenylate cyclase. *JBiol Chem.* 1982;257(20):12148–12152.
93. Bharati K, Ganguly NK. Cholera toxin: A paradigm of a multifunctional protein. *Indian J Med Res.* 2011;133(2):179–187.
94. Merritt EA, Sixma TK, Kalk KH, van Zanten BAM, Hol WGJ. Galactose-binding site in *Escherichia coli* heat-labile enterotoxin (LT) and cholera toxin (CT). *Mol Microbiol.* 1994;13(4):745–753.
95. Barra JL, Monferran CG, Balanzino LE, Cumar FA. *Escherichia coli* heat-labile enterotoxin preferentially interacts with blood group A-active glycolipids from pig intestinal mucosa and A- and B-active glycolipids from human red cells compared to H-active glycolipids. *Mol Cell Biochem.* 1992;115(1):63–70.
96. Horstman AL, Kuehn MJ. Bacterial surface association of heat-labile enterotoxin through lipopolysaccharide after secretion via the general secretory pathway. *J Biol Chem.* 2002;277(36):32538–32545.
97. Galván EM, Diema CD, Roth GA, Monferran CG. Ability of blood group A-active glycosphingolipids to Act as *Escherichia coli* heat-labile enterotoxin receptors in HT-29 Cells. *J Infect Dis.* 2004;189(9):1556–1564.

98. Zalem D, Ribeiro JP, Varrot A, Lebens M, Imberty A, Teneberg S. Biochemical and structural characterization of the novel sialic acid-binding site of *Escherichia coli* heat-labile enterotoxin LT-IIb. *BiochemiJ.* 2016;473(21):3923–3936.
99. Berenson CS, Nawar HF, Kruzel RL, Mandell LM, Connell TD. Ganglioside-binding specificities of E. coli enterotoxin LT-IIc: importance of long-chain fatty acyl ceramide. *Glycobiology.* 2013;23(1):23–31.
100. Melton-Celsa AR. Shiga toxin (Stx) classification, structure, and function. *Microbiol Spectr.* 2014;2(2):1–13.
101. Niyogi SK. Shigellosis. *J Microbiol.* 2005;43(2):133–143.
102. Endo Y, Tsurugi K. RNA N-glycosidase activity of ricin A-chain. Mechanism of action of the toxic lectin ricin on eukaryotic ribosomes. *JBiol Chem.* 1987;262(17):8128–8130.
103. Tumer NE, Li XP. Interaction of ricin and Shiga toxins with ribosomes. *Curr Top Microbiol Immunol.* 2012;357:1–18.
104. Endo Y, Tsurugi K, Yutsudo T, Takeda Y, Ogasawara T, Igarashi K. Site of action of a Vero toxin (VT2) from *Escherichia coli* O157:H7 and of Shiga toxin on eukaryotic ribosomes. *EurJ Biochem.* 1988;171(1–2):45–50.
105. Tesh VL. Activation of cell stress response pathways by shiga toxins. *Cell Microbiol.* 2012;14(1):1–9.
106. Fraser ME, Fujinaga M, Cherney MM, et al. Structure of Shiga toxin type 2 (Stx2) from *Escherichia coli* O157:H7. *JBiol Chem.* 2004;279(26):27511–27517.
107. Cilmi SA, Karalius BJ, Choy W, Smith RN, Butterton JR. Fabry disease in mice protects against lethal disease caused by Shiga toxin-expressing enterohemorrhagic *Escherichia coli*. *J Infect Dis.* 2006;194(8):1135–1140.
108. Hanashima T, Miyake M, Yahiro K, et al. Effect of Gb3 in lipid rafts in resistance to Shiga-like toxin of mutant Vero cells. *Microb Pathog.* 2008;45(2):124–133.
109. Kovbasnjuk O, Edidin M, Donowitz M. Role of lipid rafts in Shiga toxin 1 interaction with the apical surface of Caco-2 cells. *J Cell Sci.* 2001;114(22):4025–4031.
110. Mukhopadhyay S, Redler B, Linstedt AD. Shiga toxin–binding site for host cell receptor GPP130 reveals unexpected divergence in toxin-trafficking mechanisms. *Mol Biol Cell.* 2013;24(15):2311–2318.
111. Mayer CL, Leibowitz CS, Kurosawa S, Stearns-Kurosawa DJ. Shiga toxins and the pathophysiology of hemolytic uremic syndrome in humans and animals. *Toxins.* 2012;4(11):1261–1287.
112. Sandvig K, Dubinina E, Garred O, et al. Entry of Shiga toxin into cells. *Zentralbl Bakteriol.* 1993;278(2–3):296–305.
113. Natarajan R, Linstedt AD. A cycling cis-Golgi protein mediates endosome-to-Golgi traffic. *Mol Biol Cell.* 2004;15(11):4798–4806.
114. Mukhopadhyay S, Linstedt AD. Manganese blocks intracellular trafficking of shiga toxin and protects against shiga toxicosis. *Science.* 2012;335(6066):332–335.
115. Roy CJ, Song K, Sivasubramani SK, Gardner DJ, Pincus SH. Animal models of ricin toxicosis. *Curr Top Microbiol Immunol.* 2012;357:243–257.
116. Tumer NE, Li X-P. Interaction of ricin and Shiga toxins with ribosomes. *Curr Top Microbiol Immunol.* 2012;357:1–18.
117. Audi J, Belson M, Patel M, Schier J, Osterloh J. Ricin poisoning: a comprehensive review. *JAMA.* 2005;294(18):2342–2351.
118. Frankel AE, Burbage C, Fu T, Tagge E, Chandler J, Willingham MC. Ricin toxin contains at least three galactose-binding sites located in B chain subdomains 1α, 1β, and 2γ. *Biochemistry.* 1996;35(47):14749–14756.
119. Montfort W, Villafranca JE, Monzingo AF, et al. The three-dimensional structure of ricin at 2.8 A. *J Biol Chem.* 1987;262(11):5398–5403.

120. Magnusson S, Kjeken R, Berg T. Characterization of two distinct pathways of endocytosis of ricin by rat liver endothelial cells. *Exp Cell Res.* 1993;205(1): 118–125.
121. Simmons BM, Stahl PD, Russell JH. Mannose receptor-mediated uptake of ricin toxin and ricin A chain by macrophages. Multiple intracellular pathways for a chain translocation. *J Biol Chem.* 1986;261(17):7912–7920.
122. Magnússon S, Berg T. Endocytosis of ricin by rat liver cells in vivo and in vitro is mainly mediated by mannose receptors on sinusoidal endothelial cells. *Biochem J.* 1993;291 (Pt 3):749–755.
123. Alder NN, Shen Y, Brodsky JL, Hendershot LM, Johnson AE. The molecular mechanisms underlying BiP-mediated gating of the Sec61 translocon of the endoplasmic reticulum. *J Cell Biol.* 2005;168(3):389–399.
124. Byres E, Paton AW, Paton JC, et al. Incorporation of a non-human glycan mediates human susceptibility to a bacterial toxin. *Nature.* 2008;456(7222): 648–652.
125. Schauer R, Srinivasan GV, Coddeville B, Zanetta J-P, Guérardel Y. Low incidence of *N*-glycolylneuraminic acid in birds and reptiles and its absence in the platypus. *Carbohydr Res.* 2009;344(12):1494–1500.
126. Carbonetti NH. Pertussis toxin and adenylate cyclase toxin: key virulence factors of Bordetella pertussis and cell biology tools. *Future Microbiol.* 2010;5:455–469.
127. Pittman M. The concept of pertussis as a toxin-mediated disease. *Pediatr Infect Dis.* 1984;3(5):467–486.
128. Kaslow HR, Burns DL. Pertussis toxin and target eukaryotic cells: binding, entry, and activation. *FASEB J.* 1992;6(9):2684–2690.
129. Sekura RD, Fish F, Manclark CR, Meade B, Zhang YL. Pertussis toxin. Affinity purification of a new ADP-ribosyltransferase. *J Biol Chem.* 1983;258(23):14647–14651.
130. Stein PE, Boodhoo A, Armstrong GD, Cockle SA, Klein MH, Read RJ. The crystal structure of pertussis toxin. *Structure.* 1994;2(1):45–57.
131. Stein PE, Boodhoo A, Armstrong GD, et al. Structure of a pertussis toxin-sugar complex as a model for receptor binding. *Nat Struct Mol Biol.* 1994;1(9):591–596.
132. Finck-Barbancon V, Barbieri JT. Preferential processing of the S1 subunit of pertussis toxin that is bound to eukaryotic cells. *Mol Microbiol.* 1996;22(1):87–95.
133. Nencioni L, Pizza MG, Volpini G, De Magistris MT, Giovannoni F, Rappuoli R. Properties of the B oligomer of pertussis toxin. *Infect Immun.* 1991;59(12):4732–4734.
134. Millen SH, Lewallen DM, Herr AB, Iyer SS, Weiss AA. Identification and characterization of the carbohydrate ligands recognized by pertussis toxin through glycan microarray and surface plasmon resonance. *Biochemistry.* 2010;49(28):5954–5967.
135. Hausman SZ, Burns DL. Binding of pertussis toxin to lipid vesicles containing glycolipids. *Infect Immun.* 1993;61(1):335–337.
136. Brennan MJ, David JL, Kenimer JG, Manclark CR. Lectin-like binding of pertussis toxin to a 165-kilodalton Chinese hamster ovary cell glycoprotein. *J Biol Chem.* 1988;263(10):4895–4899.
137. Naylor CE, Eaton JT, Howells A, et al. Structure of the key toxin in gas gangrene. *Nat Struct Biol.* 1998;5(8):738–746.
138. Sakurai J, Nagahama M, Oda M. *Clostridium perfringens* alpha-toxin: characterization and mode of action. *J Biochem.* 2004;136(5):569–574.
139. Caplan ES, Kluge RM. Gas gangrene: review of 34 cases. *Arch Intern Med.* 1976;136 (7):788–791.
140. Oda M, Terao Y, Sakurai J, Nagahama M. Membrane-binding mechanism of *Clostridium perfringens* alpha-toxin. *Toxins.* 2015;7(12):5268–5275.
141. Oda M, Kabura M, Takagishi T, et al. *Clostridium perfringens* alpha-toxin recognizes the GM1a-TrkA complex. *J Biol Chem.* 2012;287(39):33070–33079.

142. Takagishi T, Oda M, Kabura M, et al. *Clostridium perfringens* alpha-toxin induces Gm1a clustering and Trka phosphorylation in the host cell membrane. *PLoS One.* 2015;10(4): e0120497.
143. den Akker Fv. Sarfaty S, Twiddy EM, Connell TD, Holmes RK, Hol WGJ. Crystal structure of a new heat-labile enterotoxin, LT-IIb. *Structure.* 1996;4(6):665–678.
144. Merritt EA, Sarfaty S, van den Akker F, L'Hoir C, Martial JA, Hol WG. Crystal structure of cholera toxin B-pentamer bound to receptor GM1 pentasaccharide. *Protein Sci.* 1994;3(2):166–175.
145. Murzin AG. OB(oligonucleotide/oligosaccharide binding)-fold: common structural and functional solution for non-homologous sequences. *EMBO J.* 1993;12(3): 861–867.
146. Sixma TK, Kalk KH, van Zanten BAM, et al. Refined structure of *Escherichia coli* Heat-labile Enterotoxin, a close relative of cholera toxin. *J Mol Biol.* 1993;230(3):890–918.
147. Mudrak B, Kuehn MJ. Heat-labile enterotoxin: beyond G(m1) binding. *Toxins.* 2010;2 (6):1445–1470.
148. Loosmore S, Zealey G, Cockle S, et al. Characterization of pertussis toxin analogs containing mutations in B-oligomer subunits. *Infect Immun.* 1993;61(6):2316–2324.
149. Ling H, Boodhoo A, Hazes B, et al. Structure of the Shiga-like toxin I B-pentamer complexed with an analogue of its receptor Gb3. *Biochemistry.* 1998;37(7): 1777–1788.
150. Yao J, Nellas RB, Glover MM, Shen T. Stability and sugar recognition ability of ricin-like carbohydrate binding domains. *Biochemistry.* 2011;50(19):4097–4104.
151. Sphyris N, Lord JM, Wales R, Roberts LM. Mutational analysis of the *Ricinus* lectin B-chains. Galactose-binding ability of the 2 gamma subdomain of *Ricinus communis* agglutinin B-chain. *J Biol Chem.* 1995;270(35):20292–20297.
152. Wales R, Richardson PT, Roberts LM, Lord JM. Recombinant ricin B chain fragments containing a single galactose binding site retain lectin activity. *Arch Biochem Biophys.* 1992;294(1):291–296.
153. Emsley P, Fotinou C, Black I, et al. The structures of the HC fragment of tetanus toxin with carbohydrate subunit complexes provide insight into ganglioside binding. *J Biol Chem.* 2000;275(12):8889–8894.
154. Stenmark P, Dupuy J, Imamura A, Kiso M, Stevens RC. Crystal structure of botulinum neurotoxin type A in complex with the cell surface co-receptor GT1b—insight into the toxin–neuron interaction. *PLoS Pathog.* 2008;4(8):e1000129.
155. Uppalapati SR, Kingston JJ, Qureshi IA, Murali HS, Batra HV. In silico, in vitro and in vivo analysis of binding affinity between N and C-domains of *Clostridium perfringens* alpha toxin. *PLoS One.* 2013;8(12):e82024.
156. Burke P, Schooler K, Wiley HS. Regulation of epidermal growth factor receptor signaling by endocytosis and intracellular trafficking. *Mol Biol Cell.* 2001;12(6): 1897–1910.
157. Wang X, Rahman Z, Sun P, et al. Ganglioside modulates ligand binding to the epidermal growth factor receptor. *J Invest Dermatol.* 2001;116(1):69–76.
158. Toledo MS, Suzuki E, Handa K, Hakomori S. Effect of ganglioside and tetraspanins in microdomains on interaction of integrins with fibroblast growth factor receptor. *J Biol Chem.* 2005;280(16):16227–16234.
159. Yoon S-J, Nakayama K-i. Hikita T, Handa K, Hakomori S-ii.. Epidermal growth factor receptor tyrosine kinase is modulated by GM3 interaction with N-linked GlcNAc termini of the receptor. *Proc Natl Acad Sci USA.* 2006;103(50):18987–18991.
160. Duchemin A-M, Ren Q, Mo L, Neff NH, Hadjiconstantinou M. GM1 ganglioside induces phosphorylation and activation of Trk and Erk in brain. *J Neurochem.* 2002;81 (4):696–707.

161. Gil C, Chaib-Oukadour I, Pelliccioni P, Aguilera J. Activation of signal transduction pathways involving trkA, PLCγ-1, PKC isoforms and ERK-1/2 by tetanus toxin. *FEBS Lett.* 2000;481(2):177–182.
162. Gil C, Chaib-Oukadour I, Aguilera J. C-terminal fragment of tetanus toxin heavy chain activates Akt and MEK/ERK signalling pathways in a Trk receptor-dependent manner in cultured cortical neurons. *Biochem J.* 2003;373(Pt 2):613–620.
163. Gil C, Chaib-Oukadour I, Blasi J, Aguilera J. HC fragment (C-terminal portion of the heavy chain) of tetanus toxin activates protein kinase C isoforms and phosphoproteins involved in signal transduction. *Biochem J.* 2001;356(Pt 1):97–103.
164. Yamamoto N, Matsubara T, Sato T, Yanagisawa K. Age-dependent high-density clustering of GM1 ganglioside at presynaptic neuritic terminals promotes amyloid beta-protein fibrillogenesis. *Biochim Biophys Acta.* 2008;1778(12):2717–2726.
165. Mutoh T, Tokuda A, Guroff G, Fujiki N. The effect of the B subunit of cholera toxin on the action of nerve growth factor on PC12 cells. *J Neurochem.* 1993;60(4):1540–1547.
166. Lauvrak SU, Walchli S, Iversen TG, et al. Shiga toxin regulates its entry in a Syk-dependent manner. *Mol Biol Cell.* 2006;17(3):1096–1109.
167. Utskarpen A, Massol R, van Deurs B, Lauvrak SU, Kirchhausen T, Sandvig K. Shiga toxin increases formation of Clathrin-coated pits through Syk kinase. *PLoS One.* 2010;5 (7):e10944.
168. Sonnino S, Mauri L, Chigorno V, Prinetti A. Gangliosides as components of lipid membrane domains. *Glycobiology.* 2007;17(1):1R–13R.
169. Arab S, Lingwood CA. Influence of phospholipid chain length on verotoxin/globotriaosyl ceramide binding in model membranes: comparison of a supported bilayer film and liposomes. *Glycoconj J.* 1996;13(2):159–166.
170. Falguieres T, Romer W, Amessou M, et al. Functionally different pools of Shiga toxin receptor, globotriaosyl ceramide, in HeLa cells. *FEBS J.* 2006;273(22):5205–5218.
171. Gallegos KM, Conrady DG, Karve SS, Gunasekera TS, Herr AB, Weiss AA. Shiga toxin binding to glycolipids and glycans. *PLoS One.* 2012;7(2):e30368.
172. Smith DC, Sillence DJ, Falguières T, et al. The association of shiga-like toxin with detergent-resistant membranes is modulated by glucosylceramide and is an essential requirement in the endoplasmic reticulum for a cytotoxic effect. *Mol Biol Cell.* 2006;17 (3):1375–1387.
173. Wolf AA, Jobling MG, Wimer-Mackin S, et al. Ganglioside structure dictates signal transduction by cholera toxin and association with Caveolae-like membrane domains in polarized epithelia. *J Cell Biol.* 1998;141(4):917–927.

CHAPTER TWELVE

# Gangliosides and Autoimmune Peripheral Nerve Diseases

**John A. Goodfellow, Hugh J. Willison**[1]
College of Medical, Veterinary and Life Sciences, University of Glasgow, Glasgow, Scotland
[1] Corresponding author. E-mail address: Hugh.Willison@glasgow.ac.uk

## Contents



## Abstract

A wide range of neuroimmunological diseases, referred to as autoimmune neuropathies, affect the peripheral nervous systems (PNS). The PNS is structurally diverse with complex anatomical compartments enriched in many different myelin and neuronal glycolipids, notably gangliosides. Autoimmune neuropathies are a proportion of autoimmune neuropathies mediated by autoimmune attack due to antibodies reactive with these compartmentally localized gangliosides. Antiganglioside antibodies are principally associated with the acute paralytic disease, Guillain–Barré syndrome, and are also found in several

*Progress in Molecular Biology and Translational Science*, Volume 156
ISSN 1877-1173
http://dx.doi.org/10.1016/bs.pmbts.2017.12.010

chronic autoimmune neuropathy syndromes. These antibodies may arise spontaneously from the natural autoantibody repertoire, or be induced by infections that share structurally similar glycans to gangliosides, an immunological process often referred to as molecular mimicry. The principal infection exhibiting this structural similarity is *Campylobacter jejuni*, which displays mimics of GM1, GD1a, GT1a, and other gangliosides on its surface lipo-oligosaccharide. Autoantibodies thus induced bind glycan epitopes on peripheral nerve gangliosides where they activate complement and recruit macrophages, causing structural and functional disorganization of nerve conduction. Chronic autoimmune neuropathies are also associated with naturally arising IgM antibodies directed against ganglioside epitopes present on disialylated gangliosides, that induce a sensory neuropathy, and on GM1, that induce a motor neuropathy. In a third syndrome, the so-called "anti-MAG" neuropathy, the antibodies bind a sulfated glucuronic acid epitope present on myelin-associated glycoprotein and the glycolipid sulfated glucuronyl paragloboside. This review will describe the immunological, pathological, and clinical features of these disorders in the context of our broader knowledge of the ganglioside glycobiology of the peripheral nervous system.

# 1. THE FUNCTIONAL ANATOMY OF PERIPHERAL NERVE

## 1.1 Peripheral Nerve Compartments

The peripheral nervous systems (PNS) is a highly complex structure, containing the longest axons in the human body and formed by many different classes of motor, sensory, and autonomic neurons supported by both myelin-forming and nonmyelinating Schwann cells. Autoimmune neuropathies associated with antiganglioside and antiglycolipid antibodies often target particular anatomical compartments where gangliosides congregate and are accessible to circulating factors. Motor neuron cell bodies reside within the central nervous system (CNS) in the spinal cord, but project their axons into the PNS via the ventral spinal roots and eventually terminate to form neuromuscular synapses at motor nerve terminals on muscle fibers up to 1 m away, in the case of the distal limbs. Action potential propagation over these distances is facilitated by internodal myelin and the node of Ranvier, which allows for rapid salutatory conduction.[1,2] The node of Ranvier in particular contains abundant gangliosides and is often affected in autoimmune neuropathy.[3] In the sensory arm of the PNS, the diversity of sensory fiber types and sizes, the different types of sensory ending, and the presence of the dorsal root ganglion with peripheral axons and central projections into the dorsal horn of the spinal cord contribute to an even greater level of complexity. Autonomic nerves are also regarded as PNS components. The endoneurial compartment, containing intrafascicular nerve fibers, is comprised of connective tissue rich in collagen and extracellular matrix material, and is surrounded by the perineurium. Within the endoneurial compartment lie the nerve axons enveloped by Schwann cells and their basal laminae, resident endoneurial macrophages and fibroblasts, and a vascular supply. The molecular diversity of the basal lamina includes many glycoyslated and/or glycan-binding molecules, such as laminins, dystroglycans, glycosaminoglycans, and integrins, some of which are autoantigens in neuropathy and other autoimmune diseases.[2]

## 1.2 Immunology of the Peripheral Nervous System

The PNS is also an immunologically complex structure with well-developed innate and adaptive immune function, and many sites within the PNS are subject to a wide range of inflammatory and autoimmune responses.[4,5] Much of the peripheral nervous system resides within an immunologically privileged site within the blood–nerve barrier (the PNS equivalent of the blood–-brain barrier), with the exception of the dorsal and ventral spinal roots and the sensory and motor nerve terminals that lie in sites more freely exposed to circulating factors.[6]

In the context of autoimmunity to glycan-bearing glycolipids and gangliosides, the PNS contains many glycosylated axonal and myelin molecules potentially capable of acting as antigens. The structural complexity of the PNS, combined with the limited access that autoantibodies might have to specialized compartments, such as compacted myelin, makes considerations about the relevance of such antigens as autoantibody targets a highly complex one. It is believed that the glycolipid composition of compact myelin is common to all fiber types, although it remains possible that there could be subtle differences in the content of particular antigens, including gangliosides that could affect the clinical manifestations of particular autoimmune neuropathies.[7] Peripheral nerve myelin contains many glycolipids and gangliosides that are important antigens for autoantibody responses. The myelin sheath also develops specializations associated with noncompact myelin at paranodal regions where certain glycolipids may be more readily exposed to autoantibodies than they are in compact myelin membranes.[1,7]

## 2. GLYCOLIPID ANTIGENS IN AUTOIMMUNE NEUROPATHY

Galactocerebroside (GalCer) is one of the simplest glycosphingolipids, comprising a single galactose residue linked to ceramide (monogalactosylceramide).[8] Historically, it was the first glycolipid to be identified as a key antigen in experimental autoimmune neuritis, a rodent model of autoimmune neuropathy, as well as in its CNS counterpart, experimental autoimmune encephalomyelitis.[9] Sulfatide is synthesized from GalCer following sulfation of the hydroxyl on the third carbon of galactose, and is also a prominent lipid antigen in PNS and CNS myelin. GalCer and sulfatide reside within the exofacial leaflet of the lipid bilayer, where the oligosaccharide is exposed to the extracellular environment and thus available to autoantibody binding. GalCer and sulfatide play important roles in the development and maintenance of myelin, as demonstrated by deficiency states created by targeted disruption of the UDP-galactose ceramide galactosyltransferase (CGT), the key enzyme in galactolipid biosynthesis.[10] These mice have a phenotype characterized by slow nerve conduction velocity and myelin breakdown. Deletion of the galactosylceramide sulfotransferase (CST) gene that specifically inhibits the synthesis of sulfatide, while maintaining GalCer results in a less severe phenotype than CGT deficiency, principally comprising ultrastructural abnormalities in the paranodal regions, indicating a critical role for sulfatide in the localization and maintenance of glial-axonal interactions in the nodal complex.

Peripheral nerve also contains an abundance of neolacto-series gangliosides that are localized mainly in myelin. LM1 is also termed sialosylneolactotetraosylceramide, or sialosylparagloboside, and its higher homologue sialosyllactosaminylparagloboside is referred to as Hex-LM1.[11] Although their function has not been fully elucidated, both LM1 and Hex-LM1 have been identified as autoantigens in both acute and chronic autoimmune neuropathies,[12] and in rodent models.[13] Substitution of sialic acid on the terminal saccharide chain of LM1 and Hex-LM1 for a 3-O-sulfated glucuronic acid generates sulfated glucuronyl paragloboside (SGPG) and its higher lactosaminyl homolog, sulfated glucuronyllactosaminylparagloboside (SGLPG). Both were discovered directly as a result of studying autoimmune neuropathy patients with IgM paraproteins reactive to myelin-associated glycoproiten (MAG), with which they share the immunologically cross-reactive sulfated glucuronic acid structure.[14,15] These sulfated glucuronyl glycolipids have been identified in myelin- and axon-enriched fractions of motor and sensory nerves from human subjects and have a slightly greater abundance in sensory nerve fractions compared with motor nerve, which may account for the relative predominance of sensory features in patients with anti-SGPG antibodies. The structures of the two glycolipid antigens, SGPG and SGLPG were characterized independently and revealed that the minimal carbohydrate requirement for the HNK-1 epitope resides in the terminal disaccharide residue, $SO_4$-3GlcAβ1-3Galβ1.[14,16]

Gangliosides are principally enriched in neuronal membranes but are also minor constituents of myelin.[14,16,17] With respect to identified targets for autoantibodies in autoimmune neuropathy, major gangliosides include GM1, GM2, GD1b, GD3, GD1a, GT1a, GalNAc-GD1a, and GQ1b (Fig. 1) but other gangliosides have also been identified as antibody targets. Gangliosides by definition are glycosphingolipids containing one or more sialic acid residues linked to the oligosaccharide core. Most ganglioside sialic acid in man is in the *N*-acetyl form, as opposed to the *N*-glycolylform found in many other species.[18] These sialic acid residues are often a critical part of

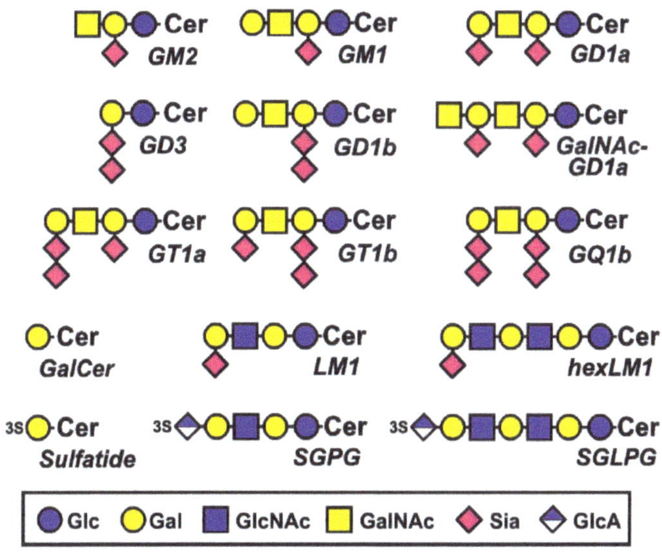


**Fig. 1** Glycolipid structures involved in human autoimmune neuropathies. Structures referred to in the text are shown using standard glycan nomenclature (https://www.ncbi.nlm.nih.gov/glycans/snfg.html). The designation "3S" indicates a sulfate group ($SO_4$) esterified to a hydroxyl on the 3-carbon of the adjacent saccharide.

the epitope involved in autoantibody binding. The amphipathic nature of gangliosides, comprising a hydrophobic ceramide moiety that is embedded in the lipid membrane and a hydrophilic oligosaccharide moiety which is exposed to the extracellular space makes the oligosaccharide ideally situated as a recognition platform for autoantibodies, and other glycan binding molecules, such as bacterial toxins.[19]

When considered as antigen targets, ganglioside localization, clustering, and distribution in the plasma membrane is an important consideration.[20] Gangliosides are concentrated in association with cholesterol in clusters referred to as lipid rafts where their close lateral interactions in the plane of the plasma membrane may enhance their ability to bind antibody with high avidity or affinity, and also affect antibody clearance by endocytosis.[21–23] Fig. 2 illustrates the role that antiganglioside antibody internalization may play in clearance of circulating pathogenic antibodies. Additionally, activated complement components and complement regulators, such as CD59 that reside in rafts may in turn modulate the pathological effects of antiganglioside antibodies.[24] The relative contribution of these factors may vary from site to site within peripheral nerve membranes, and between antiganglioside antibodies of differing immunoglobulin class and reactivity. The distribution of gangliosides in different membrane and raft domains in the PNS and the relevance that this may have to their role as antigens in autoimmune neuropathy is a subject of ongoing study. However, when considering the pathogenic relationship between the presence of an antibody and neuropathy, it is clearly important to have detailed knowledge of the glycosphingolipid composition and distribution within the PNS in both humans and experimental animals used for modeling these disorders; and furthermore, in species from which gangliosides are purified for experimental and diagnostic use. These issues are not straightforward, and a complete map of the ganglioside composition of human nerves, and a comparison between species used for experimental modeling and between individuals, would be a valuable but as yet unavailable resource.

## 3. CLINICAL SYNDROMES ASSOCIATED WITH ANTIGANGLIOSIDE ANTIBODIES

Several peripheral nerve syndromes are associated with a range of antibodies directed against one or more ganglioside species.[25] The clinical phenotype seen in affected patients can vary greatly but most of the features result from inflammatory disturbance of motor, sensory, or autonomic peripheral nerve fibers. In the Guillain–Barré syndrome (GBS) and Miller Fisher syndrome (MFS) and their variants, the disorder is due to an immunological amplification triggered by infections.[26–29] The disorder develops acutely, over days to weeks, pursuing a monophasic course which peaks within a few weeks, followed by partial or complete recovery improving over many months, concomitant with waning of the immunological response. The extent of irrecoverable axonal damage determines the long-term prognosis. In contrast to GBS, there are also a clinically diverse group of chronic disorders, usually associated with amplification of naturally occurring IgM class autoantibodies and unrelated to infectious precedents that evolve over weeks or months and follow a relapsing-remitting or slowly progressive course over many years.[30–32]

Several clinical subtypes have well established associations with particular antiganglioside antibodies.[25] Notably, anti-GQ1b ganglioside IgG antibodies are found in around 90% of patients with MFS.[33] Nonetheless, many patients with other subtypes of GBS are seronegative for antiganglioside antibodies on routine testing, and when present other antiganglioside antibody specificities do not as consistently correlate with clinical features, in comparison with the tight association seen in MFS. Consequently, the clinical utility of diagnostic testing is limited, as standard antibody testing offers low sensitivity (i.e., many patients are seronegative) but high specificity (i.e., when seropositive the result is highly indicative of a particular disorder). In recent years, there have been significant developments in antibody-detection methodologies that have allowed further characterization of antiglycolipid antibodies in these disorders.[34] A growing body of experimental data, from a variety of animal models, continues to support the hypothesis that antiganglioside antibodies are pathogenic in these disorders. The major clinical syndromes and their antiganglioside antibody associations are outlined later.

## 3.1 Guillain–Barré Syndrome

Guillain, Barré, and Strohl described a syndrome of acute, flaccid weakness with loss of tendon reflexes, high cerebrospinal fluid (CSF) protein levels and normal CSF inflammatory cell counts in 1916. The term GBS eventually became synonymous with this clinical entity and is currently used as an umbrella term encompassing a spectrum of clinical and pathological changes with various serological associations. The major clinical delineators are acute

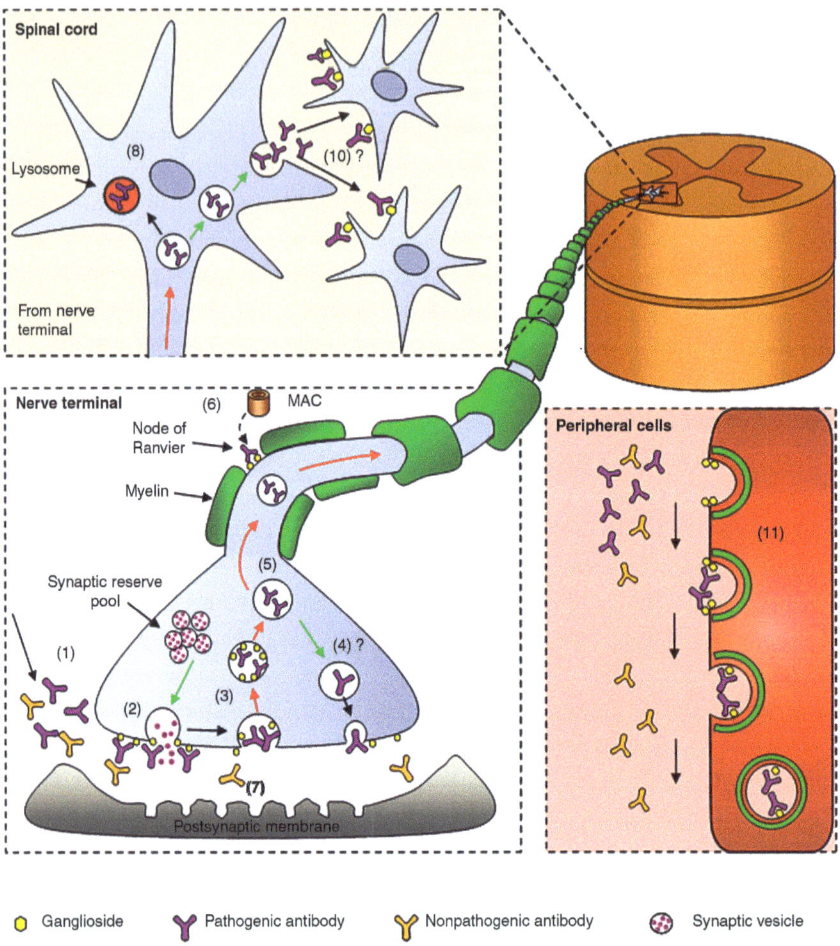


**Fig. 2** Schematic representation of anti-ganglioside antibody internalization. At the nerve terminal, pathogenic and nonpathogenic antiganglioside antibodies (that can and cannot bind their ganglioside target when embedded in the plasma membrane, respectively) arrive from the circulation (1). Vesicles containing acetylcholine fuse with the membrane during neurotransmitter release and during activity-independent vesicle recycling (2), pathogenic antibody species bind to gangliosides. During vesicle retrieval, gangliosides and their attached antibodies are internalized (3). Antibodies may be directly recycled to the plasma membrane (4), or undergo retrograde axonal transport to the motor neuron cell body in the spinal cord (5). Pathogenic antiganglioside antibody, which also binds at the node of Ranvier, cannot be significantly internalized at that site, and therefore activates complement, resulting in MAC deposition and nodal injury (6). Nonpathogenic antibodies are unbound and freely circulate in the interstitial environment of the nerve terminal and are not detectably internalized (7). Retrogradely transported antibody residing in the soma of motor neurons may be degraded in lysosomes (8), or recycled to the neuron surface (9) and

inflammatory demyelinating polyradiculoneuropathy (AIDP) and acute motor axonal neuropathy (AMAN).[28]

### 3.1.1 Acute Inflammatory Demyelinating Polyneuropathy

In AIDP, which comprises the majority of patients with GBS in Europe, North America, and Australia, the pathological changes are of multifocal segmental demyelination of peripheral nerve and nerve root motor and sensory fibers. Antibody, usually of unknown specificity and complement deposits are seen along the surfaces of myelinating Schwann cells and a lymphocytic infiltration into nerve fibers may be present. It is rare for these patients to routinely undergo diagnostic nerve biopsy; hence, this "immunopathological" diagnosis is usually inferred from the results of neurophysiological testing of peripheral nerves via nerve conduction studies/electromyography (NCS/EMG). Historically, it was considered that AIDP resulted in a number of distinct findings on NCS/EMG, including increased distal motor latencies; delayed or absent F waves; reductions in nerve conduction velocity; and temporal dispersion. Thus, these electrical studies are widely used as a surrogate marker of the underlying pathology. It is becoming increasingly recognized that strict electrophysiological diagnostic criteria can misclassify the pathological subtype or, at least with repeated testing, some patients appear to transition from one electrophysiological subtype to another.[35] At present, there is a significant debate as to the optimal electrophysiological diagnostic criteria, timing of testing and relationship of particular findings to the underlying pathology.[36] Nonetheless, NCS/EMG remains a crucial tool in diagnosing, prognosticating, and investigating GBS and are widely used as a proxy indicator as to whether an individual has predominantly demyelinating, axonal, or mixed peripheral nerve injury.

Many studies have shown an association of the AIDP variant and specific antiganglioside or antiglycolipid antibodies, although the proportion of

---

secreted, free to bind to adjacent neurons and potentially mediate pathogenic effects (10). Elsewhere in the body, circulating pathogenic antibodies may bind to their target ganglioside in the membrane of nonneuronal cells. Once bound, antibodies may be endocytosed thus cleared from the circulation (11), again leaving only non-pathogenic antibodies to circulate in the serum. These will be the preferential antibody species detected by solid-phase assays, such as ELISA. Exocytic pathways are shown by *green arrows*, whereas endocytic pathways are shown by *red arrows*. *Used with permission under the Creative Commons Attribution License from Bullens RW, O'Hanlon GM, Wagner E, et al. Complex gangliosides at the neuromuscular junction are membrane receptors for autoantibodies and botulinum neurotoxin but redundant for normal synaptic function. J Neurosci. 2002;22(16):6876–6884.*[19]

AIDP patients who can be demonstrated to have such antibodies is usually relatively low. For example, antigalactocerebroside, anti-LM1, anti-SGPG, and anti-GM2 antibodies have all been reported in small case series and await larger studies.[37] These variations in antiganglioside antibody reporting are most likely explained by methodological variations in test procedures. For example, a number of recent larger studies offer some support to earlier reports on anti-LM1 antibodies or anti-LM1-containing complexes in a subset of patients with GBS.[38]

Similarly, anti-GM1 IgG antibodies (most notably associated with AMAN, later) appear in some patients with AIDP, but only in a relatively small proportion of cases.[39] Again, some have argued that these cases are likely to be cases of AMAN at a pathological level that are misclassified on electrophysiological testing, which would leave the AIDP subtype without a reliable major serological marker. Cases of AIDP that occur in the context of cytomegalovirus (CMV) infection comprise a small subset of AIDP patients but there does appear to be a clear association in these patients with anti-GM2 IgM and quite definite electrophysiological evidence of demyelination. The pathogenic role of anti-GM2 IgM in human GBS has long been in doubt due to the observation that it is also seen in CMV infection without GBS.[29,40] However, recent studies have identified it as a major serological marker in acute canine paralysis, the equivalent of AIDP in dogs, reigniting interest in this particular antibody.[41] After decades of much effort, from numerous research teams, no single antibody has been found that consistently associates with the majority of cases of AIDP.

### 3.1.2 Acute Motor Axonal Neuropathy

The existence of a pathologically distinct subtype of GBS comprising solely of axonal injury proved a controversial concept when first proposed in the mid 1980s.[42] At that time, the widespread demyelination and lymphocytic infiltration seen in classical AIDP cases was considered the hallmark of all cases of GBS, with axonal injury being recognized, but considered a purely secondary effect of severe demyelination. Several years of careful pathological, clinical and electrophysiological reports were required to convince many that there is indeed a subgroup of patients with otherwise typical clinical features of GBS who have a primary immune attack directed against the axonal membrane rather than the myelinating Schwann cells.[43] This subtype, AMAN, and the closely related acute motor and sensory axonal neuropathy (AMSAN), are now well established as distinct pathological entities with distinct electrophysiological features on NCS/EMG and distinct associations with antiglycolipid antibodies. The serological associations in AMAN have been extensively studied.[25,44] Anti-GM1 antibodies were reported in the early descriptions of the subtype and this association has been subsequently confirmed repeatedly in many different countries in larger cohorts. Similarly, anti-GD1a antibodies appear to be consistently associated with patients with AMAN. Moreover, whereas anti-GM1 antibodies are seen in small numbers of patients classified as having AIDP according to classical electrophysiological criteria, anti-GD1a antibodies appear to be more highly specific to the AMAN variant.[45] Additional associations have been found between AMAN/AMSAN and anti-GalNAc-GD1a and anti-GM1b antibodies and more recent studies have suggested associations between AMAN and numerous antiglycolipid complex antibodies,[36,46] most recently, including anti-GM1-complexes, anti-GD1a-complexes, and anti-GA1 complexes in a large Bangladeshi cohort.[47]

## 3.2 Regional Variants of GBS

As described earlier, GBS can be classified into AIDP or AMAN on electrophysiological or pathological grounds and doing so is important for prognostication and to enable precise delineation of distinct pathogenic mechanisms. In addition to classifying cases in this manner, there are a number of highly distinct clinical presentations that vary considerably from the typical GBS cases and have given rise to the concept of "regional" or "functional variants" of GBS. Some of these can be categorized by antiganglioside antibody binding profiles.[48] These terms refer to patients who at the pathological level develop the same underlying autoimmune attack against peripheral nerves but in whom the attack is restricted to particular body regions (e.g., affecting only the facial nerves), or to nerves with a particular function (e.g., affecting only autonomic nerves) leading to distinct constellations of clinical features. Thus, some patients can also be classified on clinical grounds as having a particular regional or functional variant and some of these have distinct serological associations with antiganglioside antibodies, most likely reflecting regional distribution of gangliosides in different neuronal and glial membranes. That these cases are fundamentally part of the GBS spectrum is evidenced by an acute or subacute evolution of symptoms, typical CSF and electrophysiological findings, and that the clinical features occur as part of more typical, fully manifested GBS (e.g., facial weakness, autonomic involvement in some AIDP or AMAN cases). While rare, they inform us deeply about the immunobiology of gangliosides and autoimmunity.

### 3.2.1 *Pharyngeal-Cervical-Brachial Variant*

The pharyngeal-cervical-brachial (PCB) regional variant comprises variable manifestations of acute oropharyngeal, neck, and shoulder weakness without significant limb weakness.[49] As its original description, other case reports and series have associated it with antiganglioside antibodies, particularly anti-GT1a or anti-GD1a IgG.[50] GT1a lacks a second internal sialic acid residue that is seen in GQ1b but the molecules are otherwise identical and most anti-GQ1b antibodies in GBS patients cross-react with GT1a. However, some anti-GT1a antibodies in PCB are monospecific and do not cross-react with GQ1b, making it a relatively specific marker of this variant. Acute oropharyngeal palsy is likely an even further restricted regional variant and this has also been associated with anti-GT1a antibodies.[51,52] There is some evidence for the preferential binding of anti-GT1a antibodies to the lower cranial nerves thereby suggesting that the restricted clinical manifestations arise because of variations in regional expression, accessibility, or microenvironment of gangliosides on peripheral nerves membranes.

In typical cases of GBS, there is often a degree of ataxia; however, it is usually overshadowed by marked limb weakness and other symptoms. In some cases there is early and prominent ataxia, which leads to the initial clinical evaluation before the more fulminant typical motor weakness manifests, and in others still the ataxia remains the dominant clinical feature. These are referred to as ataxic GBS. In these later cases the link with GBS is suggested by an acute or subacute onset, distal paresthesia, areflexia, elevated CSF protein, and electrophysiological features suggesting sensory nerve demyelination. This functional variant has been associated with antibodies to GQ1b, GD1b, and related disialylatedgangliosides.[53–55] GD1b is known to be relatively enriched in sensory fibers compared with motor fibers, which may in part account for the selective injury in these patients. Ataxia is one of the core features of MFS, itself a functional variant highly associated with anti-GQ1b antibodies some of which cross-react with GD1b.

### 3.2.2 *Miller Fisher Syndrome*

Charles Miller Fisher described a number of patients with ataxia, areflexia, and ophthalmoplegia; and suggested that they were unusual manifestations of an inflammatory polyneuropathy.[56] MFS is a relatively common form of GBS in Japan, Korea, and a few other regions but only accounts for 5%–10% of cases elsewhere. In comparison to AIDP and AMAN, it is a milder illness.[57] Nonetheless, this variant has been extensively studied and modeled primarily because of the repeated and consistent finding that it is highly associated with antibodies against GQ1b, with between 80%–90% of patients testing positive with routine diagnostic techniques. Indeed, some prefer the term "anti-GQ1b syndrome" to reflect this and in order to more fully include the clinical spectrum seen in patients found to have anti-GQ1b antibodies. Anti-GQ1b antibodies are also very specific in comparison to most other antiganglioside antibodies and are only very rarely detected in patients with other conditions or in controls. Titres reach their peak at clinical presentation and drop rapidly with recovery. GQ1b has been shown biochemically to be particularly enriched in the nerve roots of the extraocular nerve roots and immunostaining with anti-GQ1b antibodies shows preferential binding at paranodal myelin and nerve terminals in these nerves but not other cranial nerves, supporting the notion that clinical features are manifestations of nerve injury restricted to regions expressing the antigen at a critical level, location, and environment.[58,59]

### *3.2.3 Bickerstaff's Brainstem Encephalitis*

Bickerstaff's brainstem encephalitis (BBE) is a rare syndrome characterized by variable degrees of ataxia, areflexia, and ophthalmoplegia, and, in distinction from MFS, impaired consciousness, extensor plantar responses, and tetraparesis.[60] Anti-GQ1b antibodies are seen in two-third of cases. Often patients initially develop the MFS-related clinical features, shortly followed by decreased consciousness or coma and tetraparesis. It can thought of as "MFS-plus" in that it has the core features of MFS plus additional, likely CNS, features, and is associated with the anti-GQ1b antibody.[61] The mechanism underlying this CNS involvement remains unknown.

In contrast to BBE which can be considered "MFS-plus," there are cases of incomplete manifestations of the features of MFS, including ataxic GBS, as described earlier, acute isolated ptosis, acute isolated internal ophthalmoplegia, and acute ophthalmoplegia (AO), which can be considered "MFS-minus." The connection with MFS is established with the observations of the same temporal course and CSF and electrophysiological features and, notably, with the finding of anti-GQ1b IgG in these cases.[62]

## 3.3 Chronic Neuropathy Syndromes

While the acute neuropathy syndromes associated with antiganglioside antibodies have an almost bewildering degree of overlap and variation between somewhat artificial nosological boundaries, the chronic syndromes are far more distinct in terms of their underlying pathological, clinical, and serological features. In general, the acute syndromes are associated with complement-fixing subclasses of IgG antiganglioside antibodies and in the chronic syndromes with IgM antibodies.

### *3.3.1 Chronic Inflammatory Demyelinating Polyradiculoneuropathy*

Chronic inflammatory demyelinating polyradiculoneuropathy (CIDP) in some respects appears to be a relapsing-remitting and less fulminant version of AIDP: pathologically there is segmental demyelination and lymphocytic inflammation; electrophysiologically there is evidence of demyelination; clinically there is areflexia and variable sensory impairment and motor weakness. The core distinction is in the more insidious progression of symptoms which, by definition, continue to progress over at least 8 weeks, and which relapse and remit over variable timescales. There are even variants of CIDP that initially develop over a shorter time period, so-called acute-onset CIDP (A-CIDP), which are often initially diagnosed as AIDP but several subsequent relapses force a reassessment and recategorization. In distinction from AIDP, there have not been any significant or consistent associations with antiganglioside antibodies, while there have been recent studies showing that a small subset of CIDP patients have IgG4 antibodies against the paranodal proteins contactin-1 (CNTN1), contactin-associated protein (Caspr) or neurofascin-155 (NF155).[63] In addition, whereas intravenous immunoglobulin (IVIg) is of proven benefit in both AIDP and CIDP, corticosteroids have also been shown to be effective in treating CIDP but not in AIDP. This suggests a fundamental difference in the immunopathology between these disorders and, given the lack of clear association with antiganglioside antibodies, CIDP will not be further discussed here.

### *3.3.2 Multifocal Motor Neuropathy*

Multifocal motor neuropathy (MMN) manifests as a slowly progressive, asymmetrical, distal predominant neuropathy affecting multiple motor fibers.[64] Usually, arms are affected first with wrist or finger drop and lower limbs affected much later on. Focal conduction block of the motor nerves with sparing of sensory fibers is the electrophysiological signature, but with time both the clinical manifestations and the electrophysiological changes become more widespread. IgM antibodies against GM1 or GM2 are seen in around half of patients using standard ELISA techniques. In patients who test negative for anti-GM1 antibodies using conventional techniques, a proportion harbor antibodies to complexes of GM1 and GalC that can only be detected using highly specialised assays.[65,66] Despite the obvious diagnostic utility of this observation this finding has not yet been adopted into widespread routine clinical practice.

### 3.3.3 Antimyelin Associated Glycoprotein Neuropathy

Anti-MAG neuropathy manifests as a chronic, slowly progressive demyelinating neuropathy mainly affecting the sensory nerves manifesting as a coarse postural tremor of the arms which progresses over time to a more widespread and disabling sensory ataxia with impaired gait and eventually often motor involvement with distal muscle weakness.[67] Pathologically, there is a selective loss of large myelinating fibers and distinct splitting of the outer myelin lamellae visible on electron microscopy, at sites where there are also IgM deposits. The "anti-MAG" antibody is an IgM paraprotein, which recognizes a carbohydrate antigen on MAG and which cross-reacts with sulfoglucuronyl glycolipid (SGPG), an acidic glycolipid which separates between GM1 and GD1a in chromatographs of human peripheral nerve, and its higher lactosaminyl homologue, SGLGP.[15,68] MAG is found mostly in the CNS, whereas SGPG is only seen in the PNS, indicating it is the most probable antigen in anti-MAG neuropathy. Detection assays tend to use MAG rather than SGPG as the antigen because anti-MAG antibodies bind at much higher affinity to MAG.[69] These antibodies have been believed to be directly pathogenic due to the observation that there are antibody and complement deposits on the split myelin, passive transfer in animal models mediates complement-dependent demyelination and myelin lamellae splitting and active immunization of cats with SGPG causes an ataxic neuropathy that resembles anti-MAG neuropathy.[70,71]

### 3.3.4 CANOMAD

The acronym CANOMAD was coined in 1996 to describe patients with a *c*hronic *a*taxic *n*europathy, *o*phthalmoplegia, Ig*M* paraprotein, cold *a*gglutinins and *d*iasialosylantibodies.[32,72] These patients develop a slowly progressive sensory neuropathy with prominent sensory ataxia and areflexia but relatively spared motor function. There is commonly also a progressive or relapsing-remitting ophthalmoplegia and bulbar palsy, and variable responsiveness to IVIg and other immunotherapies. Patients have an IgM paraprotein, which reacts with GQ1b and often other disialyated gangliosides, including GD1b, GD3, and GT1b and which can also have cold agglutinin activity. Electrophysiology reveals both demyelinating and axonal features while pathological studies have shown loss of large myelinated fibers. The ophthalmoplegia, bulbar weakness, anti-GQ1b antibody-reactivity, electrophysiological and pathological changes, and the partial response to immunotherapy are similar to that seen in MFS except in a chronic relapsing form.

## 4. METHODS FOR DETECTING ANTIGANGLIOSIDE ANTIBODIES

Antiganglioside antibodies can be measured in serum samples using a variety of different methods, each of which has distinct advantages and disadvantages and none of which is truly an independent "gold standard."[64,73,74] Commercially, versions of the enzyme-linked immunosorbant assay (ELISA) and the "dot-blot" methods are available and in widespread use. Thin-layer chromatography (TLC) of gangliosides with immuno-overlay of patient samples is only carried out in a few specialized clinical service laboratories, although remains widespread in basic glycolipid science laboratories. Similarly, glycan array techniques are highly specialized and used in a small number of specialized research facilities.[75] TLC and glycoarray techniques are very useful in experimental contexts, such as defining precise reactivities of sera or screening large numbers of samples against multiple antigens.

ELISA and dot-blot techniques are widely accepted to have significant inter- and intralaboratory variation and usually need to utilize carefully defined local cut-off values and controls.[76] Clearly, positive or negative samples usually correlate well between runs and between laboratories with a given technique but lower titre samples correlate less well. Furthermore, there is also significant variation when these two techniques are compared against each other, most likely due to differences in antigen density and the biophysical properties of the adherent surface.

One glycolipid microarray-based technique that has been developed is a further miniaturized and refined version of an earlier glycoarray technique.[47] In the microarray, picolitres of antigen, usually single and combinations of glycolipids that form heteromeric complexes (see later), are adsorbed onto a PVDF covered glass slide using a robotic micropipette. Following this a standard two-step indirect immunofluorescence approach is used to screen small volumes of sera against dozens of antigen combinations. The slide is then visualized with a microarray scanner and image analysis software is used to calculate fluorescence values for each antigen. These data can then be visualized and statistically analyzed using cluster analysis and related techniques. This method allows for screening of large cohorts of patients for antibodies against many single and heteromeric antigens with small amounts of reagents and sera that would simply not be possible with other techniques with standard volumes of patient material. Fig. 3 exemplifies a typical final

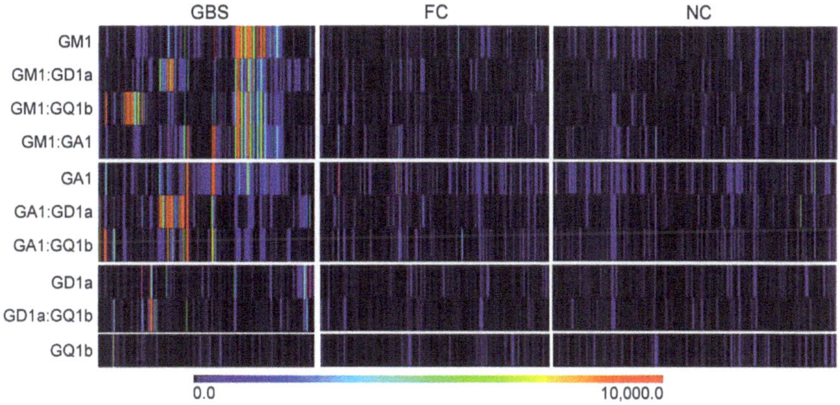


**Fig. 3** Typical dataset obtained using microarray screening technology. Heat map presentation of patient (GBS) and control sera [family controls (FC) and neurologic disease controls (NC)] reactivity with GM1, GD1a, GQ1b, and GA1 as single antigens and their six possible heteromeric complexes. Each small column displays a visual representation of antibody binding intensity against each antigen for an individual patient. Each row corresponds to a different antigen or antigen complex as labeled on the far left. Binding intensity is displayed on a rainbow color scale from *black* (no reactivity) through to *red* (highest reactivity). This particular cohort of GBS patients in Bangladesh showed a combined sensitivity of 64.3% and specificity of 77.1% against this panel of 10 targets antigens. FC and NC show minimal reactivity. *Used with permission from Kaida K, Kusunoki S. Antibodies to gangliosides and ganglioside complexes in Guillain–Barre syndrome and Fisher syndrome: mini-review. J Neuroimmunol. 2010;223 (1–2):5–12.*[44]

dataset obtained with this technique and highlights the power of the method for screening large number of samples for antibodies against a large number of antiganglioside and antiganglioside complex antibodies.

## 5. ANTIGLYCOLIPID COMPLEX ANTIBODIES

Gangliosides are relatively small and polar molecules with a hydrophobic ceramide tail that sits within the plasma membrane, and a hydrophilic head group with various combinations of glucose, galactose, GalNAc, and sialic acid that protrudes from the membrane. How the extracellular portion of a particular ganglioside is orientated at any given time and how that varies with local ionic concentrations or via the influence of interacting lipids, proteins, and other factors remains unclear.[20,77] It is often the case that the mere presence of a ganglioside in a particular membrane does not render it vulnerable to antiganglioside antibody-mediated injury due to the specific epitope being "hidden" or otherwise remaining cryptic.[78] Similarly, laboratory methods used to detect antiganglioside antibodies all must adhere, coat, or otherwise separate or immobilize gangliosides on or within a particular medium or platform in order to allow for probing of sera. Each such method is likely to result in the ganglioside taking up a particular orientation, which may favor or hinder antibody access to a particular epitope for a particular clinical sample containing an antiganglioside antibody.

In keeping with this long-appreciated concept, there has also been the equally long-standing observation that in some sera the detection of antiganglioside antibodies in vitro can be enhanced or inhibited when the target antigen is present not in isolation, but with other particular lipids, historically referred to as "accessory lipids." Twenty years ago, Pestronk *et al.* described an ELISA technique to demonstrate that some MMN sera tested negative for anti-GM1 IgM antibody when only GM1 was present, but tested positive when a GM1-containing lipid mixture was used instead.[64] Subsequently, in 2004, Kaida *et al.* used TLC immuno-overlay to demonstrate that some GBS patient sera did not bind a single ganglioside on its own, but did bind to a mixture of GD1a and GD1b, leading to the recognition of antiganglioside-complex antibodies.[46] Some antibodies bind only to a particular complex of glycolipids and not the individual glycolipids (complex-dependent), some bind to the single lipids but have their binding enhanced by complexes (complex-enhanced) or inhibited (complex-inhibited), whereas others bind at the same level irrespective of complexes (complex-independent).[34] Fig. 4 illustrates the phenomenon of complex-modified binding of Siglec-7, and antiganglioside antibody and tetanus toxin as revealed by glycoarray screening. Complex-dependent antibodies appear to be somewhat uncommon and do not account for the majority of hitherto seronegative patients, at least in the limited studies conducted to date. Nonetheless, there will be some patient serathat harbor antiganglioside antibodies, but test seronegative by standard assays which do not examine antiglycolipid complexes.[79] There may still be a small number of one or two glycolipid complexes that turn out to be the antigen in the majority of GBS patients, but what currently seems more likely is that there is a large number of distinct antiglycolipid and antiglycolipid complex antibodies that arise against a diverse range of glycolipid antigens and that this may vary by clinical and pathological subtype and by geographic region.

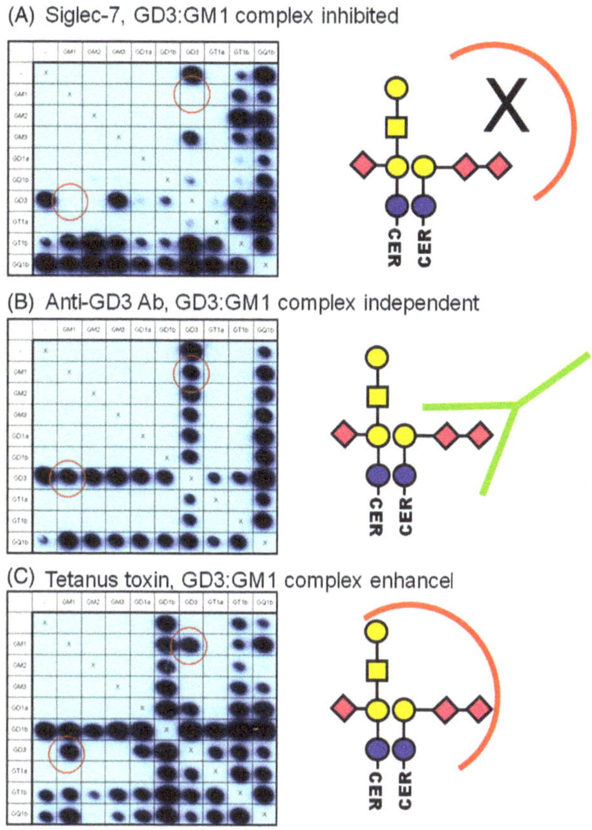


**Fig. 4** Complex inhibited, complex independent, and complex enhanced binding as demonstrated for GD3 series complexes. The arrays demonstrate antibody/toxin/siglec binding to either single ganglioside species (top row and first column in each panel) or to combinations (complexes) of two gangliosides (remainder of the panel). Adjacent to each array is a cartoon illustrating hypothetical interactions between the gangliosides and the constraints this puts on antibody interactions. The binding pattern of the ganglioside binding subunit of tetanus toxin (TeNT $H_C$), the mouse monoclonal antibody CGM3 and the Siglec fusion protein Siglec-7-Fc is markedly different with respect to GD3 series complexes. (A) The siglec-7-Fc fusion protein binds GD3 in isolation but is completely inhibited from binding in the presence of specific second gangliosides, including GM1, GM2, and partially inhibited by GD1a and GT1a (complex inhibited). (B) The mAb CGM3 binds all GD3 complexes equally (complex independent), whereas (C) TeNT $H_C$-HRP only binds to the complex of GD3:GM1, while failing to bind to either component ganglioside in isolation (complex enhanced). In addition, TeNT $H_C$ is inhibited from binding GQ1b, and to a lesser extent GT1b, by precomplexing with GM2. *Adapted with permission from Lopez PH, Comin R, Villa AM, et al. A new type of anti-ganglioside antibodies present in neurological patients. Biochimica et Biophysica Acta. 2006;1762(3):357–361.*[79]

# 6. POSTINFECTIOUS AND OTHER ORIGINS OF ANTIGANGLIOSIDE ANTIBODIES

Natural antibody repertoires that are of the IgM class and present in normal individuals found frequently bind glycolipids, including gangliosides.[80] This antibody repertoire, predominantly secreted by innate-like B cells, contains glycan binding specificities both for host and microbial glycans. Some natural antibodies are considered polyreactive, whereas others are very specific for individual glycan epitopes borne by glycolipids and gangliosides. When antiglycolipid IgM antibodies are self-reactive and highly abundant, or increase their affinity and class switch to IgG, they have the potential to induce autoimmune neuropathy.[81]

## 6.1 Campylobacter Infection

The earliest reports on GBS envisioned a toxic or infectious agent as the etiological culprit but subsequently a postinfectious, autoantibody-mediated hypothesis developed. The isolation of *Campylobacter* from human samples proved technically difficult, but subsequently this particular infectious agent came under much investigation as a potential preceding trigger for GBS and has yielded a great number of crucial observations that shape our current thinking about the pathogenesis of GBS.[82,83] Numerous other infectious agents are associated with the development of GBS.[84]

*Campylobacter jejuni* infection is by far the most extensively studied preceding infection in GBS. It has been widely shown that lipo-oligosaccharide (LOS) structures existed on *C. jejuni* that are structurally similar to gangliosides and which could induce antiganglioside antibodies. Moreover, the strains of *C. jejuni* that were associated with particular GBS subtypes carried LOS with structures that mimicked the particular ganglioside expected for that GBS subtype, based on the antiganglioside antibody associations. Thus, for example, strains of *C. jejuni* associated with AMAN (HS:19) have been shown to carry LOS structures with a structural similarity to both GM1 and GD1a.[85,86]

## 6.2 *Cytomegalovirus*

*Cytomegalovirus* is a frequent virus to be associated with GBS and associates with the AIDP variant.[87,88] It is reported to occur in around 6%–15% of cases and cases tend to occur in younger patients with more severe course, need for respiratory support, cranial nerve involvement, and delayed recovery. It has been linked to the presence of anti-GM2 IgM; however, the observation that a significant number of people with CMV infection and no neurological disease also have anti-GM2 IgM has led to doubt about whether this particular antiganglioside antibody is pathogenic. Recently, anti-GM2 IgM has been shown to be a major biomarker for the canine equivalent of GBS, thus renewing interest in its role in disease.[41]

## 6.3 *Mycoplasma*

*Mycoplasma pneumoniae* infection is associated with neurological complications, including encephalitis and GBS, where it is reported in 3%–21% of cases. Some reports indicate cross reactivity of anti-Mycoplasma antibodies and GalCer, hence its mention herein. As is seen in CMV infection, anti-GalCer antibodies are seen in patients with *M. pneumoniae* and no neurological symptoms, although the observation that in those who develop GBS the isotype more often tends to be IgG has led to the hypothesis that in *M. pneumoniae*-related GBS antibody subclass switching may be a crucial step in the development of neurological disease.[89]

## 6.4 *Haemophilus*

*Haemophilus influenzae* type b is also reported in 1%–9% of cases as a trigger of GBS, which tend to be of the axonal subtype and are associated with anti-GM1 antibodies. GM1 ganglioside-like structures have been found on an *H. influenzae* strain isolated from a patient with GBS.[90,91] In addition, *H. influenzae* strains which bear GT1a and GQ1b-like epitopes have been isolated from patients with MFS and MFS patients with serological evidence of recent *H. influenzae* infection have anti-GQ1b and anti-GT1a antibodies, suggesting it may be an important etiological trigger for MFS. Indeed, Miller Fisher isolated this organism from the sputum of one of his original MFS patients.

## 6.5 Ganglioside Administration

Ganglioside administration by injection was a popular pharmacotherapy first trailed in the 1970s and continued until the early 1990s for a number of disorders, including diabetic neuropathy, ischemic stroke, and neurodegenerative conditions. Case reports and small series of GBS occurring in patients receiving ganglioside therapy were published in the 1990s at a time when the potential pathogenic role of antiganglioside antibodies was becoming more widely recognized. There was, and remains, controversy over whether there actually is a statistically significant increased risk of GBS with ganglioside therapy. Several epidemiological studies have not indicated any increased risk and no cases were reported in trials of GM1 administration for spinal cord injury. Nonetheless most countries withdrew gangliosides from the therapeutic market.[92]

### 6.6 Vaccination

While vaccination has been widely linked to GBS for many years, based on monitoring of vaccination programs, there have been no consistent links between vaccination-related GBS and particular patterns of antiganglioside antibodies. Recently, acute Zika virus infection has been linked to GBS in French Polynesia and in Latin America; again no consistent pattern of antiganglioside antibodies has yet been detected and the etiology remains uncertain.[93]

## 7. EXPERIMENTAL MODELS OF ANTIGANGLIOSIDE ANTIBODY-MEDIATED NEUROPATHY

Dating back to the 1970s, animal models resulting from ganglioside immunization have been developed and examined. In a classic paper, Nagai immunized rabbits with gangliosides and thereby induced a paralytic syndrome, which has since been widely studied, with refinements, in Japan and elsewhere.[94] Immunization with the glycolipid, GalCer induces flaccid paralysis and sensory loss in rabbits, accompanied by pathological findings of vesicular demyelination in nerves and CNS with macrophage infiltration.[9,95] Subsequently to active immunization studies, the intraneural injection and topical application techniques were applied to rabbit anti-GalCer antiserum which produced complement mediated conduction block and demyelination in rat sciatic nerves. Similar studies have been conducted with anti-SGPG antibodies, which also react with MAG, and in humans induce the so-called "anti-MAG" neuropathy.[70] Thus, the basic principles of antiglycolipid antibody-induced injury were well established, even 40 years ago. More recently a sensory ataxic neuropathy was induced in rabbits by active immunization with GD1b ganglioside that bears close similarities to anti-GD1b antibody mediated human neuropathy.[96] Recently, in vitro data has shown that application of anti-GD1b antibody to myelinating cell culture induces both complement dependent and independent pathological effects.[97]

Anti-GM1 antibodies are a major specificity in human disease and have been well studied in experimental models, notably of AMAN in which IgG antibodies dominate and MMN, associated with anti-GM1 IgM. Affected rabbits are severely paralyzed with motor axonal degeneration and node of Ranvier injury in the ventral roots.[98] In mice, anti-GM1, anti-GD1a, and anti-GQ1b monoclonal antibodies derived from both humans and mice immunized with *C. jejuni* LOS also induce paralytic disease affecting distal axons motor nerve terminals.[99,100] The destructive pathway involves membrane attack complex pore deposition in axonal membranes with calcium ingress, calpain activation, and destructive structural changes. Inhibition of complement activation is protective and raises the possibilities of clinical trials of complement inhibiting agents.[101]

Overall animal studies, alongside human pathology studies, clearly indicate that antiganglioside antibodies are major drivers of nerve injury in a wide range of experimental paradigms, have provided crucial pathological pathway data that will hopefully be taken forward into clinical trials.

## 8. CONCLUSIONS

The antiganglioside antibody field has emerged over the last few decades as a major pathophysiological pathway for autoimmune neuropathy that informs novel treatment and diagnostic biomarker research for human disease. Further current growth focuses on identifying new autoantibodies, which may be ganglioside/lipid complexes or reactive with other nerve structures. The field has also provided information about pathways of axonal degeneration that can be further investigated. The rich variety of clinical syndromes and their associated antiganglioside autoantibodies continues to inform clinical and basic researchers and further enrich the ganglioside biology field.

## REFERENCES


1. Salzer JL, Brophy PJ, Peles E. Molecular domains of myelinated axons in the peripheral nervous system. *Glia.* 2008;56(14):1532–1540.
2. Scherer SS, Arroyo EJ. Recent progress on the molecular organization of myelinated axons. *J Peripher Nerv Syst.* 2002;7(1):1–12.
3. Sheikh KA, Deerinck TJ, Ellisman MH, Griffin JW. The distribution of ganglioside-like moieties in peripheral nerves. *Brain.* 1999;122(Pt. 3):449–460.

4. Kieseier BC, Hartung HP, Wiendl H. Immune circuitry in the peripheral nervous system. *Curr Opin Neurol.* 2006;19(5):437–445.
5. Meyer Zu HG, Hu W, Hartung HP, Lehmann HC, Kieseier BC. The immunocompetence of Schwann cells. *Muscle Nerve.* 2008;37(1):3–13.
6. Kanda T. Biology of the blood-nerve barrier and its alteration in immune mediated neuropathies. *J Neurol Neurosurg Psychiatry.* 2013;84(2):208–212.
7. Ogawa-Goto K, Abe T. Gangliosides and glycosphingolipids of peripheral nervous system myelins-a minireview. *Neurochem Res.* 1998;23(3):305–310.
8. Honke K. Biosynthesis and biological function of sulfoglycolipids. *Proc Jpn Acad Ser B Phys Biol Sci.* 2013;89(4):129–138.
9. Saida T, Saida K, Dorfman SH, et al. Experimental allergic neuritis induced by sensitization with galactocerebroside. *Science.* 1979;204(4397):1103–1106.
10. Popko B, Dupree JL, Coetzee T, Suzuki K, Suzuki K. Genetic analysis of myelin galactolipid function. *Adv Exp Med Biol.* 1999;468:237–244.
11. Svennerholm L, Bostrom K, Fredman P, et al. Gangliosides and allied glycosphingolipids in human peripheral nerve and spinal cord. *Biochim Biophys Acta.* 1994;1214 (2):115–123.
12. Susuki K, Yuki N, Hirata K, Kuwabara S. Fine specificities of anti-LM1 IgG antibodies in Guillain–Barre syndrome. *J Neurol Sci.* 2002;195(2):145–148.
13. Gu Y, Chen ZW, Siegel A, et al. Analysis of humoral immune responses to LM1 ganglioside in guinea pigs. *J Neuroimmunol.* 2012;246(1–2):58–64.
14. Chou DK, Ilyas AA, Evans JE, Costello C, Quarles RH, Jungalwala FB. Structure of sulfated glucuronyl glycolipids in the nervous system reacting with HNK-1 antibody and some IgM paraproteins in neuropathy. *J Biol Chem.* 1986;261(25):11717–11725.
15. Ilyas AA, Dalakas MC, Brady RO, Quarles RH. Sulfated glucuronyl glycolipids reacting with anti-myelin-associated glycoprotein monoclonal antibodies including IgM paraproteins in neuropathy: species distribution and partial characterization of epitopes. *Brain Res.* 1986;385(1):1–9.
16. Ariga T, Kohriyama T, Freddo L, et al. Characterization of sulfated glucuronic acid containing glycolipids reacting with IgM M-proteins in patients with neuropathy. *J Biol Chem.* 1987;262(2):848–853.
17. Ledeen RW, Yu RK. Gangliosides: structure, isolation, and analysis. *Methods Enzymol.* 1982;83:139–191.
18. Galili U. Natural anti-carbohydrate antibodies contributing to evolutionary survival of primates in viral epidemics? *Glycobiology.* 2016;26(11):1140–1150.
19. Bullens RW, O'Hanlon GM, Wagner E, et al. Complex gangliosides at the neuromuscular junction are membrane receptors for autoantibodies and botulinum neurotoxin but redundant for normal synaptic function. *J Neurosci.* 2002;22(16):6876–6884.
20. Cohen M, Varki A. Modulation of glycan recognition by clustered saccharide patches. *Int Rev Cell Mol Biol.* 2014;308:75–125.
21. Cunningham ME, McGonigal R, Meehan GR, et al. Anti-ganglioside antibodies are removed from circulation in mice by neuronal endocytosis. *Brain.* 2016;139(Pt. 6):1657–1665.
22. Fewou SN, Plomp JJ, Willison HJ. The pre-synaptic motor nerve terminal as a site for antibody-mediated neurotoxicity in autoimmune neuropathies and synaptopathies. *J Anatomy.* 2014;224(1):36–44.
23. Fewou SN, Rupp A, Nickolay LE, et al. Anti-ganglioside antibody internalization attenuates motor nerve terminal injury in a mouse model of acute motor axonal neuropathy. *J Clin Invest.* 2012;122(3):1037–1051.
24. Willison HJ, Halstead SK, Beveridge E, et al. The role of complement and complement regulators in mediating motor nerve terminal injury in murine models of Guillain–Barre syndrome. *J Neuroimmunol.* 2008;201–202:172–182.

25. Willison HJ, Yuki N. Peripheral neuropathies and anti-glycolipid antibodies. *Brain.* 2002;125(Pt. 12):2591–2625.
26. Hahn AF. Guillain–Barre syndrome. *Lancet.* 1998;352:635–641.
27. Hughes RA, Cornblath DR. Guillain–Barre syndrome. *Lancet.* 2005;366 (9497):1653–1666.
28. Willison HJ, Jacobs BC, van Doorn PA. Guillain–Barre syndrome. *Lancet.* 2016;388 (10045):717–727.
29. Yuki N. Infectious origins of, and molecular mimicry in, Guillain–Barre and Fisher syndromes. *Lancet Infect Dis.* 2001;1(1):29–37.
30. Campagnolo M, Ferrari S, Dalla Torre C, et al. Polyneuropathy with anti-sulfatide and anti-MAG antibodies: clinical, neurophysiological, pathological features and response to treatment. *J Neuroimmunol.* 2015;281:1–4.
31. Kornberg AJ, Pestronk A. Chronic motor neuropathies: diagnosis, therapy, and pathogenesis. *Ann Neurol.* 1995;37(suppl 1):S43–S50.
32. Willison HJ, O'Leary CP, Veitch J, et al. The clinical and laboratory features of chronic sensory ataxic neuropathy with anti-disialosyl IgM antibodies. *Brain.* 2001;124(Pt. 10):1968–1977.
33. Chiba A, Kusunoki S, Shimizu T, Kanazawa I. Serum IgG antibody to ganglioside GQ1b is a possible marker of Miller Fisher syndrome. *Ann Neurol.* 1992;31(6):677–679.
34. Rinaldi S, Brennan KM, Willison HJ. Heteromeric glycolipid complexes as modulators of autoantibody and lectin binding. *Prog Lipid Res.* 2010;49(1):87–95.
35. Uncini A, Kuwabara S. Nodopathies of the peripheral nerve: an emerging concept. *J Neurol Neurosurg Psychiatry.* 2015;86(11):1186–1195.
36. Kokubun N, Nishibayashi M, Uncini A, Odaka M, Hirata K, Yuki N. Conduction block in acute motor axonal neuropathy. *Brain.* 2010;133(10):2897–2908.
37. Samukawa M, Hamada Y, Kuwahara M, et al. Clinical features in Guillain–Barre syndrome with anti-Gal-C antibody. *J Neurol Sci.* 2014;337(1–2):55–60.
38. Kaida K, Kusunoki S. Immune-mediated neuropathy and anti-glycolipid antibodies. *Brain Nerve.* 2013;65(4):413–423.
39. Wakerley BR, Yuki N. Guillain-Barre syndrome. *Expert Rev Neurother.* 2015;15:847–849.
40. Ang CW, Jacobs BC, Brandenburg AH, et al. Cross-reactive antibodies against GM2 and CMV-infected fibroblasts in Guillain–Barre syndrome. *Neurology.* 2000;54 (7):1453–1458.
41. Rupp A, Galban-Horcajo F, Bianchi E, et al. Anti-GM2 ganglioside antibodies are a biomarker for acute canine polyradiculoneuritis. *J Peripher Nerv Syst.* 2013;18(1):75–88.
42. Feasby TE, Gilbert JJ, Brown WF, et al. An acute axonal form of Guillain–Barre polyneuropathy. *Brain.* 1986;109:1115–1126.
43. Griffin JW, Li CY, Ho TW, et al. Guillain-Barre-syndrome in northern china: the spectrum of neuropathological changes in clinically defined cases. *Brain.* 1995;118:577–595.
44. Kaida K, Kusunoki S. Antibodies to gangliosides and ganglioside complexes in Guillain–Barre syndrome and Fisher syndrome: mini-review. *J Neuroimmunol.* 2010;223(1–2):5–12.
45. Ho TW, Willison HJ, Nachamkin I, et al. Anti-GD1a antibody is associated with axonal but not demyelinating forms of Guillain–Barre syndrome. *Ann Neurol.* 1999;45 (2):168–173.
46. Kaida K, Morita D, Kanzaki M, et al. Ganglioside complexes as new target antigens in Guillain–Barre syndrome. *Ann Neurol.* 2004;56(4):567–571.
47. Halstead SK, Kalna G, Islam MB, et al. Microarray screening of Guillain–Barre syndrome sera for antibodies to glycolipid complexes. *Neurol Neuroimmunol Neuroinflamm.* 2016;3(6):e284.

48. Wakerley BR, Yuki N. Mimics and chameleons in Guillain–Barre and Miller Fisher syndromes. *Pract Neurol.* 2015;15(2):90–99.
49. Wakerley BR, Yuki N. Pharyngeal-cervical-brachial variant of Guillain–Barre syndrome. *J Neurol Neurosurg Psychiatry.* 2014;85(3):339–344.
50. Kim JK, Kim BJ, Shin HY, et al. Acute bulbar palsy as a variant of Guillain–Barre syndrome. *Neurology.* 2016;86(8):742–747.
51. Nagashima T, Koga M, Odaka M, Hirata K, Yuki N. Clinical correlates of serum anti-GT1a IgG antibodies. *J Neurol Sci.* 2004;219(1–2):139–145.
52. O'Leary CP, Veitch J, Durward WF, Thomas AM, Rees JH, Willison HJ. Acute oropharyngeal palsy is associated with antibodies to GQ1b and GT1a gangliosides. *J Neurol Neurosurg Psychiatry.* 1996;61(6):649–651.
53. Kaida K, Kamakura K, Ogawa G, et al. GD1b-specific antibody induces ataxia in Guillain–Barre syndrome. *Neurology.* 2008;71(3):196–201.
54. Miyazaki T, Kusunoki S, Kaida K, Shiina M, Kanazawa I. Guillain–Barre syndrome associated with IgG monospecific to ganglioside GD1b. *Neurology.* 2001;56 (9):1227–1229.
55. Willison HJ, Almemar A, Veitch J, Thrush D. Acute ataxic neuropathy with cross-reactive antibodies to GD1b and GD3 gangliosides. *Neurology.* 1994;44(12):2395–2397.
56. Fisher M. An unusual variant of acute idiopathic polyneuritis (syndrome of ophthalmoplegia, ataxia and areflexia). *New Engl J Med.* 1956;255(2):57–65.
57. Wakerley BR, Uncini A, Yuki N. Guillain–Barre and Miller Fisher syndromes: new diagnostic classification. *Nat Rev Neurol.* 2014;10(9):537–544.
58. Chiba A, Kusunoki S, Obata H, Machinami R, Kanazawa I. Ganglioside composition of the human cranial nerves, with special reference to pathophysiology of Miller Fisher syndrome. *Brain Res.* 1997;745(1–2):32–36.
59. Liu JX, Willison HJ, Pedrosa-Domellof F. Immunolocalization of GQ1b and related gangliosides in human extraocular neuromuscular junctions and muscle spindles. *Invest Ophthalmol Vis Sci.* 2009;50(7):3226–3232.
60. Koga M, Kusunoki S, Kaida K, et al. Nationwide survey of patients in Japan with Bickerstaff brainstem encephalitis: epidemiological and clinical characteristics. *J Neurol Neurosurg Psychiatry.* 2012;83(12):1210–1215.
61. Odaka M, Yuki N, Yamada M, et al. Bickerstaff's brainstem encephalitis: clinical features of 62 cases and a subgroup associated with Guillain–Barre syndrome. *Brain.* 2003;126 (Pt. 10):2279–2290.
62. Odaka M, Yuki N, Hirata K. Anti-GQ1b IgG antibody syndrome: clinical and immunological range. *J Neurol Neurosurg Psychiatry.* 2001;70(1):50–55.
63. Querol L, Devaux J, Rojas-Garcia R, Illa I. Autoantibodies in chronic inflammatory neuropathies: diagnostic and therapeutic implications. *Nat Rev Neurol.* 2017;13 (9):533–547.
64. Pestronk A, Choksi R, Blume G, Lopate G. Multifocal motor neuropathy: serum IgM binding to a GM1 ganglioside-containing lipid mixture but not to GM1 alone. *Neurology.* 1997;48(4):1104–1106.
65. Delmont E, Halstead S, Galban-Horcajo F, Yao D, Desnuelle C, Willison H. Improving the detection of IgM antibodies against glycolipids complexes of GM1 and galactocerebroside in multifocal motor neuropathy using glycoarray and ELISA assays. *J Neuroimmunol.* 2015;278:159–161.
66. Galban-Horcajo F, Fitzpatrick AM, Hutton AJ, et al. Antibodies to heteromeric glycolipid complexes in multifocal motor neuropathy. *Eur J Neurol.* 2013;20(1):62–70.
67. Nobile-Orazio E, Manfredini E, Carpo M, et al. Frequency and clinical correlates of anti-neural IgM antibodies in neuropathy associated with IgM monoclonal gammopathy. *Ann Neurol.* 1994;36:416–424.

68. Yu RK, Ariga T, Kohriyama T, Kusunoki S, Maeda Y, Miyatani N. Autoimmune mechanisms in peripheral neuropathies. *Ann Neurol.* 1990;27(suppl):S30–S35.
69. Eurelings M, Moons KG, Notermans NC, et al. Neuropathy and IgM M-proteins: prognostic value of antibodies to MAG, SGPG, and sulfatide. *Neurology.* 2001;56 (2):228–233.
70. Hays AP, Latov N, Takatsu M, Sherman WH. Experimental demyelination of nerve induced by serum of patients with neuropathy and an anti MAG IgM M protein. *Neurology.* 1987;37:242–256.
71. Willison HJ, Trapp BD, Bacher JD, Dalakas MC, Griffin JW, Quarles RH. Demyelination induced by intraneural injection of human anti MAG antibodies. *Muscle Nerve.* 1988;11:1169–1176.
72. Willison HJ, O'Hanlon GM, Paterson G, et al. A somatically mutated human anti-ganglioside IgM antibody that induces experimental neuropathy in mice is encoded by the variable region heavy chain gene, V1-18. *J Clin Invest.* 1996;97(5):1155–1164.
73. Kremer DM, Lopez PH, Mizutamari RK, Kremer LJ, Bacile EA, Nores GA. Factors defining target specificity in antibody-mediated neuropathy: density-dependent binding of anti-GD1a polyclonal IgG from a neurological patient. *J Neurosci Res.* 1997;47 (6):636–641.
74. Rongen HA, Bult A, van Bennekom WP. Liposomes and immunoassays. *J Immunol Methods.* 1997;204(2):105–133.
75. Galban-Horcajo F, Halstead SK, McGonigal R, Willison HJ. The application of glycosphingolipid arrays to autoantibody detection in neuroimmunological disorders. *Curr Opin Chem Biol.* 2014;18:78–86.
76. Willison HJ, Veitch J, Swan AV, et al. Inter-laboratory validation of an ELISA for the determination of serum anti-ganglioside antibodies. *Eur J Neurol.* 1999;6:71–77.
77. Varki A. Selectin ligands. *Proc Natl Acad Sci USA.* 1994;91(16):7390–7397.
78. Lopez PH, Zhang G, Bianchet MA, Schnaar RL, Sheikh KA. Structural requirements of anti-GD1a antibodies determine their target specificity. *Brain.* 2008;131(Pt. 7):1926–1939.
79. Lopez PH, Comin R, Villa AM, et al. A new type of anti-ganglioside antibodies present in neurological patients. *Biochimica et Bbiophysica Acta.* 2006;1762(3):357–361.
80. Defrance T, Taillardet M, Genestier L. T cell-independent B cell memory. *Curr Opin Immunol.* 2011;23(3):330–336.
81. Willison HJ, Goodyear CS. Glycolipid antigens and autoantibodies in autoimmune neuropathies. *Trends Immunol.* 2013;34:453–459.
82. Yuki N. Molecular mimicry between gangliosides and lipopolysaccharides of *Campylobacter jejuni* isolated from patients with Guillain–Barre syndrome and Miller Fisher syndrome. *J Infect Dis.* 1997;176:S150–S153.
83. Yuki N. Ganglioside mimicry and peripheral nerve disease. *Muscle Nerve.* 2007;35 (6):691–711.
84. Jacobs BC, Rothbarth PH, van der Meche FG, et al. The spectrum of antecedent infections in Guillain–Barre syndrome: a case-control study. *Neurology.* 1998;51 (4):1110–1115.
85. Yuki N. Campylobacter sialyltransferase gene polymorphism directs clinical features of Guillain–Barre syndrome. *J Neurochem.* 2007;103(suppl 1):150–158.
86. Yuki N, Susuki K, Koga M, et al. Carbohydrate mimicry between human ganglioside GM1 and *Campylobacterjejuni* lipooligosaccharide causes Guillain–Barre syndrome. *Proc Natl Acad Sci USA.* 2004;101(31):11404–11409.
87. Khalili-Shirazi A, Gregson N, Gray I, Rees J, Winer J, Hughes R. Antiganglioside antibodies in Guillain–Barre syndrome after a recent cytomegalovirus infection. *J Neurol Neurosurg Psychiatry.* 1999;66(3):376–379.

88. Visser LH, van der Meche FG, Meulstee J, et al. Cytomegalovirus infection and Guillain–Barre syndrome: the clinical, electrophysiologic, and prognostic features. Dutch Guillain–Barre Study Group. *Neurology.* 1996;47(3):668–673.
89. Kusunoki S, Shiina M, Kanazawa I. Anti-Gal-C antibodies in GBS subsequent to mycoplasma infection: evidence of molecular mimicry. *Neurology.* 2001;57(4):736–738.
90. Mori M, Kuwabara S, Miyake M, et al. *Haemophilus influenzae* has a GM1 ganglioside-like structure and elicits Guillain–Barre syndrome. *Neurology.* 1999;52(6):1282–1284.
91. Mori M, Kuwabara S, Miyake M, et al. *Haemophilus influenzae* infection and Guillain–Barre syndrome. *Brain.* 2000;123(Pt. 10):2171–2178.
92. Illa I, Ortiz N, Gallard E, Juarez C, Grau JM, Dalakas MC. Acute axonal Guillain–Barre syndrome with IgG antibodies against motor axons following parenteral gangliosides. *Ann Neurol.* 1995;38(2):218–224.
93. Sejvar JJ. Vaccines and neurologic disease. *Seminars Neurol.* 2011;31(3):338–355.
94. Nagai Y, Momoi T, Saito M, Mitsuzawa E, Ohtani S. Ganglioside syndrome, a new autoimmune neurologic disorder, experimentally induced with brain gangliosides. *Neurosci Lett.* 1976;2(2):107–111.
95. Goban Y, Saida T, Saida K, Nishitani H, Kameyama M. Ultrastructural study of central nervous system demyelination in galactocerebroside sensitized rabbits. *Lab Invest.* 1986;55:86–90.
96. Kusunoki S, Hitoshi S, Kaida K, Murayama S, Kanazawa I. Degeneration of rabbit sensory neurons induced by passive transfer of anti-GD1b antiserum. *Neurosci Lett.* 1999;273(1):33–36.
97. Clark AJ, Kaller MS, Galino J, Willison HJ, Rinaldi S, Bennett DLH. Co-cultures with stem cell-derived human sensory neurons reveal regulators of peripheral myelination. *Brain.* 2017;140(4):898–913.
98. Susuki K, Nishimoto Y, Yamada M, et al. Acute motor axonal neuropathy rabbit model: immune attack on nerve root axons. *Ann Neurol.* 2003;54(3):383–388.
99. Goodyear CS, O'Hanlon GM, Plomp JJ, et al. Monoclonal antibodies raised against Guillain–Barre syndrome-associated *Campylobacter jejuni* lipopolysaccharides react with neuronal gangliosides and paralyze muscle-nerve preparations. *J Clin Invest.* 1999;104 (6):697–708.
100. Plomp JJ, Willison HJ. Pathophysiological actions of neuropathy-related anti-ganglioside antibodies at the neuromuscular junction. *J Physiol.* 2009;587(Pt. 16):3979–3999.
101. Davidson AI, Halstead SK, Goodfellow JA, et al. Inhibition of complement in Guillain–Barre syndrome: the ICA-GBS study. *J Peripher Nerv Syst.* 2017;22(1):4–12.

CHAPTER THIRTEEN

# Gangliosides in Axon Stability and Regeneration

**Pablo H.H. Lopez[1], Bárbara Beatriz Báez**
Instituto de Investigación Médica Mercedes y Martin Ferreyra, INIMEC-CONICET-Universidad Nacional de Córdoba, Córdoba, Argentina
[1] Corresponding author. E-mail address: phhlopez@immf.uncor.edu

## Contents



## Abstract

Gangliosides are a family of sialic acid-containing glycosphingolipids highly expressed in the nervous system of vertebrates. Over the last 25 years, research has unmasked several of their neurobiological functions but the role of gangliosides in the nervous system remains not fully elucidated. Genetic disruption of genes for key enzymes involved in ganglioside biosynthesis led to the discovery of their diverse functions and highlighted the exquisite structural specificity required in this processes. In the nervous system, gangliosides regulate axonal caliber and organize ion channels at the nodes of Ranvier, a critical step to ensure fast conduction velocity of myelinated fibers. They also act as receptors for lectins located on apposing myelin membranes critical to maintain axon-glia interactions that result in cytoskeleton stabilization. After a lesion, gangliosides acting as receptors for glial-derived molecules present in the extracellular milieu can halt axon regeneration. Similarly, antiganglioside antibodies present in

*Progress in Molecular Biology and Translational Science*, Volume 156
ISSN 1877-1173
http://dx.doi.org/10.1016/bs.pmbts.2018.03.001

autoimmune neurological conditions can mimic this inhibitory effect on nerve repair. Studying the molecular details of the molecular interaction of gangliosides in *trans* with ligands present on apposing cell membranes and receptor/transducer molecules in *cis* interaction at the axolemma membrane, together with their downstream signaling pathways, represent a unique opportunity to expand our knowledge about the role of gangliosides in the nervous system.

## 1. INTRODUCTION

Gangliosides are sialic acid-containing glycosphingolipids ubiquitously distributed in all tissues but highly enriched in the nervous system from all vertebrates.[1] They are found in relatively high abundance and display broad structural complexity in peripheral nerves and brain.[2,3] The ceramide lipid moieties of gangliosides, particularly diverse in the peripheral nervous system (PNS),[4,5] are embedded in the outer leaflet of the plasma membrane, extending their complex carbohydrate structures to the extracellular space. Gangliosides are major cell surface structural determinants,[6] although the accessibility of their glycans varies among different nerve structures.[7] Their insertion in the plasma membrane allows gangliosides lateral interactions (*cis*) with other membrane components that modulate their partitioning in the membrane, which can ultimately impact their biological activity.[8,9] In addition, their glycan structures are recognized by soluble factors such as antibodies and molecules present in the apposing cell membranes via *trans* interactions.[10] Some of these interactions implicate gangliosides as anchoring receptors, but in certain circumstances these interactions can have a pronounced impact in the signaling and physiology of neurons. Gangliosides are not likely to signal on their own, but rather depend on specific transducer molecules.[8,9] Upon interaction with ligands recognizing their glycan structures, gangliosides can combine *trans* and *cis* interactions to induce a signaling event in cells.[6] In this chapter, we will focus on one of the most prominent recognition systems in which gangliosides have been implicated: the myelin–axon interaction.

Myelin is a lipid membrane extension from the myelinating cells: oligodendrocytes in the central nervous system (CNS) and Schwann cells in the PNS.[11,12] Myelin wraps most axons in vertebrates,[13] isolating them from the extracellular space. The axons are only exposed to the extracellular milieu at the nodes of Ranvier, narrow gaps depleted of myelin that allow ion currents in and out of the axon.[14] Nodes are surrounded by interdigitated noncompact myelin loops that help to establish a highly structured organization of ion channels.

The voltage-gated sodium channels clustered at the nodes, allows depolarizing currents to jump from one node to another, the mechanism for rapid "saltatory" conduction of an action potential. Interdigitated myelin loops interact with the axons via multiple receptor–ligand complexes, sealing the edge of each node to define a paranodal region, which provides the barrier impeding lateral diffusion of ion channels. A specialized segment of the axon adjacent to the paranode (opposite to the node), termed the juxtaparanode, is characterized by the presence of voltage-gated potassium channels that help return the membrane to its resting state after depolarization. Together, this complex of membrane molecules supports highly efficient and rapid action potential transmission.[14] The process of myelination has also a pronounced effect on the stability of axons, as evidenced by the presence of axonal atrophy secondary to demyelination (loss of myelin) and/or dysmyelination (alteration of myelin structure) in multiple experimental models and human neurodegenerative diseases.[15–19] In addition, after an injury of the CNS, myelin debris present the lesion site can halt axon regeneration by engaging molecules present at growth cones[20,21]; specialized structures located at the tips of axons characterized by a particular distribution of their cytoskeleton. Growth cones contain the molecular machinery to extend and guide the axons in response to signaling molecules present in the extracellular milieu.[22] Interestingly, autoantibodies present in certain neurological conditions of the PNS can mimic the effect of myelin, also stopping nerve repair.[23]

All the effects of myelin on axons described so far involve the molecular crosstalk of ligand-receptor complexes where neuronal gangliosides have a relevant role.[6,24] Understanding the way in which gangliosides manage to signal all these effects on neurons represents a critical step toward expanding our knowledge about their functional role in the nervous system and, at the same time, opens the opportunity to pharmacological intervention in neurodegenerative diseases characterized by deficient myelin-axon communication.

## 2. ROLE OF GANGLIOSIDES IN AXON CALIBER AND STRUCTURE

### 2.1 Lessons From Animal Models With Genetic Disruption of Ganglioside Biosynthesis

Human diseases associated with disrupted ganglioside biosynthesis are extremely rare and just recently identified as the cause of complex neurological symptoms initiating early after birth through childhood.[1] Therefore, most of our knowledge on ganglioside biosynthesis derives from the advent of genetic engineering in rodents.[25] In 1996, Takamiya first disrupted the biosynthetic pathway of complex ganglioside in mice by deleting the gene *B4galnt1* coding for β-1,4-N-acetylgalactosaminyltransferase 1, the enzyme responsible for the synthesis of major complex gangliosides in the nervous system; by addition of a GalNAc moiety to the terminal saccharide structures of GM3 and GD3 to produce GM2 and GD2, respectively[26] (Fig. 1). These mice were viable and displayed a normal lifespan despite lacking all major complex gangliosides. Interestingly these mutant mice maintain about the same total brain ganglioside concentration due to increase in concentration of ganglioside precursors, a hallmark from mice with genetic disruption of ganglioside biosynthesis.[27] Unexpectedly, the original finding showed only subtle neurological deficits in these mice[26]; however in a second mutant mouse line generated for this enzyme it was later identified the presence of late-onset progressive neuropathy characterized by significant Wallerian axon degeneration in CNS and PNS, decreased central myelination, and dysmyelination of peripheral nerves concurrent with a worsening of impaired hindlimb mobility in older mice[28,29] (Fig. 2). Altogether, these histological features were consistent with an impaired axon–myelin interaction. Interestingly, neuron-specific transgenic expression of *B4galnt1* gene in complex ganglioside-deficient mice almost completely corrected the neurological disorders,[30] except that this mice still exhibit minor electrophysiological abnormalities of peripheral nerves. In parallel to Takamiya's findings, gangliosides were first identified among other sialoglycans as target for myelin-associated glycoprotein (MAG),[31–33] a member of the family of lectins later named siglecs, which are characterized by their ability to bind with high specificity to glycans terminated with sialic acids. Complex gangliosides were then confirmed as axonal receptors for MAG and required to exert its inhibitory role on neurite outgrowth on cerebellar granule neurons (CGNs).[33,34] The expression of MAG (also known as Siglec-4a) remains exclusive to myelinating cells, and its location is restricted to the adaxonal layer of myelin in intimate contact with axons, besides in the PNS is also present in periaxonal myelin and Schmidt-Lanterman incisures, lateral loops, and inner and outer mesaxons. *Mag*-null mice display a phenotype characterized by late-onset axon degeneration and dysmyelination of PNS and CNS with minor alterations of myelination strikingly similar to *B4galnt1*-null mice.[35–42] The finding of MAG recognizing with high affinity gangliosides GD1a and GT1b (bearing the terminal

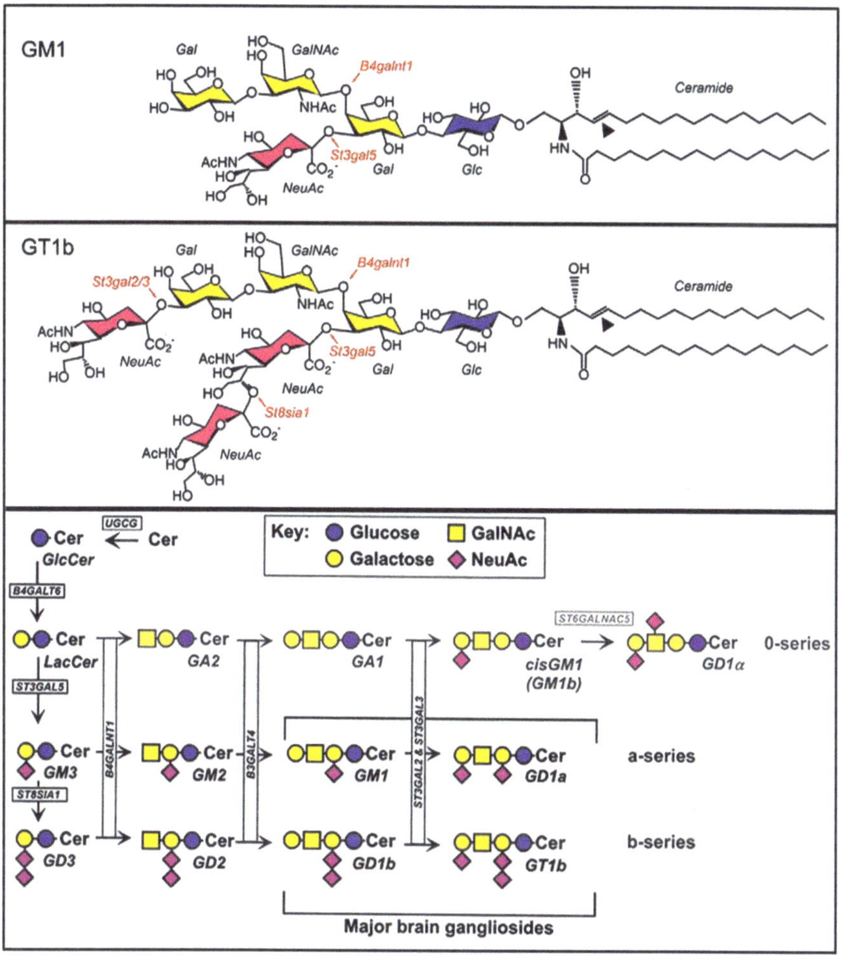


**Fig. 1** ***Ganglioside structures and their biosynthetic pathway.*** *Top*: The structure of GM1 and GT1b is depicted with the MAG-binding determinant (NeuAcα(2-3)Gal(β1-3)GalNAc) shaded. *Bottom*: Biosynthetic pathways to the major brain gangliosides. The different glycosyltransferases that appears in the text, *Ucgc*, *St3gal5*, *B4galnt1*, *St8sia1*, and *St3gal2 and St3gal3* are shown. The MAG-binding determinant is *shaded*, *St3gal5*-null mice lack normal major brain gangliosides, but instead synthesize an equivalent amount of "0" series gangliosides expressed residually in wild type mice.

NeuAcα(2-3)Galβ(1-3)GalNAc), which are missing in *B4galnt1*-null mice, lead to the hypothesis that gangliosides, acting as functional receptors for MAG, were signaling to the axons a stabilizing/protective effect emanating from myelinating cells. Consistent with this idea, generation of *St8sia1*-null mice lacking b-series gangliosides characterized by an internal disialyl residue

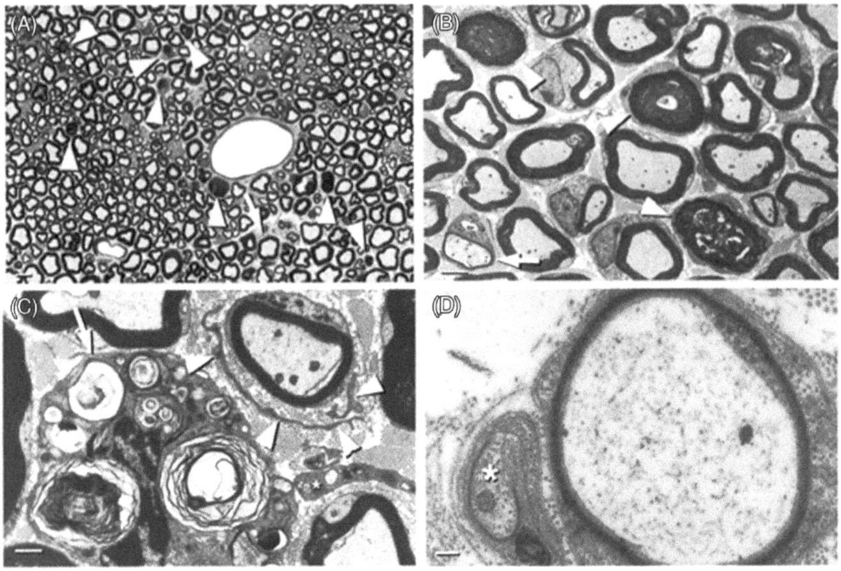


**Fig. 2** ***Pathological features including axon degeneration in sciatic nerves from 12 to 16 weeks-old B4galnt1-null mice.*** (A) Toluidine blue-stained 1 μm plastic section illustrating several myelinated fibers undergoing axonal degeneration (*arrowheads*) and a myelinated fiber surrounded by supernumerary Schwann cell processes (*arrow*). (Bar, 5 μm). (B) Low-power Electron microscopy image showing myelin figures and collapsed myelin at different stages of degeneration (*arrowheads*) and a thinly myelinated fiber surrounded by supernumerary Schwann cell processes (*arrow*) (Bar 2.5 μm). (C) Electron microscopy image showing a macrophage containing myelin debris (*arrow*) and a minor onion bulb around a myelinated fiber (*arrowheads* point to Schwann cell processes) (Bar, 1 μm). (D) Electron microscopy image showing a thinly myelinated fiber and an axonal sprout (*) invested by a Schwann cell (Bar, 200 nm). *Reproduced with permission from Sheikh KA, Sun J, Liu Y, et al. Mice lacking complex gangliosides develop Wallerian degeneration and myelination defects. Neurobiology. 1999;96:7532–7537. https://www.ncbi.nlm.nih.gov/pmc/articles/PMC22120/pdf/pq007532.pdf.*

and expressing major gangliosides GM1 and GD1a, do not display apparent neurological abnormalities or axon degeneration.[43] However, a more comprehensive analysis of neurological abnormalities could reveal reduction of some neurological functions (unpublished data, discussed by Furukawa et al., 2017). Further analysis of the *B4galnt1* and *St8sia1* double-mutant mice, which renders the nervous system with the sole expression of GM3, a very low affinity ligand for MAG, also demonstrated mild axon degeneration of peripheral nerves in aged mice consistent with the phenotype of single *B4galnt1*-null mice in one of the two double-null mice generated.[44,45] On the other hand, the generation of the *St3gal5* mutant mice, that lacks all gangliosides commonly expressed in the brain; do not display gross neurological abnormalities or problems with hindlimb mobility. Here, the expression of a rare biosynthetic pathway (only expressed residually in wild type mice), the "0" series gangliosides, which includes the potent MAG ligands GD1α and GM1b bearing the terminal NeuAcα(2-3) on an alternately sialylated core structure,[32,46] could explain the lack of gross neurological deficits beyond the recognized hearing loss.[34,46–48] In line with this, complete interruption of ganglioside biosynthesis by generation of double *St3gal5* and *B4galnt1*-null mice, which lacks the compensatory expression of "0" series gangliosides, results in axon degeneration and disturbances of axon–myelin interaction in addition to reduced brain size, apoptosis, and astrogliosis.[49] Also, deletion of the gene *Ucgc* responsible for the synthesis of the initial precursor of ganglioside biosynthesis, glucosylceramide, results in halted embryonic development during gastrulation, a limitation circumvented by the generation of conditional *Ucgc*-null mice.[50] *Ucgc* deficiency under the control of a neural promoter resulted in a robust loss of ganglioside expression in PNS and/or CNS in two different mutants generated.[51,52] One of the mutant mice characterized by almost complete ganglioside biosynthesis inhibition displays axon degeneration and dysmyelination in the PNS and excessive axonal pruning in CNS accompanied by early onset severe ataxia, and these mice die within the first month.[51] The other mutant mice reported a robust inhibition of ganglioside biosynthesis in the brain and show progressive neuropathy with dysreflexia, gait disorders, and progressive loss of Purkinje neurons.[52] Interestingly, selective conditional deletion of *Ucgc* in these cells is associated with axon degeneration and further cell death.

Pan et al. further explored the contribution of the interaction of MAG with gangliosides to the stability of axons by crossing *Mag*-null mice and *B4galnt1*-null mice in the same background.[42] Any residual MAG-ganglioside interaction in *B4galnt1-null* mice provided by GM3 would be lost in the double-null mice. Compared to wild type mice, single and double-null mice show a similar degree of degeneration of myelinated axons from spinal and ventral columns in the CNS. In contrast, morphometric analysis in the PNS revealed larger axon degeneration of myelinated fibers in the sciatic nerve from double-null mice compared to either single-null mice. This indicates that, at least in the PNS, complex gangliosides also provide stabilization and/or support to axons via MAG-independent mechanisms. In contrast to the CNS, the PNS contains an abundance of ganglioside species other than ganglio series such as the neolacto series,[5,53] which could also influence axon stabilization. Furukawa's group have described the presence of inflammation and activation of components and receptors from the complement cascade in the cerebellum and spinal cord from *B4galnt1* and *St8sia1* double-mutant mice.[54] Ganglioside deficiency in these mice is associated with disorganized rafts membranes and dysfunctions of complement-regulatory proteins. Interestingly, generation of a triple-null mice by crossing *B4galnt1* and *St8sia1* double-mutant mice with mice carrying a disrupted *C3* complement gene alleviated inflammation and neurodegeneration, which confirms a more complex scenario to the degenerative process that characterize ganglioside-deficient mice.[55]

The identification of enzymes responsible for in vivo biosynthesis of major brain gangliosides GD1a and GT1b was later reached by crossbred single *St3gal2-* and *St3gal3*-null mice, which renders the brain with a reduction of more than 95% on the content of these gangliosides.[56] These double-mutant mice were smaller that their wild type and displayed a greatly reduced life span (usually succumbed by 8 weeks). Strikingly, these mice exhibit significant reduction of brain size associated with dysmyelination, reduced myelinated axons and myelin thickness and increased axon degeneration, as well as defective nodes of Ranvier.[57] Blockage of terminal sialylation in these mice is not restricted to glycosphingolipids, as these double-null mice as well as the single-null for *St3gal3* also display reduced terminal sialylation of glycoproteins, limiting the conclusions about the impact on GD1a and GT1b expression in the nervous system. Table 1 depicts a summary with the findings of axon degeneration reported in mice with genetic disruption of ganglioside biosynthesis. Altogether, the data support a role for ganglioside signaling in axon-myelin stability.

## 2.2 In Vitro Studies Confirm the Stabilizing Role of Gangliosides

Further in vitro studies confirmed the role of gangliosides in axon-myelin stability. Nguyen et al. reported that treatment of dorsal root ganglion neurons (DRGn) explants with a soluble form of MAG (composed of its ectomain fused to the Fc region of immunoglobulins, named MAG-Fc) was able to exert protection against exposure to vincristine, a chemotherapy medication capable to induce microtubule disassembly.[58] This effect was not observed when treating cultures derived from *B4galnt1*-null mice, or when using a mutant form of MAG-Fc with a point mutation on residue R118 from its extracellular domain 1, which has been reported to be critical for its capacity to recognize sialic-acid residues including those in

**Table 1** Axon Degeneration and Nodal Integrity in Mice With Genetic Disruption of Enzymes From Ganglioside Biosynthesis.

| Enzyme Deficit | Deleted Gene | Brain Ganglioside Expression | Axon Degeneration | Nodal Alteration |
|---|---|---|---|---|
| Glucosylceramide synthase | *Ucgc* | None | ++++ | Paranode |
| GM3 synthase | *St3gal5* | α-series, including GM1b and GD1aα | – | N.D. |
| GD3 synthase | *St8sia1* | a-series, double GM1 and GD1a expression | – | –/+ |
| GM2/GD2 synthase | *B4galnt1* | GM3 and GD3 expression | ++++ | Paranode |
| GM2/GD2 synthase in non-neuronal cells | *B4galnt1* | All | – | – |
| GM2/GD2 synthase and GM3 synthase | *B4galnt1, St3gal5* | None | ++++ | Paranode |
| GM2/GD2 synthase and GD3 synthase | *B4galnt1, St8sia1* | GM3 | ++ PNS | N.D. |
| GD1a/GT1b synthase | *St3gal2, Stgal3* | GM1 and GD1b | ++++ | Paranode |

gangliosides. Protection against vincristine-induced toxicity on primary DRGn cultures was prevented by treatment with sialidase enzyme, that removes sialic acids from glycosphingolipids; or the pharmacological inhibitor P4, that inhibits glycosphingolipid biosynthesis.[59] The signaling mechanisms by which gangliosides function to protect axons remain obscure. Initial studies in DRGn explants identified that MAG-derived protection against vincristine toxicity depends on the activation of the small GTPase RhoA (a key regulator of cytoskeleton dynamics), and its associated kinase ROCK. Later the participation of the RhoA-interacting protein CRMP4, a member of the collapsing response mediator family proteins capable to bind actin and tubulin to promote its polymerization, was confirmed.[58–60] The protective effect against vincristine relies at least in part in the stabilization of tubulin cytoskeleton, but is still an open question how gangliosides manage to modulate cytoskeleton dynamics from the plasma membrane. Altogether in vivo and in vitro studies confirm that complex gangliosides present at the axolemma mediates myelin-derived axonoprotection via its interaction with MAG.

## 2.3 Gangliosides as Modulators of Axonal Structure

One the most striking observations described in the *B4galnt1*-null mice besides the presence of Wallerian-like axon degeneration was the identification of reduced caliber of myelinated axons from CNS and PNS.[28,42] This particular phenotype is also present in *Mag*-null mice, accompanied by decreased neurofilament spacing and reduced neurofilament phosphorylation[41,61]; while double-mutant mice for *B4galnt1* and *Mag* displays similar distribution of caliber in myelinated axons compared to their corresponding single-mutant mice. This reinforces the hypothesis of gangliosides as axonal receptors regulating radial growth of axons. In vitro studies using DRGn exposed to MAG provides support for the molecular mechanisms underlying modulation of axonal caliber, and includes the upregulation and phosphorylation of microtubule-associated proteins and neurofilaments associated with increased activities of extracellular-signal regulated kinases and cyclin-dependent kinase-5.[41] Future experiments should corroborate the involvement of gangliosides in these signaling pathways.

Besides providing insulation, trophic support, and stability, myelination also regulates lateral distribution of molecules along the axolemma membrane at nodes of Ranvier. Electron microscopy studies revealed that *B4galnt1* mutant mice display altered organization of nodes of Ranvier in the CNS and well as in the PNS similar to the *Mag*-null mice.[62,63] Some of the most striking observations include the presence of loose paranodal myelin loops and mislocation of potassium channels to paranodes, with a broadened nodal distribution of sodium channels, which results in reduced nerve conduction velocity preferentially affecting motor versus sensory peripheral roots. Expression of *B4galnt1* under a neural promoter in these mice prevented the alterations. Double *St3gal5* and *B4galnt1*-null mice, lacking all ganglioside receptors for MAG, also display altered interaction of paranodal loops with axons at the node.[49] In contrast, these findings are rare in *St8sia1*-null mice lacking expression of b-series gangliosides but with compensating double amounts of a-series gangliosides such as GM1 and GD1a.[64] Immunofluorescence studies in *St3gal2* and *St3gal3* double-null mice show altered organization of nodes of Ranvier characterized by reduced paranodal length and increased nodal length, paranodal protrusion and mislocation of potassium channels to paranodes.[57] Conditional mutant mice for *Ucgc* in Purkinje neurons also display disrupted paranodal loops and altered distributions of nodal ion channels at nodes of Ranvier. Nodal integrity was not studied in *St3gal5* mutant mice (see summary of nodal alterations in Table 1).

Interestingly, nodes of Ranvier are the target for anti-GM1 and anti-GD1a IgG antibodies present in the human disease Guillain Barré Syndrome (GBS), an acute autoimmune neuropathy affecting peripheral nerves in the PNS and characterized in some disease variants by the presence of high titers of circulating antiganglioside antibodies. It is accepted that these variants of the disease are triggered by a molecular mimicry mechanism where different pathogens associated with preceding gastrointestinal or respiratory infections carrying ganglioside-like oligosaccharide structures induce a robust cross-reactive humoral immune response that ultimately target axonal gangliosides.[65] A variant of the disease characterized by selective affection of motor nerves and signs of axonal pathology (AMAN) was reproduced in rabbits by passive immunization with ganglioside GM1.[66] Electron microscopy studies show that IgG antibodies bind to GM1, which is strongly expressed at the nodes of Ranvier, which results in activation of complement at the nodal axolemma with subsequent detachment of paranodal myelin loops and destruction of nodal molecular complexes including sodium channels[63,67] (Fig. 3); a finding identical to the early pathology of AMAN patients.[68] Similar findings were reported by passive transfer of GBS-associated serum containing high titers of anti-GM1 antibodies in an animal model characterized by blood-nerve barrier breakdown.[69] In an ex vivo nerve preparation, anti-GD1a antibodies disrupted the nodes by similar mechanisms.[70] Taken altogether, the evidence highlights the relevance of gangliosides in maintaining the organization of node of Ranvier, suggesting a critical role for gangliosides GM1 and GD1a as receptors for ligands at the paranodal myelin loops that favor its anchoring to the axon.

# 3. ROLE OF GANGLIOSIDES IN AXON REGENERATION

## 3.1 Exogenous Ganglioside Modulates Neurite Sprouting and Neurite Outgrowth

Early studies identified that exogenous administration of bovine brain gangliosides (BBG, in particular ganglioside GM1) promotes neurite sprouting,

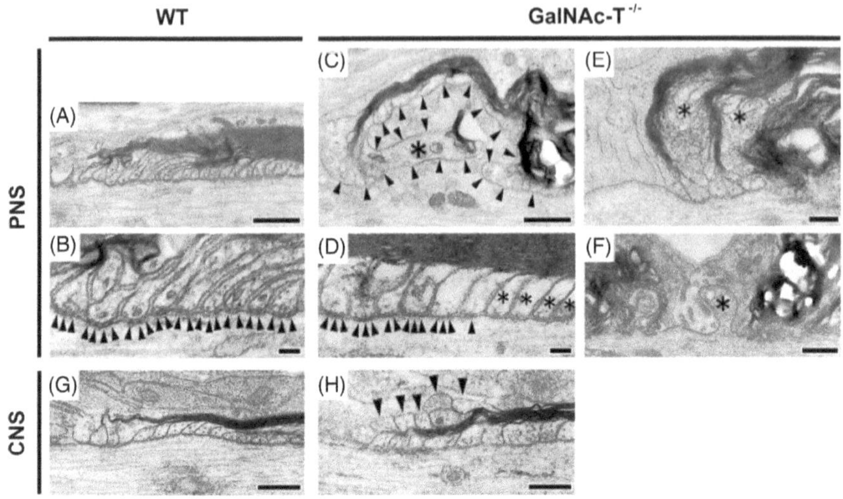


**Fig. 3** ***B4galnt1-null mice display ultrastructural abnormalities at nodes and paranodes. Electron micrographs from 12-week-old mice are depicted.*** (A) Paranodal structures in wild type (WT) sciatic nerves, further enlarged in (B) to show transverse bands, junctions between paranodal loops and axon domains (*arrowheads*). (C) Paranodal axolemma abnormally protrudes (*arrowheads*) in *B4galnt1*-null sciatic nerve. The *asterisk* indicates the area of redundant paranodal axolemma with no attachment of myelin loops. (D) In *B4galnt1*-null sciatic nerve, some paranodal myelin loops form transverse bands (*arrowheads*). No typical transverse bands are observed in the innermost paranodal loops (*asterisks*). (E) A considerable number of paranodal myelin loops (*asterisks*) do not form axo-glial junctions in *B4galnt1*-null sciatic nerve. (F) Abnormal protrusion of nodal axon (*asterisk*) in *B4galnt1*-null ventral root. (G) Paranodal structures in WT optic nerve. (H) Some lateral paranodal loops face away from the axon (*arrowheads*) in *B4galnt1*-null optic nerve. Scale bars, 0.5 μm in A, C, F, G, H; 0.1 μm in B and D; 0.2 μm in E. *Reproduced with permission from Susuki K, Rasband MN, Tohyama K, et al. Anti-GM1 antibodies cause complement-mediated disruption of sodium channel clusters in peripheral motor nerve fibers. J Neurosci 2007;27(15):3956–3967.*

differentiation and/or neurite outgrowth in primary cultures, established neuronal cell lines and neuroblastoma tumor cells following their addition to the culture medium.[71–77] For instance, in Neuro-2a neuroblastoma cells, treatment with BBG induced an remarkable increase in neurite length as well as in the number of neuritic processes.[78] In line with this, deficiency of complex gangliosides increases susceptibility to Parkinson, kainate-induced seizures, and neuronal apoptosis[79,80] while ganglioside administration exerted neuroprotection against $Ca^{+2}$ mediated excitatory amino acid-induced excitotoxicity and serum deprivation.[81–85] Even without a strong experimental support of their mechanisms of action, human trials composed of different BBG formulations were conducted (GM1 or total ganglioside extract), which led to their approval in the market for the treatment of different nerve lesions and stroke.[86–88] The formulation was withdrawn from the market in several countries after the rare description of strong neurological adverse reactions that mimics GBS.[89,90] Interestingly, BBG-associated GBS patients have high titers of IgG antiganglioside antibodies in serum with specificity for the N-glycolyl substitution in the sialic acid molecule of BBG formulations, which is absent in human nervous tissue that express the N-acetyl form.[91] This led to hypothesize that N-glycolyl BBG triggers a cross-reactive autoimmune response against human nerve gangliosides.[92]

Soluble gangliosides, present as single molecules or forming micelles in the media depending on their critical micelle concentration, reach the plasma membrane where they exert their effects by *cis* interactions with other membrane components. It is reported that ganglioside GM1 interacts with neurotrophin receptor kinases such as TrkA, TrkB, and the low affinity neurotrophin receptor $p75^{NTR}$ and GFR$\alpha$1 and RET, the protein components of the GDNF receptor.[80,84,93–100] Interestingly, the cloning of the plasma membrane sialidase NEU3 shed light into some of mechanistic properties of gangliosides at the plasma membrane.[101] Stimulation of NEU3 (the enzyme responsible for the localized hydrolysis of complex gangliosides into GM1) accelerates the formation of a single axon from multiple neurites and helps to define neuronal polarity via enhancing local TrkA activity and inhibiting RhoA signaling on hippocampal neuronal cultures.[102] On the contrary, suppressing NEU3 activity blocks axonal elongation.[101] Also, activation of NEU3 dependent on extracellular calcium influx and activation of P38/MAPK pathway (and further conversion of GD1a and GT1b to GM1) is a critical step toward active axon regeneration in neurons with intrinsic regenerative capacity in vivo (PNS) but not in neurons lacking this capacity (CNS) both in in vitro and in vivo paradigms.[103]

## 3.2 Gangliosides as Functional Receptors for Myelin-Associated Glycoprotein-Mediated Inhibition of Neurite Outgrowth in the Central Nervous System

As discussed earlier, the interest to study the role of gangliosides in the process of axon regeneration revived with the identification of gangliosides GD1a and GT1b as axonal receptors for MAG,[33,34,104,105] which was among the first identified CNS myelin-derived inhibitor of axon regeneration. MAG, isolated from the major inhibitory fraction from purified CNS myelin, conserved the vast majority of its inhibitory effect on neurite outgrowth

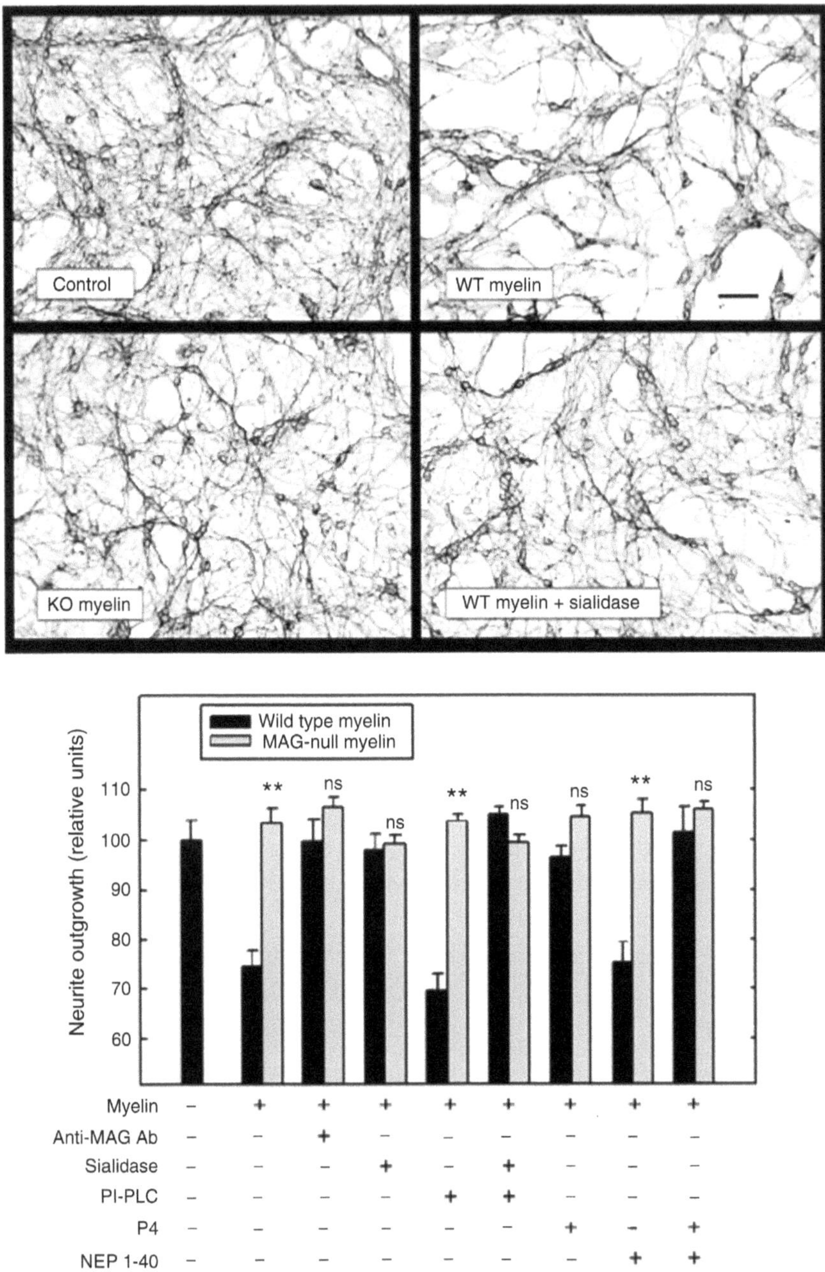


**Fig. 4** ***Axon outgrowth from cerebellar granule neurons on substrata adsorbed with extracts of wild type and Mag-null mouse myelin show that MAG-mediated inhibition is via gangliosides.*** Rat cerebellar granule neurons (CGNs) were plated on control surfaces

when tested as culture substrate for a motor neuron-like cell line.[106] This inhibitory effect was soon reproduced on different primary neuronal cultures,[107] which led to the identification of gangliosides GD1a and GT1b as functional axonal receptors for MAG in CGN[108] (see Fig. 4). In most neurons, ganglioside GT1b is part of a multimeric receptor complex recognized by MAG that includes the Nogo-66 receptor NgR1, a glycophosphatidylinositol-anchored protein acting as receptor for other myelin-derived inhibitors.[109–111] Alternatively, a second member of the NgR family, NgR2, interacts selectively with MAG to inhibit neurite outgrowth.[112] Gangliosides and NgRs transduce their inhibitory effects as part of a receptor complex via the low affinity neurotrophin receptor $p75^{NTR}$.[113–115] The complex is completed by the presence of Lingo-1, a transmembrane protein that interacts with GT1b and NgR1 and links $p75^{NTR}$ to NgR1.[116,117] GT1b can also interact with NgR1 (and probably NgR2) through a sialic acid binding FRG domain, well conserved among the NgRs members; mutations in this domain reverse the inhibitory effect of MAG and anti-GT1b antibodies on neurite outgrowth from CGN.[118] GT1b also interacts with $p75^{NTR}$ and induces its partitioning into lipids rafts, being this later effect reproduced by GD1a.[115,119] Early reports show that MAG interacts with the sialic acid of gangliosides via the Ig-like domain 1 from its ectodomain.[120] Recently, the analysis of the crystal structure of soluble MAG-Fc bound to GT1b oligosaccharide revealed novel details of its interaction with the terminal NeuAcα(2-3)Galβ(1-3) residue, identifying multiple key interaction sites in domain 1 (critical residues for recognition of sialic acid besides alanine-118 are tyrosine 65, tyrosine 127, and threonine

or the same surfaces adsorbed with detergent-extracted myelin proteins from wild type mouse brain (WT) or *Mag*-null mouse brain (KO). As indicated, cultures were treated 1 h after plating with 10 μg/mL anti-MAG antibody (to block MAG), 8 milliunits/mL sialidase (to remove sialic acids), 1 unit/mL PI-PLC (to cleave glycosylphosphaidylinositol anchors, 1 μM P4 (an inhibitor to ganglioside biosynthesis), or 1 μM NEP1–40 (an NgR blocking peptide). After 48 h, the cultures were fixed and stained with anti-GAP43 antibody to label neurites, and axon outgrowth was quantified as described[130]. Representative fluorescence micrographs are presented as reverse *gray* scale images to enhance clarity (*bar*, 50 μm). Axon outgrowth (mean ± S.E.) was quantified using image analysis and normalized with respect to the control. *Symbols* indicate statistical comparison of identically treated wild type and MAG-null cultures: **, $P \leq 0.001$; *ns*, not statistically significant. *Reproduced with permission from Mehta NR, Lopez PHH, Vyas AA, Schnaar RL. Gangliosides and Nogo receptors independently mediate myelin-associated glycoprotein inhibition of neurite outgrowth in different nerve cells. J Biol Chem. 2007;282 (38):27875–27886. doi:10.1074/jbc.M704055200.*

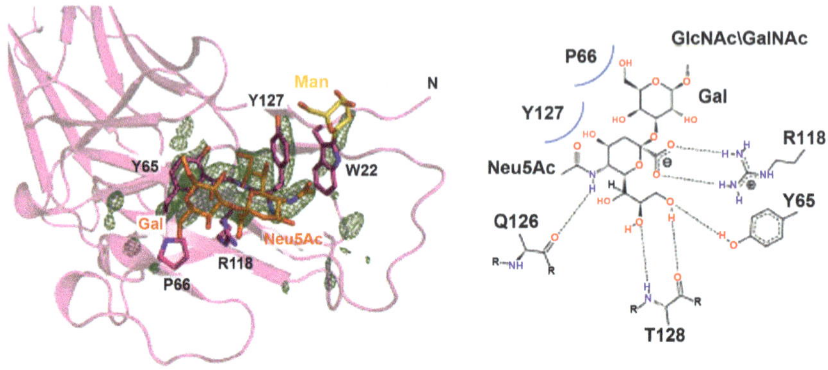


**Fig. 5** ***Structural characterization of Neu5Acα(2,3)Galβ (the terminus of GD1a and GT1b) recognition by the N-terminal Ig1 domain of MAG.*** (A) Electron density map at a contour level of 3.0 s of MAG-Fc1-3 (ectodomain of MAG lacking domains 4 and 5) at the ganglioside binding site of chain B. Image shows the unbiased electron density changes that resulted from binding of the Neu5Ac-α2,3-Gal-β1,4-GlcNAc, which includes the terminal disaccharide epitope of GT1b, and concomitant small conformational rearrangements. Residues involved in ligand engagement (sticks), as well as the C-mannosylation (*yellow*) on tryptophan W22 are shown. The first two sugars of the Neu5Ac-a2,3-Gal-b1,4-GlcNAc (*orange*) fit the density well. (B) Protein-ligand interactions with hydrogen bonds indicated by dashes and Van der Waals' contacts by *curved blue lines. Reproduced with permission from Pronker MF, Lemstra S, Snijder J, et al. Structural basis of myelin-associated glycoprotein adhesion and signalling. Nat Commun. 2016;7:13584. doi:10.1038/ncomms13584.*

128; meanwhile proline 66 support interaction with Gal, see Fig. 5).[120,121] Upon binding to gangliosides, MAG can stimulate the regulated proteolytic cleavage of $p75^{NTR}$, which releases its cytoplasmic domain allowing its interaction with Rho-GDI (a Rho-GDP dissociation inhibitor) and the subsequent activation of RhoA GTPase.[122,123] Growth cone collapse and prevention of neurite extension are later achieved by a signaling cascade including ROCK and its downstream effectors collapsin response mediator proteins 2 and 4 (CRMP2/4) and LIM domain kinase-1, which negatively modulates microtubules and actin cytoskeleton[60,124–128] (Fig. 6). CGN cultures derived from *B4galnt1*-null mice are insensitive to the inhibitory effect of MAG-Fc on neurite outgrowth, do not partition $p75^{NTR}$ into lipids rafts or activate RhoA[119]. When growing on MAG-enriched myelin adsorbed plates, inhibition of neurite outgrowth from *B4galnt1*-derived CGN is partially conserved.[107,130] This discrepancy reflects different binding capacities of soluble and adsorbed MAG to gangliosides not currently understood.[129]

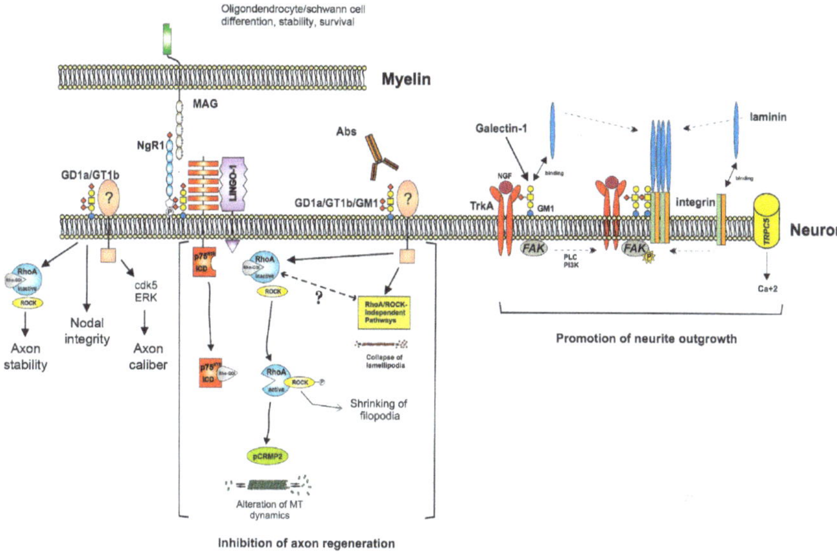


**Fig. 6** ***Model for ganglioside-mediated modulation of neurite outgrowth, axon stability and axon structure.*** The figure summarizes knowledge about the complex signaling mechanisms associated with gangliosides. Gangliosides GD1a and GT1b are part of a multimeric signaling complex at the axonal membrane recognized by MAG that include Nogo receptors 1 and 2 (NgR1 and NgR2), the neurotrophin receptor p75$^{NTR}$ and Lingo-1. A signal for MAG-mediated inhibition of neurite outgrowth is transduced through p75$^{NTR}$, which binds to a RhoA GDP-dissociation inhibitor (Rho-GDI) and induces the subsequent activation of the small GTPase RhoA and its associated kinase ROCK, which finally phosphorylates/inactivates CRMP2/4 inhibiting microtubule assembly. Disease-associated anti-ganglioside antibodies also activates RhoA/ROCK-dependent and independent pathways. Gangliosides can also signal via p75$^{NTR}$-independent pathways by using an unidentified transducer molecule. Gangliosides are also linked to stabilizing effects of MAG on neurons and are additionally associated with interactions determining Node of Ranvier integrity. Gangliosides were postulated to have a role in modulating the neuronal cytoskeleton thereby controlling axonal caliber. Laminin and Galectin-1-mediated crosslinking of GM1/$\alpha5\beta1$ integrin complex at the axolemma stimulate neurite outgrowth by triggering a signaling cascade that includes autophosphorylation of focal adhesion kinase and subsequent activation of phospholipase C, phosphoinositide-3 kinase and Ca+2 influx via voltage-independent TRPC5 channels.

Finally, MAG-induced inhibition of neurite outgrowth via gangliosides signals independently but in a synergistic manner with NgR receptors to inhibit neurite outgrowth in a cell-type-specific combination.[130,131] Despite the vast in vitro evidence about the inhibitory role of MAG on axon regeneration in the CNS, the confirmation of its in vivo role remains controversial. Among the possible causes can be identified the presence of multiple inhibitors of axon regeneration (some of them sharing axonal receptors/pathways) and/or compensatory expression of other myelin inhibitory proteins that mask its action.[132–136]

## 3.3 Gangliosides as Functional Receptors for Antibody-Mediated Inhibition of Axon Regeneration in the Peripheral Nervous System

The influence of ganglioside ligands in in vitro neuritogenesis was recognized many years ago. The use of GM1 binding reagents such as the B subunit of cholera toxin (Ct-B, which binds to GM1 with high specificity and affinity)[137] and experimentally -induced and disease-associated anti-GM1 antibodies on different cell systems support a role for GM1 in axonal growth. Some reports observed a stimulatory effect in neuritogenesis from NG-108-15 neuroblastoma cells and CGN by crosslinking GM1 at the plasma membrane with cloned Ct-B (free from adenyl cyclase activity)[138–140] and the endogenous lectin galectin-1.[141] The molecular mechanism underlying this effect is triggered by crosslinking of the GM1/$\alpha5\beta1$ integrin complex at the plasma membrane which favors its partition into lipids rafts, and in turn activates a signaling cascade that includes autophosphorylation of focal adhesion kinase and subsequent activation of phospholipase C, phosphoinositide-3 kinase and Ca+2 influx via voltage-independent TRPC5 channels[142,143] (Fig. 6). In contrast, treatment with Ct-b induced inhibition of neuritogenesis in varied cell cultures such as Neuro-2a, PC12, and N-18 neuroblastoma cells.[144–147] Likewise, mAb anti-GM1 antibodies inhibit NGF-induced sprouting and conditioned media-stimulated neurite growth of chick embryonic dorsal root ganglion cells[148] and neurite outgrowth by several stimulatory agents on Neuro-2a cells.[149]

The inhibition of axon regeneration by ganglioside crosslinking has important clinical consequences in GBS, where antiganglioside antibodies are common immune effectors. As discussed earlier, these antibodies are associated with disease onset in certain variants of the disease and the immunopathogenesis mechanism involves targeting gangliosides present at the node of Ranvier and neuromuscular junctions and further destruction via complement cascade activation and macrophage activation.[65,150–152] In addition, high titer of antigangliosides antibodies are associated with the presence of poor/delayed clinical recovery in GBS,[153–162] in most cases reflecting failure of nerve regeneration and target denervation.[163] Although initial studies found that GBS-associated IgM anti-GM1 antibodies weakly inhibited neurite sprouting in neuroblastoma cells,[146] it was later observed a robust inhibitory effect on neurite outgrowth from DRGn primary cultures by a set of IgG antigangliosides mAbs antibodies raised in ganglioside-null mice.[164–166] Neurite outgrowth inhibition by antiganglioside antibodies involves cross-linking of cell surface gangliosides and subsequent activation of RhoA/ROCK dependent and independent pathways.[166,167] The use of *B4galnt1* mutant mice in this model confirm gangliosides as in vivo targets of these antibodies, however the identity of the transducer molecule(s) associated with ganglioside signaling and the effectors of RhoA-independent pathways still remains obscure. Interestingly, one of these mAb recognizing GD1a and GT1b gangliosides (similar targets for MAG) demonstrated a robust inhibitory activity in an in vivo model of peripheral nerve regeneration associated with atrophy of target muscles,[165] an effect imitated by passive transfer with experimental and disease-associated IgG anti-GM1 antibodies as well as Ct-b.[167] Axon regeneration inhibition by anti-Gg Abs is linked to the presence of end-bulb like structures characteristic of dystrophic growth cones at the lesion site, a hallmark of halted axon regeneration in the CNS.[165] ROCK-dependent inactivation of CRMP-2 by phosphorylation at residue T555, which blocks microtubule assembly, is observed in in vivo and in vitro models and confirms the impact of RhoA/ROCK pathway in this process[167] (Fig. 6). In addition, alternative molecular mechanisms of axon regeneration inhibition caused by antiganglioside antibodies includes switching the proregenerative inflammatory environment of damaged nerves to growth inhibitory environment by engaging specific activating Fc receptors on recruited macrophages at the lesion site.[168] Overall, the data support a role for gangliosides as functional receptors for the inhibitory effect that exerts antiganglioside antibodies on nerve repair.

## 4. FUTURE PERSPECTIVES

Lessons coming from mice with disrupted ganglioside biosynthesis, in particular of gangliosides bearing the gangliotetraose core, seem to have reached a plateau in terms of understanding the role of ganglioside signaling in axon-myelin stability. Additional studies will benefit from making extensive the generation of tissue-specific conditional knockout mice for biosynthetic enzymes of gangliosides currently with limitations imposed by their phenotype. The inclusion in future studies of minor ganglioside species, specifically expressed in the PNS, could greatly enhance the overall understanding of ganglioside functions. Further dissecting the molecular identity of molecules interacting in *cis* and *trans* with gangliosides represent a critical step toward expanding our knowledge about their functional role in the nervous system and, at the same time, represents a unique opportunity for the development of novel therapeutic strategies to promote axon stabilization and improve axon regeneration in different pathological conditions of the human nervous system.

## REFERENCES


1. Schnaar RL. Gangliosides of the vertebrate nervous system. *J Mol Biol.* 2016;428 (16):3325–3336. 10.1016/j.jmb.2016.05.020.
2. Svennerholm L, Boström K, Fredman P, Månsson J-E, Rosengren B, Rynmark B-M. Human brain gangliosides: developmental changes from early fetal stage to advanced age. *Biochim Biophys Acta.* 1989;1005(2):109–117. 10.1016/0005-2760 (89)90175-6.
3. Svennerholm L, Boström K, Fredman P, et al. Gangliosides and allied glycosphingolipids in human peripheral nerve and spinal cord. *Biochim Biophys Acta.* 1994;1214 (2):115–123. http://www.ncbi.nlm.nih.gov/pubmed/7918590.
4. Ogawa-Goto K, Abe T. Gangliosides and glycosphingolipids of peripheral nervous system myelins-a minireview. *Neurochem Res.* 1998;23(3):305–310. http://www.ncbi.nlm.nih.gov/pubmed/9482242.
5. Ogawa-Goto K, Funamoto N, Abe T, Nagashima K. Different ceramide compositions of gangliosides between human motor and sensory nerves. *J Neurochem.* 1990;55 (5):1486–1493. http://www.ncbi.nlm.nih.gov/pubmed/2213006.
6. Schnaar RL, Gerardy-Schahn R, Hildebrandt H. Sialic acids in the brain: gangliosides and polysialic acid in nervous system development, stability, disease, and regeneration. *Physiol Rev.* 2014;94(2):461–518. 10.1152/physrev.00033.2013.
7. Lopez PHH, Zhang G, Bianchet MA, Schnaar RL, Sheikh KA. Structural requirements of anti-GD1a antibodies determine their target specificity. *Brain.* 2008. 10.1093/brain/awn074.
8. Regina Todeschini A, Hakomori S. Functional role of glycosphingolipids and gangliosides in control of cell adhesion, motility, and growth, through glycosynaptic microdomains. *Biochim Biophys Acta.* 2008;1780(3):421–433. 10.1016/j.bbagen.2007.10.008.
9. Sonnino S, Mauri L, Chigorno V, Prinetti A. Gangliosides as components of lipid membrane domains. *Glycobiology.* 2007;17(1):1R–13R. 10.1093/glycob/cwl052.
10. Lopez PH, Schnaar RL. Gangliosides in cell recognition and membrane protein regulation. *Curr Opin Struct Biol.* 2009;19(5):549–557. 10.1016/j.sbi.2009.06.001.
11. Simons M, Nave K-A. Oligodendrocytes: myelination and axonal support. *Cold Spring Harb Perspect Biol.* 2016;8(1):a020479. 10.1101/cshperspect.a020479.
12. Taveggia C. Schwann cells–axon interaction in myelination. *Curr Opin Neurobiol.* 2016;39:24–29. 10.1016/j.conb.2016.03.006.
13. Baraban M, Mensch S, Lyons DA. Adaptive myelination from fish to man. *Brain Res.* 2016;1641(Pt A):149–161. 10.1016/j.brainres.2015.10.026.
14. Rasband MN, Peles E. The nodes of Ranvier: molecular assembly and maintenance. *Cold Spring Harb Perspect Biol.* 2016;8(3):a020495. 10.1101/cshperspect.a020495.
15. Pan S, Chan JR. Regulation and dysregulation of axon infrastructure by myelinating glia. *J Cell Biol.* 2017;216(12):3903–3916. 10.1083/jcb.201702150.

16. Micu I, Plemel JR, Caprariello AV, Nave K-A, Stys PK. Axo-myelinic neurotransmission: a novel mode of cell signalling in the central nervous system. *Nat Rev Neurosci.* 2017;19(1):49–58. 10.1038/nrn.2017.128.
17. Nave K-A, Trapp BD. Axon-glial signaling and the glial support of axon function. *Annu Rev Neurosci.* 2008;31(1):535–561. 10.1146/annurev.neuro.30.051606.094309.
18. Verden D, Macklin WB. Neuroprotection by central nervous system remyelination: molecular, cellular, and functional considerations. *J Neurosci Res.* 2016;94 (12):1411–1420. 10.1002/jnr.23923.
19. Duncan ID, Radcliff AB. Inherited and acquired disorders of myelin: the underlying myelin pathology. *Exp Neurol.* 2016;283(Pt B):452–475. 10.1016/j.expneurol.2016.04.002.
20. Cao Z, Gao Y, Deng K, Williams G, Doherty P, Walsh FS. Receptors for myelin inhibitors: structures and therapeutic opportunities. *Mol Cell Neurosci.* 2010;43 (1):1–14. 10.1016/j.mcn.2009.07.008.
21. McKerracher L, Rosen KM. MAG, myelin and overcoming growth inhibition in the CNS. *Front Mol Neurosci.* 2015;8:51. 10.3389/fnmol.2015.00051.
22. Hur E-M, Saijilafu. Zhou F-Q. Growing the growth cone: remodeling the cytoskeleton to promote axon regeneration. *Trends Neurosci.* 2012;35(3):164–174. 10.1016/j.tins.2011.11.002.
23. Sheikh KA. Autoantobodies activate small GTPase RhoA to modulate neurite outgrowth. *Small GTPases.* 2011;2(4):233–238. 10.4161/sgtp.2.4.17115.
24. Lopez PHH. Role of myelin-associated glycoprotein (siglec-4a) in the nervous system. *Adv Neurobiol.* 2014;9:245–262. 10.1007/978-1-4939-1154-7_11.
25. Proia RL. Glycosphingolipid functions: insights from engineered mouse models. *Philos Trans R Soc Lond B Biol Sci.* 2003;358(1433):879–883. 10.1098/rstb.2003.1268.
26. Takamiya K, Yamamoto A, Furukawa K, et al. Mice with disrupted GM2/GD2 synthase gene lack complex gangliosides but exhibit only subtle defects in their nervous system. *Proc Natl Acad Sci USA.* 1996;93(20):10662–10667. http://www.ncbi.nlm.nih.gov/pubmed/8855236.
27. Sun J, Shaper NL, Itonori S, Heffer-Lauc M, Sheikh KA, Schnaar RL. Myelin-associated glycoprotein (Siglec-4) expression is progressively and selectively decreased in the brains of mice lacking complex gangliosides. *Glycobiology.* 2004;14(9):851–857. 10.1093/glycob/cwh107.
28. Sheikh KA, Sun J, Liu Y, et al. Mice lacking complex gangliosides develop Wallerian degeneration and myelination defects. *Neurobiology.* 1999;96:7532–7537. https://www.ncbi.nlm.nih.gov/pmc/articles/PMC22120/pdf/pq007532.pdf.
29. Chiavegatto S, Sun J, Nelson RJ, Schnaar RL. A functional role for complex gangliosides: motor deficits in GM2/GD2 synthase knockout mice. *Exp Neurol.* 2000;166 (2):227–234. 10.1006/exnr.2000.7504.
30. Yao D, McGonigal R, Barrie JA, et al. Neuronal expression of GalNAc transferase is sufficient to prevent the age-related neurodegenerative phenotype of complex ganglioside-deficient mice. *J Neurosci.* 2014;34(3):880–891. 10.1523/JNEUROSCI.3996-13.2014.
31. Kelm S, Pelz A, Schauer R, et al. Sialoadhesin, myelin-associated glycoprotein and CD22 define a new family of sialic acid-dependent adhesion molecules of the immunoglobulin superfamily. *Curr Biol.* 1994;4(11):965–972. http://www.ncbi.nlm.nih.gov/pubmed/7533044.
32. Collins BE, Yang LJ, Mukhopadhyay G, et al. Sialic acid specificity of myelin-associated glycoprotein binding. *J Biol Chem.* 1997;272(2):1248–1255. 10.1074/JBC.272.2.1248.
33. Yang LJ, Zeller CB, Shaper NL, et al. Gangliosides are neuronal ligands for myelin-associated glycoprotein. *Proc Natl Acad Sci USA.* 1996;93(2):814–818. http://www.ncbi.nlm.nih.gov/pubmed/8570640.

34. Collins BE, Yang LJ, Mukhopadhyay G, et al. Sialic acid specificity of myelin-associated glycoprotein binding. *J Biol Chem.* 1997;272(2):1248–1255. http://www.ncbi.nlm.nih.gov/pubmed/8995428.
35. Li C, Tropak MB, Gerlai R, et al. Myelination in the absence of myelin-associated glycoprotein. *Nature.* 1994;369(6483):747–750. 10.1038/369747a0.
36. Montag D, Giese KP, Bartsch U, et al. Mice deficient for the myelin-associated glycoprotein show subtle abnormalities in myelin. *Neuron.* 1994;13(1):229–246. http://www.ncbi.nlm.nih.gov/pubmed/7519026.
37. Fruttiger M, Montag D, Schachner M, Martini R. Crucial role for the myelin-associated glycoprotein in the maintenance of axon-myelin integrity. *Eur J Neurosci.* 1995;7(3):511–515. 10.1111/j.1460-9568.1995.tb00347.x.
38. Fujita N, Kemper A, Dupree J, et al. The cytoplasmic domain of the large myelin-associated glycoprotein isoform is needed for proper CNS but not peripheral nervous system myelination. *J Neurosci.* 1998;18(6):1970–1978. http://www.ncbi.nlm.nih.gov/pubmed/9482783.
39. Bjartmar C, Yin X, Trapp BD. Axonal pathology in myelin disorders. *J Neurocytol.* 2018;28(4–5):383–395. http://www.ncbi.nlm.nih.gov/pubmed/10739578.
40. Schachner M, Bartsch U. Multiple functions of the myelin-associated glycoprotein MAG (siglec-4a) in formation and maintenance of myelin. *Glia.* 2000;29(2):154–165. http://www.ncbi.nlm.nih.gov/pubmed/10625334.
41. Dashiell SM, Tanner SL, Pant HC, Quarles RH. Myelin-associated glycoprotein modulates expression and phosphorylation of neuronal cytoskeletal elements and their associated kinases. *J Neurochem.* 2002;81(6):1263–1272. http://www.ncbi.nlm.nih.gov/pubmed/12068074.
42. Pan B, Fromholt SE, Hess EJ, et al. Myelin-associated glycoprotein and complementary axonal ligands, gangliosides, mediate axon stability in the CNS and PNS: neuropathology and behavioral deficits in single- and double-null mice. *Exp Neurol.* 2005;195(1):208–217. 10.1016/j.expneurol.2005.04.017.
43. Okada M, Itoh Mi M, Haraguchi M, et al. b-series Ganglioside deficiency exhibits no definite changes in the neurogenesis and the sensitivity to Fas-mediated apoptosis but impairs regeneration of the lesioned hypoglossal nerve. *J Biol Chem.* 2002;277(3):1633–1636. 10.1074/jbc.C100395200.
44. Kawai H, Allende ML, Wada R, et al. Mice expressing only monosialoganglioside GM3 exhibit lethal audiogenic seizures. *J Biol Chem.* 2001;276(10):6885–6888. 10.1074/jbc.C000847200.
45. Inoue S, Egashira K, Ni W, et al. Anti-monocyte chemoattractant protein-1 gene therapy limits progression and destabilization of established atherosclerosis in apolipoprotein E-knockout mice. *Circulation.* 2002;106(21):2700–2706. http://www.ncbi.nlm.nih.gov/pubmed/12438296.
46. Collins BE, Ito H, Sawada N, Ishida H, Kiso M, Schnaar RL. Enhanced binding of the neural siglecs, myelin-associated glycoprotein and Schwann cell myelin protein, to Chol-1 (alpha-series) gangliosides and novel sulfated Chol-1 analogs. *J Biol Chem.* 1999;274(53):37637–37643. http://www.ncbi.nlm.nih.gov/pubmed/10608819.
47. Yamashita T, Hashiramoto A, Haluzik M, et al. Enhanced insulin sensitivity in mice lacking ganglioside GM3. *Proc Natl Acad Sci.* 2003;100(6):3445–3449. 10.1073/pnas.0635898100.
48. Yoshikawa M, Go S, Takasaki K, et al. Mice lacking ganglioside GM3 synthase exhibit complete hearing loss due to selective degeneration of the organ of Corti. *Proc Natl Acad Sci USA.* 2009;106(23):9483–9488. 10.1073/pnas.0903279106.
49. Yamashita T, Wu Y-P, Sandhoff R, et al. Interruption of ganglioside synthesis produces central nervous system degeneration and altered axon-glial interactions. *Proc Natl Acad Sci USA.* 2005;102(8):2725–2730. 10.1073/pnas.0407785102.

50. Yamashita T, Wada R, Sasaki T, et al. A vital role for glycosphingolipid synthesis during development and differentiation. *Proc Natl Acad Sci USA*. 1999;96(16):9142–9147. http://www.ncbi.nlm.nih.gov/pubmed/10430909.
51. Yamashita T, Allende ML, Kalkofen DN, Werth N, Sandhoff K, Proia RL. ConditionalLoxP-flanked glucosylceramide synthase allele controlling glycosphingolipid synthesis. *Genesis*. 2005;43(4):175–180. 10.1002/gene.20167.
52. Jennemann R, Sandhoff R, Wang S, et al. Cell-specific deletion of glucosylceramide synthase in brain leads to severe neural defects after birth. *Proc Natl Acad Sci*. 2005;102 (35):12459–12464. 10.1073/pnas.0500893102.
53. Svennerholm L. Designation and schematic structure of gangliosides and allied glycosphingolipids. *Progr Brain Res*. 1994;101:xi–xiv. 10.1016/S0079-6123(08)61935-4.
54. Ohmi Y, Tajima O, Ohkawa Y, et al. Gangliosides play pivotal roles in the regulation of complement systems and in the maintenance of integrity in nerve tissues. *Proc Natl Acad Sci*. 2009;106(52):22405–22410. 10.1073/pnas.0912336106.
55. Ohmi Y, Ohkawa Y, Tajima O, Sugiura Y, Furukawa K, Furukawa K. Ganglioside deficiency causes inflammation and neurodegeneration via the activation of complement system in the spinal cord. *J Neuroinflammation*. 2014;11(1):61. 10.1186/1742-2094-11-61.
56. Sturgill ER, Aoki K, Lopez PH, et al. Biosynthesis of the major brain gangliosides GD1a and GT1b. *Glycobiology*. 2012;22(10):1289–1301. 10.1093/glycob/cws103.
57. Yoo S-W, Motari MG, Susuki K, et al. Sialylation regulates brain structure and function. *FASEB J*. 2015;29(7):3040–3053. 10.1096/fj.15-270983.
58. Nguyen T, Mehta NR, Conant K, et al. Axonal protective effects of the myelin-associated glycoprotein. *J Neurosci*. 2009;29(3):630–637. 10.1523/JNEUROSCI.5204-08.2009.
59. Mehta NR, Nguyen T, Bullen JW, Griffin JW, Schnaar RL, Schnaar RL. Myelin-associated glycoprotein (MAG) protects neurons from acute toxicity using a ganglioside-dependent mechanism. *ACS Chem Neurosci*. 2010;1(3):215–222. 10.1021/cn900029p.
60. Nagai J, Goshima Y, Ohshima T. CRMP4 mediates MAG-induced inhibition of axonal outgrowth and protection against Vincristine-induced axonal degeneration. *Neurosci Lett*. 2012;519(1):56–61. 10.1016/J.NEULET.2012.05.020.
61. Yin X, Crawford TO, Griffin JW, et al. Myelin-associated glycoprotein is a myelin signal that modulates the caliber of myelinated axons. *J Neurosci*. 1998;18(6):1953–1962. http://www.ncbi.nlm.nih.gov/pubmed/9482781.
62. Pernet V, Joly S, Christ F, Dimou L, Schwab ME. Nogo-A and myelin-associated glycoprotein differently regulate oligodendrocyte maturation and myelin formation. *J Neurosci*. 2008;28(29):7435–7444. 10.1523/JNEUROSCI.0727-08.2008.
63. Susuki K, Baba H, Tohyama K, et al. Gangliosides contribute to stability of paranodal junctions and ion channel clusters in myelinated nerve fibers. *Glia*. 2007;55 (7):746–757. 10.1002/glia.20503.
64. Furukawa K, Ohmi Y, Ji S, et al. Glycolipids: Essential regulator of neuro-inflammation, metabolism and gliomagenesis. *Biochim Biophys Acta*. 2017;1861(10):2479–2484. 10.1016/j.bbagen.2017.06.007.
65. Yuki N, Susuki K, Koga M, et al. Carbohydrate mimicry between human ganglioside GM1 and *Campylobacterjejuni* lipooligosaccharide causes Guillain-Barre syndrome. *Proc Natl Acad Sci*. 2004;101(31):11404–11409. 10.1073/pnas.0402391101.
66. Yuki N, Yamada M, Koga M, et al. Animal model of axonal Guillain-Barré syndrome induced by sensitization with GM1 ganglioside. *Ann Neurol*. 2001;49(6):712–720. http://www.ncbi.nlm.nih.gov/pubmed/11409422.
67. Susuki K, Rasband MN, Tohyama K, et al. Anti-GM1 antibodies cause complement-mediated disruption of sodium channel clusters in peripheral motor nerve fibers. *J Neurosci*. 2007;27(15):3956–3967.

68. Griffin JW, Li CY, Macko C, et al. Early nodal changes in the acute motor axonal neuropathy pattern of the Guillain-Barré syndrome. *J Neurocytol.* 1996;25(1):33–51. 10.1007/bf02284784.
69. He L, Zhang G, Liu W, Gao T, Sheikh KA. Anti-ganglioside antibodies induce nodal and axonal injury via Fc receptor-mediated inflammation. *J Neurosci.* 2015;35 (17):6770–6785. 10.1523/JNEUROSCI.4926-14.2015.
70. McGonigal R, Rowan EG, Greenshields KN, et al. Anti-GD1a antibodies activate complement and calpain to injure distal motor nodes of Ranvier in mice. *Brain.* 2010;133(7):1944–1960. 10.1093/brain/awq119.
71. Roisen FJ, Bartfeld H, Nagele R, Yorke G. Ganglioside stimulation of axonal sprouting in vitro. *Science.* 1981;214(4520):577–578. http://www.ncbi.nlm.nih.gov/pubmed/7291999.
72. Leon A, Facci L, Benvegnù D, Toffano G. Morphological and biochemical effects of gangliosides in neuroblastoma cells. *Dev Neurosci.* 1982;5(1):108–114. 10.1159/000112666.
73. Facci L, Leon A, Toffano G, Sonnino S, Ghidoni R, Tettamanti G. Promotion of neuritogenesis in mouse neuroblastoma cells by exogenous gangliosides. relationship between the effect and the cell association of ganglioside $GM_1$. *J Neurochem.* 1984;42 (2):299–305. 10.1111/j.1471-4159.1984.tb02678.x.
74. Skaper SD, Katoh-Semba R, Varon S. GM1 ganglioside accelerates neurite outgrowth from primary peripheral and central neurons under selected culture conditions. *Brain Res.* 1985;355(1):19–26. http://www.ncbi.nlm.nih.gov/pubmed/4075103.
75. Dimpfel W, Moller W, Mengs U. Ganglioside-induced neurite formation in cultured neuroblastoma cells. In: Rapport M, Gorio, eds. *Gangliosides in Neurological and Neuromuscular Function, Development and Repair. New York: Raven Press; 1981:119–134.*
76. Seifert W. Gangliosides in nerve cell cultures. In: Rapport M, Gorio, eds. *Gangliosidesin Neurological and Neuromuscular Function, Development and Repair. New York: Raven Press; 1981:99–117.*
77. Leeden R, Skivanek J, Nunez J, m Sclafani J, Norton W, Farooq M. Implications of the distribution and transport of gangliosides in the nervous system. In: Rapport M, Gorio, eds. *Gangliosides in Neurological and Neuromuscular Function, Development and Repair. New York: Raven Press; 1981:211–223.*
78. Odaka M, Yuki N, Yoshino H, et al. N -glycolylneuraminic acid-containing GM1 is a new molecule for serum antibody in Guillain-Barré syndrome. *Ann Neurol.* 1998;43 (6):829–834. 10.1002/ana.410430619.
79. Wu G, Lu Z-H, Wang J, et al. Enhanced susceptibility to kainate-induced seizures, neuronal apoptosis, and death in mice lacking gangliotetraose gangliosides: protection with LIGA 20, a membrane-permeant analog of GM1. *J Neurosci.* 2005;25 (47):11014–11022. 10.1523/JNEUROSCI.3635-05.2005.
80. Hadaczek P, Wu G, Sharma N, et al. GDNF signaling implemented by GM1 ganglioside; failure in Parkinson's disease and GM1-deficient murine model. *Exp Neurol.* 2015;263:177–189. 10.1016/j.expneurol.2014.10.010.
81. Favaron M, Manev H, Alho H, et al. Gangliosides prevent glutamate and kainate neurotoxicity in primary neuronal cultures of neonatal rat cerebellum and cortex. *Proc Natl Acad Sci USA.* 1988;85:7351–7355. 10.2307/32130.
82. Manev H, Favaron M, Vicini S, Guidotti A, Costa E. Glutamate-induced neuronal death in primary cultures of cerebellar granule cells: protection by synthetic derivatives of endogenous sphingolipids. *J Pharmacol Exp Ther.* 1990;252(1).
83. Guidotti A, Manev H, Favaron M, Brooker G, Costa E. Protection by natural and semisynthetic gangliosides from $Ca^{2+}$-dependent neurotoxicity caused by excitatory amino acid (EAA) neurotransmitters. *Adv Exp Med Biol.* 1990;268:135–144.

84. Ferrari G, Anderson BL, Stephens RM, Kaplan DR, Greene LA. Prevention of apoptotic neuronal death by GM1 ganglioside. Involvement of Trk neurotrophin receptors. *J Biol Chem.* 1995;270(7):3074–3080. 10.1074/JBC.270.7.3074.
85. Wu G, Lu Z-H, Xie X, Ledeen RW. Susceptibility of cerebellar granule neurons from GM2/GD2 synthase-null mice to apoptosis induced by glutamate excitotoxicity and elevated KCl: rescue by GM1 and LIGA20. *Glycoconj J.* 2004;21(6):305–313. 10.1023/B:GLYC.0000046273.68493.f7.
86. Geisler FH, Coleman WP, Grieco G, Poonian D. Sygen Study Group.The Sygen multicenter acute spinal cord injury study. *Spine (Phila Pa 1976).* 2001;26(Suppl. 24): S87–S98. http://www.ncbi.nlm.nih.gov/pubmed/11805614.
87. Schneider JS, Gollomp SM, Sendek S, Colcher A, Cambi F, Du W. A randomized, controlled, delayed start trial of GM1 ganglioside in treated Parkinson's disease patients. *J Neurol Sci.* 2013;324(1–2):140–148. 10.1016/j.jns.2012.10.024.
88. Candelise L, Ciccone A. Gangliosides for acute ischemic stroke. *Stroke.* 2002;33(9):2336. 10.1161/01.STR.0000029272.13806.46.
89. Landi G, D'Alessandro R, Dossi BC, Ricci S, Simone IL, Ciccone A. Guillain-Barré syndrome after exogenous gangliosides in Italy. *BMJ.* 1993;307(6917):1463–1464. http://www.ncbi.nlm.nih.gov/pubmed/8281088.
90. Illa I, Ortiz N, Gallard E, Juarez C, Grau JM, Dalakas MC. Acute axonal Guillain-Barré syndrome with IgG antibodies against motor axons following parenteral gangliosides. *Ann Neurol.* 1995;38(2):218–224. 10.1002/ana.410380214.
91. Higashi H, Sasabe T, Fukui Y, Maru M, Kato S. Detection of gangliosides as N-glycolylneuraminic acid-specific tumor-associated Hanganutziu-Deicher antigen in human retinoblastoma cells. *Jpn J Cancer Res.* 1988;79(8):952–956. http://www.ncbi.nlm.nih.gov/pubmed/2460424.
92. Odaka M, Yuki N, Yoshino H, et al. N -glycolylneuraminic acid-containing GM1 is a new molecule for serum antibody in Guillain-Barré syndrome. *Ann Neurol.* 1998;43(6):829–834. 10.1002/ana.410430619.
93. Mutoh T, Tokuda A, Guroff G, Fujiki N. The effect of the B subunit of cholera toxin on the action of nerve growth factor on PC12 cells. *J Neurochem.* 1993;60(4):1540–1547. 10.1111/j.1471-4159.1993.tb03319.x.
94. Mutoh T, Tokuda A, Miyadai T, Hamaguchi M, Fujiki N. Ganglioside GM1 binds to the Trk protein and regulates receptor function. *Proc Natl Acad Sci USA.* 1995;92(11):5087–5091. http://www.ncbi.nlm.nih.gov/pubmed/7539142.
95. Mutoh T, Hamano T, Yano S, et al. Stable transfection of GM1 synthase gene into GM1-deficient NG108-15 cells, CR-72 cells, rescues the responsiveness of Trk-neurotrophin receptor to its ligand, NGF. *Neurochem Res.* 2002;27(7–8):801–806. http://www.ncbi.nlm.nih.gov/pubmed/12374216.
96. Rabin SJ, Mocchetti I. GM1 ganglioside activates the high-affinity nerve growth factor receptor trkA. *J Neurochem.* 2002;65(1):347–354. 10.1046/j.1471-4159.1995.65010347.x.
97. Panni MK, Cooper JD, Sofroniew MV. Ganglioside GM1 potentiates NGF action on axotomised medial septal cholinergic neurons. *Brain Res.* 1998;812(1–2):76–80. http://www.ncbi.nlm.nih.gov/pubmed/9813248.
98. Pitto M, Mutoh T, Kuriyama M, Ferraretto A, Palestini P, Masserini M. Influence of endogenous GM1 ganglioside on TrkB activity, in cultured neurons. *FEBS Lett.* 1998;439(1–2):93–96. 10.1016/S0014-5793(98)01344-1.
99. Abad-Rodriguez J, Robotti A. Regulation of axonal development by plasma membrane gangliosides. *J Neurochem.* 2007;103(s1):47–55. 10.1111/j.1471-4159.2007.04713.x.
100. Rabin SJ, Mocchetti I. GM1 ganglioside activates the high-affinity nerve growth factor receptor trkA. *J Neurochem.* 1995;65(1):347–354. http://www.ncbi.nlm.nih.gov/pubmed/7790879.

101. Miyagi T, Wada T, Iwamatsu A, et al. Molecular cloning and characterization of a plasma membrane-associated sialidase specific for gangliosides. *J Biol Chem.* 1999;274 (8):5004–5011. 10.1074/JBC.274.8.5004.
102. Da Silva JS, Hasegawa T, Miyagi T, Dotti CG, Abad-Rodriguez J. Asymmetric membrane ganglioside sialidase activity specifies axonal fate. *Nat Neurosci.* 2005;8 (5):606–615. 10.1038/nn1442.
103. Kappagantula S, Andrews MR, Cheah M, Abad-Rodriguez J, Dotti CG, Fawcett JW. Neu3 sialidase-mediated ganglioside conversion is necessary for axon regeneration and is blocked in CNS axons. *J Neurosci.* 2014;34(7):2477–2492. 10.1523/ JNEUROSCI.4432-13.2014.
104. Kelm S, Pelz A, Schauer R, et al. Sialoadhesin, myelin-associated glycoprotein and CD22 define a new family of sialic acid-dependent adhesion molecules of the immunoglobulin superfamily. *Curr Biol.* 1994;4(11):965–972. 10.1016/S0960-9822(00) 00220-7.
105. Collins BE, Kiso M, Hasegawa A, et al. Binding specificities of the sialoadhesin family of I-type lectins. Sialic acid linkage and substructure requirements for binding of myelin-associated glycoprotein, Schwann cell myelin protein, and sialoadhesin. *J Biol Chem.* 1997;272(27):16889–16895. http://www.ncbi.nlm.nih.gov/pubmed/9201997.
106. McKerracher L, David S, Jackson DL, Kottis V, Dunn RJ, Braun PE. Identification of myelin-associated glycoprotein as a major myelin-derived inhibitor of neurite growth. *Neuron.* 1994;13(4):805–811. http://www.ncbi.nlm.nih.gov/pubmed/7524558.
107. Mukhopadhyay G, Doherty P, Walsh FS, Crocker PR, Filbin MT. A novel role for myelin-associated glycoprotein as an inhibitor of axonal regeneration. *Neuron.* 1994;13 (3):757–767. http://www.ncbi.nlm.nih.gov/pubmed/7522484.
108. Vyas AA, Patel HV, Fromholt SE, et al. Gangliosides are functional nerve cell ligands for myelin-associated glycoprotein (MAG), an inhibitor of nerve regeneration. *Proc Natl Acad Sci.* 2002;99(12):8412–8417. 10.1073/pnas.072211699.
109. Liu BP, Fournier A, GrandPré T, Strittmatter SM. Myelin-associated glycoprotein as a functional ligand for the Nogo-66 receptor. *Science.* 2002;297(5584):1190–1193. 10.1126/science.1073031.
110. Domeniconi M, Cao Z, Spencer T, et al. Myelin-associated glycoprotein interacts with the Nogo66 receptor to inhibit neurite outgrowth. *Neuron.* 2002;35(2):283–290. http://www.ncbi.nlm.nih.gov/pubmed/12160746.
111. Laurén J, Hu F, Chin J, Liao J, Airaksinen MS, Strittmatter SM. Characterization of myelin ligand complexes with neuronal Nogo-66 receptor family members. *JBiolChem.* 2007;282(8):5715–5725. 10.1074/jbc.M609797200.
112. Venkatesh K, Chivatakarn O, Lee H, et al. The Nogo-66 receptor homolog NgR2 is a sialic acid-dependent receptor selective for myelin-associated glycoprotein. *J Neurosci.* 2005;25(4):808–822. 10.1523/JNEUROSCI.4464-04.2005.
113. Wong ST, Henley JR, Kanning KC, Huang K, Bothwell M, Poo M. A p75NTR and Nogo receptor complex mediates repulsive signaling by myelin-associated glycoprotein. *Nat Neurosci.* 2002;5(12):1302–1308. 10.1038/nn975.
114. Wang KC, Kim JA, Sivasankaran R, Segal R, He Z. p75 interacts with the Nogo receptor as a co-receptor for Nogo, MAG and OMgp. *Nature.* 2002;420 (6911):74–78. 10.1038/nature01176.
115. Yamashita T, Higuchi H, Tohyama M. The p75 receptor transduces the signal from myelin-associated glycoprotein to Rho. *J Cell Biol.* 2002;157(4):565–570. 10.1083/ jcb.200202010.
116. Saha N, Kolev MV, Semavina M, Himanen J, Nikolov DB. Ganglioside mediate the interaction between Nogo receptor 1 and LINGO-1. *Biochem Biophys Res Commun.* 2011;413(1):92–97. 10.1016/j.bbrc.2011.08.060.

117. Mi S, Lee X, Shao Z, et al. LINGO-1 is a component of the Nogo-66 receptor/p75 signaling complex. *Nat Neurosci.* 2004;7(3):221–228. 10.1038/nn1188.
118. Williams G, Wood A, Williams E-J, et al. Ganglioside inhibition of neurite outgrowth requires nogo receptor function. *J Biol Chem.* 2008;283(24):16641–16652. 10.1074/jbc.M802067200.
119. Fujitani M, Kawai H, Proia RL, Kashiwagi A, Yasuda H, Yamashita T. Binding of soluble myelin-associated glycoprotein to specific gangliosides induces the association of p75NTR to lipid rafts and signal transduction. *J Neurochem.* 2005;94(1):15–21. 10.1111/j.1471-4159.2005.03121.x.
120. Tang S, Shen YJ, DeBellard ME, et al. Myelin-associated glycoprotein interacts with neurons via a sialic acid binding site at ARG118 and a distinct neurite inhibition site. *J Cell Biol.* 1997;138(6):1355–1366. http://www.ncbi.nlm.nih.gov/pubmed/9298990.
121. Pronker MF, Lemstra S, Snijder J, et al. Structural basis of myelin-associated glycoprotein adhesion and signalling. *Nat Commun.* 2016;7:13584. 10.1038/ncomms13584.
122. Domeniconi M, Zampieri N, Spencer T, et al. MAG induces regulated intramembrane proteolysis of the p75 neurotrophin receptor to inhibit neurite outgrowth. *Neuron.* 2005;46(6):849–855. 10.1016/J.NEURON.2005.05.029.
123. Niederöst B, Oertle T, Fritsche J, McKinney RA, Bandtlow CE. Nogo-A and myelin-associated glycoprotein mediate neurite growth inhibition by antagonistic regulation of RhoA and Rac1. *J Neurosci.* 2002;22(23):10368–10376. http://www.ncbi.nlm.nih.gov/pubmed/12451136.
124. Mimura F, Yamagishi S, Arimura N, et al. Myelin-associated glycoprotein inhibits microtubule assembly by a Rho-kinase-dependent mechanism. *J Biol Chem.* 2006;281(23):15970–15979. 10.1074/jbc.M510934200.
125. Hsieh SH., Ferraro GB, Fournier AE. Myelin-associated inhibitors regulate cofilin phosphorylation and neuronal inhibition through LIM kinase and slingshot phosphatase. *J Neurosci.* 2006;26(3):1006–1015. 10.1523/JNEUROSCI.2806-05.2006.
126. Arimura N, Inagaki N, Chihara K, et al. Phosphorylation of collapsin response mediator protein-2 by Rho-kinase. Evidence for two separate signaling pathways for growth cone collapse. *J Biol Chem.* 2000;275(31):23973–23980. 10.1074/jbc.M001032200.
127. Arimura N, Menager C, Kawano Y, et al. Phosphorylation by Rho kinase regulates CRMP-2 activity in growth cones. *Mol Cell Biol.* 2005;25(22):9973–9984. 10.1128/MCB.25.22.9973-9984.2005.
128. Khazaei MR, Girouard M-P, Alchini R, et al. Collapsin response mediator protein 4 regulates growth cone dynamics through the actin and microtubule cytoskeleton. *J Biol Chem.* 2014;289(43):30133–30143. 10.1074/jbc.M114.570440.
129. Al-Bashir N, Mellado W, Filbin MT. Sialic acid is required for neuronal inhibition by soluble MAG but not for membrane bound MAG. *Front Mol Neurosci.* 2016;9:21. 10.3389/fnmol.2016.00021.
130. Mehta NR, Lopez PHH, Vyas AA, Schnaar RL. Gangliosides and Nogo receptors independently mediate myelin-associated glycoprotein inhibition of neurite outgrowth in different nerve cells. *J Biol Chem.* 2007;282(38):27875–27886. 10.1074/jbc.M704055200.
131. Venkatesh K, Chivatakarn O, Sheu S-S, Giger RJ. Molecular dissection of the myelin-associated glycoprotein receptor complex reveals cell type–specific mechanisms for neurite outgrowth inhibition. *J Cell Biol.* 2007;177(3):393–399. 10.1083/jcb.200702102.
132. Cafferty WBJ, Duffy P, Huebner E, Strittmatter SM. MAG and OMgp synergize with Nogo-A to restrict axonal growth and neurological recovery after spinal cord trauma. *J Neurosci.* 2010;30(20):6825–6837. 10.1523/JNEUROSCI.6239-09.2010.

133. Lee JK, Geoffroy CG, Chan AF, et al. Assessing spinal axon regeneration and sprouting in Nogo-, MAG-, and OMgp-deficient mice. *Neuron.* 2010;66(5):663–670. 10.1016/j.neuron.2010.05.002.
134. Dickendesher TL, Baldwin KT, Mironova YA, et al. NgR1 and NgR3 are receptors for chondroitin sulfate proteoglycans. *Nat Neurosci.* 2012;15(5):703–712. 10.1038/nn.3070.
135. Bartsch U, Montag D, Bartsch S, Schachner M. Multiply myelinated axons in the optic nerve of mice deficient for the myelin-associated glycoprotein. *Glia.* 1995;14(2):115–122. 10.1002/glia.440140206.
136. Wong EV, David S, Jacob MH, Jay DG. Inactivation of myelin-associated glycoprotein enhances optic nerve regeneration. *J Neurosci.* 2003;23(8):3112–3117. http://www.ncbi.nlm.nih.gov/pubmed/12716917.
137. Fishman PH. Role of membrane gangliosides in the binding and action of bacterial toxins. *J Membr Biol.* 1982;69(2):85–97. 10.1007/BF01872268.
138. Wu G, Ledeen RW. Stimulation of neurite outgrowth in neuroblastoma cells by neuraminidase: putative role of GM1 ganglioside in differentiation. *J Neurochem.* 1991;56(1):95–104. 10.1111/j.1471-4159.1991.tb02567.x.
139. Wu G, Lu ZH, Nakamura K, Spray DC, Ledeen RW. Trophic effect of cholera toxin B subunit in cultured cerebellar granule neurons: modulation of intracellular calcium by GM1 ganglioside. *J Neurosci Res.* 1996;44(3):243–254. 10.1002/(SICI)1097-4547(19960501)44:3<243::AID-JNR5>3.0.CO;2-G.
140. Fang Y, Wu G, Xie X, Lu Z-H, Ledeen RW. Endogenous GM1 ganglioside of the plasma membrane promotes neuritogenesis by two mechanisms. *Neurochem Res.* 2000;25(7):931–940. https://link.springer.com/content/pdf/10.1023/A:1007596223484.pdf
141. Wu G, Lu Z-H, André S, Gabius H-J, Ledeen RW. Functional interplay between ganglioside GM1 and cross-linking galectin-1 induces axon-like neuritogenesis via integrin-based signaling and TRPC5-dependent $Ca^{2+}$ influx. *J Neurochem.* 2016;136(3):550–563. 10.1111/jnc.13418.
142. Fang Y, Xie X, Ledeen RW, Wu G. Characterization of cholera toxin B subunit-induced $Ca^{2+}$ influx in neuroblastoma cells: evidence for a voltage-independent GM1 ganglioside-associated $Ca^{2+}$ channel. *J Neurosci Res.* 2002;69(5):669–680. 10.1002/jnr.10333.
143. Wu G, Lu Z-H, Obukhov AG, Nowycky MC, Ledeen RW. Induction of calcium influx through TRPC5 channels by cross-linking of GM1 ganglioside associated with α5β1 integrin initiates neurite outgrowth. *J Neurosci.* 2007;27(28):7447–7458. 10.1523/JNEUROSCI.4266-06.2007.
144. Masco D, Van de Walle M, Spiegel S. Interaction of ganglioside GM1 with the B subunit of cholera toxin modulates growth and differentiation of neuroblastoma N18 cells. *J Neurosci.* 1991;11(8):2443–2452. http://www.ncbi.nlm.nih.gov/pubmed/1651376.
145. Wu G, Nakamura K, Ledeen RW. Inhibition of neurite outgrowth of neuroblastoma neuro-2a cells by cholera toxin B-subunit and anti-GM1 antibody. *Mol Chem Neuropathol.* 1994;21(2–3):259–271. 10.1007/BF02815354.
146. O'Hanlon GM, Hirst TR, Willison HJ. Ganglioside GM1 binding toxins and human neuropathy-associated IgM antibodies differentially promote neuritogenesis in a PC12 assay. *Neurosci Res.* 2003;47(4):383–390. http://www.ncbi.nlm.nih.gov/pubmed/14630342.
147. Wu G, Ledeen RW. Chapter 7 Gangliosides as modulators of neuronal calcium. *Prog Brain Res.* 1994;101:101–112. 10.1016/S0079-6123(08)61942-1.
148. Spirman N, Sela B-A, Schwartz M. Antiganglioside antibodies inhibit neuritic outgrowth from regenerating goldfish retinal explants. *J Neurochem.* 1982;39(3):874–877. 10.1111/j.1471-4159.1982.tb07974.x.

149. Spoerri PE, Rapport MM, Mahadik SP, Roisen FJ. Inhibition of conditioned media-mediated neuritogenesis of sensory ganglia by monoclonal antibodies to GM1 ganglioside. *Brain Res.* 1988;469(1–2):71–77. http://www.ncbi.nlm.nih.gov/pubmed/3401809.
150. Uncini A, Susuki K, Yuki N. Nodo-paranodopathy: beyond the demyelinating and axonal classification in anti-ganglioside antibody-mediated neuropathies. *Clin Neurophysiol.* 2013;124(10):1928–1934. 10.1016/j.clinph.2013.03.025.
151. Willison HJ, Yuki N, Willison HJ. Peripheral neuropathies and anti-glycolipid antibodies. *Brain.* 2002;125(Pt 12):2591–2625. https://www-ncbi-nlm-nih-gov.ezp.welch.jhmi.edu/pubmed/12429589.
152. Van Den Berg B, Walgaard C, Drenthen J, Fokke C, Jacobs BC, Van Doorn PA. Guillain-Barré syndrome: pathogenesis, diagnosis, treatment and prognosis. *Nat Rev Neurol.* 2014. 10.1038/nrneurol.2014.121.
153. Jacobs BC, van Doorn PA, Tio-Gillen AP, et al. *Campylobacterjejuni* infections and anti-GM1 antibodies in guillain-barré syndrome. *Ann Neurol.* 1996;40(2):181–187. 10.1002/ana.410400209.
154. Carpo M, Pedotti R, Allaria S, et al. Clinical presentation and outcome of Guillain-Barré and related syndromes in relation to anti-ganglioside antibodies. *J Neurol Sci.* 1999;168(2):78–84. http://www.ncbi.nlm.nih.gov/pubmed/10526187.
155. Press R, Matá S, Lolli F, Zhu J, Andersson T, Link H. Temporal profile of anti-ganglioside antibodies and their relation to clinical parameters and treatment in Guillain-Barré syndrome. *J Neurol Sci.* 2001;190(1–2):41–47. 10.1016/S0022-510X(01)00580-9.
156. Ilyas AA, Mithen FA, Dalakas MC, Chen ZW, Cook SD. Antibodies to acidic glycolipids in Guillain-Barré syndrome and chronic inflammatory demyelinating polyneuropathy. *J Neurol Sci.* 1992;107(1):111–121. http://www.ncbi.nlm.nih.gov/pubmed/1578228.
157. Simone IL, Annunziata P, Maimone D, Liguori M, Leante R, Livrea P. Serum and CSF anti-GM1 antibodies in patients with Guillain-Barré syndrome and chronic inflammatory demyelinating polyneuropathy. *JNeurolSci.* 1993;114(1):49–55. http://www.ncbi.nlm.nih.gov/pubmed/8433097.
158. Bech E, Orntoft TF, Andersen LP, Skinhøj P, Jakobsen J. IgM anti-GM1 antibodies in the Guillain-Barré syndrome: a serological predictor of the clinical course. *JNeuroimmunol.* 1997;72(1):59–66. http://www.ncbi.nlm.nih.gov/pubmed/9003245.
159. Annunziata P, Figura N, Galli R, Mugnaini F, Lenzi C. Association of anti-GM1 antibodies but not of anti-cytomegalovirus. *Campylobacter jejuni* and *Helicobacter pylori* IgG, with a poor outcome in Guillain-Barré syndrome. *J Neurol Sci.* 2003;213(1–2):55–60. http://www.ncbi.nlm.nih.gov/pubmed/12873755.
160. Koga M, Yuki N, Hirata K, Morimatsu M, Mori M, Kuwabara S. Anti-GM1 antibody IgG subclass: a clinical recovery predictor in Guillain-Barré syndrome. *Neurology.* 2003;60(9):1514–1518. 10.1212/01.WNL.0000061615.77865.83.
161. Gregson NA, Koblar S, Hughes RA. Antibodies to gangliosides in Guillain-Barré syndrome: specificity and relationship to clinical features. *Q J Med.* 1993;86(2):111–117. http://www.ncbi.nlm.nih.gov/pubmed/8464986.
162. Kuwabara S, Asahina M, Koga M, Mori M, Yuki N, Hattori T. Two patterns of clinical recovery in Guillain-Barré syndrome with IgG anti-GM1 antibody. *Neurology.* 1998;51(6):1656–1660. http://www.ncbi.nlm.nih.gov/pubmed/9855518
163. Brown WF, Feasby TE. Conduction block and denervation in Guillain-Barré polyneuropathy. *Brain.* 1984;107(Pt 1):219–239. http://www.ncbi.nlm.nih.gov/pubmed/6697157.
164. Lunn MP, Johnson LA, Fromholt SE, et al. High-affinity anti-ganglioside IgG antibodies raised in complex ganglioside knockout mice: reexamination of GD1a

immunolocalization. *J Neurochem.* 2000;75(1):404–412. http://www.ncbi.nlm.nih.gov/pubmed/10854286.
165. Lehmann HC, Lopez PHH, Zhang G, et al. Passive immunization with anti-ganglioside antibodies directly inhibits axon regeneration in an animal model. *J Neurosci.* 2007;27(1):27–34. 10.1523/JNEUROSCI.4017-06.2007.
166. Zhang G, Lehmann HC, Manoharan S, et al. Anti-ganglioside antibody-mediated activation of RhoA induces inhibition of neurite outgrowth. *J Neurosci.* 2011;31(5):1664–1675. 10.1523/JNEUROSCI.3829-10.2011.
167. Rozés Salvador V, Heredia F, Berardo A, et al. Anti-glycan antibodies halt axon regeneration in a model of Guillain Barrè Syndrome axonal neuropathy by inducing microtubule disorganization via RhoA-ROCK-dependent inactivation of CRMP-2. *Exp Neurol.* 2016;278:42–53. 10.1016/j.expneurol.2016.01.016.
168. Zhang G, Bogdanova N, Gao T, et al. Fcγ receptor-mediated inflammation inhibits axon regeneration (Costigan M, ed). *PLoS One.* 2014;9(2):e88703. 10.1371/journal.pone.0088703.

CHAPTER FOURTEEN

# Ganglioside-Mediated Assembly of Amyloid β-Protein: Roles in Alzheimer's Disease

**Katsumi Matsuzaki*, Koichi Kato[†,‡], Katsuhiko Yanagisawa[§,1]**

* Kyoto University, Kyoto, Japan
† Nagoya City University, Nagoya, Japan
‡ Okazaki Institute for Integrative Bioscience and Institute for Molecular Science, National Institutes of Natural Sciences, Okazaki, Japan
§ Center for Advanced Medicine for Dementia, National Center for Geriatrics and Gerontology, Obu, Japan
1 Corresponding author. E-mail address: katuhiko@ncgg.go.jp

## Contents



## Abstract

Assembly and deposition of amyloid β-protein (Aβ) is an early and invariable pathological event of Alzheimer's disease (AD), a chronic neurodegenerative disease

*Progress in Molecular Biology and Translational Science*, Volume 156
ISSN 1877-1173
http://dx.doi.org/10.1016/bs.pmbts.2017.10.005

affecting the neurons in the brain of aging population. Thus, clarification of the molecular mechanism underlying Aβ assembly is crucial not only for understanding the pathogenesis of AD, but also for developing disease-modifying remedies. In 1995, ganglioside-bound Aβ (GAβ), with unique molecular characteristics, including its altered immunoreactivity and its conspicuous ability to accelerate Aβ assembly, was discovered in an autopsied brain showing early pathological changes of AD. Based on these findings, it was hypothesized that GAβ is an endogenous seed for amyloid fibril formation in the AD brain. A body of evidence that supports the GAβ hypothesis has been growing for over 20 years as follows. First, the conformational changes of Aβ from a random coil to an α-helix, and then to a β-sheet in the presence of ganglioside were validated by several techniques. Second, the seed activity of GAβ to accelerate the assembly of soluble Aβ into amyloid fibrils was confirmed by various in vitro and in vivo experiments. Third, it was found that the Aβ binding to ganglioside to form GAβ occurs under limited conditions, which were provided by the lipid environment surrounding ganglioside. Fourth, the region-specific Aβ deposition in the brain appeared to be dependent on the presence of the lipid environment that was in favor of GAβ generation. In this chapter, further progress of the study of ganglioside-mediated Aβ assembly, especially from the aspects of physicochemistry, structural biology, and neuropathology, is reviewed.

## 1. INTRODUCTION

Assembly of the soluble amyloid β-protein (Aβ), which is a physiological cleavage product of the amyloid precursor protein, into insoluble amyloid fibrils is an early and invariable pathological event in the brain of patients with Alzheimer's disease (AD).[1] Thus, elucidation of the molecular mechanism underlying Aβ assembly in the brain is one of the most fundamental issues regarding the pathogenesis of AD. Aβ may assemble owing to the rise in soluble Aβ level caused by increased production in the brain and/or decreased clearance from the brain. However, there is no conclusive proof that the soluble Aβ level in the brain is high enough to induce spontaneous Aβ assembly, which can be induced at several orders of magnitude higher Aβ levels in vitro than in the brain. More importantly, the increase in soluble Aβ level, if any, cannot be responsible for the region-specific Aβ deposition in the brain; some regions such as the precuneus are vulnerable while other regions such as the calcarine cortex are resistant to Aβ deposition. This region-specific Aβ deposition is further noticeable in the case of some Aβ mutants, which predominantly deposit in blood vessel walls but do so only minimally in the brain parenchyma and vice versa.[2,3] Thus, it seems reasonable that local factors induce Aβ assembly and deposition in the brain.

In 1995, a novel Aβ species, which was characterized by its tight binding to GM1 ganglioside, was discovered in the brain of a patient with early pathological changes of AD.[4] Interestingly, the subsequent characterization of the ganglioside-bound Aβ (GAβ) revealed its altered immunoreactivity distinct from those of soluble and insoluble Aβs, and its conspicuous ability to accelerate the assembly of soluble Aβ into amyloid fibrils[4,5]: these unique characteristics of GAβ are now attributable to the conformational changes of Aβ, which are induced by the binding of Aβ to ganglioside.[6–8] On the basis of these findings, together with the findings obtained from other studies,[9–12] it was hypothesized that GAβ is an endogenous seed for amyloid fibril formation in AD brains (GAβ hypothesis).[5]

Subsequently, various in vitro and in vivo studies were performed to explore the interaction between Aβ and ganglioside and provided evidence indicating a pivotal role of ganglioside in the initiation of Aβ assembly as follows. First, the conformational changes of Aβ from a random coil to an α-helix, and then to a β-sheet in the presence of ganglioside were validated by several techniques.[13–16] Second, the seed activity of GAβ to accelerate the assembly of soluble Aβ into amyloid fibrils was confirmed by in vitro and in vivo experiments.[17–25] Third, it was found that the Aβ binding to ganglioside to form GAβ occurs under strictly limited conditions, which were provided by the specific lipid microenvironment surrounding ganglioside.[26,27] Fourth, the region-specific deposition of mutant Aβs appeared to be dependent on the presence of a particular ganglioside species required by each Aβ mutant.[18,20]

The aim of this chapter is to describe recent progress in the study of ganglioside-mediated Aβ assembly, mainly after the publication of a previous review article by Matsuzaki *et al.*[28] Another aim is to emphasize that the pathological interaction between an amyloidogenic protein and ganglioside may trigger the development of various human amyloidoses including AD.

## 2. PHYSICOCHEMICAL ASPECTS OF Aβ INTERACTION WITH MEMBRANE GANGLIOSIDES

### 2.1 Aβ in Solution

The amyloid β peptide fragment representing the first 40 amino acids [Aβ-(1–40)] assumes a random coil conformation in aqueous solution under physiological conditions, as revealed by circular dichroism (CD).[6,29] However, NMR measurements suggested that monomeric Aβ is in

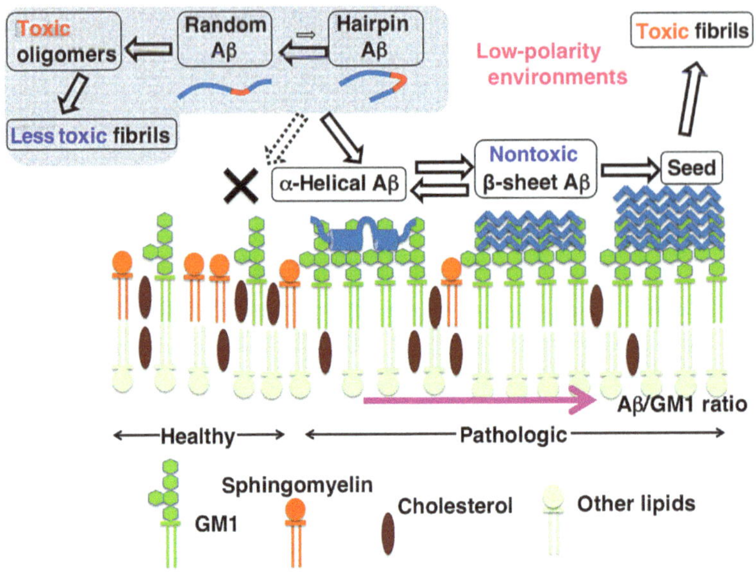


**Fig. 1** Schematic representation of amyloid β-protein (Aβ)–GM1 interactions. Aβ is in equilibrium between a random conformation and a hairpin structure with a turn at around Asp23–Ala30. The equilibrium is extremely shifted toward the random coil structure. Aβ does not bind to membranes as long as GM1 is uniformly distributed in the membrane, corresponding to the healthy state. Once GM1 forms a cluster, Aβ recognizes it, assuming an α-helix-rich conformation. At Aβ/GM1 ratios between ~0.013 and ~0.044, the helical form and aggregated β-sheets (10–20 mer) coexist. However, the β-structure is nontoxic and stable. At higher Aβ densities, the β-structure is converted to a second, seed-prone β-structure, which recruits monomers from the aqueous phase to form toxic amyloid fibrils containing antiparallel β-sheets. Low-polarity environments provided by GM1 clusters trigger the formation of this unique fibril. In contrast, in aqueous phase assembly *(dotted area)*, oligomers are more toxic than fibrils, which are composed of in-register parallel β-sheets. Toxic oligomers may be off-pathway products.

equilibrium between at least two conformations, among which the populations of minor conformations are extremely small.[30] The region Asp23–Ala30 experiences a drastic change in chemical environment upon the conformational transition. A plausible model is that a minor transient conformation has a folded structure with a turn at around Asp23–Ala30 (Fig. 1).

## 2.2 Interaction of Aβ With Ganglioside-Containing Membranes

Aβ-(1–40) selectively binds to ganglioside clusters, the formation of which is facilitated by cholesterol.[7] Molecular dynamics (MD) simulations suggested that string-like clusters of sugar groups of GM1 ganglioside are formed when

GM1 is mixed with raft-forming cholesterol and sphingomyelin (SM), but not with phosphatidylcholine, indicating the importance of membrane environments surrounding GM1.[31] Cholesterol has an important role to stabilize other lipids for interaction with each other and numerous hydrogen bonds are formed among glycan portions of GM1 clusters.

Conformational transitions of Aβ upon binding to GM1 clusters were analyzed in detail by a single-value decomposition analysis of CD spectra.[29] Aβ-(1–40) was mixed with various concentrations of raft-like liposomes composed of GM1/cholesterol/SM (4:3:3) and CD spectra were measured. The CD spectra were composed of contributions from free Aβ in solution, as well as membrane-bound Aβ. The latter spectra were calculated by independently determining the concentration of free Aβ by ultracentrifugation. The eight obtained spectra were analyzed to determine how many basis spectra were needed to reconstitute the spectra. We found three basis spectra, that is, spectra for an α-helix–rich species (~50% helicity), a β-sheet–rich species, and spectra originating from an exciton coupling between phenylalanine (Phe) side chains. From this analysis, the concentrations of the α-helix–rich and β-sheet–rich forms could be obtained under eight Aβ/GM1 ratios. The results are illustrated in Fig. 1. There were three stages, and the stability of Aβ was also examined. At lower Aβ densities in the membrane (Aβ/GM1 molar ratio ~0.013), only the α-helix–rich species existed. At intermediate Aβ densities (~0.013 < Aβ/GM1 < ~0.044), the helical form and aggregated β-sheets (10–20mer) coexisted. However, the β-structure was stable and did not form larger aggregates. At Aβ/GM1 ratios above ~0.044, the β-structure was converted to a second, seed-prone β-structure, which recruited monomers from the aqueous phase to form amyloid fibrils. The fibrils formed on GM1-containing membranes were different from those formed in aqueous solution.[26] The former were thicker (~12 nm wide) than the latter (~7 nm), and, more importantly, stickier and cytotoxic to neuronal cells.

We hypothesized that low-polarity environments provided by GM1 clusters are important for the formation of these unique fibrils.[32] Using Aβ conjugated with the polarity-sensitive dye 7-diethylaminocoumarin-3-carbonyl at the N-terminus, we estimated the polarity of Aβ-(1–40) bound to GM1 clusters, which was similar to that in a mixture of 80% 1,4-dioxane and 20% water (v/v). In this solution, amyloid fibrils with similar morphology and cytotoxicity to those formed in GM1 clusters were generated, supporting our hypothesis. Fourier-transform infrared spectroscopy revealed that amyloid fibrils formed in aqueous solution are composed of parallel β-sheets as also revealed by solid-state NMR,[33] whereas fibrils formed in 80% dioxane, as well as in GM1 clusters, contain antiparallel β-sheets, as characterized by major bands at around 1629 $cm^{-1}$ and minor bands at around 1695 $cm^{-1}$. The formation of in-register parallel β-sheets is unfavorable in low-polarity environments because electrostatic repulsion between charged amino acid side chains in adjacent β-strands is manifested. The presence of antiparallel β-sheets has been reported to correlate closely with the cytotoxicity of various amyloid fibrils and soluble oligomers.[34,35]

We found that aromatic π–π interactions play a crucial role in the assembly of Aβ both in aqueous solution and GM1-containing membranes.[36] Essentially, interactions between side chains are important for fibril formation because they are arranged in a highly ordered fashion within fibrillar structures. Indeed, a Phe–Phe exciton coupling was observed during assembly in GM1-containing membranes (vide supra).[29] Not only Aβ, but also other amyloidogenic proteins, frequently contain aromatic residues. To reveal the effects of interactions involving π-electrons on the assembly of Aβ we substituted cyclohexylalanine (Cha) for each or all Phe residues at the 4th, 19th, and 20th positions of Aβ-(1–40). Cha is sterically similar to Phe, but lacks π-electrons and is more hydrophobic than Phe. Thus, the introduction of Cha does not weaken hydrophobic interactions. Each Cha substitution significantly inhibited fibril formation by Aβ both in aqueous solutions and GM1-containing membranes, indicating a pivotal role of aromatic interactions. The Cha substitutions do not exert significant effects on the binding of Aβ to GM1-containing membranes, although MD simulations suggested that CH–π and/or OH–π interactions are involved in the binding of Aβ to GM1.[37] Interestingly, the Aβ analog with three Cha residues effectively retarded Aβ fibrillation, suggesting that small peptides containing Cha could be promising candidates for disease modulators.

## 2.3 Comparison Between Aβ-(1–40) and Aβ-(1–42)

Aβ-(1–40) and Aβ-(1–42) are two major forms of Aβ present at an approximate ratio of 9:1 in the brain. An increased level of more amyloidogenic and neurotoxic Aβ-(1–42) is related to the early onset of AD.[38] Aβ-(1–42) exhibits lipid specificity similar to that of Aβ-(1–40) and it selectively binds to GM1 clusters. Unexpectedly, the binding affinity of Aβ-(1–42) for GM1-containing membranes is close to that of Aβ-(1–40), although the longer version has two additional hydrophobic amino acids at the C-terminus.[39]

Recently, similar results were reported for GM1-containing nanodiscs.[40] Membrane-bound Aβ-(1–42) also undergoes an α-helix-to-β-sheet transition depending on the Aβ-to-GM1 ratio. Its amyloidogenic activity is more than 10-fold higher than that of Aβ-(1–40).[39]

### 2.4 Comparison Between Human Aβ-(1–40) and Rodent Aβ-(1–40)

Aged rodents rarely develop the characteristic lesions of AD.[41] Rodent Aβ-(1–40) differs from its human counterpart only in the three substitutions of Arg-to-Gly, Tyr-to-Phe, and His-to-Arg at positions 5, 10, and 13, respectively. Both Aβs form similar mature amyloid fibrils in aqueous solution. However, their interactions with GM1-containing membranes are different.[42] The rodent protein binds to GM1-containing membranes 3 times more strongly than the human version does. Rodent Aβ also undergoes an α-helix-to-β-sheet transition, but its α-helix and β-sheet forms are not identical to those of human Aβ. On both GM1-containing membranes and cell membranes, human Aβ-(1–40) forms toxic, mature amyloid fibrils, whereas rodent Aβ-(1–40) produces less toxic protofibrils that are not stained by the amyloid-specific dye Congo red. These findings can explain the lack of cerebral Aβ amyloid deposition in rodents.

## 3. ASSEMBLY OF Aβ ON NEURONAL CELLS AND CYTOTOXICITY

Aβ binds to GM1-rich and cholesterol-rich membrane domains on neuronal PC12 cells and forms amyloids. Removal of cholesterol prevents cells from amyloidogenesis and reduces cytotoxicity by Aβ.[43,44] Assembly of Aβ on neuronal cells and cytotoxicity were examined in more detail by antibody labeling and multicolor imaging techniques.[39] Human SH-SY5Y neuroblastoma cells expressing GM1 were incubated with Aβ-(1–40) and Aβ-(1–42). The accumulation of Aβ, the formation of amyloid fibrils, the formation of oligomers, and cell viability were detected using the mouse anti-Aβ monoclonal antibody 6E10, the rabbit anti-Aβ-oligomer antibody A11, the amyloid-staining dye Congo red, and the cell viability marker calcein AM, respectively. Fluorescent dye-labeled secondary antibodies were used to visualize 6E10 and A11. Thus, Aβ-treated cells were stained in four colors and visualized by a confocal laser scanning microscope equipped with a spectral detector. A spectral unmixing procedure was utilized to separate each component in the spectral images. When cells were incubated with 5 μM Aβ-(1–42) at 37°C, the accumulation of Aβ on cell membranes was detected within 0.5 h and increased with incubation time (Fig. 2A–F). Congo red–positive spots appeared later at 24 h and became prominent with time. Some cells were surrounded by amyloids at 48 h; however, they were still alive. Cell death was observed only after substantial amounts of amyloid fibrils had been accumulated. In contrast, no A11-positive spots were detected during the entire period. A11 is known to stain toxic oligomers.[45] These results indicate that the formation of fibrils could lead to cell death, in striking contrast to the Aβ assembly in aqueous solution, in which certain types of oligomers are more toxic than fibrils.[46]

When cells were incubated with 5 μM Aβ-(1–40), neither fibrils nor cell death was detected even after 72 h (Fig. 2G). Incubation with a 10-fold higher concentration of the protein, however, resulted in the appearance of Congo red–positive spots at 48 h and cell death at 72 h (data not shown). These results were in accordance with those found in liposomal study.[39] In summary, Aβ-(1–40) and Aβ-(1–42) behave similarly in terms of cell interaction, although the former needs 10-fold higher concentrations to induce neurotoxicity.

## 4. NMR CHARACTERIZATION OF Aβ AND GANGLIOSIDE MEMBRANES

### 4.1 Conformational Analyses of Gangliosides by NMR Spectroscopy

The structure of GM1 ganglioside has long been studied by using NMR spectroscopy.[47] Recently, a paramagnetism-assisted NMR technique in conjunction with MD simulation has been applied for conformational analysis of a series of ganglioside oligosaccharides including GM1 pentasaccharide in aqueous solution[48–50] (Fig. 3). In this approach, dynamic conformational ensembles derived from MD simulations were validated based on NMR information regarding atomic arrangements with respect to a paramagnetic probe introduced at the reducing terminus of the oligosaccharide. The NMR-validated MD simulations indicate that GM1 pentasaccharide and GM2 tetrasaccharide share similarities in the conformational spaces of their common parts. In contrast, GM3 trisaccharide

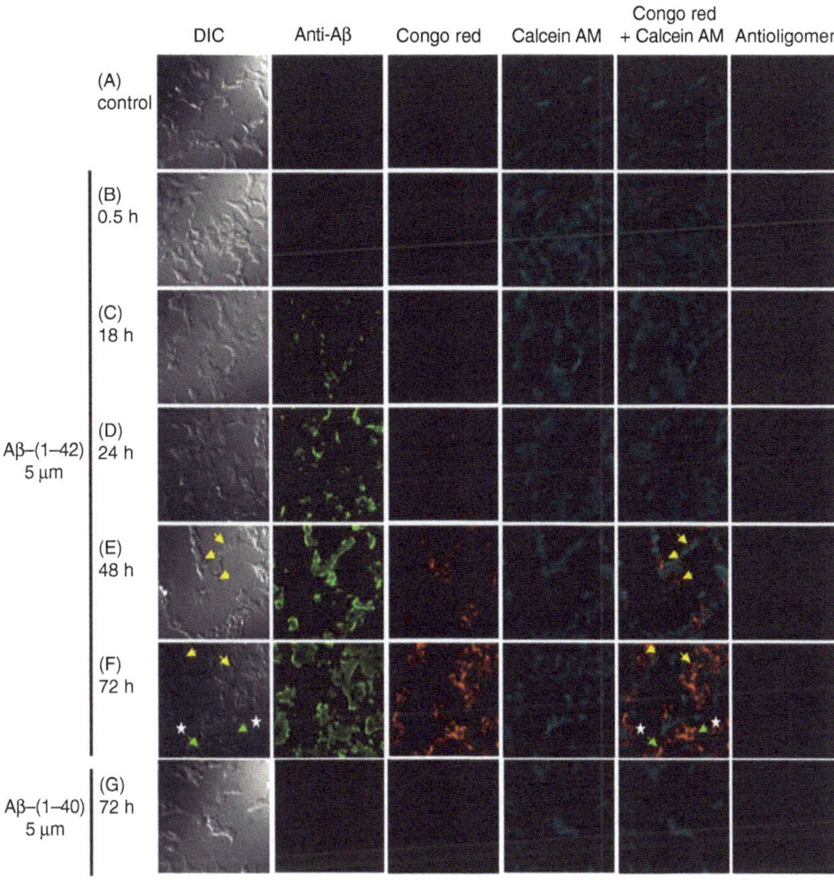


**Fig. 2** Assembly of Aβ-(1–42) and Aβ-(1–40) on cell membranes. SH-SY5Y cells were incubated with 5 μM Aβ-(1–42) for 0.5 (B), 18 (C), 24 (D), 48 (E), or 72 h (F). The accumulation of Aβ-(1–42) on cell membranes, the formation of amyloids, the formation of oligomers, and cell death were simultaneously visualized with 6E10, Congo red, A11, and calcein-AM, respectively. The accumulation of Aβ-(1–42) was detected at 0.5 h, whereas fibrils were only observed after 24 h. Live cells show calcein fluorescence. *Arrows* indicate living cells with fibrils on the cell membrane and *arrows with stars* denote dead cells with fibrils on the cell membrane. No A11-positive spots were detected at any time point. When cells were incubated with 5 μM Aβ-(1–40) for 72 h, amyloid was not detected, although a small amount of Aβ-(1–40) bound to the cell membrane was observed (G). *From Ogawa M, Tsukuda M, Yamaguchi T, Ikeda K, Okada T, Yano Y, Hoshino M, Matsuzaki K. Ganglioside-mediated aggregation of amyloid β-proteins (Aβ): comparison between Aβ-(1-42) and Aβ-(1-40). J Neurochem. 2011;116:851–857.*[39]

(A)

(B)

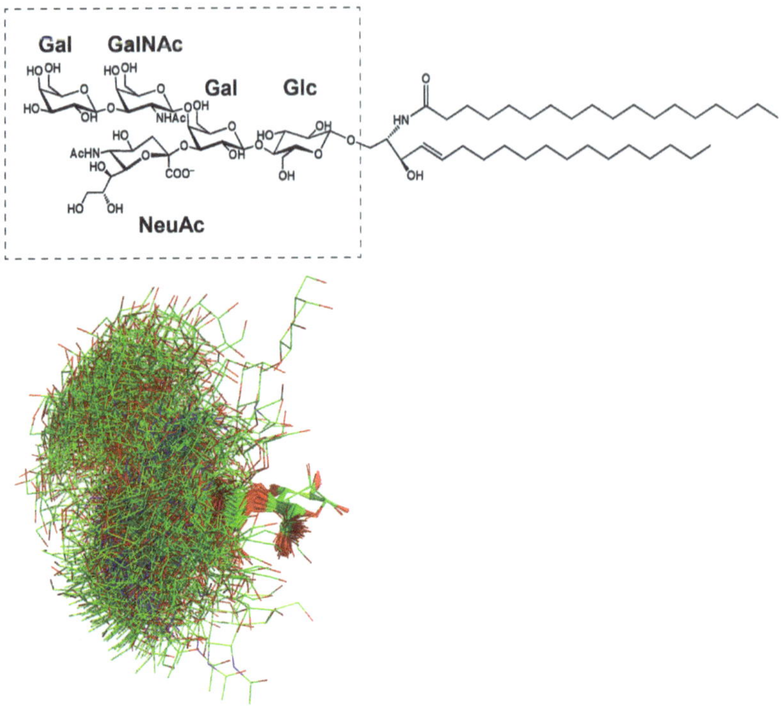


**Fig. 3** (A) Structure of GM1 ganglioside with a *broken-line square* indicating the pentasaccharide moiety. (B) Superimposition of 200 conformers extracted from the NMR-validated results of replica exchange molecular dynamics (MD) simulation of GM1 pentasaccharide. *Parts of this figure were reproduced from Yagi-Utsumi M, Kato K. Structural and dynamic views of GM1 ganglioside. Glycoconj J. 2015;32:105–112.*[47]

exhibits a distinct conformational ensemble in terms of the NeuAc-Gal glycosidic linkage, which is conformationally restricted by the GalNAc residue in the branched GM1 and GM2 glycans. Solution NMR analysis using small bicelles embedding the gangliosides revealed that GM1 and GM2, but not GM3, interact with α-synuclein, another protein that aggregates to cause neuronal dysfunction, through its N-terminal segment, indicating that the branching glycans are a prerequisite for specific interactions with this intrinsically disordered protein.[51]

Crystallographic data suggest that GM1-binding toxins and viruses select specific conformations among those found in the conformational ensemble of GM1 pentasaccharide.[47] The membrane-embedded conformation of GM1 has been characterized by NMR spectroscopy using water-soluble and membrane-anchored paramagnetic probes.[52] The NMR data combined with MD simulations revealed that the outermost Gal and NeuAc residues are structurally flexible and accessible for binding to toxins and viruses.

## 4.2 Conformational States of Aβ on Gangliosidic Clusters

In contrast to the lectin-like stoichiometric GM1-binding of the toxins and viruses, the interactions of ganglioside clusters with amyloidogenic proteins, including Aβ, are characterized as transient, multiple binding processes offering catalytic environments for conformational alteration of the proteins. Although solution NMR methods have been applied to characterize dynamic interactions of Aβ with various proteins including molecular chaperones,[53–55] appropriately designed artificial ganglioside clusters based on micelles and lipid bilayers are necessary for detailed NMR analyses of the enigmatic interactions between Aβ and GM1. An intriguing example is a recently developed self-assembled spherical complex that displays a GM1 pentasaccharide cluster but lacks the hydrophobic lipid moieties.[56] This artificial glycocluster captures Aβ-(1–40) primarily through interacting with its N-terminal segment, giving insights into ganglioside-binding specificity of Aβ during the initial encounter complex formation.

Aβ-(1–40) bound to aqueous micelles consisting of lyso-GM1 or GM1 have been structurally characterized using solution NMR techniques.[15,16,57] The data indicate that Aβ lies on the carbohydrate–lipid interface of these gangliosidic clusters with an up-and-down topology in which the two α-helical segments, His14-Val24 and Ile31-Val36, along with the C-terminal dipeptide, Val39-Val40, are inserted into the hydrophobic interior while leaving the remaining parts unstructured and exposed to the solvent environment. The segments in the hydrophobic milieu eventually form β-hairpin in the fibril. It has been hypothesized that this restricted topology facilitates specific Aβ–Aβ interactions at the hydrophilic/hydrophobic interface in membranous environments. GM1 micelles, which are larger and have lower surface curvature in comparison with lyso-GM1 micelles, are capable of accommodating more Aβ molecules. Under conditions where the Aβ density on the GM1 micelles is high, specific Aβ–Aβ interactions are promoted through their C-terminal segments coupled with the formation of a Thioflavin T–reactive β-structure.[57] Recently reported solid-state NMR data indicate that Aβ-(1–40) bound to 1,2-dimyristoyl-*sn*-glycero-3-phosphocholine multilamellar vesicles exhibits dichotomous conformation with the disordered N-terminal segment followed by the stable C-terminal β-structure[58] (Fig. 4A). These data suggest that Aβ fibrillization in the membrane environment is triggered

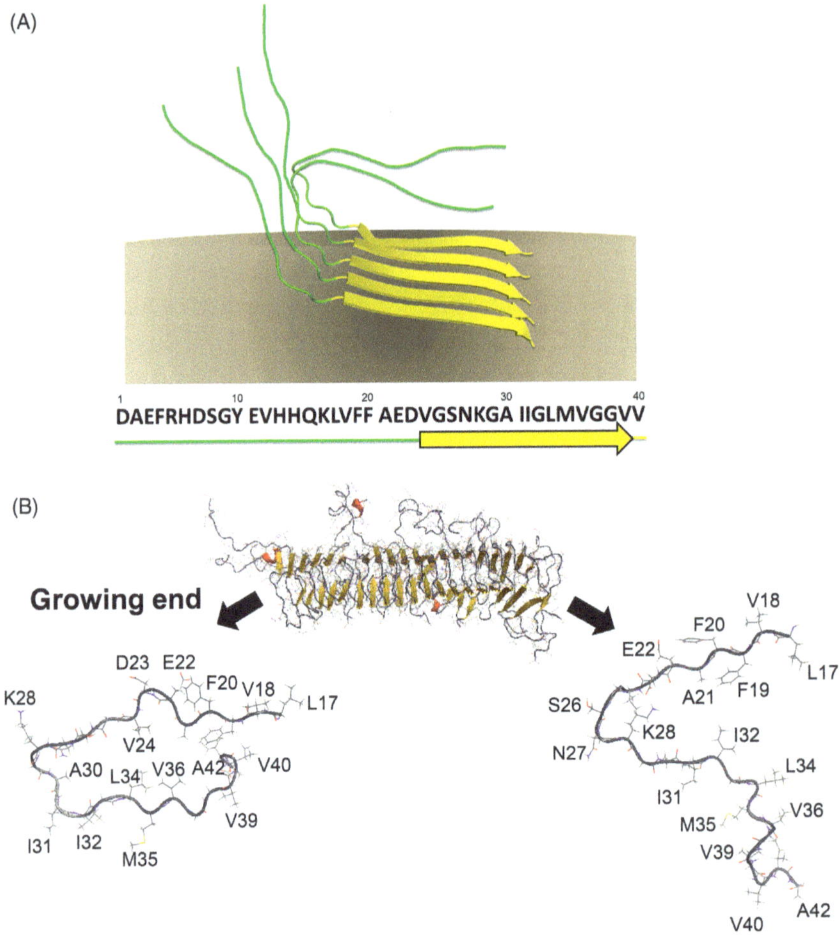


**Fig. 4** (A) Structural model of Aβ-(1–40) bound to DMPC bilayers characterized by solid-state NMR analyses together with amino acid sequence of Aβ-(1–40) indicating location of β-strand *(arrow)*. (B) Final conformation of one all-atom MD simulation of fibril composed of 20 Aβ-(1–42) molecules. Side views of the Aβ-(1–42) monomers at the growing end (left) and the opposing end (right). The N-terminal 16 residues are not shown in the side views. *Part (A) From Yagi-Utsumi M, Kato K, Nishimura K. Membrane-induced dichotomous conformation of amyloid β with the disordered N-terminal segment followed by the stable C-terminal β structure. PLoS One. 2016;11:e0146405*[58]*; Part (B) Parts of this figure were reproduced from Okumura H, Itoh S.G. Structural and fluctuational difference between two ends of Aβ amyloid fibril: MD simulations predict only one end has open conformations. Sci Rep. 2016;6:38422.*[59]

by α-to-β conformational transition initiated from their C-terminal segments assembled into parallel β-structures.

It is possible that hereditary mutations have impacts on these on-membrane molecular events. For example, the Flemish-type mutation (A21G) disrupts the first α-helix at His14-Val24 in Aβ-(1–40) bound to lyso-GM1 micelles, rendering the unstructured N-terminal segment tethered through the residual C-terminal helix.[60] This mutation precludes the initial nucleation process of amyloid formation by Aβ-(1–40), suggesting that the mutational perturbation to the membrane binding properties of Aβ also affects its nucleation behavior during fibrillization.

## 4.3 Integrative Approach for Characterizing Molecular Recognition of Aβ

As the anti-GAβ monoclonal antibody (4396C) cross-reacts with Aβ fibril specifically at the growing end,[5] detailed structural characterization of the putative conformational epitope recognized by this antibody commonly displayed on GAβ and the growing fibril end will provide clues for the design and development of novel drugs targeting AD. Recently, several independent groups have reported NMR-derived high-resolution structural models of Aβ-(1–42) fibrils, in which Aβ-(1–42) molecules adopt an S-shaped conformation forming a salt bridge between Lys28 and the C-terminal carboxylate, in contrast to the U-shaped conformation of Aβ-(1–40) in its fibrils.[61–65] Recently, Qiang *et al.* performed solid-state NMR analysis of fibrils of Aβ-(1–40) and Aβ-(1–42) prepared by seeded growth from extracts derived from brain cortex of AD patients, thereby demonstrating correlation between AD phenotype and structural polymorphism of the fibrils.[66]

Based on the NMR-derived structural model of an Aβ fibril, the dynamical structure of the fibril was explored by all-atom MD simulation.[59] The results revealed that Aβ at the growing end adopts a β-hairpin conformation with less fluctuation as compared with the opposing terminus, which is characterized by dynamic Aβ conformation (Fig. 4B). These results provide a structural basis for the unidirectional extension of Aβ fibril and offer insight into the GAβ structure recognized by mAb 4396C.

An integrative approach is thus necessary for detailed structural characterization of the GAβ motif by combining NMR spectroscopy with other biophysical methods, including recently developed cryoelectron microscopy, high-speed atomic force microscopy, and MD simulation techniques.[67–70]

# 5. PATHOLOGICAL SIGNIFICANCE OF GANGLIOSIDE-MEDIATED Aβ ASSEMBLY

## 5.1 GAβ Generation in Alzheimer's Animal Models

GAβ was discovered in a human brain with early pathological changes of AD[4]; however, GAβ generation in the brain was not experimentally confirmed until mAb 4396C was developed.[5] To date, taking advantage of the conformation-specific immunoreactivity of mAb 4396C, GAβ generation has been corroborated in various Alzheimer's animal models as follows. First, GAβ generation was examined in nonhuman primate brains, which invariably show amyloid deposition at ages over 25.[71] Notably, GAβ immunoreactivity was readily observed as a granular pattern in neurons of the brains of aged monkeys, suggesting that GAβ is generated and accumulated in specific organelles with age.[5] This possibility was further examined in brains obtained from animals of the same species at different ages and it was concluded that GAβ accumulation indeed occurs with age exclusively in endosomes.[72] Second, GAβ generation was examined in transgenic mouse brains, which were designed to express the genes responsible for familial AD and to develop marked amyloid deposition.[73] In this study, in good agreement with the GAβ discovery in the human brains with early-stage but not advanced-stage AD, GAβ was readily detected in the brains of young mice before amyloid deposition became apparent.[73]

## 5.2 GAβ Generation is Responsible for Region-Specific Amyloid Deposition in Human Brains

In addition to the question on how Aβ starts to assemble into amyloid fibrils in the brain, another crucial question is why does amyloid deposit exclusively in a region-specific manner. To answer this question, an experiment has recently been performed focusing on two contrasting regions of the human brain; the precuneus as a vulnerable region and the calcarine cortex as a resistant region to amyloid deposition.[74] The cortices of these two regions were carefully analyzed to determine whether the membrane microenvironment that facilitates GAβ generation is favorably formed in the precuneus. Notably, Aβ assembly into amyloid fibrils on the reconstituted membranes of the lipids extracted from synaptic plasma membranes (SPMs) isolated from the precuneus, but not the calcarine cortex, was markedly enhanced through GAβ generation (Fig. 5).[74] Lipid chemical analysis of the SPMs obtained

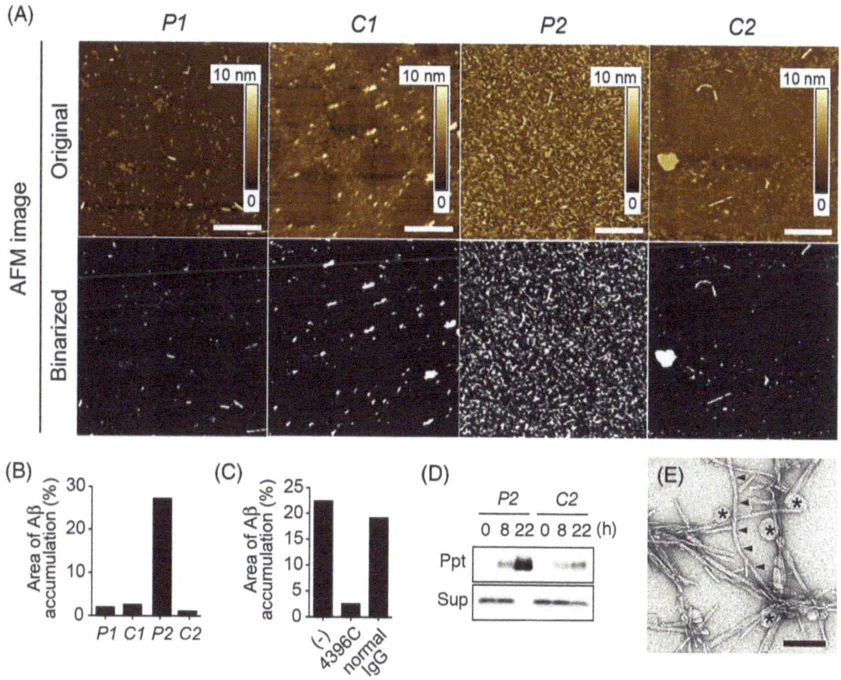


**Fig. 5** Accelerated Aβ assembly in the presence of lipids extracted from the amyloid-bearing precuneus. (A) Representative atomic force microscope images (AFM) (upper, original; lower, binarized) of Aβ mixtures incubated on the reconstituted membranes prepared from lipids extracted from synaptic plasma membranes (SPMs); P1 from the amyloid-free precuneus, C1 from calcarine cortex from the brain providing P1, P2 from the amyloid-bearing precuneus, C2 from calcarine cortex from the brain providing P2. (B–C) Area of Aβ assembly on the membrane. (D) Western blots of insoluble (Ppt) and soluble (Sup) Aβ mixtures following incubation. (E) An electron microscope image of the mixture incubated with liposomes *(asterisks)* prepared from P2. Typical amyloid fibrils were observed *(arrowheads)*. Scale: Part (A) 500 nm; Part (E) 100 nm. *From Oikawa N, Matsubara T, Fukuda R. Imbalance in fatty-acid-chain length of gangliosides triggers Alzheimer amyloid deposition in the precuneus. PLoS One. 2015;10 (3):e0121356.*

from the precuneus revealed that enhanced amyloid fibril formation was attributable to an increase in the ratio of the level of GD1b-ganglioside with long (C20:0) fatty acid chains to that with short (C18:0) chains.[74] It remains to be elucidated how the expression of certain ganglioside species is altered in particular brain regions such as the precuneus; however, this study suggests that the local glycolipid microenvironment can be responsible for the region-specific amyloid deposition in AD.

## 5.3 GAβ Generation in the Presence of Alzheimer's Disease Risk Factors

### *5.3.1 APOE Genotype*

The *APOE* genotype is a strong genetic risk factor for AD development; the epsilon 4 allele of *APOE* increases whereas the epsilon 2 allele decreases the risk of the disease.[75] Given that the major physiological function of apolipoprotein E (ApoE), which is encoded by *APOEgene*, is cholesterol transport in the brain, and that the cholesterol content in ganglioside-containing lipid bilayers is a critical factor for GAβ generation,[6] the cholesterol content in SPMs was previously examined in *APOE*–knock-in mouse brains[76] and also recently in human brains.[77] These studies suggest that the cholesterol content in SPMs changes in an *APOE* genotype–dependent manner, which is closely related to the formation of a specific membrane microenvironment that is favorable or inhibitory for GAβ generation.

### *5.3.2 Endosomal–Lysosomal Dysfunction*

Cataldo *et al.* previously found that Rab5-positive abnormally enlarged early endosomes appeared in neurons before amyloid deposition in AD brains, suggesting that the endocytic pathway is impaired in a very early stage of AD.[78] This possibility was examined using cultured cells, which corroborated the previous finding that perturbation of the endocytic pathway induces enlargement of early endosomes with ganglioside accumulation, leading to amyloid deposition on the cell surface through GAβ generation.[79,80] Furthermore, mouse models of lysosomal dysfunction were examined, and intraneuronal GAβ accumulation was clearly observed.[81] This line of evidence suggests that endosomal–lysosomal dysfunction, which likely induces abnormal accumulation of ganglioside and GAβ generation in endosomes, is one of the causative factors for amyloid deposition in AD brains.

### *5.3.3 Diabetes Mellitus*

Evidence is accumulating to indicate that diabetes mellitus (DM) is a risk factor for AD development, and various possibilities have been raised to explain how DM accelerates AD development. Recently, brains of a nonhuman primate with type 2 DM were examined to clarify how DM facilitates the amyloid pathology.[82] Notably, amyloid deposition was readily observed in animals with DM earlier than those without DM, and the intraneuronal accumulation of GAβ and Rab5 was also observed.[82] Although clarification of the exact mechanism awaits further studies, these findings suggest that DM also perturbs the endocytic pathway and in turn facilitates amyloid deposition through GAβ generation.

### 5.4 GAβ as a Possible Target of Clinical Intervention for Alzheimer's Disease

GAβ-induced initiation of Aβ assembly in the brain may justify the targeting of GAβ to suppress amyloid deposition. This strategy was previously examined by intraperitoneal administration of mAb 4396C into transgenic AD mouse models, and as expected, the amyloid deposition in the brain was suppressed significantly.[83] Given that the Aβs at the end of growing fibrils likely share the same structure as GAβ, as shown by immunoelectron microscopy using mAb 4396C,[5] GAβ is a promising target in developing safe and efficient measures for suppression of amyloid deposition.

## 6. CONCLUSIONS AND PERSPECTIVES

Aβ likely starts to assemble into amyloid fibrils in the brain not spontaneously but in the presence of a favorable lipid microenvironment containing ganglioside through GAβ generation. Concerning GAβ generation, several issues remain to be elucidated as follows. First, the lipid chemical features of a membrane microenvironment that induce GAβ formation need to be clarified. It should be studied how lipids localized in the microenvironment fine-tune ganglioside conformation as previously shown with cholesterol.[84] Second, it should be clarified how various risk factors for AD development, including aging, expression of the *APOE* allele epsilon 4, and DM, are uniformly related to the formation of the lipid microenvironment. Third, it is challenging but crucial to investigate whether genetic causes of AD, including presenilin mutations, can be a driving force to induce the membrane microenvironment. Finally, it is intriguing to investigate whether the assembly of a given amyloidogenic protein other than Aβ into amyloid fibrils is also facilitated in the presence of gangliosides, as previously suggested.[85–88] As described in this chapter, a series of studies on ganglioside-mediated Aβ assembly may provide new insights into the pathogenesis of various human amyloidosis including AD.

## ACKNOWLEDGMENTS

We thank Dr. Maho Yagi-Utsumi and Hisashi Okumura (Institute for Molecular Science) for useful discussion. This work was supported in part by JSPS/MEXT KAKENHI Grant in-Aid for Scientific Research on Innovative Areas (JP25102001 and JP25102008 to K.K.), the Nanotechnology Platform Program of MEXT, and the Okazaki ORION project.

## REFERENCES


1. Selkoe DJ. Alzheimer's disease: genotypes, phenotypes, and treatments. *Science.* 1997;275:630–631.
2. Levy E, Carman MD, Fernandez-Madrid IJ, et al. Mutation of the Alzheimer's disease amyloid gene in hereditary cerebral hemorrhage, Dutch type. *Science.* 1990;248: 1124–1126.
3. Nilsberth C, Westlind-Danielsson A, Eckman CB, et al. The 'Arctic' APP mutation (E693G) causes Alzheimer's disease by enhanced Aβ protofibril formation. *Nat Neurosci.* 2001;4:887–893.
4. Yanagisawa K, Odaka A, Suzuki N, Ihara Y. GM1 ganglioside-bound amyloid β-protein (Aβ): a possible form of preamyloid in Alzheimer's disease. *Nat Med.* 1995;1: 1062–1066.
5. Hayashi H, Kimura N, Yamaguchi H, et al. A seed for Alzheimer amyloid in the brain. *J Neurosci.* 2004;24:4894–4902.
6. Matsuzaki K, Horikiri C. Interactions of amyloid β-peptide (1-40) with ganglioside-containing membranes. *Biochemistry.* 1999;38:4137–4142.
7. Kakio A, Nishimoto SI, Yanagisawa K, Kozutsumi Y, Matsuzaki K. Cholesterol-dependent formation of GM1 ganglioside-bound amyloid β-protein, an endogenous seed for Alzheimer amyloid. *J Biol Chem.* 2001;276:24985–24990.
8. Kakio A, Nishimoto S, Yanagisawa K, Kozutsumi Y, Matsuzaki K. Interactions of amyloid β-protein with various gangliosides in raft-like membranes: importance of GM1 ganglioside-bound form as an endogenous seed for Alzheimer amyloid. *Biochemistry.* 2002;41:7385–7390.
9. McLaurin J, Chakrabartty A. Membrane disruption by Alzheimer β-amyloid peptides mediated through specific binding to either phospholipids or gangliosides. Implications for neurotoxicity. *J Biol Chem.* 1996;271:26482–26489.
10. Choo-Smith LP, Surewicz WK. The interaction between Alzheimer amyloid β (1-40) peptide and ganglioside GM1-containing membranes. *FEBS Lett.* 1997;402:95–98.
11. Choo-Smith LP, Garzon-Rodriguez W, Glabe CG, Surewicz WK. Acceleration of amyloid fibril formation by specific binding of Aβ-(1-40) peptide to ganglioside-containing membrane vesicles. *J Biol Chem.* 1997;272:22987–22990.
12. McLaurin J, Franklin T, Fraser PE, Chakrabartty A. Structural transitions associated with the interaction of Alzheimer beta-amyloid peptides with gangliosides. *J Biol Chem.* 1998;273:4506–4515.
13. Williamson MP, Suzuki Y, Bourne NT, Asakura T. Binding of amyloid β-peptide to ganglioside micelles is dependent on histidine-13. *Biochem J.* 2006;397:483–490.
14. Kato K, Sasakawa H, Kamiya Y, et al. 920 MHz ultra-high field NMR approaches to structural glycobiology. *Biochim Biophys Acta.* 2008;1780:619–625.
15. Yagi-Utsumi M, Kameda T, Yamaguchi Y, Kato K. NMR characterization of the interactions between lyso-GM1 aqueous micelles and amyloid β. *FEBS Lett.* 2010;584: 831–836.
16. Utsumi M, Yamaguchi Y, Sasakawa H, Yamamoto N, Yanagisawa K, Kato K. Up-and-down topological mode of amyloid β-peptide lying on hydrophilic/hydrophobic interface of ganglioside clusters. *Glycoconj J.* 2009;26:999–1006.
17. Yamamoto N, Igbabvoa U, Shimada Y, et al. Accelerated Aβ aggregation in the presence of GM1-ganglioside-accumulated synaptosomes of aged apoE4-knock-in mouse brain. *FEBS Lett.* 2004;569:135–139.
18. Yamamoto N, Hirabayashi Y, Amari M, et al. Assembly of hereditary amyloid β-protein variants in the presence of favorable gangliosides. *FEBS Lett.* 2005;579:2185–2190.
19. Yuyama K, Yamamoto N, Yanagisawa K. Chloroquine-induced endocytic pathway abnormalities: Cellular model of GM1 ganglioside-induced Aβ fibrillogenesis in Alzheimer's disease. *FEBS Lett.* 2006;580:6972–6976.

20. Yamamoto N, Van Nostrand WE, Yanagisawa K. Further evidence of local ganglioside-dependent amyloid β-protein assembly in brain. *Neuroreport*. 2006;17:1735–1737.
21. Yamamoto N, Fukata Y, Fukata M, Yanagisawa K. GM1-ganglioside-induced Aβ assembly on synaptic membranes of cultured neurons. *Biochim Biophys Acta*. 2007;1768: 1128–1137.
22. Yamamoto N, Matsubara T, Sato T, Yanagisawa K. Age-dependent high-density clustering of GM1 ganglioside at presynaptic neuritic terminals promotes amyloid β-protein fibrillogenesis. *Biochim Biophys Acta*. 2008;1778:2717–2726.
23. Yuyama K, Yamamoto N, Yanagisawa K. Accelerated release of exosome-associated GM1 ganglioside (GM1) by endocytic pathway abnormality: another putative pathway for GM1-induced amyloid fibril formation. *J Neurochem*. 2008;105:217–224.
24. Yuyama K, Yanagisawa K. Late endocytic dysfunction as a putative cause of amyloid fibril formation in Alzheimer's disease. *J Neurochem*. 2009;109:1250–1260.
25. Oikawa N, Yamaguchi H, Ogino K, et al. Gangliosides determine the amyloid pathology of Alzheimer's disease. *Neuroreport*. 2009;20:1043–1046.
26. Okada T, Ikeda K, Wakabayashi M, Ogawa M, Matsuzaki K. Formation of toxic Aβ(1-40) fibrils on GM1 ganglioside-containing membranes mimicking lipid rafts: polymorphisms in Aβ(1-40) fibrils. *J Mol Biol*. 2008;382:1066–1074.
27. Matsubara T, Iijima K, Yamamoto N, Yanagisawa K, Sato T. Density of GM1 in nanoclusters is a critical factor in the formation of a spherical assembly of amyloid β-protein on synaptic plasma membranes. *Langmuir*. 2013;29:2258–2264.
28. Matsuzaki K, Kato K, Yanagisawa K. Aβ polymerization through interaction with membrane gangliosides. *Biochim Biophys Acta*. 2010;1801:868–877.
29. Ikeda K, Yamaguchi T, Fukunaga S, Hoshino M, Matsuzaki K. Mechanism of amyloid β-protein aggregation mediated by GM1 ganglioside clusters. *Biochemistry*. 2011;50: 6433–6440.
30. Yamaguchi T, Matsuzaki K, Hoshino M. Transient formation of intermediate conformational states of amyloid β-peptide revealed by heteronuclear magnetic resonance spectroscopy. *FEBS Lett*. 2011;585:1097–1102.
31. Mori K, Mahmood MI, Neys S, Matsuzaki K, Hoshino T. Formation of GM1 ganglioside clusters on the lipid membrane containing sphingomyelin and cholesterol. *J Phys Chem B*. 2012;116:5111–5121.
32. Fukunaga S, Ueno H, Yamaguchi T, Yano Y, Hoshino M, Matsuzaki K. GM1 cluster mediates formation of toxic Aβ fibrils by providing hydrophobic environments. *Biochemistry*. 2012;51:8121–8131.
33. Petkova AT, Ishii Y, Balbach JJ, et al. A structural model for Alzheimer's β-amyloid fibrils based on experimental constraints from solid state NMR. *Proc Natl Acad Sci USA*. 2002;99:16742–16747.
34. Stroud JC, Liu C, Teng PK, Eisenberg D. Toxic fibrillar oligomers of amyloid-β have cross-β structure. *Proc Natl Acad Sci USA*. 2012;109:7717–7722.
35. Berthelot K, Ta HP, Géan J, Lecomte S, Cullin C. In vivo and in vitro analyses of toxic mutants of HET-s: FTIR antiparallel signature correlates with amyloid toxicity. *J Mol Biol*. 2011;412:137–152.
36. Genji M, Yano Y, Hoshino M, Matsuzaki K. Aromaticity of phenylalanine residues is essential for amyloid formation by Alzheimer's amyloid β-peptide. *Chem Pharm Bull*. 2017;65:668–673.
37. Hoshino T, Mahmood MI, Mori K, Matsuzaki K. Binding and aggregation mechanism of amyloid β-peptides onto the GM1 ganglioside-containing lipid membrane. *J Phys Chem B*. 2013;117:8085–8094.
38. Karran E, Mercken M, De Strooper B. The amyloid cascade hypothesis for Alzheimer's disease: an appraisal for the development of therapeutics. *Nat Rev Drug Disc*. 2011;10: 698–712.

39. Ogawa M, Tsukuda M, Yamaguchi T, et al. Ganglioside-mediated aggregation of amyloid β-proteins (Aβ): comparison between Aβ-(1-42) and Aβ-(1-40). *J Neurochem.* 2011;116:851–857.
40. Thomaier M, Gremer L, Dammers C, Fabig J, Neudecker P, Willbold D. High-affinity binding of monomeric but not oligomeric amyloid-β to ganglioside GM1 containing nanodiscs. *Biochemistry.* 2016;55:6662–6672.
41. Selkoe DJ. Biochemistry of altered brain proteins in Alzheimer's disease. *Annu Rev Neurosci.* 1989;12:463–490.
42. Ueno H, Yamaguchi T, Fukunaga S, et al. Comparison between the aggregation of human and rodent amyloid β-proteins in GM1 ganglioside clusters. *Biochemistry.* 2014;53:7523–7530.
43. Wakabayashi M, Okada T, Kozutsumi Y, Matsuzaki K. GM1 ganglioside-mediated accumulation of amyloid β-protein on cell membranes. *Biochem Biophys Res Commun.* 2005;328:1019–1023.
44. Wakabayashi M, Matsuzaki K. Formation of amyloids by Aβ-(1-42) on NGF-differentiated PC12 cells: roles of gangliosides and cholesterol. *J Mol Biol.* 2007;371: 924–933.
45. Liu P, Reed MN, Kotilinek LA, et al. Quaternary structure defines a large class of amyloid-β oligomers neutralized by sequestration. *Cell Rep.* 2015;11:1760–1771.
46. Lee SJ, Nam E, Lee HJ, Savelieff MG, Lim MH. Towards an understanding of amyloid-β oligomers: characterization, toxicity mechanisms, and inhibitors. *Chem Soc Rev.* 2017;46:310–323.
47. Yagi-Utsumi M, Kato K. Structural and dynamic views of GM1 ganglioside. *Glycoconj J.* 2015;32:105–112.
48. Yamamoto S, Zhang Y, Yamaguchi T, Kameda T, Kato K. Lanthanide-assisted NMR evaluation of a dynamic ensemble of oligosaccharide conformations. *Chem Commun.* 2012;48:4752–4754.
49. Zhang Y, Yamamoto S, Yamaguchi T, Kato K. Application of paramagnetic NMR-validated molecular dynamics simulation to the analysis of a conformational ensemble of a branched oligosaccharide. *Molecules.* 2012;17:6658–6671.
50. Zhang Y, Yamaguchi T, Satoh T, et al. Conformational dynamics of oligosaccharides characterized by paramagnetism-assisted NMR spectroscopy in conjunction with molecular dynamics simulation. *Adv Exp Med Biol.* 2015;842:217–230.
51. Yamaguchi T, Uno T, Uekusa Y, Yagi-Utsumi M, Kato K. Ganglioside-embedding small bicelles for probing membrane-landing processes of intrinsically disordered proteins. *Chem Commun.* 2013;49:1235–1237.
52. Demarco ML, Woods RJ, Prestegard JH, Tian F. Presentation of membrane-anchored glycosphingolipids determined from molecular dynamics simulations and NMR paramagnetic relaxation rate enhancement. *J Am Chem Soc.* 2010;132:1334–1338.
53. Yagi-Utsumi M, Kunihara T, Nakamura T, et al. NMR characterization of the interaction of GroEL with amyloid β as a model ligand. *FEBS Lett.* 2013;587:1605–1609.
54. Libich DS, Fawzi NL, Ying J, Clore GM. Probing the transient dark state of substrate binding to GroEL by relaxation-based solution NMR. *Proc Natl Acad Sci USA.* 2013;110:11361–11366.
55. Kitago Y, Nagae M, Nakata Z, et al. Structural basis for amyloidogenic peptide recognition by sorLA. *Nat Struct Mol Biol.* 2015;22:199–206.
56. Sato S, Yoshimasa Y, Fujita D, et al. A self-assembled spherical complex displaying a gangliosidic glycan cluster capable of interacting with amyloidogenic proteins. *Angew Chem Int Ed Engl.* 2015;54:8435–8439.
57. Yagi-Utsumi M, Matsuo K, Yanagisawa K, Gekko K, Kato K. Spectroscopic characterization of intermolecular interaction of amyloid β promoted on GM1 micelles. *Int J Alzheimers Dis.* 2010;2011:925073.

58. Yagi-Utsumi M, Kato K, Nishimura K. Membrane-induced dichotomous conformation of amyloid β with the disordered N-terminal segment followed by the stable C-terminal β structure. *PLoS One.* 2016;11:e0146405.
59. Okumura H, Itoh SG. Structural and fluctuational difference between two ends of Aβ amyloid fibril: MD simulations predict only one end has open conformations. *Sci Rep.* 2016;6:38422.
60. Yagi-Utsumi M, Dobson CM. Conformational effects of the A21G Flemish mutation on the aggregation of amyloid β peptide. *Biol Pharm Bull.* 2015;38:1668–1672.
61. Colvin MT, Silvers R, Ni QZ, et al. Atomic resolution structure of monomorphic Aβ42 amyloid fibrils. *J Am Chem Soc.* 2016;138:9663–9674.
62. Walti MA, Ravotti F, Arai H, et al. Atomic-resolution structure of a disease-relevant Aβ(1-42) amyloid fibril. *Proc Natl Acad Sci USA.* 2016;113:E4976–E4984.
63. Xiao Y, Ma B, McElheny D, et al. Aβ(1-42) fibril structure illuminates self-recognition and replication of amyloid in Alzheimer's disease. *Nat Struct Mol Biol.* 2015;22:499–505.
64. Bertini I, Gonnelli L, Luchinat C, Mao J, Nesi A. A new structural model of Ab40 fibrils. *J Am Chem Soc.* 2011;133:16013–16022.
65. Tycko R. Physical and structural basis for polymorphism in amyloid fibrils. *Protein Sci.* 2014;23:1528–1539.
66. Qiang W, Yau WM, Lu JX, Collinge J, Tycko R. Structural variation in amyloid-β fibrils from Alzheimer's disease clinical subtypes. *Nature.* 2017;541:217–221.
67. Schmidt M, Sachse C, Richter W, Xu C, Fandrich M, Grigorieff N. Comparison of Alzheimer Aβ(1-40) and Aβ(1-42) amyloid fibrils reveals similar protofilament structures. *Proc Natl Acad Sci USA.* 2009;106:19813–19818.
68. Schmidt M, Rohou A, Lasker K, et al. Peptide dimer structure in an Aβ(1-42) fibril visualized with cryo-EM. *Proc Natl Acad Sci USA.* 2015;112:11858–11863.
69. Okumura H, Itoh SG. Amyloid fibril disruption by ultrasonic cavitation: nonequilibrium molecular dynamics simulations. *J Am Chem Soc.* 2014;136:10549–10552.
70. Watanabe-Nakayama T, Ono K, Itami M, Takahashi R, Teplow DB, Yamada M. High-speed atomic force microscopy reveals structural dynamics of amyloid β1-42 aggregates. *Proc Natl Acad Sci USA.* 2016;113:5835–5840.
71. Nakamura S, Nakayama H, Goto N, Ono F, Sakakibara I, Yoshikawa Y. Histopathological studies of senile plaques and cerebral amyloidosis in cynomolgus monkeys. *J Med Primatol.* 1998;27:244–252.
72. Kimura N, Yanagisawa K. Endosomal accumulation of GM1 ganglioside-bound amyloid β-protein in neurons of aged monkey brains. *Neuroreport.* 2007;18:1669–1673.
73. Hong S, Ostaszewski BL, Yang T, et al. Soluble Aβ oligomers are rapidly sequestered from brain ISF in vivo and bind GM1 ganglioside on cellular membranes. *Neuron.* 2014;82:308–319.
74. Oikawa N, Matsubara T, Fukuda R, et al. Imbalance in fatty-acid-chain length of gangliosides triggers Alzheimer amyloid deposition in the precuneus. *PLoS One.* 2015;10:e0121356.
75. Holtzman DM, Herz J, Bu G. Apolipoprotein E and apolipoprotein E receptors: normal biology and roles in Alzheimer disease. *Cold Spring Harb Perspect Med.* 2012;2:a006312.
76. Hayashi H, Igbavboa U, Hamanaka H, et al. Cholesterol is increased in the exofacial leaflet of synaptic plasma membranes of human apolipoprotein E4 knock-in mice. *Neuroreport.* 2002;13:383–386.
77. Oikawa N, Hatsuta H, Murayama S, Suzuki A, Yanagisawa K. Influence of APOE genotype and the presence of Alzheimer's pathology on synaptic membrane lipids of human brains. *J Neurosci Res.* 2014;92:641–650.
78. Cataldo AM, Peterhoff CM, Troncoso JC, Gomez-Isla T, Hyman BT, Nixon RA. Endocytic pathway abnormalities precede amyloid beta deposition in sporadic

Alzheimer's disease and Down syndrome: differential effects of APOE genotype and presenilin mutations. *Am J Pathol.* 2000;157:277–286.

79. Yuyama K, Yamamoto N, Yanagisawa K. Chloroquine-induced endocytic pathway abnormalities: Cellular model of GM1 ganglioside-induced Aβ fibrillogenesis in Alzheimer's disease. *FEBS Lett.* 2006;580:6972–6976.
80. Yuyama K, Yanagisawa K. Late endocytic dysfunction as a putative cause of amyloid fibril formation in Alzheimer's disease. *J Neurochem.* 2009;109:1250–1260.
81. Keilani S, Lun Y, Stevens AC, et al. Lysosomal dysfunction in a mouse model of Sandhoff disease leads to accumulation of ganglioside-bound amyloid-β peptide. *J Neurosci.* 2012;32:5223–5236.
82. Okabayashi S, Shimozawa N, Yasutomi Y, Yanagisawa K, Kimura N. Diabetes mellitus accelerates Aβ pathology in brain accompanied by enhanced GAβ generation in nonhuman primates. *PLoS One.* 2015;10:e0117362.
83. Yamamoto N, Yokoseki T, Shibata M, Yamaguchi H, Yanagisawa K. Suppression of Aβ deposition in brain by peripheral administration of Fab fragments of anti-seed antibody. *Biochem Biophys Res Commun.* 2005;335:45–47.
84. Yahi N, Aulas A, Fantini J. How cholesterol constrains glycolipid conformation for optimal recognition of Alzheimer's β amyloid peptide (Aβ1-40). *PLoS One.* 2010;5: e9079.
85. Gellermann GP, Appel TR, Tannert A, et al. Raft lipids as common components of human extracellular amyloid fibrils. *Proc Natl Acad Sci USA.* 2005;102:6297–6302.
86. Wang SS, Good TA, Rymer DL. The influence of phospholipid membranes on bovine calcitonin secondary structure and amyloid formation. *Protein Sci.* 2005;14:1419–1428.
87. Miura T, Yoda M, Takaku N, Hirose T, Takeuchi H. Clustered negative charges on the lipid membrane surface induce β-sheet formation of prion protein fragment 106-126. *Biochemistry.* 2007;46:11589–11597.
88. Wakabayashi M, Matsuzaki K. Ganglioside-induced amyloid formation by human islet amyloid polypeptide in lipid rafts. *FEBS Lett.* 2009;583:2854–2858.

CHAPTER FIFTEEN

# Gangliosides, α-Synuclein, and Parkinson's Disease

**Robert W. Ledeen[1], Gusheng Wu**
Department of Pharmacology, Physiology & Neuroscience, Rutgers New Jersey Medical School, Newark, NJ, United States
[1] Corresponding author. E-mail address: ledeenro@njms.rutgers.edu

## Contents



## Abstract

This review addresses the role of α-synuclein (αSyn) in the etiopathology of Parkinson's disease (PD), with emphasis on its interaction with GM1 ganglioside. We begin with a brief review of some of the milestone discoveries that helped to elucidate PD neuropathology, including the fibrous inclusions of Lewy that characterize the degenerating dopaminergic neurons of the substantia nigra and the presence of αSyn as a major constituent of these Lewy bodies and neurites. This enabled Braak *et al.* to define the progressive nature of PD in developing their staging hypothesis which described the topographically predictable sequence of neuropathological changes giving rise to prodromal nonmotor symptoms that precede the classical motor dysfunctions. We recount recent studies demonstrating strong, specific binding of αSyn to GM1 that serves to inhibit fibril formation and the key role of *N*-acetylation of αSyn in enhancing GM1 binding and specificity. The consequences of insufficient GM1 are illustrated in a newly presented mouse model of PD based on partial deletion of this ganglioside due to heterologous disruption of *B4galnt1* (GM2/GD2 synthase), such mice presenting accurate recapitulation of the PD phenotype. A key feature of these mice was marked elevation of αSyn aggregates which accompanied motor impairment, both aggregates and motor dysfunction being corrected by GM1 replacement therapy. Such therapy was achieved with high dosage of GM1 and more effectively with lower doses

*Progress in Molecular Biology and Translational Science*, Volume 156
ISSN 1877-1173
http://dx.doi.org/10.1016/bs.pmbts.2017.12.009

of LIGA20, a membrane permeable analog of GM1. The accuracy of this mouse model was emphasized by the finding that various central nervous system and noncentral nervous system tissues from PD patients manifested similar GM1 deficiency as the *B4galnt1*$^{+/-}$ mouse. A mechanism is proposed whereby the GM1 deficiency detected in PD patients gives rise to αSyn aggregation and facilitation by the latter in blocking glial cell-derived neurotrophic factor neuroprotection.

## 1. INTRODUCTION

This year (2017) marks the 200th anniversary of James Parkinson's landmark essay on "the shaking palsy" describing a condition subsequently named after him and initially viewed as a prototypical movement disorder. That perception has gradually evolved to the current recognition of Parkinson's disease (PD) as a multisystem disorder with numerous nonmotor features involving several portions of the central, peripheral, autonomic, and enteric nervous systems.[1,2] Moreover, it is progressive in development with earliest symptoms (e.g., gastrointestinal, cardiovascular, olfactory) often appearing a decade or more before overt motor dysfunction. A small fraction (5%–10%) of PD cases are familial or genetic in origin, whereas the large majority, termed sporadic or idiopathic, are known only to have age as the major risk factor. Genetic and/or environmental factors are speculated to have a role, but as yet no cogent theory has emerged that reconciles the disparate data into a compelling narrative concerning the etiology of sporadic PD. Accumulating evidence is pointing to such a role for gangliosides, which have been seriously considered as possible therapeutic agents but only recently as initiator of PD pathophysiology. This pertains especially to GM1 which, along with GD1a, its catabolic precursor, exhibits progressive decline with advancing age leading eventually to neuropathological dysfunction in several key areas. Two such functions have commanded special attention in regard to PD: α-synuclein (αSyn) aggregation into Lewy inclusions and neuroprotective signaling via neurotrophic factors (NTFs). αSyn now occupies center stage in current thinking on PD pathophysiology. It has been referred to as the "Rosetta stone" of PD and related synucleinopathies[3] owing largely to its prominence as a misfolded, aggregated molecule within the Lewy bodies and Lewy neurites that define the affected neurons. Neuroprotection via NTFs, which is essential for maintaining the viability of mature neurons,[4] is also gaining recognition as a GM1-dependent phenomenon based on mandatory association of this ganglioside with NTF receptors. These and related ganglioside phenomena will be reviewed here in relation to PD pathophysiology along with encouraging findings, suggesting GM1 therapeutic efficacy.

## 2. MILESTONES IN ELUCIDATION OF PARKINSON'S DISEASE NEUROPATHOLOGY

To understand the putative roles of gangliosides in PD etiology and therapy, it is useful to review the general evolution of the PD field, especially in regard to αSyn considering its intimate association with GM1 in normal brain. Approximately 100 years elapsed following Parkinson's initial essay before it was discovered that severe loss of pigmented neurons in the substantia nigra pars compacta (SNpc) together with the presence of Lewy bodies are the most consistent pathological lesions.[5] These fibrous inclusions were first described by Lewy in the dorsal motor nucleus of the vagus nerve, the nucleus basalis of Meynert, and some thalamic nuclei of PD patients.[6] A major breakthrough in understanding their origin was the discovery by Spillantini *et al.*[7] that αSyn, a ubiquitous neuronal protein, is a major component of these aggregated bodies. Following the discovery that dopamine (DA) is a neurotransmitter[8] and that 80% of the brain content occurs in the striatum,[9] Carlsson[10] proposed that DA deficiency in that region might account for the motor dysfunction in PD. Initial attempts to replace the missing DA were unsuccessful due to the inability of catecholamines to cross the blood–brain barrier, but the DA precursor, 3,4-dihydroxyphenylalanine (DOPA), was found to enter the brain readily and to be decarboxylated in situ to DA by aromatic amino acid decarboxylase.[11] L-DOPA, the active enantiomer, was approved by the US FDA for the treatment of PD and remains the "gold standard" for PD therapy in its initial stages. Current L-DOPA preparations include inhibitors of aromatic amino acid decarboxylase to prevent peripheral metabolism of DA. In some patients with advanced PD where chronic exposure to L-DOPA leads to motor complications, alternative treatments such as deep brain stimulation are being successfully applied,[12] although the mechanism of such treatment is not well understood. The progressive nature of PD was well delineated by Braak *et al.*,[13,14] who utilized the sequential appearance of Lewy bodies and Lewy neurites in specific areas of the CNS and PNS to develop their staging hypothesis. This demonstrated a topographically predictable sequence in six stages involving prodromal nonmotor symptoms of PD which can include lesions in the olfactory, autonomic, limbic, and sensorimotor systems; these can and often do antedate classical motor symptoms by many years.[15,16] Later stages involving lesions in the neocortex with consequent cognitive dysfunction were also defined by Braak staging.

## 3. α-SYNUCLEIN AND PARKINSON'S DISEASE

αSyn occurs in conjunction with closely related β- and γ-isoforms, but is the only form implicated in PD. It is a cytosolic protein that is abundant in neurons as well as certain other cell types such as erythrocytes and endothelial cells, though its function in the latter types is uncertain. In neurons it is present in the nucleus and axon terminals, the latter being its site of highest concentration where it influences the distribution of SNARE proteins and the clustering of synaptic vesicles.[17] It is a relatively small protein of 140 amino acids which, in its pathological manifestations, initially forms oligomers followed by small protofibrils and eventually large, insoluble fibrils that comprise the Lewy aggregates. It has a propensity to bind to membrane lipids which tends to promote its α-helical conformation, but αSyn can also exist in cytosol unbound to lipids in an α-helical oligomeric state that is more resistant to aggregation than the unfolded monomer.[18–20]This concept of a helically folded tetramer, although supported by in vivo cross-linking studies,[20] is somewhat controversial in view of the counter-proposal that αSyn exists predominantly as disordered monomer (see later). An important advance came with the revelation that all detectable αSyn in vivo contains an acetyl group attached to the α-amino group of the first N-terminal amino acid[18,21,22] which has been implicated in stabilizing helical conformations of αSyn[23] including folded, oligomeric αSyn.[24] This posttranslational modification seems important for lipid interaction[25,26] and conceivably it is failure of such interaction that could lead to pathological αSyn aggregation. Preventing such misfolding has been postulated as an important function of GM1 (see later). In relation to αSyn function, mice lacking αSyn showed altered DA release with reduction in striatal DA and an attenuation of DA-dependent locomotor response to amphetamine, suggesting αSyn behaves as a presynaptic, activity-dependent regulator of DA neurotransmission.[27] On the contrary, overexpression of αSyn resulted in reduced DA release.[28] It was thus proposed that αSyn attenuates neurotransmitter release in a manner possibly involving multimerization at the synapse with redistribution of SNARE proteins and clustering of synaptic vesicles.[29,30] αSyn was shown to associate as a peripheral membrane protein with synaptic vesicles,[31] although it was described as behaving almost entirely as a soluble protein in brain extracts.[32] The localization of αSyn at the axon terminal was found to be highly dynamic, characterized by rapid exchange of the protein between neighboring synapses,[33] perhaps thus accounting in part for its reported prion-like behavior. The evidence indicated αSyn dissociates from the synaptic vesicle membrane after exocytosis, behaving like a soluble protein that rapidly diffuses away from the synapse.

*SNCA* is the gene encoding αSyn, and six dominantly inherited missense mutations have been described as the cause of familial PD. These tend to replicate the pathophysiology of sporadic PD, although some have an earlier age of onset than the classical idiopathic form.[3] Elevation of unchanged αSyn is sufficient to cause disease, as seen in cases involving duplication of the *SNCA* gene that induces a familial form of PD similar to sporadic PD in relation to symptoms and age of onset.[34] However, triplication of *SNCA* gives rise to a more severe phenotype with earlier onset and prominent cognitive dysfunction.[35] Mouse models based on overexpression of normal or mutant αSyn are proving useful in PD research,[36,37] although some of the initial models lacked significant neurodegeneration in DA cells of the SNpc. This was partially solved by transgenic mouse lines that express carboxy-terminally truncated human αSyn on an αSyn-null background under the control of rat tyrosine hydroxylase promoter[38]; truncation of the carboxy-terminal region promoted αSyn aggregation as did expression of human αSyn on a mouse *SNCA*-null background.[39] Overexpression of human wild-type αSyn in nigral DA neurons led to impaired neurotransmission as reflected in reduced DA release and early dysfunction of the DA transporter.[40] This rodent model was believed to replicate human PD pathology more closely than models generated with neurotoxins which, although reproducing the canonical loss of midbrain DA neurons, fail to recapitulate the progressive, time-dependent feature of the pathological process and the development of α-synucleinopathy.[41,42] However, it was observed that in these αSyn-overexpressing rodent models the nigral DA neurons failed to respond to neuroprotective signaling by glial cell line-derive neurotrophic factor (GDNF),[43] upon which DA neuron survival depends (see later). This directed attention to nuclear function where it was revealed that αSyn disrupts GDNF signaling in such neurons by downregulation of the transcription factor Nurr1 along with its downstream target Ret, the tyrosine hydroxylase component of the GDNF receptor.[44] Protection of nigral DA neurons against αSyn toxicity was restored by overexpression of Nurr1. The latter study also demonstrated that αSyn targeted to the nucleus promotes toxicity by interacting with histone H3 and reducing its acetylation in αSyn-overexpressing animals. In addition to this Nurr1-related loss of GDNF signaling, a similar effect was shown to result from a reduced level of GM1 ganglioside, suggesting functional interplay between GM1 and αSyn (see later).

## 4. GM1 AND α-SYNUCLEIN INTERACTION IN PARKINSON'S DISEASE

The initial studies demonstrating that lipid binding inhibits αSyn fibril formation employed vesicles containing phospholipids of varying composition and reported a strong correlation between induction of α-helix conformers through such association and the inhibition of fibril formation.[45] This led to the further observation that such interaction is particularly strong between αSyn and GM1 compared with other gangliosides and lipids.[46] Helical folding through GM1 association was shown to be more thermodynamically favorable with the *N*-acetylated form of αSyn which also enhanced its specificity for this ganglioside.[47] These in vitro findings imply in vivo association of GM1 with αSyn, although evidence for this is mainly indirect; Fortin *et al.*[48] reported that αSyn localizes primarily to the nerve terminal and behaves as a soluble, monomeric protein in brain extracts, yet demonstrated association of αSyn with intracellular structures of HeLa cells that double label for components of lipid rafts. A functional connection between αSyn and GM1 suggesting such association was revealed by studying mice with homozygous disruption of *B4galnt1*, a necessary enzyme in the biosynthetic pathway to GM1.[49] Such mice, which are totally deficient in GM1 and the other ganglio-series gangliosides,[50,51] were shown to manifest elevated, aggregated αSyn in nigral neurons of the SNpc (Fig. 1). This occurred together with additional neuropathological lesions characteristic of PD including loss of tyrosine hydroxylase-expressing neurons in the SNpc and depletion of striatal DA.[49] These mice also suffered motor dysfunction as previously reported for similar mice.[52] These various CNS-associated lesions, including elevated αSyn, were abrogated by treatment with relatively high doses of GM1 and more effectively with much lower doses of LIGA20,

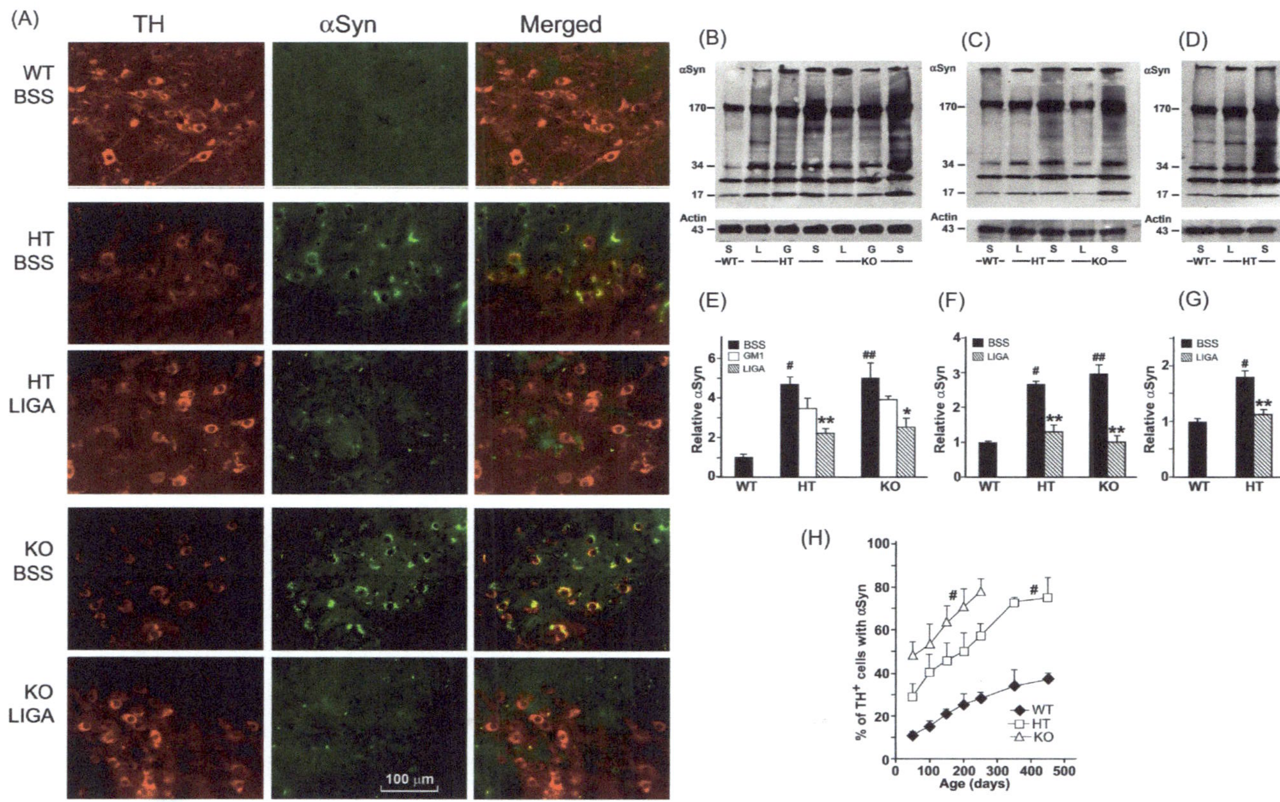

(A)
TH
αSyn
Merged
WT BSS
HT BSS
HT LIGA
KO BSS
KO LIGA
100 μm
(B)
(C)
(D)
αSyn
170
34
17
Actin
43
S L G S L G S
WT HT KO
S L S L S
S L S
(E)
(F)
(G)
BSS
GM1
LIGA
Relative αSyn
WT
HT
KO
(H)
% of TH+ cells with αSyn
Age (days)
WT
HT
KO

◀ **Fig. 1** Inverse relationship of GM1 and α-synuclein. (A) Immunohistochemical evidence for elevation and aggregation of αSyn in SNpc of 200-day-old mice with heterozygous (HT) or homozygous (KO) disruption of *B4galnt1* compared with wild-type (WT) mice. Mice were treated with basic salt solution (BSS) or LIGA20 (LIGA) administered 3×/week for 5 weeks. Antityrosine hydroxylase (TH) Ab revealed loss of DA neurons. LIGA20 reduced αSyn elevation/aggregation and restored TH expression in both HT and KO mice. (B–D) Detection of αSyn aggregation by Western blot. WT, HT, and KO mice at 35 (B), 200 (C), and 350 (D) days of age were treated with BSS (S), GM1 (G), or LIGA20 (L) 3×/week for 5 weeks. Anti-αSyn Ab revealed aggregated αSyn from 34 to 170 kD. Actin was run in parallel. (E–G) Densitometric quantification of αSyn polymers in (B–D), respectively. Mean ± SEM, two-tailed Student's *t* test. $^{\#}P < 0.01$; $^{\#\#}P < 0.001$ versus BSS-treated WT. *P = 0.05; **P < 0.01 versus BSS-treated mice with same genotype. (H) Age-dependent increase in $TH^+$ neurons with αSyn aggregation in HT and KO mice. Mean ± SEM, $n = 5$ mice. Two-way ANOVA; $^{\#}P < 0.01$ compared with WT; P < 0.001 for age-dependent increase in each genotype. *From Wu G, Lu Z-H, Kulkarni N, Ledeen RW. Deficiency of ganglioside GM1 correlates with Parkinson's disease in mice and humans. J Neurosci Res. 2012;90:1997–2008,*[53] *with permission from Wiley Periodicals, Inc.*

a membrane-permeable synthetic analog of GM1 (Fig. 2). This pointed to GM1 as the operative ganglioside whose deficiency induced the observed effects. Of particular interest was the finding that monoallelic disruption of this gene (*B4galnt1*$^{+/-}$) involving mice with approximately half the normal amount of GM1 showed virtually the same motor dysfunctions and neuropathological lesions as the biallelic (null) mice and responded similarly to LIGA20 treatment.[53] Importantly, these genetically altered mice also showed non-CNS lesions characteristic of PD including αSyn accumulations in the gastrointestinal and cardiovascular systems.[54]

These findings with the heterozygous *B4galnt1*$^{+/-}$ mice prompted more detailed study of actual PD tissues, the first of which revealed significant reduction of GM1 in the DA neurons of the SNpc from PD brains.[53] It was also noted that virtually all cells of that brain region showed significant GM1 depletion, the first of several indications that the GM1 deficit in sporadic PD is systemic in nature. Similar GM1 deficiency was reported for gray matter of the occipital cortex,[55] and also for a number of extra-CNS tissues, including colon, heart, skin, and peripheral blood mononuclear cells.[54] This provided evidence that the systemic deficiency of GM1 (and/or GD1a) approximates in magnitude the deficiencies found in the *B4galnt1*$^{+/-}$ mice and were thus consistent with the designation of this mouse as a true pathophysiological model of idiopathic PD.[53,55] These findings further supported the hypothesis that GM1 deficiency underlies the multisystem nature of sporadic PD with

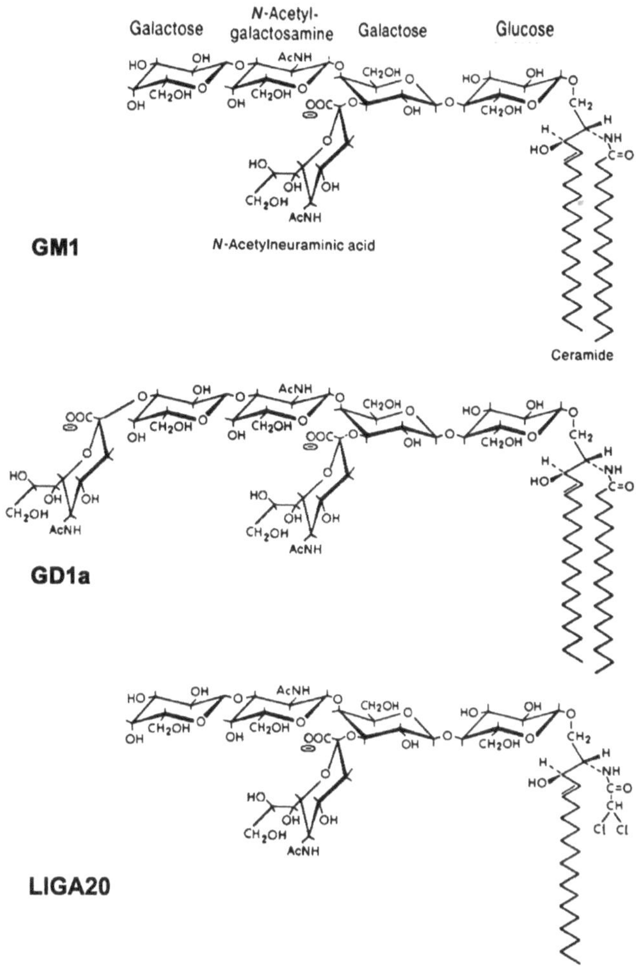


**Fig. 2** Structures of LIGA20 and a-series gangliosides: GM1 and GD1a. GM1 and LIGA20 have identical oligosaccharide moieties, but differ in the hydrophobic ceramide structure. The stearoyl unit in GM1 joined in amide linkage to sphingosine is replaced by dichloroacetyl in LIGA20, which renders the analog membrane permeable, thereby accounting for its greater neuroprotective potency when administered peripherally. GD1a is a catabolic precursor of GM1 (via Neu3 sialidase) and the two function as a unit.

numerous nonmotor symptoms that include non-DA lesions.[56] Moreover, subnormal levels of GM1 in non-neural cells such as peripheral blood mononuclear cells point to a systemic pathophysiology that extends beyond the CNS/PNS, conceivably epigenetic or possibly even genetic in origin.

The finding of specific, high-affinity binding between GM1 and αSyn observed in vitro and proposed to occur in vivo can be seen as representing a stoichiometric balance between soluble αSyn and the pool of colocalized GM1 required to maintain it in its oligomeric, nonaggregated state. Such GM1 would be expected to represent a relatively small nonmembranous pool of the type previously identified as the soluble component of GM1 associated with cytosolic proteins.[57,58] Disruption of this proposed homeostatic balance could result from subnormal levels of GM1 of the type demonstrated in several PD tissues and the *B4galnt1*$^{+/-}$ and also from overexpression of αSyn as in the abovementioned PD mouse models. In such conditions the portion of αSyn not associated with GM1 would be a candidate to form the β-sheet-rich fibrils that become Lewy aggregates and also to enter the nucleus to suppress the Nurr1 transcription factor necessary for expression of the GDNF Ret receptor. Fig. 3 indicates some of the expected consequences arising over time from insufficient GM1 in nigral and certain other neurons of the CNS and PNS.

The question then arises as to how the subnormal levels of GM1 observed in sporadic PD tissues come into being. At least a partial answer is suggested in the work of Svennerholm who showed that (a) brain GM1 levels vary significantly among individuals of the same age, and (b) the a-series gangliosides (GM1, GD1a) decline progressively with age.[59] This decrease is most striking for GD1a,[60] the catabolic precursor of GM1 whose level has a major influence on that of GM1 by functioning as a reserve pool via the coassociated NEU3 sialidase.[61] Individuals endowed with GM1/GD1a in the low normal range, for example, might start life with an adequate level, but reach a point in later years when GM1/GD1a falls below the threshold necessary to maintain all soluble αSyn in its tetrameric α-helical nonaggregating conformation, thereby leading to Lewy aggregates. As the population ages and GM1/GD1a continues to decline the number and percentage of persons with this condition will multiply. Additional factors besides aging might contribute to suppression of GM1/GD1a including epigenetic mechanisms that influence GM1 synthesis.[62] A prime example of this is the recent finding that GM1 occurs in the chromatin of neurons where it promotes its own further synthesis through epigenetic activation of *B4galnt1*,[63] perhaps explaining some of the neuroregenerative activity of this ganglioside. Interestingly, the latter study reported involvement of histone H3 acetylation on the *B4galnt1* promoter, the same histone whose acetylation is reduced by nuclear αSyn in blocking Nurr1 expression.[44]

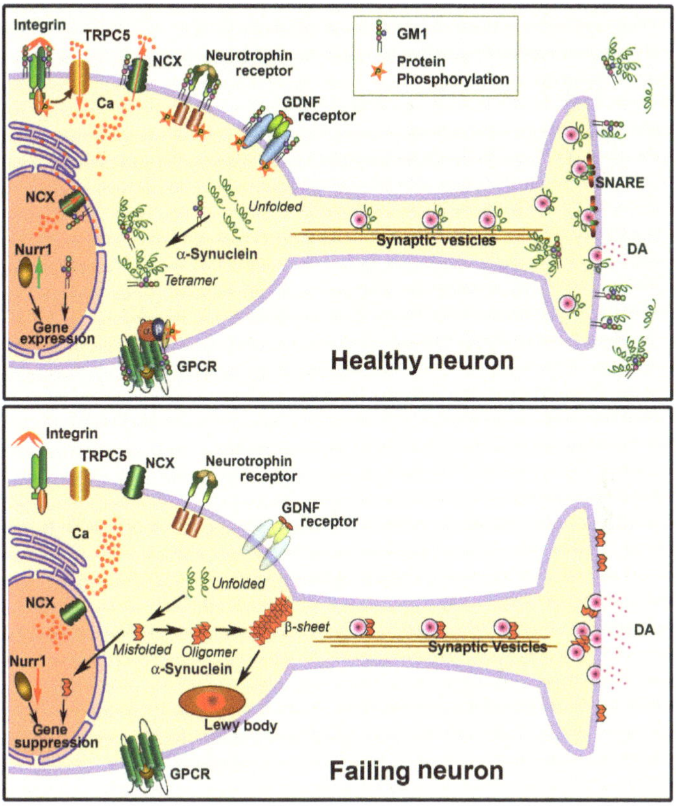


**Fig. 3** Neuronal functions dependent on GM1 and mechanism underlying the observed inverse relation between GM1 and α-synuclein. Some of the major neuronal processes that require GM1-protein associations are depicted (see Refs. 64 and 65 for more complete description). The viability of DA neurons depends especially on effective neuroprotective signaling by GDNF through its Ret/GFRα1/GM1 receptor, shown here as diminished due to Nurr1 suppression and failed signaling due to insufficient GM1. αSyn was described by some investigators as occurring in α-helical tetrameric conformation,[18] and to associate with synaptic vesicles as a peripheral membrane protein[31] while behaving as a soluble protein in brain extracts.[32] αSyn was also described as a presynaptic regulator of DA neurotransmission influencing SNARE proteins and the clustering of synaptic vesicles.[27–30] It was said to undergo rapid exchange between neighboring synapses. Also, shown is our hypothesized consequence of disequilibrium between soluble αSyn and soluble GM1, healthy neurons (upper) having GM1 associated with all αSyn tetramer to maintain nonaggregating α-helical conformation, and failing neurons (lower) with some αSyn unassociated with GM1 to form Lewy bodies and enter nucleus to suppress Nurr1 expression and downstream formation of Ret.[44]

## 5. NEUROPROTECTION AND OTHER NEURONAL FUNCTIONS OF GM1

GM1 ganglioside has been viewed as a master modulator in the nervous system in light of the truly numerous functions it regulates in developing and mature neurons. As elaborated in recent reviews,[64,65] several neuronal functions can become gradually compromised as GM1 levels recede with aging and/or epigenetic influences. A prime example is $Ca^{+2}$ transport at both the plasma and nuclear membranes, efflux being effected by GM1 complexed with one or another sodium–calcium exchanger which expels $Ca^{+2}$ from the nucleoplasm or cytoplasm.[66] Calcium influx can also be regulated via GM1 cross-linking with galectin-1 which triggers $Ca^{+2}$ influx through TRPC5 channels[67] as an initial phase of axonogenesis.[68] Several G-protein-coupled receptors are modulated by GM1, an interesting case being the δ-opioid receptor which is converted from inhibitory to excitatory mode through association with GM1.[69] Most relevant in regard to neurodegeneration are NTFs whose receptors require GM1 association to facilitate lifelong survival of neurons. An initial indication of this was the finding that GM1 occurs with high-affinity binding to TrkA, the tyrosine kinase receptor for nerve growth factor, and promotes tyrosine autophosphorylation together with neurite elongation.[70,71] Of interest was the observation that expression of TrkA in cells lacking GM1 elicited failure of this receptor to appear on the cell surface, a defect that was corrected by stable transfection of GM1 synthase.[72] This suggested a trafficking role involving intracellular association of GM1 with TrkA during microdomain assembly. The receptor for brain-derived neurotrophic factor, TrkB, showed similar association with GM1 which regulated its activity.[73] In regard to PD, a similar GM1 association has been reported for the receptor for glial GDNF, an NTF essential for preserving the viability of catecholaminergic (and perhaps other) neuronal types.[74] This involved binding of GM1 to the receptor complex comprising Ret, the tyrosine kinase component, and GFRα, a GPI-anchored coreceptor, cohesion of these two proteins being severely impaired in neurons totally or even partially devoid of GM1.[55] As with TrkA, this suggested a trafficking role with respect to Ret, which is recruited to lipid rafts by GFRα1 following GDNF binding. Efficacy of GDNF signaling, as indicated by IHC of phosphorylated Ret, was severely impaired in the SNpc from PD patients and also in the SNpc of both the monoallelic (*B4galnt1*$^{+/-}$) and biallelic (*B4galnt1*$^{-/-}$) mice; GDNF signaling (phosphorylated Ret levels) was restored in both mouse genotypes with LIGA20.[55] In another system, GM1 addition to striatal slices in situ increased Ret activity in a concentration- and time-dependent manner with GM1 concentration dependent increases in the binding of endogenous GDNF.[75]

All three of these NTFs have been studied in relation to cause and/or possible cure of PD, especially GDNF which, along with neurturin, is a distant member of the TGF-β superfamily. The latter two have been administered as NTFs to the SNpc or striatal regions of brain or as viral or cell-based therapies.[76] There have been indications of potential therapeutic benefit in some cases,[77] but on the whole the results have been variable and largely unsuccessful. Unfortunately, such studies have given little consideration to the possibility that the variable outcomes might relate, at least in part, to erratic GDNF receptor activity, perhaps involving varying levels of residual GM1. Consequently, it could prove useful in such studies to consider coadministration of GM1 along with the NTF. GM1 alone is receiving serious consideration for PD therapy (see later).

## 6. GM1 AND PARKINSON'S DISEASE THERAPY

Ganglioside therapy in various forms, but principally as direct peripheral application of GM1, has been shown to exert significant benefit to the nigrostriatal DA system in a variety of animal models. In rodents these lesions have included surgical transection,[78] and neurotoxin-induced degeneration via 1-methyl-4-phenyl,1,2,3,6-tetrahydropyridine (MPTP)[79,80] and 6-hydroxydopamine.[81] Intraperitoneal administration of GM1 prevented depletion of striatal DA and nigrostriatal neuron loss. LIGA20, the above-mentioned membrane-permeable analog of GM1, proved even more efficacious than GM1 in the mouse MPTP model and had the important advantage of oral bioavailability.[82] Subhuman primate models, principally MPTP-treated monkeys, also benefited from GM1 in terms of both motor and cognitive functioning.[83,84] This correlated with small but statistically significant increased density of tyrosine hydroxylase-positive fibers and increased striatal DA.[83] Elevation of brain GM1 has been achieved by other means such as intraventricular sialidase administration which converted some of the ganglio-series gangliosides (GD1a, GD1b, GT1b) to GM1; this proved partially neuroprotective in MPTP-treated mice as seen in the sparing of striatal DA and SNpc DA neurons.[85] Elevation of GM1 was also achieved genetically through deletion of GD3-synthase in mice prior to MPTP administration, which significantly reduced the loss of striatal DA and tyrosine hydroxylase-expressing neurons in the SNpc compared with wild-type mice.[86] It was noted that therapeutic benefit might also have derived in this model from elimination of GD3, a proapoptotic ganglioside.

These promising preclinical results with GM1 gave way to clinical trials by Schneider's group beginning with a phase I safety study.[87] This was followed by a 5-year open-label study in which PD patients, following an initial loading dose of 1000 mg GM1 i.v., were given two subcutaneous injections of 100 mg GM1 per day. This resulted in reduced, slow linear projection of motor dysfunction based on UPDRS motor scores, and in improved activities of daily living scores. Significantly, after 5 years of treatment the motor disability scores were lower than at baseline.[88] This was followed by a randomized, controlled, delayed-start phase II trial that demonstrated GM1 superiority to placebo in reducing motor impairment and slowing symptom progression over a 2-year period; as before, UPDRS motor scores were improved at trial end of 120 weeks compared with baseline, suggesting that " . . . GM1 may have symptomatic and potentially disease modifying effects" on PD.[89] An additional study by this group involving a subset of patients from the delayed-start trial obtained positron emission tomography imaging data to demonstrate significant slowing by GM1 of binding potential loss of the DA transporter with [$^{11}$C]methylphenidate.[90]

These several indications of GM1 therapeutic proficiency in preclinical as well as clinical studies appear consonant with its proposed function as a form of replacement therapy in PD patients who, on the basis of other studies, would manifest subnormal levels of GM1. One might speculate on the potentially greater benefit that might accrue to a structurally modified form of GM1 that is better able to enter the brain and, importantly, to gain access to intraneuronal loci associated with GM1-dependent functions. A promising development along this line was introduction of the LIGA series of "semisynthetic" derivatives of GM1 in which the stearoyl fatty acid of the ceramide moiety was replaced by shorter chain fatty acids.[91] The mechanism underlying the superiority of these GM1 analogs in providing neuroprotection was eventually attributed to their greater membrane permeability compared with GM1, the abovementioned LIGA20, for example, affording in vivo protection against kainate-induced seizures in ganglio-series-deficient mice with demonstration of in vivo neuronal entry.[92] Although LIGA20 itself may not be a promising candidate for clinical trials owing to potential toxicity due to the dichloroacetyl group, it is to be hoped that other GM1 analogs may soon be subjected to testing in this manner.

## 7. CONCLUDING REMARKS AND FUTURE PROSPECTS

The essential place of gangliosides in maintaining a healthy, cohesive nervous system is well illustrated in the disastrous consequences of their genetic deletion in human inherited disorders. That applies, for example, to loss of the ganglio-series in the early-onset spastic paraplegia involving disruption of *B4galnt1*,[93] and as described earlier,[49] mice with this lesion are similarly, though less drastically, affected. Deletion of the ganglio-series along with GM3 and GD3 by mutation of *St3gal5* (GM3 synthase) is even more devastating to humans[94] and mice.[95,96] This condition was discovered in the Amish religious community[97] which also manifests the world's highest prevalence of PD—a point of interest in regard to GM1 since first degree relatives of afflicted children would likely be heterozygous carriers of GM1 deficiency. As summarized in this review, the *B4galnt1*$^{+/-}$ mouse PD model based on partial, systemic depletion of GM1 provides an accurate recapitulation of behavioral and neuropathological lesions of sporadic PD which, in patients, is paralleled by similarly reduced levels of GM1 in brain and extra-CNS regions. The specific nature of these lesions is directly traceable to the established functions of GM1 in the nervous system (Fig. 3), an especially essential one being constituent of NTF receptors. This applies especially to GDNF and its Ret/GFRα1 receptor upon which survival of DA (and other catecholaminergic) neurons depends. In regard to αSyn, the aptly termed "Rosetta stone of PD" refers to its prominence in Lewy inclusions which constitute the hallmark neuropathological symptoms of PD, and we have proposed that the inverse correlation of αSyn and GM1 levels points to pathological interplay between these two. We believe this highlights the singular ability of GM1 to specifically bind and maintain αSyn in its α-helical, nonaggregating conformation. Additional work is required to establish the nature of such association in vivo and test the hypothesis that GM1 is an essential constituent of the proposed tetrameric/oligomeric form of αSyn which is necessary to prevent αSyn fibrillation. One must bear in mind the alternate claim that αSyn exists predominantly as disordered monomer[98] that would bear no special relationship to GM1, although evidence has been presented for the in vivo presence of the tetrameric form[20] which would presumably be associated with GM1. The possibility that the portion of αSyn that lacks α-helical folding and forms Lewy inclusions is due to insufficient GM1 seems an idea worth exploring to further consider the pathological consequences of subnormal GM1. In regard to PD therapy, various mechanisms to elevate endogenous GM1 other than peripheral application might command attention including epigenetic modulation[63] that could include the fascinating new area of gut microbiota that was recently studied in relation to a PD model.[99] Efforts to clinically remedy the systemic GM1 deficiency in PD will likely continue in the pharmacological mode following the pioneering efforts that have shown encouraging results.[88,89] Finally, the use of membrane-permeable GM1 analogs with superior ability to access intraneuronal GM1 functional sites may hold considerable therapeutic promise, considering also their oral bioavailability and capacity for systemic dissemination to all sites affected by sporadic PD.

## REFERENCES

1. Jellinger KA. Synuclein deposition and non-motor symptoms in Parkinson disease. *J Neurol Sci.* 2011;310:107–111.
2. Lang AE. A critical appraisal of the premotor symptoms of Parkinson's disease: potential usefulness in early diagnosis and design of neuroprotective trials. *Mov Disord.* 2011;26:775–783.
3. Goedert M, Jakes R, Spillantini MG. The synucleinopathies: twenty years on. *J Parkinsons Dis.* 2017;7:S53–S71.
4. Aron L, Klein R. Repairing the parkinsonian brain with neurotrophic factors. *Trends Neurosci.* 2011;34:88–100.
5. Tretiakoff K, Tretiakoff K. Contribution a l'etude de l'etude de l'anatomiepathologique du locus niger de Soemmering avec quelques deductions relatives a la pathogenie des troubles du tonus musculaire et de la maladie de Parkinson. These, Universite de Paris; 1919
6. Lewy F, Paralysis agitans. I. Pathologische anatomie. Lewandowsky M, Abelsdorff G, eds. *Handbuch der Neurologie.* Vol. 3. Berlin: Springer Verlag; 1912: 920–933.
7. Spillantini MG, Schmidt ML, Lee VM, Trojanowski JQ, Jakes R, Goedert M. Alpha-synuclein in Lewy bodies. *Nature.* 1997;388:839–840.
8. Carlsson A, Lindqvist M, Magnusson T, Waldeck B. On the presence of 3-hydroxytyramine in brain. *Science.* 1958;127:471.
9. Bertler A, Rosengren E. Occurrence and distribution of dopamine in brain and other tissues. *Experientia.* 1959;15:10–11.
10. Carlsson A. The occurrence, distribution and physical role of catecholamines in the nervous system. *Pharmacol Rev.* 1959;11:490–493.
11. Birkmayer W, Hornykiewicz O. Der L-Dioxyphenylalanin (=DOPA) Effect bei der Parkinson Akinese. *Wien Klin Wschr.* 1961;73:787–788.
12. Poewe W, Seppi K, Tanner CM, et al. Parkinson disease. *Nat Rev Dis Primers.* 2017;3:17013.
13. Braak H, Del Tredici K, Rub U, de Vos RAI, JansenSteur ENH, Braak E. Staging of brain pathology related to sporadic Parkinson's disease. *Neurobiol Aging.* 2003;24:197–211.
14. Braak H, de Vos RAI, Bohl J, Del Tredici K. Gastric α-synuclein immunoreactive inclusions in Meissner's and Auerbach's plexuses in cases staged for Parkinson's disease-related brain pathology. *Neurosci Lett.* 2006;396:67–72.
15. Braak H, Bohl JR, Muller CM, et al. Stanley Fahn Lecture 2005: the staging procedure for the inclusion body pathology associated with sporadic Parkinson's disease reconsidered. *Mov Disord.* 2006;21:2042–2051.

16. Pellicano C, Benincasa D, Pisani V, et al. Prodromal non-motor symptoms of Parkinson's disease. *Neuropsychiatr Dis Treat.* 2007;3:145–152.
17. Bendor JT, Logan TP, Edwards RH. The function of α-synuclein. *Neuron.* 2013;79:1044–1066.
18. Bartels T, Choi JG, Selkoe DJ. α-Synuclein occurs physiologically as a helically folded tetramer that resists aggregation. *Nature.* 2011;477:107–110.
19. Wang W, Perovic I, Chittuluru J, et al. A soluble α-synuclein construct forms a dynamic tetramer. *Proc Natl Acad Sci USA.* 2011;108:17797–17802.
20. Dettmer U, Newman AJ, Luth ES, et al. In vivo cross-linking reveals principally oligomeric forms of α-synuclein and β-synuclein in neurons and non-neural cells. *J Biol Chem.* 2013;288:6371–6385.
21. Anderson JP, Walker DE, Goldstein JM, et al. Phosphorylation of Ser-129 is the dominant pathological modification of alpha-synuclein in familial and sporadic Lewy body disease. *J Biol Chem.* 2006;281:29739–29752.
22. Ohrfelt A, Zetterberg H, Andersson K, et al. Identification of novel α-synuclein isoforms in human brain tissue by using an online nanoLC-ESI-FTICR-MS method. *Neurochem Res.* 2011;36:2029–2042.
23. Maltsev AS, Ying J, Bax A. Impact of N-terminal acetylation of α-synuclein on its random coil and lipid binding properties. *Biochemistry.* 2012;51:5004–5013.
24. Trexler AJ, Rhoades E. N-terminal acetylation is critical for forming α-helical oligomer of α-synuclein. *Protein Sci.* 2012;21:601–605.
25. Vamvaca K, Volles MJ, Lansbury Jr PT. The first terminal amino acids of alpha-synuclein are essential for alpha-helical structure formation in vitro and membrane binding in yeast. *J Mol Biol.* 2009;389:413–424.
26. Bartels T, Ahlstrom LS, Leftin A, et al. The N-terminus of the intrinsically disordered protein α-synuclein triggers membrane binding and helix folding. *Biophys J.* 2010;99:2116–2124.
27. Abelovich A, Schmitz Y, Farinas I, et al. Mice lacking α-synuclein display functional deficits in the nigrostriatal dopamine system. *Neuron.* 2000;25:239–252.
28. Nemani VM, Lu W, Berge V, et al. Increased expression of α-synuclein reduces neurotransmitter release by inhibiting synaptic vesicle re-clustering after endocytosis. *Neuron.* 2010;65:66–79.
29. Wang L, Das U, Scott DA, et al. α-Synuclein multimers cluster synaptic vesicle and attenuate recycling. *Curr Biol.* 2014;24:2319–2326.
30. Garcia-Reitbock P, Anichtchik O, Bellucci A, et al. SNARE protein redistribution and synaptic failure in a transgenic mouse model of Parkinson's disease. *Brain.* 2010;133:2032–2044.
31. Maroteaux L, Campanelli JT, Scheller RH. Synuclein: a neuron-specific protein localized to the nucleus and presynaptic nerve terminal. *J Neurosci.* 1988;8:2804–2815.
32. Maroteaux L, Scheller RH. The rat brain synucleins; family of proteins transiently associated with neuronal membrane. *Brain Res Mol Brain Res.* 1991;11:335–343.
33. Fortin DL, Nemani VM, Voglmaier SM, et al. Neural activity controls the synaptic accumulation of α-synuclein. *J Neurosci.* 2005;25:10913–10921.
34. Chartier-Harlin MC, Kachergus J, Roumier C, et al. α-Synuclein locus duplication as a cause of familial Parkinson's disease. *Lancet Neurol.* 2004;364:1167–1169.
35. Singleton AB, Farrer M, Johnson J, et al. α-Synuclein locus triplication causes Parkinson's disease. *Science.* 2003;302:841.
36. Magen I, Chesselet MF. Genetic mouse models of Parkinson's disease: the state of the art. *Prog Brain Res.* 2010;184:53–87.
37. Dawson TM, Ko HS, Dawson VL. Genetic animal models of Parkinson's disease. *Neuron.* 2010;66:646–661.

38. Tofaris GK, Garcia Reitbock PG, Humby T, et al. Pathological changes in dopaminergic nerve cells of the substantia nigra and olfactory bulb in mice transgenic for truncated human α-synuclein (1-120): implications for Lewy body disorders. *J Neurosci.* 2006;26:3942–3950.
39. Fares MB, Maco B, Oueslati A, et al. Induction of de novo α-synuclein fibrilization in a neuronal model for Parkinson's disease. *Proc Natl Acad Sci USA.* 2016;113:E912–E921.
40. Decressac M, Mattsson B, Lundblad M, Weikop P, Bjorklund A. Progressive neurodegenerative and behavioural changes induced by AAV-mediated overexpression of α-synuclein in midbrain dopamine neurons. *Neurobiol Dis.* 2012;45:939–953.
41. Decressac M, Mattsson B, Bjorklund A. Comparison of the behavioural and histological characteristics of the 6-OHDA and α-synuclein rat models of Parkinson's disease. *Exp Neurol.* 2012;235:306–315.
42. Cannon JR, Greenamyre. Neurotoxic in vivo models of Parkinson's disease recent advances. *Prog Brain Res.* 2010;184:17–33.
43. Decressac M, Ulusoy A, Mattson B, et al. GDNF fails to exert neuroprotection in a rat α-synuclein model of Parkinson's disease. *Brain.* 2011;134:2302–2311.
44. Decressac M, Kadkhodaei B, Mattsson B, et al. α-Synuclein-induced down-regulation of Nurr1 disrupts GDNF signaling in Nigral dopamine neurons. *Sci Transl Med.* 2012;4:1–15.
45. Zhu M, Fink AL. Lipid binding inhibits alpha-synuclein fibril formation. *J Biol Chem.* 2003;278:16873–16877.
46. Martinez Z, Zhu M, Han S, Fink AL. GM1 specifically interacts with α-synuclein and inhibits fibrillation. *Biochemistry.* 2007;46:1868–1877.
47. Bartels T, Kim NC, Luth ES, Selkoe DJ. N-Alpha-acetylation of α-synuclein increases its helical folding propensity, GM1 binding specificity and resistance to aggregation. *PloS One.* 2014;9:103727.
48. Fortin DL, Troyer MD, Nakamura K, et al. Lipid rafts mediate the synaptic localization of α-synuclein. *J Neurosci.* 2004;24:6715–6723.
49. Wu G, Lu Z-H, Kulkarni N, Amin R, Ledeen RW. Mice lacking major brain gangliosides develop parkinsonism. *Neurochem Res.* 2011;36:1706–1714.
50. Liu Y, Wada R, Kawai H, et al. A genetic model of substrate deprivation therapy for a glycosophingoliid storage disorder. *J Clin Invest.* 1999;103:497–505.
51. Takamiya K, Yamamoto A, Furukawa K, et al. Mice with disrupted GM2/GD2 synthase gene lack complex gangliosides but exhibit only subtle defects in their nervous system. *Proc Natl Acad Sci USA.* 1996;93:10662–10667.
52. Chiavegatto S, Sun J, Nelson RJ, Schnaar RL. A functional role for complex gangliosides: motor deficits in GM2/GD2 synthase knockout mice. *Exp Neurol.* 2000;166:227–234.
53. Wu G, Lu Z-H, Kulkarni N, Ledeen RW. Deficiency of ganglioside GM1 correlates with Parkinson's disease in mice and humans. *J Neurosci Res.* 2012;90:1997–2008.
54. Wu G, Lu Z-H, Ledeen RW, et al. In preparation.
55. Hadaczek P, Wu G, Sharma N, et al. GDNF signaling implemented by GM1 ganglioside; failure in Parkinson's disease and GM1-deficient murine model. *Exp Neurol.* 2015;263:177–189.
56. Halliday GM, Barker RA, Rowe DB, eds. In: *Non-dopamine Lesions in Parkinsons Disease. Oxford, New York: Oxford University Press; 2011.*
57. Ledeen RW, Skrivanek JA, Tirri LJ, et al. Gangliosides of the neuron: localization and origin. *Adv Exp Med Biol.* 1976;71:83–103.
58. Sonnino S, Ghidoni R, Fiorilli A, et al. Cytosolic gangliosides of rat brain: their fractionation into protein-bound complexes of different ganglioside compositions. *J Neurosci Res.* 1984;12:193–204.

59. Svennerholm L, Bostrom K, Jungbjer B, Olsson L. Membrane lipids of adult human brain: lipid composition of frontal and temporal lobe in subjects of age 20 to 100 years. *J Neurochem.* 1994;63:1802–1811.
60. Svennerholm L, Bostrom K, Fredman P, et al. Human brain gangliosides: developmental changes from early fetal stage to advanced age. *Biochim Biophys Acta.* 1989;1005:109–117.
61. Miyagi T, Yamaguchi K. Mammalian sialidases: physiological and pathological roles in cellular functions. *Glycobiology.* 2012;22:880–896.
62. Tsai Y-T, Yu RK. Epigenetic activation of mouse ganglioside synthase genes: implications for neurogenesis. *J Neurochem.* 2014;128:101–110.
63. Itokazu Y, Tsai Y-T, Yu RK. Epigenetic regulation of ganglioside expression in neural stem cells and neuronal cells. *Glycoconj J.* 2017;34:749–756.
64. Aureli M, Mauri L, Ciampa M, et al. GM1 ganglioside: past studies and future potential. *Mol Neurobiol.* 2016;53:1824–1842.
65. Ledeen R, Wu G. The multi-tasked life of GM1 ganglioside, a true factotum of nature. *Trends Biochem Sci.* 2015;40:407–418.
66. Xie X, Wu G, Lu Z-H, Ledeen RW. Potentiation of a sodium-calcium exchanger in the nuclear envelope by nuclear GM1 ganglioside. *J Neurochem.* 2002;81:1185–1195.
67. Wu G, Lu ZH, Obukhov AG, et al. Induction of calcium influx through TRPC5 channels by cross-linking of GM1 ganglioside associated with alpha5beta1 integrin initiates neurite outgrowth. *J Neurosci.* 2007;27:7447–7458.
68. Wu G, Lu Z-H, André S, et al. Functional interplay between ganglioside GM1 and cross-linking galectin-1 induces axon-like neuritogenesis via integrin-based signaling and TRPC5-dependent $Ca^{2+}$ influx. *J Neurochem.* 2016;136:550–563.
69. Wu G, Lu Z-H, Ledeen RW. Interaction of δ-opioid receptor with GM1 ganglioside: conversion from inhibitory to excitatory mode. *Brain Res Mol Brain Res.* 1997;44:341–346.
70. Mutoh T, Tokuda A, Miyadai T, et al. Ganglioside GM1 binds to the Trk protein and regulates receptor function. *Proc Natl Acad Sci USA.* 1995;92:5087–5091.
71. Rabin SJ, Mocchetti I. GM1 ganglioside activates the high-affinity nerve growth factor receptor TrkA. *J Neurochem.* 1995;65:347–354.
72. Mutoh T, Hamano T, Yano S, et al. Stable transfection of GM1 synthase gene into GM1-deficient NG108-15 cells, CR-72 cells, rescues the responsiveness of Trk-neurotrophin receptor to its ligand, NGF. *Neurochem Res.* 2002;27:801–806.
73. Pitto M, Mutoh T, Kuriyama M, et al. Influence of endogenous GM1 ganglioside on TrkB activity in cultured neurons. *FEBS Lett.* 1998;439:93–96.
74. Pascual A, Hidalgo-Figueroa M, Piruat JI, et al. Absollute requirement of GDNF for adult catecholaminergic neuron survival. *Nat Neurosci.* 2008;11:755–761.
75. Newburn EN, Duchemin AM, Neff NH, Hadjiconstantinou M. GM1 ganglioside enhances Ret signaling in striatum. *J Neurochem.* 2014;130:541–554.
76. Staudt MD, Di Sebastiano AR, Xu H, et al. Advances in neurotrophic factor and cell-based therapies for Parkinson's disease: a mini-review. *Gerontology.* 2016;62:371–380.
77. Patel NK, Pavese N, Javed S, et al. Benefits of putaminal GDNF infusion in Parkinson disease are maintained after GDNF cessation. *Neurology.* 2013;81:1176–1178.
78. Toffano G, Savoini G, Moroni F, et al. GM1 ganglioside stimulates the regeneration of dopaminergic neurons in the central nervous system. *Brain Res.* 1983;261:163–166.
79. Hadjiconstantinou M, Rossetti ZL, Paxton RC, Neff NH. Administration of GM1 ganglioside restores the dopamine content in striatum after chronic treatment with MPTP. *Neuropharmacology.* 1986;25:1075–1077.
80. Schneider JS, Yuwiler A. GM1 ganglioside treatment promotes recovery of striatal dopamine concentrations in the mouse model of MPTP-induced parkinsonism. *Exp Neurol.* 1989;105:177–183.
81. Tilson HA, Harry GJ, Nanry K, et al. Ganglioside interactions with the dopaminergic system of rats. *J Neurosci Res.* 1988;19:88–93.

82. Schneider JS, DiStefano L. Response of the damaged dopamine system to GM1 and semisynthetic gangliosides: effects of dose and extent of lesion. *Neuropharmacology*. 1995;34.489–493.
83. Schneider JS, Pope A, Simpson K, et al. Recovery from experimental parkinsonism in primates with GM1 ganglioside treatment. *Science*. 1992;256:843–846.
84. Herrero MT, Perez-Otano I, Oset C, et al. GM-1 ganglioside promotes the recovery of surviving midbrain dopaminergic neurons in MPTP-treated monkeys. *Neuroscience*. 1993;56:965–972.
85. Schneider JS, Seyfried TN, Choi H-S, Kidd SK. Intraventricular sialidase administration enhances GM1 ganglioside expression and is partially neuroprotective in a mouse model of Parkinson's disease. *PloS One*. 2015;10:e0143351.
86. Akkhawattanangkul Y, Maiti P, Xue Y, et al. Targeted deletion of GD3 synthase protects against MPTP-induced neurodegeneration. *Genes Brain Behav*. 2017;16:522–536.
87. Schneider JS, Roeltgen DP, Rothblat DS, et al. GM1 ganglioside treatment of Parkinson's disease: an open pilot study of safety and efficacy. *Neurology*. 1995;45:1149–1154.
88. Schneider JS, Sendek S, Daskalakis C, Cambi F. GM1 ganglioside in Parkinson's disease: results of a five-year open study. *J Neurol Sci*. 2010;292:45–51.
89. Schneider JS, Gollomp SM, Sendek S, et al. A randomized, controlled, delayed start trial of GM1 ganglioside in treated Parkinson's disease patients. *J Neurol Sci*. 2013;324:140–148.
90. Schneider JS, Cambi F, Gollomp SM, et al. GM1 ganglioside in Parkinson's disease: pilot study of effects on dopamine transporter binding. *J Neurol Sci*. 2015;356:118–123.
91. Manev H, Favaron M, Vicini S, et al. Glutamate-induced neuronal death in primary cultures of cerebellar granule cells: protection by synthetic derivatives of endogenous sphingolipids. *J Pharmacol Exp Ther*. 1990;252:419–427.
92. Wu G, Lu Z-H, Wang J, et al. Enhanced susceptibility to kainate-induced seizures, neuronal apoptosis, and death in mice lacking gangliotetraose gangliosides: protection with LIGA20, a membrane-permeant analog of GM1. *J Neurosci*. 2005;25:11014–11022.
93. Boukhris A, Schule R, Loureiro JL, et al. Alteration of ganglioside biosynthesis responsible for complex hereditary spastic paraplegia. *Am J Hum Genet*. 2013;93:118–123.
94. Simpson MA, Cross H, Proukakis C, et al. Infantile-onset symptomatic epilepsy syndrome caused by a homozygous loss-of-function mutation of GM3 synthase. *Nat Genet*. 2004;36:1225–1229.
95. Yoshikawa M, Go S, Takasaki K, et al. Mice lacking ganglioside GM3 synthase exhibit complete hearing loss due to selective degeneration of the organ of Corti. *Proc Natl Acad Sci USA*. 2009;106:9483–9488.
96. Niimi K, Nishioka C, Miyamoto T, et al. Impairment of neuropsychological behaviors in ganglioside GM3-knockout mice. *Biochem Biophys Res Commun*. 2011;406:524–528.
97. Tell S. Medical genetic research on the Amish: from genetic tourism to community health centers. *Einstein J Biol Med*. 2012;28:18–24.
98. Fauvet B, Mbefo MK, Fares MB, et al. (-Synuclein in central nervous system and from erythrocytes, mammalian cells, and Escherichia coli exists predominantly as disordered moniomer. *J Biol Chem*. 2012;287:15345–15364.
99. Sampson TR, Debellus JW, Thron T, et al. Gut microbiota regulate motor deficits and neuroinflammation in a model of Parkinson's disease. *Cell*. 2016;167:1469–1480.

# INDEX

## C



## D



## E

## H



## I

## T



## V

Kolibri 360 GmbH
Ferdinand-Jühlke-Straße 7
99095 Erfurt, Deutschland
produktsicherheit@kolibri360.de